Handbook of Radioembolization

IMAGING IN MEDICAL DIAGNOSIS AND THERAPY

Series Editors: Andrew Karellas and Bruce R. Thomadsen

Published titles

IMAGING IN MEDICAL DIAGNOSIS AND THERAPY

Series Editors: Andrew Karellas and Bruce R. Thomadsen

Published titles

Forthcoming titles

Handbook of
Radioembolization

Physics, Biology, Nuclear Medicine, and Imaging

Edited by

Alexander S. Pasciak, PhD
J. Mark McKinney, MD
Yong C. Bradley, MD

CRC Press
Taylor & Francis Group
Boca Raton London New York

CRC Press is an imprint of the
Taylor & Francis Group, an **informa** business

A TAYLOR & FRANCIS BOOK

CRC Press
Taylor & Francis Group
6000 Broken Sound Parkway NW, Suite 300
Boca Raton, FL 33487-2742

First issued in paperback 2019

ISBN-13: 978-1-4987-4201-6 (hbk)
ISBN-13: 978-0-367-87468-1 (pbk)

Library of Congress Cataloging-in-Publication Data

Names: Pasciak, Alexander S., editor. | McKinney, J. Mark, editor. | Bradley, Yong C., editor.
Title: Handbook of radioembolization : physics, biology, nuclear medicine, and imaging / edited by Alexander S. Pasciak, J. Mark McKinney, and Yong C. Bradley.
Other titles: Imaging in medical diagnosis and therapy.
Description: Boca Raton : Taylor & Francis, 2017. | Series: Imaging in medical diagnosis and therapy | Includes bibliographical references and index.
Identifiers: LCCN 2016032428 | ISBN 9781498742016 (hardcover : alk. paper)
Subjects: | MESH: Liver Neoplasms--radiotherapy | Liver Neoplasms--radionuclide imaging | Embolization, Therapeutic | Yttrium Radioisotopes--therapeutic use | Tomography, Emission-Computed
Classification: LCC RC280.L5 | NLM WI 735 | DDC 616.99/4360757--dc23
LC record available at https://lccn.loc.gov/2016032428

Visit the Taylor & Francis Web site at
http://www.taylorandfrancis.com

and the CRC Press Web site at
http://www.crcpress.com

Dedication

To Alina...

Dedication

Contents

Series preface

The advances in the science and technology of medical imaging and radiation therapy are more profound and rapid than ever before since their inception over a century ago. Further, these disciplines are increasingly cross-linked as imaging methods become more widely used to plan, guide, monitor, and assess treatments in radiation therapy. Today, the technologies of medical imaging and radiation therapy are so complex and so computer-driven that it is difficult for the persons (physicians and technologists) responsible for their clinical use to know exactly what is happening at the point of care, when a patient is being examined or treated. The persons best equipped to understand the technologies and their applications are medical physicists, and these individuals are assuming greater responsibilities in the clinical arena to ensure that what is intended for the patient is actually delivered in a safe and effective manner.

The growing responsibilities of medical physicists in the clinical arenas of medical imaging and radiation therapy are not without their challenges, however. Most medical physicists are knowledgeable in either radiation therapy or medical imaging and expert in one or a small number of areas within their discipline. They sustain their expertise in these areas by reading scientific articles and attending scientific talks at meetings. In contrast, their responsibilities increasingly extend beyond their specific areas of expertise. To meet these responsibilities, medical physicists periodically must refresh their knowledge of advances in medical imaging or radiation therapy, and they must be prepared to function at the intersection of these two fields. How to accomplish these objectives is a challenge.

At the 2007 annual meeting of the American Association of Physicists in Medicine in Minneapolis, this challenge was the topic of conversation during a lunch hosted by Taylor & Francis Publishers and involving a group of senior medical physicists (Arthur L. Boyer, Joseph O. Deasy, C.-M. Charlie Ma, Todd A. Pawlicki, Ervin B. Podgorsak, Elke Reitzel, Anthony B. Wolbarst, and Ellen D. Yorke). The conclusion of this discussion was that a book series should be launched under the Taylor & Francis banner, with each volume in the series addressing a rapidly advancing area of medical imaging or radiation therapy of importance to medical physicists. The aim would be for each volume to provide medical physicists with the information needed to understand technologies driving a rapid advance and their applications to safe and effective delivery of patient care.

Each volume in the series is edited by one or more individuals with recognized expertise in the technological area encompassed by the book. The editors are responsible for selecting the authors of individual chapters and ensuring that the chapters are comprehensive and intelligible to someone without such expertise. The enthusiasm of volume editors and chapter authors has been gratifying and reinforces the conclusion of the Minneapolis luncheon that this series of books addresses a major need of medical physicists.

Imaging in Medical Diagnosis and Therapy would not have been possible without the encouragement and support of the series manager, Luna Han, of Taylor & Francis Publishers. The editors and authors, and most of all I, are indebted to her steady guidance of the entire project.

William Hendee
Founding Series Editor
Rochester, Minnesota

Preface

Primary and metastatic liver cancers are the fifth most commonly diagnosed cancer and the second most common cause of cancer-related mortality in men worldwide, with slightly lower rates in women. Environmental factors based on geographic location carry a profound impact in primary liver cancers. In some developing countries, liver cancer is the most common form of cancer with higher rates of incidence secondary to viral hepatitis or exposure to aflatoxin B1. Unfortunately, liver cancer often carries a poor prognosis as surgical options are commonly limited by multifocality combined with other factors, such as underlying liver disease.

External beam radiation therapy (EBRT) is often a preferred treatment modality for many surgically unresectable cancers, especially in concert with chemotherapy. However, for many years, its utility in the treatment of primary and secondary liver cancer has been limited by the high radiosensitivity of normal liver tissue. As is discussed in the chapters of this book, 30–40 Gy represents a maximum tolerable dose to normal liver from EBRT, above which radiation-induced liver disease and liver failure are potentially fatal complications. This fact has completely eliminated the use of conventional whole-liver EBRT as a potential treatment option for liver cancer. Newer advancements in image-guided intensity modulated radiation therapy are greatly improving the prospects of EBRT as a viable treatment modality; however, the high radiosensitivity of normal liver tissue, worsened by underlying liver disease in many patients, remains a challenge.

Despite technological advancements in EBRT, the most common form of radiation therapy for the treatment of primary and metastatic liver cancer is radioembolization, sometimes referred to as selective internal radiation therapy or SIRT.

Radioembolization is a brachytherapy treatment delivered as part of a minimally invasive fluoroscopically guided intervention. In this procedure, millions of microscopic embolic spheres containing calibrated activities of either yttrium-90 (^{90}Y) or holmium-166 (^{166}Ho) are infused into the right or left hepatic artery where they embolize both tumor tissue and, to some extent, normal liver tissue. Because the hepatic artery primarily perfuses the tumor, greater concentrations of radioactive microspheres are trapped in the tumor compared to the normal liver. The relative difference in microsphere concentration in tumor compared to normal liver tissue following a successful radioembolization therapy can range anywhere from a factor of 2 to a factor of 15, providing the potential for sparing of healthy liver tissue compared to conventional EBRT. However, there are other advantages. While a 40 Gy absorbed dose to normal liver tissue from EBRT could potentially cause liver failure, it is well below the toxic threshold for a single-session treatment using radioembolization, which is greater than 80 Gy. This unique paradox is due to differences in irradiated tumor volume, dose rate, and other factors, such as the heterogeneous microscopic dose distribution that results from radioembolization. This microscopic absorbed dose heterogeneity, combined with the regenerative propensity of healthy liver tissue, vastly reduces the toxicity of radioembolization and is a hallmark of the technique's utility.

Given the clear inherent benefits of the radioembolization treatment, its use as a treatment option for primary and metastatic liver cancer is advancing extremely rapidly. While the methodology behind radioembolization has been relatively stable over the past 10 years, it is our belief that this treatment is on the cusp of some rapid changes that will increase both its efficacy and the breadth of its clinical use. Our prediction is based on the following:

- Commercial manufacturers of ^{90}Y radioembolization products are rapidly seeking approval for new hepatic treatment indications both in the United States and worldwide. Over the next several years, on-label indications may match what many leading institutions are currently performing regularly as off-label treatment with radioembolization.
- New techniques in posttreatment quantitative imaging will vastly expand the field's understanding of the dose–response relationships associated with radioembolization. These include the following: ^{90}Y positron emission tomography/computed tomography (PET/CT), quantitative bremsstrahlung single-photon emission computed tomography (SPECT)/CT, and increasing use of directly imageable ^{166}Ho microspheres. Worldwide clinical trials are currently being planned to collect the data necessary to determine these dose–response relationships using advanced imaging techniques.
- Alternatives to ^{90}Y radioembolization, such as ^{166}Ho radioembolization, are currently under clinical use in some parts of the world. Alternative isotopes such as this provide certain advantages over ^{90}Y, including effective imaging with magnetic resonance imaging (MRI) and SPECT, and will undoubtedly lead to expansion of the field of radioembolization in the near future.
- Extrahepatic usage of radioembolization is currently being investigated in clinical trials, including use for treatment of primary renal cell carcinoma.
- Multiple clinical trials are underway to assess the utility of adjuvant ^{90}Y radioembolization

with systemic chemotherapy, radio frequency, cryo- and IRE percutaneous ablative techniques with promising preliminary results.

While this is by no means an exhaustive list, it is still highly suggestive that the field of radioembolization is poised for rapid advancement in the near future. Although generally considered a third- or fourth-line palliative therapy for some forms of metastatic liver cancer, some leading institutions have moved radioembolization to a second-line treatment in combination with chemotherapy by taking advantage of new information and treatment planning techniques.

Many of the recent advancements in radioembolization are related more to radiation biology, nuclear medicine, and the physics of the treatment rather than the vascular aspects. As such, a book focusing on these topics is appropriate, especially in light of the expected near-term growth and advancement of the field. We expect that with the expanded use of radioembolization, many individuals who have little prior experience with the procedure may pick up this book. While they come from different backgrounds—medical physicists, radiation oncologists, nuclear medicine radiologists, interventional radiologists, and health physicists—all have a necessary role to play in the execution of a well-planned radioembolization therapy. We suggest that regardless of background, all individuals begin with Chapter 1, which expertly summarizes all aspects of the procedure in its entirety. The remaining chapters in this book fill in the details of radioembolization treatments as a currently valuable therapeutic method with many clinically relevant examples as well as some ideas that may aid the advancement of the field.

Alexander S. Pasciak, J. Mark McKinney,
and Yong C. Bradley

Editors

Alexander S. Pasciak, PhD, earned a BS in electrical engineering at the University of Washington, MS in health physics and PhD in nuclear engineering at Texas A&M University. In 2010, Dr. Pasciak completed a 2-year diagnostic medical physics residency at the University of Texas MD Anderson Cancer Center after which he worked for 5 years as a diagnostic medical physicist at the University of Tennessee Medical Center in Knoxville, Tennessee. Dr. Pasciak maintains a position as an associate professor of radiology at the University of Tennessee and is concurrently pursuing his MD degree at the Johns Hopkins University School of Medicine in Baltimore, Maryland. Dr. Pasciak is active in multiple research endeavors in the fields of interventional radiology and medical physics, and he has published papers in high impact journals. Dr. Pasciak has published 35 articles in peer-reviewed medical journals, presented 62 research abstracts at national meetings, has written 6 book chapters, and has filed 2 patents. He currently serves as principal investigator on three externally funded research grants involving radioembolization. Dr. Pasciak is certified by the American Board of Radiology (ABR) in diagnostic medical physics and is Mammography Quality Standards Act (MQSA) qualified.

J. Mark McKinney, MD, earned a medical degree at Loma Linda University School of Medicine in California where he completed a diagnostic radiology residency and interventional radiology fellowship. Dr. McKinney joined the Mayo Clinic in 1993. While at Mayo Clinic in Jacksonville, Florida he developed the interventional radiology fellowship, mentored residents, made numerous presentations, and authored more than 60 peer-reviewed articles, abstracts, and book chapters. From 2008 to 2012 Dr. McKinney was Chair of Radiology at the University of Tennessee Medical Center and initiated the University of Tennessee interventional oncology radioembolization program. Dr. McKinney returned to Mayo Clinic in 2012 and is serving as Chair of Radiology. Dr. McKinney is an associate professor of radiology in the Mayo Clinic College of Medicine. Dr. McKinney is a recent past president of the Association of Program Directors in Radiology and is involved in the new interventional radiology residency program.

Yong C. Bradley, MD, is an ABNM and ABR-certified nuclear medicine subspecialist at the University of Tennessee Medical Center in Knoxville, Tennessee. He completed his residency training in radiology at Tripler Army Medical Center in Honolulu, Hawaii, and subsequently completed a 2-year fellowship in nuclear medicine at Brooke Army Medical Center in San Antonio, Texas. Dr. Bradley is a former chief of radiology and nuclear medicine and molecular imaging at the Brooke Army Medical Center, Wilford Hall Medical Center, and San Antonio Military Medical Center. Dr. Bradley has been interested in cancer imaging and therapy for over 20 years. He has been involved in positron emission tomography (PET) and PET/computed tomography (CT) imaging since 1998, when he was first introduced to PET along with radioimmunotherapy. Presently, his interest has been concentrated in liver radioembolization.

Contributors

Oreste Bagni
National Institute of Ionizing Radiation Metrology
Rome, Italy

Guido Bonomo
European Institute of Oncology (IEO)
Milan, Italy

Francesca Botta
European Institute of Oncology (IEO)
Milan, Italy

Austin C. Bourgeois
University of Tennessee Graduate School
of Medicine
Knoxville, Tennessee

Ray Bradford
Mayo Clinic
Jacksonville, Florida

Yong C. Bradley
University of Tennessee Graduate School
of Medicine
Knoxville, Tennessee

Marta Cremonesi
European Institute of Oncology (IEO)
Milan, Italy

Marco D'Arienzo
National Institute of Ionizing Radiation Metrology
Rome, Italy

William D. Erwin
University of Texas MD Anderson Cancer Center
Houston, Texas

Mahila Ferrari
European Institute of Oncology (IEO)
Milan, Italy

Luca Filippi
Santa Maria Goretti Hospital
Latina, Italy

Gregory T. Frey
Mayo Clinic
Jacksonville, Florida

Marcelo S. Guimaraes
Medical University of South Carolina
Charleston, South Carolina

Ram Gurajala
Cleveland Clinic
Cleveland, Ohio

Christopher Hannegan
Medical University of South Carolina
Charleston, South Carolina

Yung Hsiang Kao
Royal Melbourne Hospital
Melbourne, Australia

S. Cheenu Kappadath
University of Texas MD Anderson Cancer Center
Houston, Texas

Karunakaravel Karuppasamy
Cleveland Clinic
Cleveland, Ohio

Edward Kim
Mount Sinai School of Medicine
New York, New York

Susan D. Kost
Cleveland Clinic
Cleveland, Ohio

Marnix G.E.H. Lam
University Medical Center
Utrecht, The Netherlands

Safet Lekperic
Mount Sinai School of Medicine
New York, New York

J. Mark McKinney
Mayo Clinic
Jacksonville, Florida

Roberto Orecchia
European Institute of Oncology (IEO)
Milan, Italy

Franco Orsi
European Institute of Oncology (IEO)
Milan, Italy

Alexander S. Pasciak
University of Tennessee Graduate School
of Medicine
Knoxville, Tennessee

Ricardo Paz-Fumagalli
Mayo Clinic
Jacksonville, Florida

David M. Sella
Mayo Clinic
Jacksonville, Florida

Sankaran Shrikanthan
Cleveland Clinic
Cleveland, Ohio

Shyam M. Srinivas
Cleveland Clinic
Cleveland, Ohio

Lidia Strigari
Regina Elena National Cancer Institute (IFO)
Rome, Italy

Daniel Y. Sze
Stanford University Medical Center
Stanford, California

Joseph Titano
Mount Sinai School of Medicine
New York, New York

Andor F. Van Den Hoven
University Medical Center
Utrecht, The Netherlands

Stephan Walrand
Catholic University of Louvain
Louvain, Belgium

Naichang Yu
Cleveland Clinic
Cleveland, Ohio

PART 1

Introduction

1

Introduction to hepatic radioembolization

ANDOR F. VAN DEN HOVEN, DANIEL Y. SZE, AND MARNIX G.E.H. LAM

1.1 GENERAL INTRODUCTION

1.1.1 WHAT IS RADIOEMBOLIZATION?

Radioembolization is a therapy during which radioactive microspheres are administered through a microcatheter placed in the hepatic arterial vasculature to irradiate liver tumors from within. This therapy is based on the principle that liver tumors are almost exclusively vascularized by the hepatic artery, whereas the healthy liver tissue receives the majority of its blood supply from the portal vein. Therefore, following the administration in the hepatic artery, microspheres will be carried preferentially toward the distal arterioles in and around tumors. Clusters of microspheres are formed inside and in the periphery of tumors, where they emit high-energy β-radiation to induce cell death, while relatively sparing the healthy liver tissue (Braat et al., 2015). Radioembolization is a minimally invasive, image-guided, locoregional alternative, or adjunct to more conventional therapies such as surgery, systemic chemotherapy, and external beam radiation therapy for patients with liver-dominant malignancy. The advantages of this treatment are the targeted delivery of a very high radiation-absorbed dose to tumors, with limited systemic side effects and hepatotoxicity (Kennedy, 2014).

The efficacy and safety of radioembolization have been proven in patients with primary liver tumors such as hepatocellular carcinoma (HCC) (Hilgard et al., 2010) and intrahepatic cholangio-carcinoma (ICC) (Mouli et al., 2013), as well as in metastatic liver tumors from various primary tumors, with colorectal cancer (CRC) (Kennedy et al., 2015), breast cancer (BrC), neuroendocrine tumors (NET) (Devcic et al., 2014), and uveal melanoma (Xing et al., 2014) being the most common. Typically, radioembolization is performed as a stand-alone treatment in salvage patients with liver-dominant disease, but several clinical trials are currently evaluating its role in earlier lines of treatment and in combination with systemic therapy or other locoregional treatments such as radiofrequency ablation.

"Radioembolization" is used as an umbrella term for the treatment of liver tumors with varying disease extents ranging from a single focal subsegmental liver tumor to extensive disseminated or infiltrative disease, which can be hypo- to hypervascular in nature, situated in livers that are relatively healthy, cirrhotic, partially resected, transplanted, or heavily pretreated with systemic or intra-arterial chemotherapy. These situations pose various challenges and require other approaches with regard to safety precautions, treatment planning and dose calculation, microsphere type usage, and catheter positioning during administration. Furthermore, treatment techniques and strategies are dependent on operator experience and preferences and may differ considerably among practices.

Research continues to provide new insights into how to optimize radioembolization treatment, and new indications continue to arise. Among the latest introductions are radiation segmentectomy as a potentially curative technique to eradicate focal solitary liver tumors (Riaz et al., 2011), downstaging of unresectable disease to enable potentially curative surgical resection or transplantation (Braat et al., 2014), and radiation lobectomy to induce contralateral hypertrophy as an alternative to portal vein embolization in surgical candidates (Gaba et al., 2009; Vouche et al., 2013). Additional information on these techniques is presented in Chapter 6. Applying radioembolization principles to the treatment of solid tumors in organs other than the liver has also been provisionally explored, but falls outside the scope of this book.

1.1.2 A BRIEF HISTORY OF RADIOEMBOLIZATION

Several earlier studies and discoveries have set the backdrop for the clinical development of radioembolization as a technique to treat liver tumors. These investigations showed that large quantities of glass microspheres could be safely administered intra-arterially in animal experiments (Prinzmetal and Ornitz, 1948), that radioactive gold-covered charcoal particles administered intravenously or yttrium oxide particles administered via a pulmonary artery catheter could be used to treat lung cancer patients successfully (Muller and Rossier, 1951), and that liver tumors, even ones that reached the liver via the portal circulation, were preferentially vascularized by the hepatic artery when they exceeded about 50 μm in diameter (Bierman et al., 1951). The first report on radioembolization was published in 1960 by the American surgeon Edgar D. Grady and his colleagues, affiliated with Piedmont Hospital and Georgia Institute of Technology in Atlanta, GA, USA (Grady et al., 1960). Subsequent preclinical and clinical investigations by Kim et al. (1962), Caldarola et al. (1964), Blanchard et al. (1965a), and Ariel (1965) followed shortly thereafter. However, technical aspects such as the method to access the hepatic vasculature, the site of administration, safety precautions, size and material of the particles, and the radioactive isotope and the amount of activity to be infused still needed to be refined in the years to follow.

Experiments with New Zealand rabbits demonstrated that injection of radioactive microspheres via the hepatic artery established preferential tumor targeting, whereas injection via the portal vein did not (Blanchard et al., 1965a), which echoed early clinical results in humans (Grady, 1979). However, it proved challenging to catheterize the hepatic artery in both animals and humans. Access methods included antegrade catheterization of the celiac artery via brachial artery access, retrograde catheterization through femoral arteriotomy with the use of a balloon below the level of the celiac artery, and catheterization of the hepatic artery by accessing the gastroepiploic artery during laparotomy.

After trial and error it was learned that additional safety precautions were required, since extrahepatic deposition of radioactive microspheres (in the gastrointestinal tract or lungs) as well as too much radiation exposure of the healthy liver tissue could result in life-threatening complications (Blanchard et al., 1965b). Therefore, routine "skeletonization" (a surgical term used to describe isolation of the main vascular trunk by ligating all side branches) of the hepatic artery, as well as injection and imaging of radiolabelled albumin particles before treatment to simulate the therapeutic microsphere distribution, was advocated and eventually became standard of practice (Grady, 1979; Ariel and Padula, 1982).

Initially, glass microspheres of 50–100 µm diameter were used. Soon, however, it was recognized that smaller resin microspheres (15–30 µm) were easier to keep in suspension and would still not pass through the capillaries. After several years of experimentation with other isotopes such as Phosphorus-32 (^{32}P) (Caldarola et al., 1964; Grady et al., 1975), Yttrium-90 (^{90}Y) established its dominance. Reported benefits of ^{90}Y included a pure high-energy yield of tumoricidal β-radiation (max energy of 2.28 MeV), a short soft-tissue penetration (max 11 mm), and a 64-h half-life, which limited potential safety hazards for persons in close proximity to a treated patient. Early reports did, however, acknowledge the importance of imaging the posttreatment microsphere distribution and the limited possibilities inherent to the use of ^{90}Y (Grady et al., 1963; Ariel, 1965). The secondary bremsstrahlung γ-ray produced by β-activity could be detected with a Geiger–Muller survey meter or a scintillation crystal probe. Ariel even added Ytterbium-169 (^{169}Yb; γ-ray 52–310 keV; $T_{1/2}$ 32 days) to the microspheres as a radiation source for imaging with a γ-camera (Ariel, 1965).

Determining the optimal treatment activity (pretreatment dosimetry) has been a challenge from the start (Blanchard et al., 1965b). It was already recognized that the intrahepatic microsphere distribution is highly heterogeneous after treatment, but imaging methods available at that time precluded the assessment of the tissue mass exposed to radiation. Therefore, treatment activity could not be adapted to effective tumor-absorbed dose and safe healthy liver-absorbed dose values. Instead, the required treatment activity was calculated based on a target whole liver-absorbed dose of 5000 rad (50 Gy), which had been demonstrated as a safe dose in animal experiments. Doses were prescribed based on the formula that per gram of liver tissue 1 mCi (37 MBq) would be required to deliver an absorbed dose of 182 rad (1.82 Gy) (Grady, 1979).

The first efficacy reports were case series reporting posttreatment survival and the clinical condition of patients with primary or metastatic liver cancer. These results were generally promising, and some cases showed unprecedented disease control, but these reports were written prior to the availability of computed tomography, magnetic resonance imaging, and quantitative ultrasonography. Patients with inoperable disease had no good alternatives at that time, since the effectiveness of systemic chemotherapy and external beam radiation therapy remained disappointing. In 1989, Gray et al. published the first prospective trial results on radioembolization demonstrating an objective treatment response, defined as a decline of carcinoembryonic antigen (CEA) levels after treatment in 9/10 treated patients with colorectal cancer liver metastases (Gray et al., 1989). In the next two decades, only a few prospective studies followed patients with primary liver cancer and colorectal liver metastases (Lau et al., 1994; Rosler et al., 1994; Gray et al., 2001). Among these studies was the first randomized controlled trial, which demonstrated that the addition of radioembolization to regional hepatic arterial chemotherapy (floxuridine) in salvage patients with colorectal cancer liver metastases resulted in significantly improved tumor response.

Eventually, ^{90}Y-microspheres received Conformité Européenne (CE) mark in the European Union and U.S. Food and Drug Administration (FDA) approval in the United States for the treatment of HCC and metastatic colorectal cancer, which in turn led to a broader availability of radioembolization to patients and a renewed scientific interest.

The past two decades have been characterized by an enormous growth in the widespread use of radioembolization to treat salvage patients, with either primary or metastatic liver cancer. It is increasingly acknowledged that, as long as the liver disease is the survival-limiting factor in the patients' prognosis, radioembolization treatment is expected to be beneficial in patients with all kinds of liver-dominant tumor types. Patient selection, workup, treatment technique, and analyses of treatment toxicity and response have all been vastly improved. Modern imaging techniques including multidetector contrast-enhanced computed tomography (CT), magnetic resonance imaging (MRI), and C-arm cone beam CT now allow for a detailed assessment of tumor location, tumor characteristics, and individual hepatic arterial anatomy before treatment. This enables the operator to set a feasible individualized treatment strategy with the aim to achieve adequate tumor targeting, while minimizing the chance of treatment-related complications. The advent of nuclear medicine imaging techniques such as single photon emission computed tomography (SPECT)/computed tomography (CT) and positron emission tomography (PET)/CT, as well as the development of non-^{90}Y microspheres such as Holmium-166 (^{166}Ho) microspheres, has enabled imaging of the particle distribution and quantification of radiation-absorbed doses. It is now possible to identify an unfavorable particle distribution early on when the treatment plan can still be modified. Tumor response assessment is also becoming less observer dependent with all the possibilities that functional MRI and 18-fluoro-2-deoxyglucose positron emission tomography (^{18}F-FDG-PET) imaging have to offer.

The challenges for the near future will be to clarify which patients will benefit most from radioembolization, to improve methods for treatment activity calculations, to maximize treatment efficacy, to reduce treatment-related toxicity, to standardize treatment technique, to enhance our understanding of relevant particle-fluid dynamics, radiobiology, and systemic treatment effects, to explore combination therapies, and to strengthen scientific evidence by proving superiority over conventional and emerging therapies in large-scale phase III randomized controlled trials. These topics will be discussed in more detail in Chapter 15.

1.1.3 INDICATIONS FOR RADIOEMBOLIZATION

At this moment, the indication for radioembolization as a stand-alone therapy for patients with liver metastases is primarily based on unresectable, liver-dominant metastases refractory to standard systemic therapy. The standard for systemic therapy differs per primary tumor type and per geographical location, and may include cytotoxic chemotherapeutic agents as well as targeted small molecules, monoclonal antibodies, and immunomodulators. The prevailing principle is that no other therapy should be available with more convincing scientific evidence of effectiveness. Patients with contraindications to or unacceptable toxicity from systemic therapy are also eligible. Since large randomized controlled studies are currently investigating the role of radioembolization combined with systemic therapy in the first- and second-line treatment of colorectal cancer liver metastases, radioembolization may potentially be performed earlier in the treatment cycle in the future.

In patients with HCC, radioembolization is generally reserved for patients with intermediate and early advanced disease stages (Braat et al., 2015). These are patients with large multinodular tumors (>3, ≥3 cm), with or without macrovascular invasion, sufficient liver function (Child–Pugh A–B), and an acceptable clinical condition [World Health Organization (WHO) performance status score 0–2], corresponding to Barcelona Clinic Liver Cancer staging system stages B–C (Forner et al., 2014). Some patients may have already failed chemoembolization and/or systemic treatment with sorafenib, but radioembolization is offered as an alternative to chemoembolization in some practices, even for earlier stage disease.

Treatment with radioembolization should be considered relatively aggressive, and must be technically feasible and clinically tolerable. Additional important criteria for patient selection are summarized in Table 1.1. It should be noted that indications and contraindications are subject to change over time as clinical experience, both positive and negative, accumulate over the years.

Table 1.1 Eligibility criteria and contraindications for radioembolization

| | Contraindications | |
Eligibility criteria	Relative	Absolute
Good clinical condition		WHO/ECOG PS >2[a]
Life expectancy ≥3 months		Life expectancy <3 months
Adequate vital functions	Mild laboratory abnormalities	Severe laboratory abnormalities indicating critical liver, renal, or bone-marrow failure
Adequate portal venous liver vascularization	Portal vein thrombosis[b] Prior portal vein embolization or main portal vein occlusion	
Adequate functional hepatic reserve	Cirrhotic liver (Child–Pugh score >B7), total bilirubin >2.0 mg/dL (=34.2 µmol/L)	Uncompensated liver failure
Liver tumor burden <70%		Liver tumor burden ≥70%
Undisturbed biliary system	Biliary stent or prior sphincterotomy, choledochojejunostomy, or hepaticojejunostomy[c]	Active cholangitis
Adequate arterial access	Celiac axis and superior mesenteric artery occluded	Occluded intrahepatic arterial network
Successful preparatory angiographic procedure		Uncorrectable gastrointestinal microsphere deposition; expected lung dose >30 Gy (or 50 Gy cumulative)
Interval since last dose of systemic therapy ≥4 weeks	Interval since last dose of systemic therapy <4 weeks	Active use of antiangiogenic agents (bevacizumab, aflibercept)

Note: The most important eligibility criteria and associated contraindications for radioembolization are presented in this table. A distinction between relative and absolute contraindications is made. ECOG PS, World Health Organization/Eastern Cooperative Oncology Group Performance Score.

[a] Corresponds to Karnofsky score <50.
[b] No contraindication in HCC.
[c] May require antibiotic prophylaxis to prevent liver abscesses.

1.1.4 COMPARISON OF RADIOEMBOLIZATION AND EXTERNAL BEAM RADIATION THERAPY PRINCIPLES

Radioembolization is also referred to as selective internal radiation therapy (SIRT), indicating that it is a preferentially tumor targeted form of brachytherapy that differs considerably from external beam radiation therapy (EBRT). Both therapies aim to achieve tumor cell death through radiation-induced apoptosis and proliferative capacity inhibition (Eriksson and Stigbrand, 2010). However, differences in the technical methods to target the tumor have an important impact on the radiation distribution, as well as the dose rate and the possibility to fractionate the treatment dose, which in turn determines inherent strengths and limitations of both treatment modalities.

Conventional whole-liver EBRT is no longer considered a viable treatment option. The healthy liver tissue is very sensitive to radiation, and whole-liver-absorbed doses exceeding 30 Gy are associated with an increased risk of potentially fatal radiation-induced liver failure (Emami et al., 1991; Fuss et al., 2004; Sharma, 2014). This phenomenon is even more pronounced in patients with cirrhotic liver disease, or a prior history of hepatotoxic systemic therapy. However, the introduction of image-guided, conformal, intensity modulated, stereotactic body radiation therapy (SBRT) has made it possible to achieve highly preferential tumor targeting, thereby sparing much of the adjacent and remote healthy liver tissue (Fuss et al., 2004; Sharma, 2014). A safety margin of up to 2 cm surrounding the tumor is required, which can result in some hepatotoxicity. Three-dimensional images of the liver are acquired during a simulation procedure, allowing drafting of a detailed therapy plan. Patients subsequently undergo multiple treatment sessions, during which photon beams produced by a linear accelerator apply small fractions of the total radiation dose. The greatest advantage of EBRT is that the radiation can be actively targeted with precision, yielding a predictable dose and treatment effect. The most important disadvantages are that it is not feasible to treat patients with a high tumor burden, and the maximum radiation dose delivered is still limited by potential hepatotoxicity.

Radioembolization, on the contrary, depends on a hemodynamic mechanism for tumor targeting. The interplay between catheter positioning, particle-fluid dynamics, and tumor vascularization ultimately determines the dose delivered to tumors and to normal liver. Pathological examination of explanted livers that were treated with radioembolization showed that the posttreatment microsphere distribution was highly heterogeneous. Furthermore, the microspheres tended to cluster preferentially in the periphery of the tumors with a concentration up to 200 times greater than in the tumor core or in healthy liver tissue (Campbell et al., 2001; Kennedy et al., 2004). Therefore, the greatest benefits of radioembolization are that very high tumor-absorbed doses can be achieved and that treatment feasibility is less dependent on tumor burden. However, inherent disadvantages are that tumor targeting can turn out to be suboptimal in patients with relatively hypovascular tumors, yielding highly variable tumor response rates and liver toxicity, and that the accurate prediction of the therapeutic microsphere distribution is a great challenge.

Additional detailed discussion on the hepatic radiation biology of radioembolization and EBRT are presented on macroscopic and microscopic levels in Chapters 8 and 9, respectively.

1.2 TYPES OF MICROSPHERES AND RADIONUCLIDES USED

Different types of radioactive microspheres are commercially available. The type of microspheres can be divided based on the embedded radioactive isotope (^{90}Y or ^{166}Ho) or microsphere material (resin, glass, or poly-L-lactic acid). These microspheres all have different production processes, physical characteristics, and methods of use. The most important characteristics of the different microsphere types are summarized in Table 1.2.

1.2.1 YTTRIUM-90 MICROSPHERES

^{90}Y is a nearly pure (99.99%) β-emitter. It has a half-life of 64.1 h and decays to stable Zirconium-90 (^{90}Zn). With maximum beta particle (β$^-$) energy of 2.28 MeV—resulting in an energy release of 49.67 J/GBq—and a range in water or soft-tissue of 2.5 mm (mean) and 11 mm maximum, it is a suitable radionuclide to treat cancer with an appropriate safety profile. Radioactive ^{90}Y can either be produced by neutron irradiation of stable Yttrium-89 (^{89}Y) or by chemical separation from the parent isotope Strontium-90 (^{90}Sr), a fission product of uranium (Walker, 1964). Imaging of the radiation emission from ^{90}Y is a challenge due to the absence of γ-radiation emission. SPECT images can only be acquired by the detection of bremsstrahlung, secondary γ-radiation produced by slowing of the beta particles in tissue (Figure 1.1a), a modality with very limited spatial resolution. Actually, ^{90}Y has a minor branch to the first excited state of ^{90}Zn at 1.76 MeV (0$^+$–0$^+$ transition). As a result, once in every 32 million (31.86 × 10^6) decays, an electron–positron (β$^-$/β$^+$) pair is created. This process is called internal-pair production and enables positron emission detection with PET at high ^{90}Y-activities (Figure 1.1a) (D'Arienzo, 2013). SPECT and PET imaging of ^{90}Y are discussed further in Chapters 10 and 11.

Table 1.2 Microsphere characteristics

Isotope	Yttrium-90 (^{90}Y)		Holmium-166 (^{166}Ho)
Half-life	64.1 h		26.8 h
Decay product	Zirconium-90 (^{90}Zn)		Erbium-166 (^{166}Er)
Radiation emission	β (max 2.28 MeV)		β (max 1.74 and 1.85), γ (max 81 and 1.38 keV)
Energy per activity	49.67 J/GBq		15.87 J/GBq
Tissue penetration	2.5 mm mean, 11 mm maximum		2.5 mm mean, 8.4 mm maximum
Imaging	PET (internal-pair production) SPECT (bremsstrahlung)		SPECT (γ-imaging) MRI (R2* mapping)
Material	Glass (ceramic)	Resin	PLLA
Product name	TheraSphere®	SIR-Spheres®	QuiremSpheres®
Size	20–30 μm	32.5 ± 5 μm	20–50 μm
Density	3.3 g/cc	1.6 g/cc	1.4 g/cc
Spheres per vial	$1.2–8 \times 10^6$	$40–80 \times 10^6$	33×10^6
Specific activity per sphere	2500 Bq	40–70 Bq	450 Bq
Max activity per dose	20 GBq	3 GBq	15 GBq
Dosimetry method recommended by manufacturer	MIRD based	BSA method	MIRD based
Embolic effect	Low	Moderate	Moderate

Note: The characteristics of the glass and resin yttrium-90 microspheres and holmium-166 microspheres are summarized in this table. PLLA, poly-L-lactic acid.

Glass and resin ^{90}Y-microspheres currently have near worldwide commercial availability. In 2002, resin ^{90}Y-microspheres received FDA approval for the treatment of metastatic colorectal cancer when given in conjunction with floxuridine, and received CE mark for the treatment of inoperable liver tumors. Glass ^{90}Y-microspheres received FDA approval under a humanitarian device exemption in 1999 for the treatment of patients with HCC, including those with branch portal vein thrombosis, and received CE mark in 2005 for the treatment of hepatic neoplasms.

1.2.1.1 Glass microspheres

Glass ^{90}Y-microspheres (TheraSphere®, Nordion Inc. for BTG International, Ottawa, ON, Canada) are produced by incorporating ^{89}Y oxide into the glass matrix of the microsphere and subsequent activation by neutron bombardment in a nuclear reactor facility (Wollner et al., 1988). Compared with the other microsphere types, glass ^{90}Y-microspheres have a relatively high density, and a high specific activity per sphere (2500 Bq/sphere). Therefore,

10–20 times less particles need to be injected than with resin microspheres to administer the same treatment activity. As a consequence, the embolic effect is much smaller during injection, so the entire treatment dose can be injected at once with a lower risk of stasis and particle reflux.

1.2.1.2 Resin microspheres

The production process of resin ^{90}Y-microspheres (SIR-Spheres®, Sirtex Medical Limited, North Sydney, Australia) is different; in this type of microsphere, ^{90}Y cations in solution are chemically incorporated onto the bland microsphere surface by binding to the carboxylic group of the acrylic polymer matrix (Gulec and Siegel, 2007; Giammarile et al., 2011). Resin microspheres have a much lower density than glass microspheres, which could potentially result in a more distal distribution in the tumor vasculature (Jernigan et al., 2015). Furthermore, the relatively low specific activity requires injection of a higher number of microspheres, approximately 20-80 million. Since this involves a greater embolic effect, stasis of blood

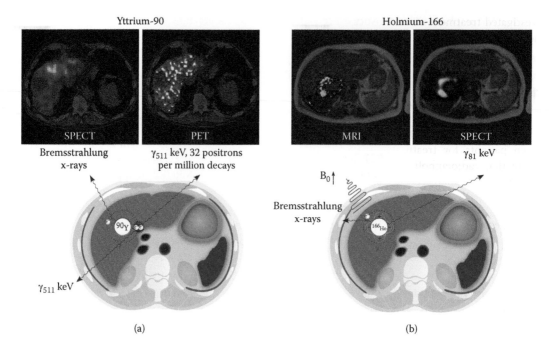

Figure 1.1 Schematic illustration of the physical properties and decay of ^{90}Y **(a)** and ^{166}Ho **(b)** in the liver. The *upper row* displays examples of the microsphere biodistribution in the liver using ^{90}Y bremsstrahlung SPECT (a, *upper left*), ^{90}Y PET (a, *upper right*) ^{166}Ho-MRI (b, *upper left*), and ^{166}Ho-SPECT (b, *upper right*). (With kind permission from Springer Science+Business Media. *Cardiovasc Intervent Radiol*, Radioembolization dosimetry: The road ahead, 38, 2014, Smits et al.)

flow may occur during administration. Therefore, resin ^{90}Y-microspheres must be administered carefully by hand injection in smaller aliquots, with intervening angiography to reevaluate pace of flow and degree of stasis. Glass and resin microspheres may be used in different or similar tumor types and disease extents, but it remains controversial how the differences in distribution patterns impact treatment efficacy.

1.2.2 HOLMIUM-166 MICROSPHERES

The isotope ^{166}Ho emits both high-energy β-radiation and low-energy γ-radiation. It has a shorter half-life than ^{90}Y (26.8 h) and decays with a relatively high dose rate to the stable element erbium-166 (^{166}Er). ^{166}Ho emits β-radiation at two energy levels, maximum 1.74 MeV (48.7%) and 1.85 MeV (50%), with a maximum soft-tissue range of 8.4 mm. The resulting energy release is much lower (15.87 J/GBq) than with ^{90}Y; therefore, a larger administered treatment activity is required to achieve the same radiation-absorbed

dose in liver tissue (Prince et al., 2014a). The biodistribution of ^{166}Ho-microspheres can be visualized on SPECT (Figure 1.1b), using the low-energy γ-radiation (81 keV, 6.2%; 1.38 keV, 0.93%), and with magnetic resonance imaging, utilizing the paramagnetic properties of ^{166}Ho (Figure 1.1b) (Smits et al., 2013a).

^{165}Ho acetylacetonate poly(L-lactic acid) particles are manufactured through a solvent evaporation process, and subsequently activated in a nuclear reactor facility to create radioactive ^{166}Ho acetylacetonate poly(L-lactic acid) microspheres (Nijsen et al., 2001; Zielhuis et al., 2006). The density, size, specific activity per sphere, number of injected ^{166}Ho-microspheres, and embolic effect are comparable with resin ^{90}Y-microspheres. ^{166}Ho-microspheres received CE mark approval, but no FDA approval yet. The transition toward commercial availability in Europe is ongoing. The safety of ^{166}Ho-radioembolization has been demonstrated in a phase I dose escalation study in patients with unresectable chemorefractory liver metastases from various primary tumor types, and the results of a phase II clinical trial that investigated treatment efficacy are expected to be released soon.

1.3 PRETREATMENT WORKUP

A thorough pretreatment workup is essential to screen patients for treatment eligibility, and to ensure that radioembolization treatment is performed as safely and effectively as possible. A standard workup includes laboratory and clinical investigations, pretreatment cross-sectional imaging, a preparatory angiographic procedure, simulation scout dose imaging, and pretreatment activity calculation.

1.3.1 LABORATORY AND CLINICAL INVESTIGATIONS

Laboratory and clinical investigations are used to assess vital functions and general clinical status, and to record baseline values as a reference for toxicity assessments.

The laboratory investigations should include parameters to assess hepatobiliary function (albumin, total bilirubin, gamma-glutamyl transferase, alkaline phosphatase, alanine aminotransferase, aspartate aminotransferase), renal function (creatinine, estimated glomerular filtration ratio), coagulation status (platelet count, prothrombin time, activated partial thromboplastin time, thrombin time, internationalized normalized ratio), and hematological function (hemoglobin and hematocrit, white blood cell count). Additional parameters to consider are tumor markers (carcinoembryonic antigen [CEA], alpha-fetoprotein [AFP], chromogranin-A [CgA], etc., depending on the tumor type and biomarker secretion), and indicators for an acute infection (C-reactive protein) or cell tissue damage (lactate dehydrogenase).

Deviations from reference values are expected, considering the severity of disease in most patients. However, severe deviations (Common Terminology Criteria for Adverse Events or CTCAE grades 3–4) should be considered indicators of organ dysfunction, which may render radioembolization treatment unsafe.

Previous treatments, chronic diseases, recent periods of acute illness, current medication use, and allergies (especially for contrast agents) should be recorded during clinical investigation. The general clinical status can be summarized by performance status (as defined by the WHO/Eastern Cooperative Oncology Group): 0—asymptomatic, 1—symptomatic but completely ambulatory, 2—symptomatic <50% of time in bed, 3—symptomatic >50% of time in bed but not bedbound, 4—bedbound, 5—dead. A performance status >2 is considered an exclusion criterion for radioembolization. Occasionally, a poorer performance status may reflect toxicity from other therapies that is expected to improve after discontinuation of those therapies, a potential exception to the guidelines.

1.3.2 PRETREATMENT IMAGING: LIVER CT/MRI AND ¹⁸F-FDG-PET

Cross-sectional pretreatment imaging is used for the evaluation of the liver parenchyma, vasculature, and presence and extent of extrahepatic disease. CT, magnetic resonance imaging (MRI), and whole-body ¹⁸F-FDG-PET all play an important role.

For the evaluation of the liver parenchyma—characterization and localization of liver tumors and their relation with surrounding vessels and biliary ducts—MRI is superior to CT in terms of soft-tissue contrast. Dynamic contrast-enhanced sequences can be used to assess tumor hypervascularity and washout in a larger number of phases, while diffusion weighted and T2-weighted imaging provide options for high-sensitivity tumor detection. Imaging with CT is faster, cheaper, higher spatial resolution, and less susceptible to motion artifacts, but this comes at the cost of ionizing radiation and iodinated contrast agent burden. When using CT, multiphasic images (arterial, portal venous, equilibrium phase) should be acquired and adequate timing of these phases is especially important to depict tumor types with different degrees of arterial vascularization. A late arterial phase and an equilibrium phase are recommended for the imaging of hypervascular tumors, while relatively hypovascular tumors are often best visualized on a portal venous phase (Rengo et al., 2011).

An arterial phase and a portal venous phase are required for vascular assessment. The arterial phase can be used to evaluate accessibility of the hepatic artery (through the celiac axis and superior mesenteric artery), to reveal the individual hepatic arterial anatomy including variants and parasitized arteries, and to identify tumor-feeding branches. The portal venous phase allows for evaluation of the portal and hepatic veins for patency and tumor invasion, and it confirms segmentation of the liver.

The importance of oncological ^{18}F-FDG-PET imaging is increasingly acknowledged. ^{18}F-FDG-PET has a high sensitivity for the detection of liver metastases, especially in patients with tumor types that are expected to be PET avid (cholangiocarcinoma, liver metastases from CRC, BrC, and uveal melanoma) (Tsurusaki et al., 2014). Furthermore, ^{18}F-FDG-PET shows more extrahepatic lesions than CT during radioembolization workup in patients with metastatic colorectal cancer (mCRC), which can lead to considerable changes in management (Rosenbaum et al., 2013). In patients with HCC and NET liver metastases, the role of ^{18}F-FDG-PET remains secondary, due to a less reliable tumor avidity on ^{18}F-FDG-PET. Other tracers such as gallium-68 tetraazacyclododecane tetraacetic acid–octreotate (DOTATATE) are currently in development to improve the sensitivity of PET or scintigraphy in these tumor types.

The liver tumor burden can be determined on all three imaging modalities, and programs are being developed to segment the liver automatically based on thresholds for contrast enhancement or metabolic activity. Some tumors replace liver parenchyma and some add to the total liver volume. Guidelines warn against treating patients with liver tumor burden >50% or >70% to ensure enough healthy liver tissue reserve to tolerate radioembolization; a measurement of volume and function of liver tissue would be more valid, but is not currently standard.

1.3.3 ASSESSING THE INDIVIDUAL HEPATIC ARTERIAL ANATOMY

The functional anatomy of the liver is based on the branching pattern of the portal vein, which usually parallels the hepatic artery and biliary ducts. According to the Couinaud model of segmental anatomy, eight (or nine in the Couinaud–Bismuth model) liver segments can be distinguished with a distinct vascularization and biliary drainage. An avascular plane, called the portal scissura, separates the functional left and right liver. Cantlie's line—an imaginary line drawn anteriorly from the middle of the gallbladder fossa to the inferior vena cava posteriorly—or the course of the middle hepatic vein can be used to indicate this division on cross-sectional imaging. The right hemi-liver is divided into an anterior and posterior sector by the right portal scissura, as indicated by the course of the right hepatic vein. The level of the portal bifurcation marks the distinction of superior and inferior segments in these sectors (segments 5/8 anterior, and segments 6/7 posterior). The left hemi-liver is similarly divided into an anterior and posterior sector by the left portal scissura, as indicated by the course of the left hepatic vein. The fissure for the falciform ligament divides segment 3 and 4 in the left anterior sector, whereas segment 2 forms the only segment of the left posterior sector. Segment 4 can also be divided into cranial and caudal subsegments 4a and 4b. Segment 1 is situated between the portal bifurcation and the inferior vena cava. This independent segment receives small branches from the left and right portal vein (Bismuth, 1982; Majno et al., 2014).

Deep within the liver, hepatic arterial branches, portal–venous branches and biliary ducts run alongside one another, enclosed in the Glissonian sheath. The hepatic arterial anatomy is nevertheless more complex and variable than the portal–venous anatomy, because anatomical variants of the hepatic artery can occur on three different levels: the origin of hepatic arterial branches, the branching pattern of the hepatic arterial tree, and the segmental territory vascularized by the individual hepatic arterial branches (van den Hoven et al., 2015). Anatomical variants of the portal vein, on the contrary, are mainly limited to variants in branching order (van Leeuwen et al., 1994).

The standard arterial anatomy of the adult liver is described as a common hepatic artery (CHA) originating from a celiac trifurcation that gives off the gastroduodenal artery branch (GDA) and then continues as the proper hepatic artery (PHA), dividing into the left (LHA) and right (RHA) hepatic arteries, vascularizing segments 2–4 and 5–8, respectively. However, approximately half of radioembolization candidates have a variant to this configuration (van den Hoven et al., 2015).

An explanation of why variants of the hepatic arterial anatomy are so common lies in the embryological development. During early development, three main arteries exist: an embryological LHA (eLHA) originating from the left gastric artery (LGA), an embryological middle hepatic artery (eMHA) from the celiac axis, and an embryological right hepatic artery (eRHA) from the superior mesenteric artery (SMA). The eLHA and eRHA should eventually regress, so that the eMHA forms a separate LHA and RHA that take over the vascularization of the entire liver (Wang et al., 2010). However, one (or both) of the embryological arteries often persists, resulting in the presence of a hepatic artery with an aberrant origin, a so-called aberrant hepatic artery. Logically, most aberrant hepatic arteries originate from the LGA or SMA. Aberrant hepatic arteries can be divided into replaced hepatic arteries that are the main supply to the right or left hepatic lobe (or both in the case of a replaced CHA), and accessory hepatic arteries that only supply a part of the lobe. In the latter case, the remaining part of the liver lobe is vascularized by a normally derived counterpart of the aberrant LHA/RHA.

A variant in the order of arterial branching is present in up to 20% of patients. In these patients, an early branching LHA/RHA may be present as (part of) the LHA or RHA that originates proximal to the GDA. Another variant is a trifurcation (or quadrifurcation) of the CHA. This means that the PHA is absent, so that the CHA branches into a GDA, LHA, RHA (and MHA).

Variants in the segmental vascularization pattern include variants in the origin of the arteries vascularizing segments 1 and 4, as well as unexpected vascular territories of aberrant hepatic arteries. The arterial feeding branch(es) to segment 4 may originate from the LHA, RHA, both, or from a separate origin of the CHA/PHA. Segment 1 may in addition be vascularized by a branch originating from the segment 4 artery.

All of the previous outlined anatomical variants can coincide, leading to complex and unexpected individual hepatic arterial configurations and segmental vascularization patterns. Failing to identify aberrant hepatic arteries may result in incomplete treatment, whereas incorrect judgment of the segmental vascularization pattern may result in under- or overdosing of the treatment activity (van den Hoven et al., 2014b). Thus, it is essential to assess the individual hepatic arterial anatomy in each patient.

Assessment on pretreatment arterial enhanced CT or MRI enables identification and characterization of anatomical variants, while providing guidance for subsequent catheterization during the preparatory angiography. Reading the pretreatment CT or MRI should be done systematically, scanning all potential sources for hepatic arterial branches,

and following them up to their segmental territory of vascularization, with attention for the order of branching. The fissure for the ligamentum venosum and the portocaval space should be screened with special attention, since these are the locations where the majority of aberrant LHAs and aberrant or early branching RHAs, respectively, course before entering the liver (van den Hoven et al., 2014b, 2015). Chapter 3 discusses hepatic arterial anatomy in the context of treatment planning in additional detail.

1.3.4 PREPARATORY ANGIOGRAPHY AND INTRAPROCEDURAL IMAGING

Preparatory angiography is performed before treatment to map the arterial anatomy, to allow prophylactic or redistributive coil embolization of arterial branches if necessary to determine optimal catheter positioning and to administer a simulation scout tracer. Intra-arterial access is gained through transcutaneous puncture, using the Seldinger technique, either through a transfemoral or transradial approach. After securing the access site, a preshaped catheter is used to enter the source of the hepatic arterial vasculature (usually the celiac axis). A standard end-hole microcatheter is advanced over an atraumatic microguidewire for further selectively catheterization.

Intraprocedural imaging with digital subtraction angiography (DSA) and C-arm cone beam CT (CBCT) plays a pivotal role during the preparatory angiography. DSA provides two-dimensional images of the vasculature at a high spatial resolution, providing a high-resolution projectional map to guide the catheterization. Furthermore, cinematic DSA images can be acquired during high rate contrast administration with a power injector, thereby depicting arterial flow rates, tumor blushes, and altered dynamics.

CBCT is a relatively new imaging modality that enables the acquisition of three-dimensional CT-like images, depicting contrast-enhanced vessels in relation to their surrounding soft-tissue structures, by rotation of the C-arm mounted flat-panel detector around the patient. The timing between contrast injection and start of the scan can be adjusted to influence whether the images mainly depict contrast enhancement of the arterial tree (early arterial phase), the liver and tumor parenchyma (late arterial phase, Figure 1.2a and b), or both (van den Hoven et al., 2015b). The hepatic arterial anatomy is mapped by a combination of DSA and CBCT. "Complete" hepatic arteriography should be performed to depict comprehensively all arteries supplying any portion of the liver (Liu et al., 2005; Lewandowski et al., 2007; Salem et al., 2007; van den Hoven et al., 2014b).

"Complete" hepatic angiography requires identification of parasitized extrahepatic arteries (Figure 1.3). These are tumor-feeding branches that have been recruited from arteries outside the hepatic vasculature (including prominent phrenic, intercostal, omental, internal mammary, adrenal, renal capsular, gastric, and pancreaticoduodenal arteries) by stimulation of neovascularization and are found in 17% of radioembolization patients. These branches can often be embolized, causing intrahepatic arteries to take over the entire blood supply to the tumor, to ensure complete tumor coverage during treatment (Abdelmaksoud et al., 2011a). Failure to recognize these arteries results in incomplete treatment.

Since extrahepatic deposition of radioactive microspheres may cause serious complications, it should be assessed whether side branches of the hepatic arterial vasculature pose a risk. The GDA, right gastric artery (RGA), and supraduodenal arteries (SDAs) deserve special attention. Nontarget deposition into the cystic artery and the falciform artery is less likely to be life threatening. The catheter can often be positioned distal to these side branches without sacrificing tumor coverage, but this may require splitting of doses. If that is not possible, prophylactic embolization with coils or vascular plugs may be performed. Routine embolization of the GDA was previously a standard of care but has become controversial, since this may actually induce the hypertrophy of new, very small hepatofugal vessels during the interval between the preparatory angiography and treatment (Abdelmaksoud et al., 2010; Lam et al., 2013a; Samuelson et al., 2013). Alternatively, administration of yttrium microspheres using an antireflux catheter or occlusion balloon catheter may provide a solution (Ahmadzadehfar et al., 2011; Prince et al., 2014b).

Choosing the optimal catheter position(s) requires attention for the risk of extrahepatic shunting by evaluating the distance between catheter tip and patent gastrointestinal side branches, by evaluating tumor coverage by making sure that the catheter is placed proximal to all tumor-feeding branches, and by evaluating flow dynamics by recognizing preferential flow directions. In

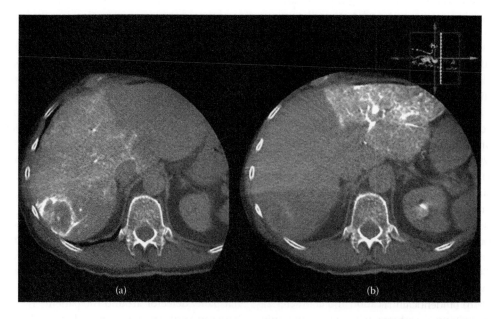

(a)   (b)

Figure 1.2 C-arm CT images demonstrating the arterial perfusion territory of the right **(a)** and left **(b)** hepatic artery. These late arterial phase images were separately acquired, with contrast injection by a microcatheter placed in the right and left hepatic artery, respectively. Both the healthy liver parenchyma and the tumor in the posterior part of the right liver lobe show contrast enhancement. Note how the vascular territories of both arteries complement each other.

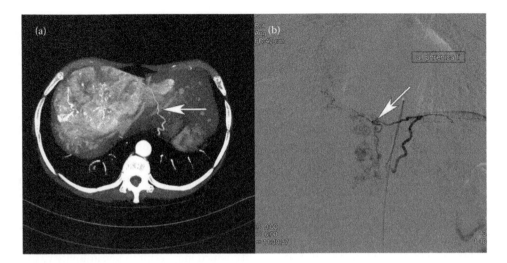

Figure 1.3 Parasitized left inferior phrenic artery. The arterial phase of the pretreatment CT **(a)** shows a prominent left inferior phrenic artery coursing adjacent to the upper part of a large hypervascular tumor mass (arrow). The suspected presence of a parasitized extrahepatic artery was confirmed during angiography. Digital subtraction angiography acquired after selective catheterization of the left inferior phrenic artery **(b)** demonstrates intrahepatic tumor blush (arrow).

general, it is considered impractical to use more than two to three injection positions. Coil embolization of intrahepatic arterial branches (e.g., the S4A, or an accessory LHA) may be used as a means to reduce the number of required injection positions, since this will induce redistribution of blood flow through intrahepatic collateral vessels (Bilbao et al., 2010; Abdelmaksoud et al., 2011b).

At the end of the preparatory angiography, it is advisable to do a test injection with contrast agent from the intended treatment positions. A gentle hand injection simulating the administration injection may reveal problematic flow dynamics resulting in a disproportional microsphere distribution over tumor-bearing liver segments, while a late arterial phase CBCT may show unintentional gastrointestinal shunting or lack of target segment perfusion (van den Hoven et al., 2015b).

After confirming adequacy of the catheter position(s), a simulation tracer test dose of scout particles is injected. Circa 30–150 MBq technetium-99m (^{99m}Tc)-macroaggregated albumin (MAA) is used as a surrogate for ^{90}Y-microspheres, while a scout dose (250 MBq) of identical ^{166}Ho-microspheres is available for ^{166}Ho-microspheres treatment. DSA images of the exact catheter position during injection should be stored.

The access site is closed by manual compression until hemostasis, or by using a closure device.

1.3.5 IMAGING OF THE SCOUT DOSE DISTRIBUTION

Currently, imaging of the scout dose distribution is mainly performed for safety reasons, to quantitate hepatopulmonary shunting and to rule out unintentional extrahepatic deposition. Planar scintigraphy can be used to calculate the lung shunt fraction after administration of ^{99m}Tc-MAA. For resin microspheres, the manufacturer recommends to lower the treatment activity in patients with a lung shunt fraction of 10%–20%, and to refrain from treatment in patients with a lung shunt fraction of >20%. This is a convenient and simple recommendation based on whole liver treatment of average sized persons, but with some critical flaws. First, lung shunt fraction is less important than the estimated lung-absorbed dose. The latter should be kept below 30 Gy (or 50 Gy cumulatively in patients who undergo sequential treatments). Second, planar scintigraphy-based assessment gives a less accurate estimation of the lung-absorbed dose than SPECT/CT, oftentimes averaging hepatic dome activity into right lung base activity (Yu et al., 2013). Third, it has been demonstrated that the use of ^{99m}Tc-MAA itself leads to an overestimation of the true liver-to-lung shunting, probably due to differences in particle characteristics, the broad range of ^{99m}Tc-MAA particle size, and circulating free pertechnetate (Elschot et al.,

2014; Smits et al., 2014). These points are discussed in greater detail in Chapter 4.

SPECT/CT is considered the gold standard for detection of unintentional extrahepatic activity in Europe. However, its relatively low spatial resolution and potential occurrence of misregistration between the SPECT data set and the low-dose CT data set can make the distinction between intra- and extrahepatic activity accumulation challenging. Therefore, many centers now base their assessment of extrahepatic perfusion primarily on CBCT.

The intrahepatic scout dose distribution should ideally be a good predictor for the treatment distribution, since determining the particle distribution in tumorous and nontumorous tissue (T/N ratio) would enable a patient-tailored treatment strategy. However, it has been demonstrated that the intrahepatic distribution of ^{99m}Tc-MAA is not a reliable predictor for the ^{90}Y-microsphere distribution, especially not in patients with relatively hypovascular liver tumors (Wondergem et al., 2013). This may be explained by the fact that the difference in particle characteristics, including different particle size, density, shape, and number, strongly affects the particle distribution (Van de Wiele et al., 2012). In patients with markedly hypervascular HCCs, preferential tumoral blood flow is so strong that these differences in particle characteristics have less influence, making the ^{99m}Tc-MAA distribution a more reliable predictor (Garin et al., 2012; Garin, 2015).

Using a scout dose of particles identical to the treatment microspheres, such as with ^{166}Ho-microspheres, may provide a more accurate prediction of the intrahepatic therapeutic microsphere distribution. Further investigation is required to confirm this hypothesis, because differences in catheter positioning between the scout procedure and the therapy procedure may also play an important role. The embolic effect of a scout dose could theoretically change flow patterns, and even the beginning and end of a partially embolic therapeutic administration could have different dynamics (Prince et al., 2015; van den Hoven et al., 2015a).

1.3.6 PRETREATMENT ACTIVITY CALCULATIONS

Different methods for pretreatment activity calculations (pretreatment dosimetry) have been proposed. The empirical method is the simplest

method. According to this method, a fixed treatment activity of ^{90}Y between 2 and 3 GBq (depending on the tumor burden) is prescribed. This method is no longer advocated, because it has been associated with unacceptable clinical toxicity (Smits et al., 2014).

The body surface area (BSA) method adjusts the prescribed activity for the patient's BSA and the fractional tumor burden. It assumes a correlation between BSA and liver weight, but this is not necessarily true for patients with liver cancer. As a consequence, small patients with a relatively large liver will be undertreated, potentially resulting in progressive disease, while tall patients with a small cirrhotic liver will receive too much activity, which may result in hepatotoxicity (Lam et al., 2014). In the vast majority of cases, however, it is a safe method to use, but it may be overly conservative for tumor-enlarged livers. The manufacturer of resin ^{90}Y-microspheres recommends the use of this method.

The method based on medical internal radiation dose (MIRD) is recommended by the manufacturer of glass ^{90}Y-microspheres. With this method, the prescribed activity is determined by calculating the activity required to achieve a desired absorbed dose (usually 80–120 Gy), assuming a homogeneous intrahepatic microsphere distribution throughout the treated portion of the liver and a known yield of 50 Gy per GBq ^{90}Y per kilogram of liver tissue. Although this method accounts for liver mass, it neglects the inter- and intrapatient variability of intrahepatic microsphere distribution between tumorous and nontumorous tissue (T/N ratio). It does not account for fractional or total tumor burden. The activity calculation method for ^{166}Ho-microspheres is also a MIRD method, with a fixed target whole liver-absorbed dose of 60 Gy (maximum tolerable dose derived from a phase I trial) and the ^{166}Ho-specific energy yield of 15.87 Gy per GBq per kilogram.

An anatomic partition model could be the most scientifically sound method for pretreatment activity calculations. This method extends the MIRD calculation by incorporating the T/N microsphere uptake ratio, the weight of the healthy liver tissue, and the maximum tolerable healthy liver-absorbed dose. Partition models are currently applicable for patients with few and well-defined tumors where a T/N ratio can be estimated. However, partitioning the liver into only two discreet volumes is at best inaccurate and in many cases impossible when disease is diffuse, infiltrative, and/or hypovascular. Physiologic and functional imaging methods are being developed to address this patient population.

Radioembolization treatment planning and dosimetry techniques are discussed in greater detail in Chapter 5.

1.4 TREATMENT

Most centers perform the preparatory angiography and treatment sessions as outpatient procedures, but some patients may require admission to the hospital before and/or after undergoing treatment, since clinical observation and drug therapy may be required. The treatment procedure is typically scheduled ~1–2 weeks after a successful preparatory angiography, but in some circumstances, can even be performed on the same day as the preparatory angiography.

1.4.1 MEDICATION AND PERIPROCEDURAL CARE

Monitoring of vital signs, with a heart rate monitor, electrocardiogram (ECG) registration, pulse oximetry, and periodic blood pressure measurements, is considered standard. The use of proton pump inhibitors or H2 receptor antagonists during the first week before treatment and a month after treatment is recommended as a prophylaxis for gastrointestinal ulcer formation. Furthermore, corticosteroids given for a few days or weeks may help to mitigate the expected symptoms of postembolization syndrome. Intravenous analgesia and sedation should also be considered during the procedure to control pain and anxiety (Mahnken et al., 2013). Heparin (up to 50 IU/kg) may be administered intra-arterially or intravenously at the beginning of the procedure for thrombosis prophylaxis. Intra-arterial administration of a vasodilator (nitroglycerine or a calcium antagonist) is indicated if vasospasm occurs at any time during the procedure. Patients treated for NET may require aggressive somatostatin analog prophylaxis against carcinoid crisis from sudden hormonal release. Patients with enterobiliary reflux from prior surgery, endoscopic retrograde cholangiopancreatography (ERCP), sphincterotomy,

biliary drainage, or biliary stent are at high risk for abscess formation and should be administered a long course of prophylactic antibiotics before and after the procedure.

1.4.2 TREATMENT TECHNIQUE

During treatment, radioactive microspheres should be administered at the same catheter position as during the preparatory angiography unless contraindications were discovered from the scout dose simulation. In patients with bilobar disease, treatment of both lobes can either be performed at once or in two-staged treatment sessions. The latter may be advisable in patients with a high tumor burden to allow a more careful approach with regard to toxicity.

It is advisable to perform a last check with DSA and/or CBCT before the administration, to make sure that no new hepatico-enteric collaterals have been recruited in the interval between the preparatory angiography and treatment, and that the catheter is indeed placed in the same position.

Although the administration systems of the different microspheres slightly differ from each other, they all have a vial that contains the radioactive microspheres in an acrylic container, an afferent line that allows the operator to inject solution into the vial, and an efferent line that connects the vial with the microcatheter. After setting up the administration system, ensuring that no air is present in any of the lines, administration of the microspheres can be started.

A 5% glucose solution has replaced the formerly recommended sterile water for injection of resin microspheres. Sterile water presumably leads to temporary changes in blood osmolality, causing hemolysis, vascular endothelial damage, vasospasm, and premature stasis of blood flow during injection. Recently, it has been demonstrated that using isotonic 5% glucose water instead reduces the occurrence of stasis, reduces periprocedural pain, and improves the percentage delivered activity (Ahmadzadehfar et al., 2015). Saline and iodinated contrast should not be used to avoid the risk of displacement of the yttrium-90 from resin microspheres via an ion exchange mechanism.

The injection technique differs significantly between glass ^{90}Y-microspheres and resin ^{90}Y-microspheres or ^{166}Ho-microspheres due to the significantly lower number of injected microspheres. Glass microspheres can be injected in a single bolus, without angiographic monitoring, since stasis of blood flow or reflux is not expected. A typical injection with glass microspheres may be completed in less than 5 min. When using resin ^{90}Y-microspheres or ^{166}Ho-microspheres, injection should be, however, performed carefully, using short pulsatile pressure strokes on the syringe (ideally a small syringe of 3–5 cc). The blood flow velocity should be regularly checked, since the administration should be paused in slow flow conditions, and stopped entirely when stasis occurs to prevent particle reflux. A change in preferential blood flow direction can sometimes be observed during the administration, which may be explained by stasis in some branches and subsequent redistribution of blood flow toward other branches. The treatment is completed if the entire activity is administered, or when the administration is stopped prematurely for stasis, which can take up to 30 min in total.

To determine the net administered activity, the vial, administration lines, and catheter may be checked for remnant activity with a dose calibrator. The process of pre- and posttreatment assay of administered activity and residual is described in Chapter 7.

1.4.3 CATHETER TYPES AND PARTICLE-FLUID DYNAMICS

Two types of administration catheters are commercially available for radioembolization procedures. The standard end-hole microcatheter remains the standard default catheter for all embolotherapies in interventional radiology. This may be explained by the low cost, simplicity of use, atraumatic character, and wide range of available sizes and flexibility that allow for catheterization of small, tortuous vessels. However, a disadvantage is that the catheter has no fixed support in the vessel lumen. Therefore, the microcatheter may deviate toward a vessel wall during injection of microspheres, which can lead to streaming and preferential deposition in daughter branches on the side of deviation.

A microcatheter specifically designed for embolotherapy, the Surefire Infusion System (Surefire Medical Inc., Westminster, Colorado), has been developed to prevent reflux during radioembolization. The tip of this antireflux catheter (ARC) dynamically expands radially to make contact with the vessel wall during administrations to prevent particle reflux without blocking antegrade blood flow. Using this ARC makes embolization of side branches proximal to the administration site unnecessary (Fischman et al., 2014; Morshedi et al., 2014; van den Hoven et al., 2014a). Earlier studies also used occlusion balloons for administration, with similar objectives.

Using the ARC may in theory also affect particle distribution in two ways. First, the fixed position of the ARC may improve the predictive value of the treatment simulations with a scout dose particle (^{99m}Tc-MAA or ^{166}Ho) by preventing catheter tip deviation. Second, the design of the ARC has a complex effect on blood flow and particle outflow dynamics, which may ultimately improve tumor targeting (Pasciak et al., 2015; van den Hoven et al., 2015a).

1.4.4 IMAGING OF THE THERAPEUTIC MICROSPHERE DISTRIBUTION

Imaging of the therapeutic microsphere distribution is increasingly performed. Bremsstrahlung SPECT/CT was formerly the only technique available to depict the ^{90}Y distribution after treatment, and was limited by poor spatial resolution. More recently, internal-pair production-based PET/CT has largely replaced bremsstrahlung SPECT/CT due to its superior spatial resolution and lower scatter (Padia et al., 2013; Elschot et al., 2013b). The biodistribution of ^{166}Ho-microspheres can also be imaged by SPECT, using the primary γ-radiation emission, as well as by MRI, by measuring the R2* dephasing effects induced by the paramagnetic microspheres (Elschot et al., 2013a; Smits et al., 2013a; van de Maat et al., 2013).

Confirming an adequate intrahepatic treatment distribution as well as lack of extrahepatic activity distribution seems a logical part of treating patients with radioembolization. However, this has not yet become standard of care. One of the reasons is lack of financial reimbursement. Furthermore, low signal-to-noise ratio on these images may limit the accuracy of both visual and quantitative assessments of the ^{90}Y-activity on PET/CT (Pasciak et al., 2014). Also, the clinical consequences of finding an unfavorable microsphere distribution on posttreatment imaging remain uncertain and irreversible. Nevertheless,

^{90}Y-PET/CT and ^{166}Ho-SPECT/CT should be incorporated into study protocols of future clinical trials on radioembolization to gather data about dosimetry and treatment outcomes. While such data could beneficially impact radioembolization in the future, the methods by which these data are gathered must be carefully controlled as described in Chapter 15.

1.4.5 DOSE–RESPONSE RELATIONSHIP

Several studies have shown an association between tumor-absorbed dose and tumor response and overall survival after radioembolization. Therefore, the goal is to strive for a high tumor-absorbed dose (Figure 1.4). However, a clear threshold or goal for effective tumor-absorbed dose has not been found.

Most dose–response investigations have been performed in patients treated with HCC with glass ^{90}Y-microspheres (Strigari et al., 2010; Riaz et al., 2011; Garin et al., 2012, 2015; Kao et al., 2013; Mazzaferro et al., 2013; Eaton et al., 2014; Srinivas et al., 2014). These dose–response relationships may not apply to other cell types and microsphere types (Flamen et al., 2008; Lam et al., 2013b, 2015; Demirelli et al., 2015). The difference in tumor biology among various tumors may require a different amount of radiation to induce tumor cell death, similar to findings in EBRT (Lausch et al., 2013). Furthermore, due to activity and distribution differences between resin and glass ^{90}Y-microspheres, generally higher tumor-absorbed doses are reported for glass microspheres, without a significantly different biological effect (Cremonesi et al., 2014).

Some authors report that an absorbed dose of 120 Gy is commonly considered tumoricidal for HCC (Riaz et al., 2011), whereas colorectal cancer liver metastases may require at least 70 Gy (Srinivas et al., 2014). Yet, a recent literature review showed that the range of reported effective tumor-absorbed dose values is extremely wide, with 66–495 Gy for resin microspheres and 163–1214 Gy for glass microspheres. The differences in study population and methods for quantifying tumor dose and response in these studies certainly contribute to this uncertainty. Even within an individual liver, two studies independently demonstrated

that there is high intraindividual dose-distribution heterogeneity and that inadequate treatment of at least one tumor is fairly common in patients with colorectal cancer liver metastases (Flamen et al., 2008; Smits et al., 2013a). Dose–response relationships for radioembolization will be discussed in greater detail in Chapter 8.

If effective tumor-absorbed dose thresholds can be established for the various microspheres and tumors, and imaging can monitor actual distribution of therapeutic dose, an iterative administration technique should be feasible. Additional treatment could be given to tumors that appear to have received inadequate initial treatment. If feasible, such a strategy should translate into improved outcomes.

1.5 TREATMENT-RELATED LABORATORY AND CLINICAL TOXICITY

1.5.1 COMPLAINTS DURING TREATMENT

Microsphere administration may sometimes induce complaints such as abdominal discomfort, pain, nausea, and vomiting. This can be accompanied by a vasovagal reaction. The exact cause of these symptoms remains unclear, but clinical observation suggests that complaints often occur simultaneously with stasis of blood flow, indicating that the embolic effect of therapy is causing symptoms. The symptoms are self-limited and can last for days but usually resolve on the day of treatment with supportive medical therapy.

1.5.2 LABORATORY TOXICITY

Abnormal laboratory values (liver function tests, complete blood count) should be expected for several weeks after treatment. Mild-to-severe (CTCAE grades 3–4) laboratory toxicity may even occur in one-third of patients without signs of associated clinical toxicity (Smits et al., 2013b). Therefore, laboratory investigations are not suited to discriminate between patients with and without a normal reaction to treatment. Abnormal liver function test (especially bilirubin level) in combination with ascites, however, is an alarming finding that may indicate radioembolization-induced liver disease (REILD).

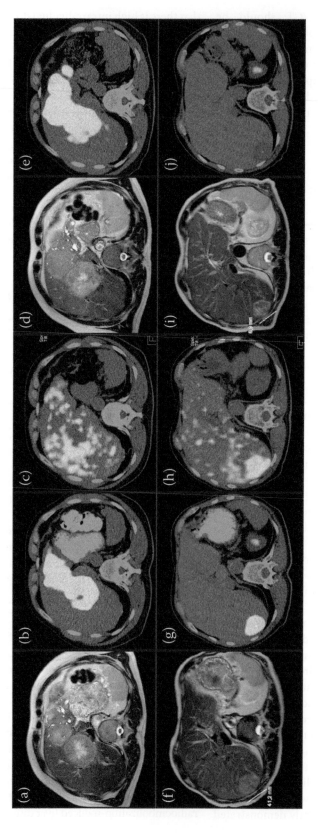

Figure 1.4 Clinical examples of colorectal carcinoma liver metastasis patients with adequate and inadequate tumor targeting. The *upper row* **(a–e)** shows a patient with a large tumor mass on pretreatment T2-weighted (T2W) MRI (a) and ^{18}F-FDG-PET (b) that extends from the central liver to the left liver lobe, which did not show a sufficient tumor-absorbed dose on ^{90}Y-PET/CT (c). At 1-month posttreatment, progressive disease was noted on the MRI (d) and ^{18}F-FDG-PET (h). The patient in the *lower row* **(f–j)**, on the contrary, had a metastasis in the posterior sector of the right liver lobe (f and g) that did have a sufficient tumor-absorbed dose on ^{90}Y-PET/CT (h). The posttreatment MRI (i) at 1-month follow-up showed reduction in size and ^{18}F-FDG-PET (j) demonstrated a complete metabolic response.

1.5.2.1 Postembolization syndrome

The most common symptoms after radioembolization are fatigue, abdominal pain, nausea, vomiting, low-grade fever, and anorexia/cachexia. These symptoms constitute the so-called postradioembolization syndrome, which is related to complaints after other embolotherapies (Riaz et al., 2009). These symptoms are considered an expected reaction to radioembolization treatment and they are usually self-limited and manageable with standard supportive medical treatment. Failure to recognize this may unfortunately result in unnecessary extension of patient hospitalization.

1.5.2.2 Treatment complications

Treatment complications are uncommon, but can occur by accumulation of radioactive microspheres outside the liver or overexposure of the healthy liver tissue to radiation.

Extrahepatic activity in the gallbladder wall, pancreas, and gastrointestinal tract may, respectively, result in radiation-induced cholecystitis, pancreatitis, or gastric/duodenal ulceration. Radiation-induced cholecystitis and pancreatitis are often subclinical in presentation, but may in some cases cause severe morbidity. Radiation-induced ulcers are particularly cumbersome because they originate from the serosal surface instead of the mucosal surfaces as in normal peptic ulcers, impairing the healing process and complicating surgical intervention. Detecting and correcting extrahepatic shunting before treatment can prevent these complications (Riaz et al., 2009).

Accumulation of too much radioactive microspheres in the lung causes tissue damage due to the combined effects of ischemia and radiation. Cases of radiation pneumonitis have been described in the literature, many of them fatal, often associated with hepatic venous and/or portal venous tumor invasion. To prevent this, estimated lung-absorbed doses should be kept <30 Gy (or 50 Gy cumulatively) (Leung et al., 1995). Additional exposure to pulmonary toxins such as chemotherapy may also compound the risk.

Exposing the healthy liver tissue to high radiation-absorbed doses during radioembolization may lead to the development of REILD within the first 2 months after treatment. Clinically, it is characterized by jaundice, weight gain, ascites, and a distinct rise in bilirubin, while transaminases and alkaline phosphatase are only mildly increased. Pathologically, REILD is characterized by sinusoidal congestion, venule occlusion by sloughed necrotic endothelium, and eventually fibrosis of the liver. If severe, REILD can be fatal in the acute phase, but some patients with milder REILD may develop chronic hepatic insufficiency and complications of portal hypertension. The radiation tolerance of the liver tissue depends on the involved volume, previous exposure to hepatotoxins including many common systemic chemotherapeutics, and underlying cirrhosis (Gil-Alzugaray et al., 2013). The diagnosis and management of complications following radioembolization are discussed in greater detail in Chapter 14.

1.6 TUMOR RESPONSE ASSESSMENT

A variety of methods can be used to assess tumor response after radioembolization treatment. These methods can be divided into morphologic and physiologic tumor response assessment methods. Morphologic or anatomical assessment methods only evaluate the treatment effect on tumor size/volume as visualized on cross-sectional imaging. Physiologic or functional assessment methods assess the treatment effect on tumor biology, including arterial vascularization, water diffusivity, and glucose uptake (Figure 1.5). All of these methods strive to provide valuable prognostic information using a reproducible method. In general, morphologic tumor assessment is simple, standardized, relatively subjective, variably reproducible, and time consuming. Physiologic tumor assessments, on the other hand, are complex, objective, and partly automated, but reproducibility is heavily dependent on the technique for image acquisition, reconstruction, and analysis. Unfortunately, only few comparative studies have been performed to assess which methods work best in radioembolization patients.

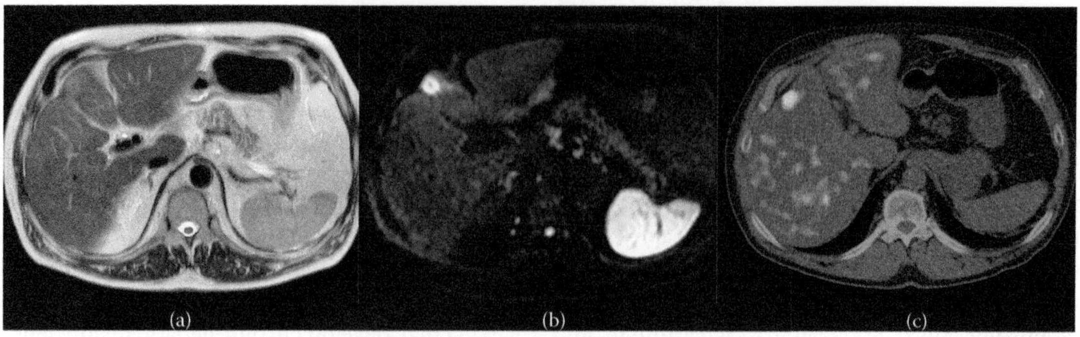

Figure 1.5 Early detection of tumor recurrence on functional imaging. This patient with colorectal cancer liver metastasis had a good response after initial treatment with radioembolization, but developed tumor recurrence in segment 4, 9 months after treatment. The metastasis is hardly visible on a standard (T2W) MRI sequence **(a)**, but was detected on diffusion-weighted (DW)-MRI **(b)** and on ^{18}F-FDG-PET **(c)**. The patient received additional treatment of segment 4, based on this finding, and responded well to treatment again.

1.6.1 ANATOMICAL TUMOR RESPONSE ASSESSMENT

WHO published the first response criteria in 1981. These criteria use the sum of the product (SPD) of tumor diameters (longest diameter in the axial imaging plane and the one perpendicular to it) to evaluate whether therapy led to significant tumor shrinkage. Response was classified into complete response (CR, –100%), partial response (PR, –50% to –99%), stable disease (SD, –49% to 25%), and progressive disease (PD, >25%), based on the percentage of change in SPD.

The Response Evaluation Criteria in Solid Tumors (RECIST) system was introduced to simplify the evaluation, and was found to correlate highly with the WHO system, and version 1.1 is now the most widely used response classification system in solid tumors. It uses the sum of the longest one-dimensional tumor diameters in two target lesions on cross-sectional imaging to classify response into the same categories as the WHO criteria (CR –100%, PR –99% to –30%, SD –30% to 20%, PD >20%). Target lesion response can also be extended to liver response and overall (whole-body) response by incorporating information about the size change in nontarget lesions and the appearance of new lesions in or outside the liver (Tirkes et al., 2013). Studies generally report either the best response during follow-up or the objective response rate (CR + PR) or disease control rate (CR + PR + SD) at a specific time of follow-up.

There are no guidelines on the required time intervals for follow-up imaging, but most oncology studies adapt follow-up to chemotherapy treatment cycles (e.g., in intervals of 3 weeks) and interventional radiology treatments often use 1, 2, and/or 3 months posttreatment. If scans are performed with short intervals (i.e., <6 weeks), parameters such as time to progression (TTP) or progression-free survival (PFS) may be used (Llovet et al., 2008).

RECIST has been adopted as the standard for response assessment by both research institutions and regulatory authorities. It functions as a surrogate endpoint for overall survival, provides important prognostic information (especially in therapies where tumor shrinkage is expected such as cytotoxic chemotherapy), is relatively simple and reproducible, and does not require state-of-the-art imaging facilities. However, its validity is questioned for therapies where response to therapy is not always associated with lesion shrinkage, such as radioembolization, ablation, and some systemic treatments such as immunotherapy.

1.6.2 FUNCTIONAL TUMOR RESPONSE ASSESSMENT

Modern imaging technology has enabled a more physiological approach to tumor response assessment. Modified RECIST (mRECIST) has been developed to address the issue that a change in arterial tumor vascularization might be more representative than a change in entire tumor size in patients with HCC treated with locoregional therapy. Similar to RECIST, the longest diameter is measured but only of the viable (arterially enhanced) portion of the tumor. Lack of arterial tumor enhancement on follow-up imaging, even if the total lesion size is unchanged or even increased, is considered a complete response (CR) (Lencioni and Llovet, 2010). The other response categories are defined similarly as with RECIST based on the enhanced tumor diameter measurement.

More quantitative approaches are offered by dynamic contrast-enhanced MRI (DCE-MRI). With DCE-MRI, serial images are acquired before, during, and after contrast agent administration, allowing for contrast kinetic modeling with time-signal intensity curves (Choyke et al., 2003). MRI can also be used to assess functional tumor response by diffusion-weighted imaging (DWI). The principle behind this imaging technique is that the water diffusivity is restricted in tumors, as opposed to healthy liver tissue. Diffusivity can be quantified by calculating apparent diffusion coefficients (ADC). An increase of ADC after treatment indicates treatment response (Barabasch et al., 2015).

^{18}F-FDG-PET is also increasingly used to assess treatment response. Standardized uptake values (SUV) can be used to quantify the selective uptake of the ^{18}F-isotope labeled glucose analog in malignant tissues (Larson et al., 1999). A combination of SUV and metabolic tumor volume—called tumor lesion glycolysis or metabolic product—may be especially interesting, since this reflects the total glucose turnover in a tumor. Many options are available when acquiring, reconstructing, and analyzing PET data, yet none of these methods has proven to be clearly superior to the others, and they may lead to different results (Boellaard, 2011). An attempt to standardize response analysis, PET Response Criteria in Solid Tumors (PERCIST) has not resulted in the same widespread adaptation as with the RECIST criteria.

Both the anatomical and functional assessments of tumor response following radioembolization are discussed in greater detail in Chapter 14.

1.7 CONCLUSION

In recent decades, radioembolization has evolved into a safe and effective liver cancer therapy. Despite major advances brought by modern imaging technology and clinical experience, improvement of patient workup, treatment technique, toxicity, and response assessment is an ongoing process. In the following chapters, these subjects will be discussed in further detail.

REFERENCES

Abdelmaksoud, M.H.K., Hwang G.L., Louie J.D. et al. (2010). Development of new hepaticoenteric collateral pathways after hepatic arterial skeletonization in preparation for yttrium-90 radioembolization. J Vasc Interv Radiol 21:1385–1395. Available at http://dx.doi.org/10.1016/j.jvir.2010.04.030.

Abdelmaksoud, M.H.K., Louie, J.D., Kothary, N. et al. (2011a). Embolization of parasitized extrahepatic arteries to reestablish intrahepatic arterial supply to tumors before yttrium-90 radioembolization. J Vasc Interv Radiol 22:1355–1362. Available at http://www.ncbi.nlm.nih.gov/pubmed/21961979 [Accessed August 4, 2014].

Abdelmaksoud, M.H.K., Louie, J.D., Kothary, N. et al. (2011b). Consolidation of hepatic arterial inflow by embolization of variant hepatic arteries in preparation for yttrium-90 radioembolization. J Vasc Interv Radiol 22:1364–1371.e1. Available at http://www.ncbi.nlm.nih.gov/pubmed/21961981 [Accessed August 5, 2014].

Ahmadzadehfar, H., Meyer, C., Pieper, C.C. et al. (2015). Evaluation of the delivered activity of yttrium-90 resin microspheres using sterile water and 5% glucose during administration. Eur J Nucl Med Mol Imaging Res 5:54. Available at http://www.ejnmmires.com/content/5/1/54.

Ahmadzadehfar, H., Möhlenbruch, M., Sabet, A. et al. (2011). Is prophylactic embolization of the hepatic falciform artery needed before radioembolization in patients with 99mTc-MAA accumulation in the anterior abdominal wall? *Eur J Nucl Med Mol Imaging* 38:1477–1484. Available at http://www.ncbi.nlm.nih.gov/pubmed/21494857 [Accessed August 4, 2014].

Ariel, I.M. (1965). Treatment of inoperable primary pancreatic and liver cancer by the intra-arterial administration of radioactive isotopes (Y90 radiating microspheres. *Ann Surg* 162:267–278.

Ariel, I.M., Padula, G. (1982). Treatment of asymptomatic metastatic cancer to the liver from primary colon and rectal cancer by the intraarterial administration of chemotherapy and radioactive isotopes. *J Surg Oncol* 20:151–156.

Barabasch, A., Kraemer, N.A., Ciritsis, A. et al. (2015). Diagnostic accuracy of diffusion-weighted magnetic resonance imaging versus positron emission tomography/computed tomography for early response assessment of liver metastases to Y90-radioembolization. *Invest Radiol* 50:409–415.

Bierman, H.R., Byron, R.L., Kelley, K.H., Grady, A. (1951). Studies on the blood supply of tumors in man. III. Vascular patterns of the liver by hepatic arteriography in vivo. *J Natl Cancer Inst* 12:107–131.

Bismuth, H. (1982). Surgical anatomy and anatomical surgery of the liver. *World J Surg* 6: 3–9. Available at http://www.ncbi.nlm.nih.gov/pubmed/7090393.

Blanchard, R.J., Grotenhuis, I., Lafave, J.W., Perry, J.F. (1965a). Blood supply to hepatic V2 carcinoma implants as measured by radioactive microspheres. *Proc Soc Exp Biol Med* 118:465–468.

Blanchard, R.J., Lafave, J.W., Kim, Y.S. et al. (1965b). Treatment of patients with advanced cancer utilizing Y90 microspheres. *Cancer* 18:375–380.

Boellaard, R. (2011). Need for standardization of 18F-FDG PET/CT for treatment response assessments. *J Nucl Med* 52(Suppl 2):93S–100S. Available at http://www.ncbi.nlm.nih.gov/pubmed/22144561 [Accessed July 17, 2014].

Braat, A.J.A.T., Huijbregts, J.E., Molenaar, I.Q. et al. (2014). Hepatic radioembolization as a bridge to liver surgery. *Front Oncol* 4:1–13. Available at http://journal.frontiersin.org/article/10.3389/fonc.2014.00199/abstract.

Braat, A.J.A.T., Smits, M.L.J., Braat, M.N.G.J. et al. (2015). 90Y hepatic radioembolization: An update on current practice and recent developments. *J Nucl Med* 56:1079–1087. Available at http://jnm.snmjournals.org/cgi/doi/10.2967/jnumed.115.157446.

Caldarola, L., Rosa, U., Badellino, F. et al. (1964). Preparation of 32-P labelled resin microspheres for radiation treatment of tumours by intra-arterial injection. *Minerva Nucl* 55:169–174.

Campbell, A.M., Bailey, I.H., Burton, M.A. (2001). Tumour dosimetry in human liver following hepatic yttrium-90 microsphere therapy. *Phys Med Biol* 46:487–498. Available at http://www.ncbi.nlm.nih.gov/pubmed/11229728.

Choyke, P.L., Dwyer, A.J., Knopp, M.V. (2003). Functional tumor imaging with dynamic contrast-enhanced magnetic resonance imaging. *J Magn Reson Imaging* 17:509–520. Available at http://www.ncbi.nlm.nih.gov/pubmed/12720260. [Accessed August 4, 2014].

Cremonesi, M., Chiesa, C., Strigari, L. et al. (2014). Radioembolization of hepatic lesions from a radiobiology and dosimetric perspective. *Front Oncol* 4:210. Available at http://journal.frontiersin.org/Journal/10.3389/fonc.2014.00210/. [Accessed September 10, 2014].

D'Arienzo, M. (2013). Emission of β+ particles via internal pair production in the 0+ – 0+ transition of 90Zr: Historical background and current applications in nuclear medicine imaging. *Atoms* 1:2–12. Available at http://www.mdpi.com/2218-2004/1/1/2/.

Demirelli, S., Erkilic, M., Oner, A.O. et al. (2015). Evaluation of factors affecting tumor response and survival in patients with primary and metastatic liver cancer treated with microspheres. *Eur J Gastroenterol Hepatol*: 1. Available at http://content.wkhealth.com/linkback/openurl?sid=WKPTLP:landingpage&an=00042737-900000000-98792.

Devcic, Z., Rosenberg, J., Braat, A.J. et al. (2014). The efficacy of hepatic 90Y resin radioembolization for metastatic neuroendocrine tumors: A meta-analysis. *J Nucl Med* 55:1404–1410. doi: 10.2967/jnumed.113.135855. Available at http://www.ncbi.nlm.nih.gov/pubmed/25012459 [Accessed July 18, 2014].

Eaton, B.R., Kim, H.S., Schreibmann, E. et al. (2014). Quantitative dosimetry for yttrium-90 radionuclide therapy: Tumor dose predicts fluorodeoxyglucose positron emission tomography response in hepatic metastatic melanoma. *J Vasc Interv Radiol* 25:288–295. Available at http://dx.doi.org/10.1016/j.jvir.2013.08.021.

Elschot, M., Nijsen, J.F.W., Lam, M.G.E.H. et al. (2014). (^{99}m)Tc-MAA overestimates the absorbed dose to the lungs in radio-embolization: A quantitative evaluation in patients treated with ^{166}Ho-microspheres. *Eur J Nucl Med Mol Imaging* 41:1965–1975. Available at http://www.ncbi.nlm.nih.gov/pubmed/24819055 [Accessed January 8, 2015].

Elschot, M., Smits, M.L.J., Nijsen, J.F.W. et al. (2013a). Quantitative Monte Carlo-based holmium-166 SPECT reconstruction. *Med Phys* 40:112502. Available at http://www.ncbi.nlm.nih.gov/pubmed/24320461 [Accessed August 4, 2014].

Elschot, M., Vermolen, B.J., Lam, M.G.E.H. et al. (2013b). Quantitative comparison of PET and bremsstrahlung SPECT for imaging the in vivo yttrium-90 microsphere distribution after liver radioembolization. *PLoS One* 8:e55742. Available at http://www.pubmedcentral.nih.gov/articlerender.fcgi?artid=3566032&tool=pmcentrez&rendertype=abstract [Accessed July 24, 2014].

Emami, B., Lyman, J., Brown, A. et al. (1991). Tolerance of normal tissue to therapeutic irradiation. *Int J Radiat Oncol Biol Phys* 21:109–122.

Eriksson, D., Stigbrand, T. (2010). Radiation-induced cell death mechanisms. *Tumor Biol.* 31: 363–372.

Fischman, A.M., Ward, T.J., Patel, R.S. et al. (2014). Prospective, randomized study of coil embolization versus surefire infusion system during yttrium-90 radioembolization with resin microspheres. J Vasc Interv Radiol. 25:1709–1716. doi: 10.1016/j.jvir.2014.08.007.

Flamen, P., Vanderlinden, B., Delatte, P. et al. (2008). Multimodality imaging can predict the metabolic response of unresectable colorectal liver metastases to radioembolization therapy with yttrium-90 labeled resin microspheres. *Phys Med Biol* 53:6591–6603. Available at http://www.ncbi.nlm.nih.gov/pubmed/18978442 [Accessed August 5, 2014].

Forner, A., Gilabert, M., Bruix, J., Raoul, J.-L. (2014). Treatment of intermediate-stage hepatocellular carcinoma. *Nat Rev Clin Oncol* 11:525–535. Available at http://dx.doi.org/10.1038/nrclinonc.2014.122 [Accessed August 6, 2014].

Fuss, M., Salter, B.J., Herman, T.S., Thomas, C.R. (2004). External beam radiation therapy for hepatocellular carcinoma: Potential of intensity-modulated and image-guided radiation therapy. *Gastroenterology* 127:206–217.

Gaba, R.C., Lewandowski, R.J., Kulik, L.M. et al. (2009). Radiation lobectomy: Preliminary findings of hepatic volumetric response to lobar yttrium-90 radioembolization. *Ann Surg Oncol* 16:1587–1596. Available at http://www.ncbi.nlm.nih.gov/pubmed/19357924 [Accessed August 4, 2014].

Garin, E. (2015). Radioembolization with 90Y-loaded microspheres: High clinical impact of treatment simulation with MAA-based dosimetry. *Eur J Nucl Med Mol Imaging* 42:1189–1191. Available at http://link.springer.com/10.1007/s00259-015-3073-y.

Garin, E., Lenoir, L., Rolland, Y. et al. (2012). Dosimetry based on 99mTc-macroaggregated albumin SPECT/CT accurately predicts tumor response and survival in hepatocellular carcinoma patients treated with 90Y-loaded glass microspheres: Preliminary results. *J Nucl Med* 53:255–263.

Garin, E., Rolland, Y., Edeline, J. et al. (2015). Personalized dosimetry and intensification concept with 90Y-loaded glass microsphere radioembolization induce prolonged overall survival in hepatocelluar carcinoma patients with portal vein thrombosis. *J Nucl Med* 56:339–346. Available at http://www.ncbi.nlm.nih.gov/pubmed/25678490.

Giammarile, F., Bodei, L., Chiesa, C. et al. (2011). EANM procedure guideline for the treatment of liver cancer and liver metastases with intra-arterial radioactive compounds. *Eur J Nucl Med Mol Imaging* 38:1393–1406. Available at http://www.ncbi.nlm.nih.gov/pubmed/ [Accessed July 17, 2014].

Gil-Alzugaray, B., Chopitea, A., Iñarrairaegui, M. et al. (2013). Prognostic factors and prevention of radioembolization-induced liver disease. *Hepatology* 57:1078–1087. Available at http://doi.wiley.com/10.1002/hep.26191.

Grady, E.D. (1979). Internal radiation therapy of hepatic cancer. *Dis Colon Rectum* 22:371–375.

Grady, E.D., Nolan, T.R., Crumbley, A.J. et al. (1975). Internal hepatic radiotherapy: II. Intra-arterial radiocolloid therapy for hepatic tumors. *Am J Roentgenol Radium Ther Nucl Med* 124:596–599.

Grady, E.D., Sale, W., Nicolson, W.P., Rollins, L.C. (1960). Intra-arterial radioisotopes to treat cancer. *Am Surg* 26:678–684.

Grady, E.D., Sale, W.T., Rollins, L.C. (1963). Localization of radioactivity by intravascular injection of large radioactive particles. *Ann Surg* 157:97–114.

Gray, B., Van Hazel, G., Hope, M. et al. (2001). Randomised trial of SIR-Spheres plus chemotherapy vs. chemotherapy alone for treating patients with liver metastases from primary large bowel cancer. *Ann Oncol* 12:1711–1720. Available at http://www.ncbi.nlm.nih.gov/pubmed/11843249 [Accessed August 2, 2014].

Gray, B.N., Burton, M.A., Kelleher, D.K. et al. (1989). Selective internal radiation (SIR) therapy for treatment of liver metastases: Measurement of response rate. *J Surg Oncol* 42:192–196.

Gulec, S.A., Siegel, J.A. (2007). Posttherapy radiation safety considerations in radiomicrosphere treatment with 90Y-microspheres. *J Nucl Med* 48:2080–2086.

Hilgard, P., Hamami, M., Fouly, A.E.L. et al. (2010). Radioembolization with yttrium-90 glass microspheres in hepatocellular carcinoma: European experience on safety and long-term survival. *Hepatology* 52:1741–1749.

Jernigan, S.R., Osborne, J.A., Mirek, C.J., Buckner, G. (2015). Selective internal radiation therapy: Quantifying distal penetration and distribution of resin and glass microspheres in a surrogate arterial model. *J Vasc Interv Radiol* 26:897–904. Available at http://www.ncbi.nlm.nih.gov/pubmed/25891507.

Kao, Y.-H., Steinberg, J.D., Tay, Y.S. et al. (2013). Post-radioembolization yttrium-90 PET/CT - part 2: Dose-response and tumor predictive dosimetry for resin microspheres. *Eur J Nucl Med Mol Imaging Res* 3:57. Available at http://www.pubmedcentral.nih.gov/articlerender.fcgi?artid=3733999&tool=pmcentrez&rendertype=abstract [Accessed August 4, 2014].

Kennedy, A. (2014). Radioembolization of hepatic tumors. *J Gastrointest Oncol* 5:178–189. Available at http://www.pubmedcentral.nih.gov/articlerender.fcgi?artid=4074949&tool=pmcentrez&rendertype=abstract [Accessed July 23, 2014].

Kennedy, A.S., Ball, D., Cohen, S.J. et al. (2015). Multicenter evaluation of the safety and efficacy of radioembolization in patients with unresectable colorectal liver metastases selected as candidates for 90 Y resin microspheres. *J Gastrointest Oncol* 90:134–142.

Kennedy, A.S., Nutting, C., Coldwell, D. et al. (2004). Pathologic response and microdosimetry of (90)Y microspheres in man: Review of four explanted whole livers. *Int J Radiat Oncol Biol Phys* 60:1552–1563. Available at http://www.ncbi.nlm.nih.gov/pubmed/15590187 [Accessed August 1, 2014].

Kim, Y.S., Lafave, J.W., Maclean, L.D. (1962). The use of radiating microspheres in the treatment of experimental and human malignancy. *Surgery* 52:220–231.

Lam, M.G.E.H., Banerjee, A., Goris, M.L. et al. (2015). Fusion dual-tracer SPECT-based hepatic dosimetry predicts outcome after radioembolization for a wide range of tumour cell types. *Eur J Nucl Med Mol Imaging* 42:1192–1201. Available at http://www.ncbi.nlm.nih.gov/pubmed/25916740.

Lam, M.G.E.H., Banerjee, S., Louie, J.D. et al. (2013a). Root cause analysis of gastroduodenal ulceration after yttrium-90 radioembolization. *Cardiovasc Intervent Radiol* 36:1536–1547. Available at http://www.ncbi.nlm.nih.gov/pubmed/23435742 [Accessed August 4, 2014].

Lam, M.G.E.H., Goris, M.L., Iagaru, A.H. et al. (2013b). Prognostic utility of 90Y radioembolization dosimetry based on fusion 99mTc-macroaggregated albumin-99mTc-sulfur colloid SPECT. *J Nucl Med* 54:2055–2061. Available at http://www.ncbi.nlm.nih.gov/pubmed/24144563 [Accessed August 4, 2014].

Lam, M.G.E.H., Louie, J.D., Abdelmaksoud, M.H.K. et al. (2014). Limitations of body surface area-based activity calculation for radioembolization of hepatic metastases in colorectal cancer. *J Vasc Interv Radiol* 25:1085–1093. Available at http://www.ncbi.nlm.nih.gov/pubmed/24457263 [Accessed August 4, 2014].

Larson, S.M., Erdi, Y., Akhurst, T. et al. (1999). Tumor treatment response based on visual and quantitative changes in global tumor glycolysis using PET-FDG imaging. The visual response score and the change in total lesion glycolysis. *Clin Positron Imaging* 2:159–171. Available at http://www.ncbi.nlm.nih.gov/pubmed/14516540.

Lau, W.Y., Leung, W.T., Ho, S. et al. (1994). Treatment of inoperable hepatocellular carcinoma with intrahepatic arterial yttrium-90 microspheres: A phase I and II study. *Br J Cancer* 70:994–999.

Lausch, A., Sinclair, K., Lock, M. et al. (2013) Determination and comparison of radiotherapy dose responses for hepatocellular carcinoma and metastatic colorectal liver tumours. *Br J Radiol* 86:20130147. Available at http://www.ncbi.nlm.nih.gov/pubmed/23690438.

Lencioni, R., Llovet, J.M. (2010). Modified RECIST (mRECIST) assessment for hepatocellular carcinoma. *Semin Liver Dis* 30:52–60. Available at http://www.ncbi.nlm.nih.gov/pubmed/20175033.

Leung, T.W., Lau, W.Y., Ho, S.K. et al. (1995). Radiation pneumonitis after selective internal radiation treatment with intraarterial 90yttrium-microspheres for inoperable hepatic tumors. *Int J Radiat Oncol Biol Phys* 33:919–924.

Lewandowski, R.J., Sato, K.T., Atassi, B. et al. (2007). Radioembolization with 90Y microspheres: Angiographic and technical considerations. *Cardiovasc Intervent Radiol* 30:571–592. Available at http://www.ncbi.nlm.nih.gov/pubmed/17516113 [Accessed August 4, 2014].

Liu, D.M., Salem, R., Bui, J.T. et al. (2005). Angiographic considerations in patients undergoing liver-directed therapy. *J Vasc Interv Radiol* 16:911–935. Available at http://www.ncbi.nlm.nih.gov/pubmed/16002500 [Accessed August 4, 2014].

Llovet, J.M., Di Bisceglie, A.M., Bruix, J. et al. (2008). Design and endpoints of clinical trials in hepatocellular carcinoma. *J Natl Cancer Inst* 100:698–711. Available at http://www.ncbi.nlm.nih.gov/pubmed/18477802 [Accessed July 11, 2014].

Mahnken, A.H., Spreafico, C., Maleux, G. et al. (2013). Standards of practice in transarterial radioembolization. *Cardiovasc Intervent Radiol* 36:613–622. Available at http://www.ncbi.nlm.nih.gov/pubmed/23511991 [Accessed August 4, 2014].

Majno, P., Mentha, G., Toso, C. et al. (2014). Anatomy of the liver: An outline with three levels of complexity—A further step toward tailored territorial liver resections. *J Hepatol* 60:654–662. Available at http://www.ncbi.nlm.nih.gov/pubmed/24211738 [Accessed October 13, 2014].

Mazzaferro, V., Sposito, C., Bhoori, S. et al. (2013). Yttrium-90 radioembolization for intermediate-advanced hepatocellular carcinoma: A phase 2 study. *Hepatology* 57:1826–1837.

Morshedi, M.M., Bauman, M., Rose, S.C., Kikolski, S.G. (2014). Yttrium-90 resin microsphere radioembolization using an antireflux catheter: An alternative to traditional coil embolization for nontarget protection. *Cardiovasc Intervent Radiol* 38:381–388. Available at http://www.ncbi.nlm.nih.gov/pubmed/24989143 [Accessed July 18, 2014].

Mouli, S., Memon, K., Baker, T. et al. (2013). Yttrium-90 radioembolization for intrahepatic cholangiocarcinoma: Safety, response, and survival analysis. *J Vasc Interv Radiol* 24:1227–1234. Available at http://www.pubmedcentral.nih.gov/articlerender.fcgi?artid=3800023&tool=pmcentrez&rendertype=abstract [Accessed August 4, 2014].

Muller, J.H., Rossier, P.H. (1951). A new method for the treatment of cancer of the lungs by means of artificial radioactivity. *Acta Radiol* 35:449–468. Available at http://www.tandfonline.com/doi/full/10.3109/00016925109136677.

Nijsen, J.F.W., Van Steenbergen, M.J., Kooijman, H. et al. (2001). Characterization of poly(L-lactic acid) microspheres loaded with holmium acetylacetonate. *Biomaterials* 22:3073–3081.

Padia, S.A., Alessio, A., Kwan, S.W. et al. (2013). Comparison of positron emission tomography and bremsstrahlung imaging to detect particle distribution in patients undergoing yttrium-90 radioembolization for large hepatocellular carcinomas or associated portal vein thrombosis. *J Vasc Interv Radiol* 24:1147–1153. Available at http://www.ncbi.nlm.nih.gov/pubmed/23792126 [Accessed August 4, 2014].

Pasciak, A.S., Bourgeois, A.C., McKinney, J.M. et al. (2014). Radioembolization and the dynamic role of (90)Y PET/CT. *Front Oncol* 4:38. Available at http://www.pubmedcentral.nih.gov/articlerender.fcgi?artid=3936249&tool=pmcentrez&rendertype=abstract [Accessed August 4, 2014].

Pasciak, A.S., Mcelmurray, J.H., Bourgeois, A.C. et al. (2015). The impact of an antireflux catheter on target volume particulate distribution in liver-directed embolotherapy: A pilot study. *J Vasc Interv Radiol* 26:660–669. Available at http://www.ncbi.nlm.nih.gov/pubmed/25801854.

Prince, J.F., Smits, M.L.J., Krijger, G.C. et al. (2014a). Radiation emission from patients treated with holmium-166 radioembolization. *J Vasc Interv Radiol* 25:1956–1963.e1. Available at http://www.ncbi.nlm.nih.gov/pubmed/25311966 [Accessed January 8, 2015].

Prince, J.F., van den Hoven, A.F., van den Bosch, M.A.A.J. et al. (2014b). Radiation-induced cholecystitis after hepatic radioembolization: Do we need to take precautionary measures? *J Vasc Interv Radiol* 25:1717–1723. Available at http://linkinghub.elsevier.com/retrieve/pii/S1051044314006447.

Prince, J.F., van Rooij, R., Bol, G.H. et al. (2015). Safety of a scout dose preceding hepatic radioembolization with 166Ho microspheres. *J Nucl Med* 56:817–823. Available at http://jnm.snmjournals.org/cgi/doi/10.2967/jnumed.115.155564.

Prinzmetal, M., Ornitz, E.M. (1948). Arteriovenous anastomoses in liver, spleen, and lungs. *Am J Physiol* 152:48–52.

Rengo, M., Bellini, D., De Cecco, C.N. et al. (2011). The optimal contrast media policy in CT of the liver. Part II: Clinical protocols. *Acta Radiol* 52:473–480.

Riaz, A., Gates, V.L., Atassi, B. et al. (2011). Radiation segmentectomy: A novel approach to increase safety and efficacy of radioembolization. *Int J Radiat Oncol Biol Phys* 79:163–171. Available at http://www.ncbi.nlm.nih.gov/pubmed/20421150 [Accessed August 4, 2014].

Riaz, A., Lewandowski, R.J., Kulik, L.M. et al. (2009). Complications following radioembolization with yttrium-90 microspheres: A comprehensive literature review. *J Vasc Interv Radiol* 20:1121–1130; quiz 1131. Available at http://www.ncbi.nlm.nih.gov/pubmed/19640737 [Accessed July 16, 2014].

Rosenbaum, C.E.N.M., van den Bosch, M.A.A.J., Veldhuis, W.B. et al. (2013). Added value of FDG-PET imaging in the diagnostic workup for yttrium-90 radioembolisation in patients with colorectal cancer liver metastases. *Eur Radiol* 23:931–937. Available at http://www.ncbi.nlm.nih.gov/pubmed/23111818 [Accessed August 4, 2014].

Rosler, H., Triller, J., Baer, H.U. et al. (1994). Superselective radioembolization of hepatocellular carcinoma: 5-year results of a prospective study. *Nuklearmedizin* 33:206–214.

Salem, R., Lewandowski, R.J., Sato, K.T. et al. (2007). Technical aspects of radioembolization with 90Y microspheres. *Tech Vasc Interv Radiol* 10:12–29. Available at http://www.ncbi.nlm.nih.gov/pubmed/17980315 [Accessed August 4, 2014].

Samuelson, S.D., Louie, J.D., Sze, D.Y. (2013). N-butyl cyanoacrylate glue embolization of arterial networks to facilitate hepatic arterial skeletonization before radioembolization. *Cardiovasc Intervent Radiol* 36:690–698. Available at http://www.ncbi.nlm.nih.gov/pubmed/23070102.

Sharma, H. (2014). Role of external beam radiation therapy in management of hepatocellular carcinoma. *J Clin Exp Hepatol* 4:S122–S125. Available at http://linkinghub.elsevier.com/retrieve/pii/S0973688314002989.

Smits, M.L.J., Elschot, M., Sze, D.Y. et al. (2014). Radioembolization dosimetry: The road ahead. *Cardiovasc Intervent Radiol*. Available at http://www.ncbi.nlm.nih.gov/pubmed/25537310 [Accessed December 30, 2014].

Smits, M.L.J., Elschot, M., van den Bosch, M.A.A.J. et al. (2013a). In vivo dosimetry based on SPECT and MR imaging of 166Ho-microspheres for treatment of liver malignancies. *J Nucl Med* 54:2093–2100. Available at http://www.ncbi.nlm.nih.gov/pubmed/24136931 [Accessed August 4, 2014].

Smits, M.L.J., van den Hoven, A.F., Rosenbaum, C.E.N.M. et al. (2013b). Clinical and laboratory toxicity after intra-arterial radioembolization with 90Y-microspheres for unresectable liver metastases. *PLoS One* 8:e69448. Available at http://journals.plos.org/plosone/article?id=10.1371/journal.pone.0069448.

Srinivas, S.M., Natarajan, N., Kuroiwa, J. et al. (2014). Determination of radiation absorbed dose to primary liver tumors and normal liver tissue using post-radioembolization 90Y PET. *Front Oncol* 4:1–12. Available at http://journal.frontiersin.org/journal/10.3389/fonc.2014.00255/full.

Strigari, L., Sciuto, R., Rea, S. et al. (2010). Efficacy and toxicity related to treatment of hepatocellular carcinoma with 90Y-SIR spheres: Radiobiologic considerations. *J Nucl Med* 51:1377–1385.

Tirkes, T., Hollar, M.A., Tann, M. et al. (2013). Response criteria in oncologic imaging: Review of traditional and new criteria. *Radiographics* 33:1323–1341 Available at http://eutils.ncbi.nlm.nih.gov/entrez/eutils/elink.fcgi?dbfrom=pubmed&id=24025927&retmode=ref&cmd=prlinks\npapers2://publication/doi/10.1148/rg.335125214.

Tsurusaki, M., Okada, M., Kuroda, H. et al. (2014). Clinical application of 18F-fluorodeoxyglucose positron emission tomography for assessment and evaluation after therapy for malignant hepatic tumor. *J Gastroenterol* 49:46–56. Available at http://link.springer.com/10.1007/s00535-013-0790-5.

van de Maat, G.H., Seevinck, P.R., Elschot, M. et al. (2013). MRI-based biodistribution assessment of holmium-166 poly(L-lactic acid) microspheres after radioembolisation. *Eur Radiol* 23:827–835. Available at http://www.pubmedcentral.nih.gov/articlerender.fcgi?artid=3563959&tool=pmcentrez&rendertype=abstract [Accessed August 4, 2014].

Van de Wiele, C., Maes, A., Brugman, E. et al. (2012). SIRT of liver metastases: Physiological and pathophysiological considerations. *Eur J Nucl Med Mol Imaging* 39:1646–1655. Available at http://www.ncbi.nlm.nih.gov/pubmed/22801733 [Accessed July 16, 2014].

van den Hoven, A.F., Lam, M.G.E.H., Jernigan, S. et al. (2015a). Innovation in catheter design for intra-arterial liver cancer treatments results in favorable particle-fluid dynamics. *J Exp Clin Cancer Res* 34:74. Available at http://www.jeccr.com/content/34/1/74.

van den Hoven, A.F., Prince, J.F., de Keizer, B. et al. (2015b). Use of C-Arm Cone beam CT during hepatic radioembolization: Protocol optimization for extrahepatic shunting and parenchymal enhancement. *Cardiovasc Intervent Radiol.* Available at http://link.springer.com/10.1007/s00270-015-1146–1148.

van den Hoven, A.F., Prince, J.F., Samim, M. et al. (2014a). Posttreatment PET-CT-confirmed intrahepatic radioembolization performed without coil embolization, by using the antireflux surefire infusion system. *Cardiovasc Intervent Radiol* 37:523–528. Available at http://www.ncbi.nlm.nih.gov/pubmed/23756882 [Accessed August 5, 2014].

van den Hoven, A.F., Smits, M.L.J., de Keizer, B. et al. (2014b). Identifying aberrant hepatic arteries prior to intra-arterial radioembolization. *Cardiovasc Intervent Radiol.* Available at http://www.ncbi.nlm.nih.gov/pubmed/24469409 [Accessed August 5, 2014].

van den Hoven, A.F., van Leeuwen, M.S., Lam, M.G.E.H., van den Bosch, M.A.A.J. (2015). Hepatic arterial configuration in relation to the segmental anatomy of the liver; observations on MDCT and DSA relevant to radioembolization treatment. *Cardiovasc Intervent Radiol:* 38:100–111. Available at http://www.ncbi.nlm.nih.gov/pubmed/24603968.

van Leeuwen, M.S., Fernandez, M.A., van, Es.H.W. et al. (1994). Variations in venous and segmental anatomy of the liver: Two- and three-dimensional MR imaging in healthy volunteers. *AJR Am J Roentgenol* 162:1337–1345. Available at http://www.ncbi.nlm.nih.gov/pubmed/8191995 [Accessed August 5, 2014].

Vouche, M., Lewandowski, R.J., Atassi, R. et al. (2013). Radiation lobectomy: Time-dependent analysis of future liver remnant volume in unresectable liver cancer as a bridge to resection. *J Hepatol* 59:1029–1036. Available at http://www.ncbi.nlm.nih.gov/pubmed/23811303. Accessed August 4, 2014.

Walker, L.A. (1964). Radioactive yttrium 90: A review of its properties, biological behavior, and xlinical uses. *Acta Radiol Ther Phys Biol* 2:302–314.

Wang, S., He, X., Li, Z. et al. (2010). Characterization of the middle hepatic artery and its relevance to living donor liver transplantation. *Liver Transplant* 16:736–741. Available at http://onlinelibrary.wiley.com/doi/10.1002/lt.22082/abstract.

Wollner, I., Knutsen, C., Smith, P. et al. (1988). Effects of hepatic arterial yttrium 90 glass microspheres in dogs. *Cancer* 61:1336–1344.

Wondergem, M., Smits, M.L.J., Elschot, M. et al. (2013). 99mTc-macroaggregated albumin poorly predicts the intrahepatic distribution of 90Y resin microspheres in hepatic radioembolization. *J Nucl Med* 54:1294–1301. Available at http://www.ncbi.nlm.nih.gov/pubmed/23749996 [Accessed July 10, 2014].

Xing, M., Prajapati, H.J., Dhanasekaran, R. et al. (2014). Selective internal yttrium-90 radioembolization therapy (90Y-SIRT) versus best supportive care in patients with unresectable metastatic melanoma to the liver refractory to systemic therapy. *Am J Clin Oncol* 00:1. Available at http://www.ncbi.nlm.nih.gov/pubmed/25089529.

Yu, N., Srinivas, S.M., Difilippo, F.P. et al. (2013). Lung dose calculation with SPECT/CT for [90]yittrium radioembolization of liver cancer. *Int J Radiat Oncol Biol Phys* 85:834–839. Available at http://www.ncbi.nlm.nih.gov/pubmed/22871239.

Zielhuis, S.W., Nijsen, J.F.W., de Roos, R. et al. (2006). Production of GMP-grade radioactive holmium loaded poly(L-lactic acid) microspheres for clinical application. *Int J Pharm* 311:69–74. Available at http://www.ncbi.nlm.nih.gov/pubmed/16439073 [Accessed August 4, 2014].

PART 2

Patient Selection and Treatment Planning

2

Treatment options for patients with primary and secondary liver cancer: An overview of invasive, minimally invasive, and noninvasive techniques

RICARDO PAZ-FUMAGALLI, DAVID M. SELLA, AND GREGORY T. FREY

2.1 INTRODUCTION

Liver cancer occurs when cellular proliferation within the liver escapes normal control mechanisms, exhibits aggressive behavior, and reaches or invades other body parts causing deterioration of the patient's well-being and shortening of the patient's life expectancy. It may start in the liver (primary) or reach the liver from another origin (secondary or metastatic). The signs and symptoms of liver cancer are generally determined by the tumor size, number, location, proximity to vulnerable structures, rate of growth, production of substances, pattern of spread, and underlying condition of the patient's organ systems. Oncologists study cancer populations rather than individuals and express outcomes in statistical terms that reflect cancer control or freedom from cancer (overall survival, median survival, cancer-specific survival, disease-free survival, progression-free survival, time-to-progression) or can reflect improvement, maintenance, or deterioration of quality of life.

Cancer treatment aims to maintain or improve quality of life and lifespan and can be extremely complex due to the wide variety of tumor types, the presence or absence of underlying liver disease, and a multitude of available therapies. Treatment can be systemic (whole body) or locoregional (organ- or volume-specific) and can have a curative intent or be meant to lessen the impact of the disease (palliation). Treatment objectives can cross over; occasionally, a treatment given with palliative intent can later enable a curative approach.

This book is focused specifically on one form of liver cancer treatment: yttrium-90 (^{90}Y) radioembolization. However, because of the multidisciplinary nature of radioembolization, many of the individuals involved in a radioembolization program may not be very familiar with other treatments for hepatic malignancy. Those who directly or indirectly contribute to the care of a patient receiving ^{90}Y radioembolization should have a basic understanding of the therapies a patient has received or may receive in the future. To frame the context in which the role of radioembolization can be appreciated, this chapter provides an overview of standard-of-care treatment modalities for primary and secondary liver cancer and discusses the strengths, weaknesses, and contraindications of the therapeutic options stratified by tumor type.

2.2 APPROACHES TO THE PATIENT WITH HEPATIC MALIGNANCY

When approaching the patient with hepatic cancer, one must first determine whether the process is primary or metastatic. Hepatocellular carcinoma (HCC) is the most common primary liver tumor (Ferlay et al., 2010). Colorectal cancer is a common secondary tumor of the liver. Other liver malignancies commonly treated with locoregional therapies include neuroendocrine tumor (NET) metastases, breast cancer metastases, and intrahepatic cholangiocarcinoma (ICC). Second, the synthetic function of the patient's liver should be considered. Preserved liver function is important because the antitumor effects of various therapies can be counteracted by treatment-induced

liver failure (Lewandowski and Davenport, 2015). Various models and staging systems can be utilized to predict overall performance status, disease burden, hepatic reserve, and prognosis. The model for end-stage liver disease (MELD) and Child–Pugh scores take into account various chemical and clinical factors including creatinine, total bilirubin, serum albumin, prothrombin time, ascites, and encephalopathy to assess prognosis of chronic liver disease/cirrhosis. The Eastern Cooperative Oncology Group (ECOG) scale of performance status is used for measuring how a disease impacts a patient's daily living abilities. It determines a patient's level of functioning in terms of daily activity, physical ability, and self-care ability. Finally, one must consider the goals of therapy. When designing a treatment plan, one must take into consideration factors such as the extent of disease, both intra- and extrahepatic, as well as previous therapies. The treatment of liver malignancies can be complex, and a multispecialty team may include medical oncologists, hepatologists, surgeons, radiologists, and radiation oncologists.

2.3 SYSTEMIC THERAPIES

Systemic therapies act throughout the body. The most important categories include cytotoxic chemotherapy, hormonal therapy, and targeted therapy.

Cytotoxic chemotherapy interferes with the different steps in the cell cycle and directly targets and kills cancer cells but it is also harmful to normal cells. There are numerous cytotoxic pharmacologic mechanisms and agents. Alkylating agents interact with DNA (cyclophosphamide, cisplatin, oxaliplatin), antimetabolites interfere with DNA precursors and cellular metabolism (5-fluorouracil, gemcitabine), and antitumor antibiotics interfere with DNA activity (mitomycin, doxorubicin). Topoisomerase inhibitors (irinotecan) and mitotic inhibitors (vincristine, paclitaxel, docetaxel) represent other mechanisms. Cytotoxic chemotherapy is limited by its nonspecificity, as it is toxic to both cancer and normal cells, resulting in the adverse effects and toxicities observed during therapy.

Hormonal therapy is a broad category that includes inhibitors of hormone synthesis (letrozole), hormone receptor antagonists (tamoxifen), and hormone supplements (estrogens, progestins, androgens, and somatostatin analogs such as octreotide). Hormonal therapy has an anticancer effect but is also used to alleviate the symptoms of hormone- and peptide-secreting tumors.

Targeted therapy promises to minimize the problems that cytotoxic chemotherapy causes because it interferes with specific molecules that participate in the genesis of cancer and tumor growth rather than interfering nonspecifically in cell multiplication, which is a function shared with normal tissues. These agents are often called biologicals, and the most commonly used in liver cancer include tyrosine kinase inhibitors (sorafenib, sunitinib, erlotinib, imatinib), serine/threonine kinase inhibitors (everolimus), and monoclonal antibodies [bevacizumab has an antiangiogenesis effect by blocking vascular endothelial growth factor A (VEGF-A), and cetuximab inhibits epidermal growth factor receptor (EGFR)]. Immunotherapy is a variant of targeted therapy that has the objective of activating the immune system to identify and reject the cancer cells. Monoclonal antibody-based therapy is the most common and successful form of cancer immunotherapy. Cytokines (interleukin, interferon) and cellular therapy (cancer vaccines) are other forms of immunotherapy.

2.4 LOCOREGIONAL THERAPIES

Locoregional therapy of liver cancer is directed to a well-defined zone and lacks any activity outside of the volume defined by the treatment plan. The therapy can be surgical, minimally invasive guided by imaging, or noninvasive in the form of external radiation therapy. Surgical options include tumor resection, intraoperative thermal ablation, and liver transplantation. Image-guided therapy is most commonly done percutaneously by means of needles or probes inserted into the body and directed with computed tomography (CT), ultrasound, magnetic resonance imaging (MRI), or fluoroscopy, or it can be administered into the hepatic arterial blood flow after catheterization. Radiation therapy for liver tumors is most commonly given in tightly focused external beams that use stereotactic techniques or delivered during catheterization into the hepatic arterial circulation to achieve selective internal radiation therapy.

2.4.1 SURGERY

The objective of liver surgery is to remove the malignant tissue surrounded by a margin of tumor-free liver (resection) or the entire liver (transplantation). The success and safety are determined by proper patient selection based on the patient's health, tumor characteristics, the anatomy of liver in relation to the malignancy, the quality and volume of the liver remnant after the resection is complete, and adherence to established criteria for transplantation.

The type of resection is chosen after determining that the future liver remnant is large enough (usually >30% of total liver volume), that the liver functional reserve is adequate, and that there is the possibility of achieving a surgical margin negative for residual cancer. The surgery can follow the segmental anatomy of the liver and varies in extent from segmentectomy, to lobectomy, and to extended hepatectomy. The resection can also be nonanatomic, which is often referred to as wedge resection (Kishi et al., 2009).

If the expected liver remnant is too small, it is possible to induce growth of the remnant by interrupting portal vein perfusion of the liver portion to be removed. Preoperative portal vein embolization (PVE) and the so-called "associating liver partition and portal vein ligation for staged hepatectomy" (ALPPS) surgical procedure are the two procedures designed to achieve this effect (Shindoh et al., 2013a, 2013b). The physiologic response in the liver remnant is called the atrophy–hypertrophy complex (Kim et al., 2008). PVE is performed by interventional radiologists using multiple techniques to achieve complete occlusion of the portal vein branches expected to be removed (Guiu et al., 2013). PVE is safe and has a low complication rate (Abulkhir et al., 2008; Ratti et al., 2010). However, questions have been raised about the potential of tumor growth following embolization related to alterations in blood flow patterns and increases in local growth factors (Simoneau et al., 2012). Another method to induce hypertrophy of the functional liver remnant is lobar radioembolization with Y-90 microspheres (Fernandez-Ros et al., 2014; Garlipp et al., 2014; Teo and Goh, 2015).

Liver transplantation is well established as the treatment of choice for HCC, and HCC accounts for the vast majority of transplantation for malignancy, but metastatic NETs, epithelioid hemangioendothelioma, and cholangiocarcinoma are occasionally treated with transplantation. Liver transplantation is particularly valuable for HCC because it simultaneously treats the hepatic malignancy and the underlying cirrhosis and/or liver insufficiency (Eghtesad and Aucejo, 2014).

2.4.2 TUMOR ABLATION

Tumor ablation intends to destroy tissue in the body without removal, leaving a zone of necrosis that heals into a scar over time (Figure 2.1). Tissue destruction can be achieved with extreme temperature by inducing heating or freezing (radiofrequency ablation or RFA, laser ablation, microwave ablation [MW], high-intensity focused ultrasound, cryoablation), with injection of chemicals such as alcohol and concentrated acetic acid and by disrupting the cell membrane at the molecular level with electrical fields (irreversible electroporation or IRE) (Ahmed, 2014). For percutaneous ablation,

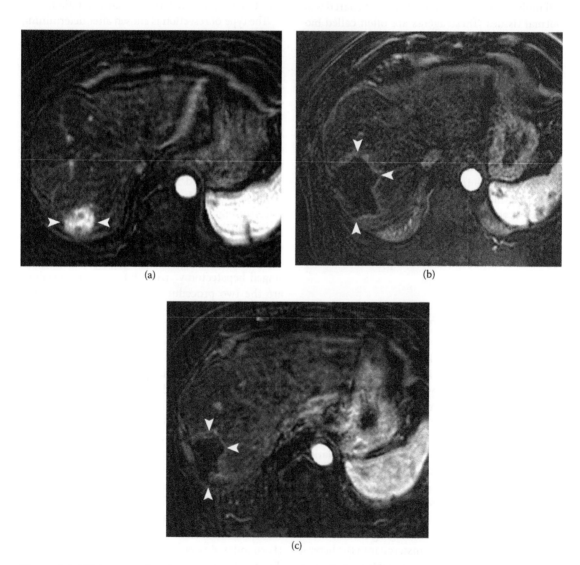

(a)

(b)

(c)

Figure 2.1 MR images after intravenous contrast processed with subtraction technique in a 67-year-old man with hepatocellular carcinoma. **(a)** Single intensely enhancing hepatocellular carcinoma in segment VII of the liver (arrowheads). **(b)** One month after percutaneous radiofrequency ablation, the zone of ablation is sharply defined without any residual abnormal contrast enhancement to indicate residual tumor (arrowheads). **(c)** At 76 years of age, 9 years after the ablation, the treated volume has decreased (arrowheads) and there is no abnormal enhancement to indicate active tumor.

probes (thermal ablation, IRE) or needles (chemical injection) are inserted through skin punctures and directed to the treatment zone guided with imaging, typically CT or ultrasound, but other imaging methods can be used. Tumor ablation can also be guided surgically, whether laparoscopically or through a laparotomy. The number and distribution of the probes or needles will define a specific kill volume. Tumor ablation can be performed alone or in combination with other treatments including chemotherapy, surgery, or embolization (Wells et al., 2015). Obesity, under-lying liver disease, and tumor size and location can affect both the results and frequency of complications with tumor ablation (Komorizono et al., 2003; Livraghi et al., 2003; Teratani et al., 2006).

2.4.2.1 Radiofrequency ablation

Radiofrequency ablation applies to target tissues alternating current at a high frequency, similar to electrocautery. The term radiofrequency is misleading because this method does not apply electromagnetic energy in the form of radio waves; rather, it refers to the frequency of the alternating current, typically 460–500 kHz, that falls within the frequency range of radio. As the current alternates, dipolar molecules and ions such as sodium and potassium move quickly to align with the current, inducing frictional heat. Temperatures can rise well above the boiling point of water. Temperatures >50°C are quickly lethal at the cellular level.

The probes function as electrodes in the electrical circuit that runs through the targeted tissues. The size of the ablation volume and extent of cell death with RFA are limited by the electrical conductivity of the tissue, patterns of heat convection and conduction, and the presence of "heat sinks." Flowing blood in medium to large vessels can cool nearby tissues during RFA and prevent reaching lethal temperatures. RFA is relatively slow and is unreliable for ablation volumes with diameters >5 cm; however, it is quite reliable for volumes ≤3 cm (Hong and Georgiades, 2010). Of all thermal ablation methods, RFA is the most studied and most widely reported in the medical literature.

2.4.2.2 Microwave ablation

MW probes are antennae that broadcast electromagnetic waves with frequencies from 900 to 2450 MHz. The probe is tuned to the natural frequency of water, and similarly to RFA, vigorous motion occurs at the molecular level during application of microwaves, producing frictional heat. MW has several advantages over RFA. It is faster, can achieve higher temperatures, is less susceptible to heat-sink effect, and does not depend on electrical conductivity so it is not encumbered by poorly conducting tissues such as bone or air-filled lung. The devices are capable of operating multiple probes simultaneously. Large ablation volumes ≥5 cm can be reached more reliably than with RFA (Simon et al., 2005).

2.4.2.3 Cryoablation

The clinically available cryoablation probes are designed to circulate a gas, such as argon, into an expansion chamber at the tip of the probe, which causes profound cooling (Joule–Thompson effect). The adjacent tissues freeze and the frozen volume expands over time. Because cells suffer harm both from freezing as well as thawing, a commonly utilized sequence requires freezing for 10 min, thawing for 8 min, and refreezing for 10 min. Tissue is reliably devitalized with this type of protocol at temperatures below –20°C. Therefore, frozen tissues at the edges of the ice ball survive, and the kill zone is smaller than the size of the freeze, as shown with intraprocedural CT.

The x-ray beam attenuation of ice is less than water, making the frozen volume very clearly visible with CT. Cryoablation is, therefore, advantageous when sharp visibility of the ablation edge is needed to monitor and protect nearby vulnerable structures from injury. Multiple probes used simultaneously permit sculpting larger volumes of ablation than are achievable with heat-based systems. Cryoablation causes less pain that RFA or MW and is less destructive of the connective tissue structure. For these reasons, it is better suited for locations where substantial postprocedural pain can be expected and when attempting to preserve the integrity of an adjacent structure.

Because cryotherapy does not have a cauterizing effect like RFA or MW and the probe diameter tends to be larger, there is greater potential for postprocedural hemorrhage, which limits the application of percutaneous cryotherapy for liver tumors. Caution must be exercised when creating large ice balls because of the threat of cryoshock, a severe consequence of therapy that can be fatal, and because of the threat of tissue fracture with hemorrhage (Erinjeri and Clark, 2010).

2.4.2.4 Irreversible electroporation

Irreversible electroporation units apply high-voltage current across a cellular membrane, which creates small pores, destabilizes the cellular membrane, and causes cell death (Gehl, 2003). IRE requires general anesthesia and complete muscle relaxation. Because it does not have a significant effect on connective tissues, it is the least likely of the ablation methods to damage vessels, bile ducts, the gallbladder, or the bowel (Silk et al., 2014). Patients who benefit the most from IRE have limited disease in the central portion of the liver where resection and other ablative modalities cannot be done safely (Lencioni et al., 2015; Scheffer et al., 2015). Experience with IRE at this time is preliminary, and there are scant data focused on outcomes.

2.4.3 TRANSARTERIAL THERAPY

The hepatic arteries provide another route for locoregional therapy. Both primary and metastatic liver cancers derive blood supply almost completely from the hepatic arterial circulation. In contrast, the functional liver tissue receives 60%–75% of blood flow from the portal vein. Transarterial therapies exploit this differential perfusion. Consequently, the tumor receives a highly concentrated therapeutic dose while, for the most part, sparing the functional liver tissue. The most common transarterial options include inert particle embolization, chemoembolization, and radioembolization.

2.4.3.1 Inert particle embolization

Administration of small particles (usually 45–750 μm in size) into the hepatic arteries causes microvascular occlusion and interrupts the blood supply. The malignant disease quickly develops large zones of ischemia and coagulative necrosis (Figure 2.2). Particle embolization can be very effective for treatment of HCC and metastatic NET but is less effective with cholangiocarcinoma and metastatic disease of other histologic types (Brown et al., 1999; Maluccio et al., 2008).

2.4.3.2 Chemoembolization

Transcatheter arterial chemoembolization (TACE) treats liver cancer with a combination of local delivery of tumorcidal chemicals in a highly concentrated form while inducing tumor ischemia and maintaining very low concentrations of systemic chemotherapeutic agents (Lencioni, 2010). The conventional TACE (cTACE) mixture is based on ethiodized oil used either alone, in an emulsion with chemotherapeutic agents or in combination with absolute ethanol (Figure 2.3). Doxorubicin is the most commonly used chemotherapy agent in the United States; however, epirubicin, cisplatin, and mitomycin C are also utilized. The mixture is delivered through an arterial catheter placed as selectively as possible into the arteries feeding the tumor, usually followed by embolization particles that induce ischemic necrosis of the tumor and prevent washout of the drug.

Drug-eluting bead TACE (DEB-TACE) involves the delivery of microspheres loaded with chemotherapy and provides a sustained release of drug and tumor vessel occlusion following intra-arterial administration (Figure 2.4). The two agents utilized most often are doxorubicin for HCC or virtually any other tumor type drug-eluting beads with doxorubicin (DEB-DOX) or drug-eluting beads with irinotecan for colorectal metastases (DEB-IRI). DEB-TACE utilizes catheter-based techniques similar to cTACE.

After chemoembolization, most patients experience a postembolization syndrome of right upper quadrant pain, nausea, fever, and loss of appetite that is self-limited over a period of several days. Fatigue often occurs, can be profound, and can last a few weeks. Reported complications include bile duct injury, liver abscess (particularly after biliary intervention), tumor rupture, and nontarget injury with necrosis in vascular territories supplying the bowel, gallbladder, and diaphragm. Some studies have demonstrated fewer adverse events and an improved pharmacokinetic profile with DEB-TACE compared with cTACE (Liapi and Geschwind, 2010; Molvar and Lewandowski, 2015).

2.4.3.3 Radioembolization

The radioembolization procedure will be presented in detail elsewhere in this text. Briefly, radioembolization refers to the delivery of glass or resin

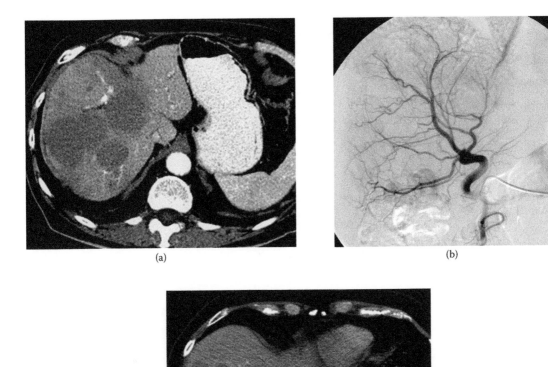

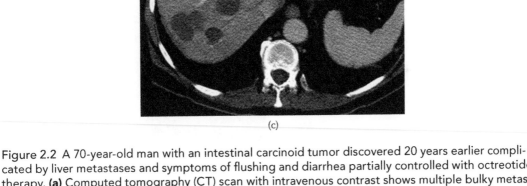

Figure 2.2 A 70-year-old man with an intestinal carcinoid tumor discovered 20 years earlier complicated by liver metastases and symptoms of flushing and diarrhea partially controlled with octreotide therapy. **(a)** Computed tomography (CT) scan with intravenous contrast shows multiple bulky metastases in the liver with only modest enhancement. **(b)** Hepatic arteriography showed distortion of the hepatic artery branches and only subtle enhancement of the disease. **(c)** CT scan follow-up after 6 months confirmed complete loss of the modest enhancement in most lesions and marked decrease in the size of the tumors.

microspheres loaded with Y-90 radioactive isotope into the liver circulation by means of catheterization. Like TACE, it takes advantage of the differential arterial flow between tumor and normal tissues to deliver a high dose to the malignancy with minimization of the radiation dose to the normal liver (Lencioni, 2010). Because hepatic artery occlusion is not necessarily the goal in radioembolization, the typical postembolization syndrome is usually avoided or minimized. Patient eligibility for radioembolization is similar to that of chemoembolization. Contraindications include established liver insufficiency, unsuitable anatomy that places the patient at risk for nontarget radiation exposure to the gastrointestinal tract, and hepatopulmonary shunting estimated to exceed the maximum allowable radiation lung doses as shown on a Tc-99m macroaggregated albumin nuclear scan done after administration of the radionuclide into the hepatic artery.

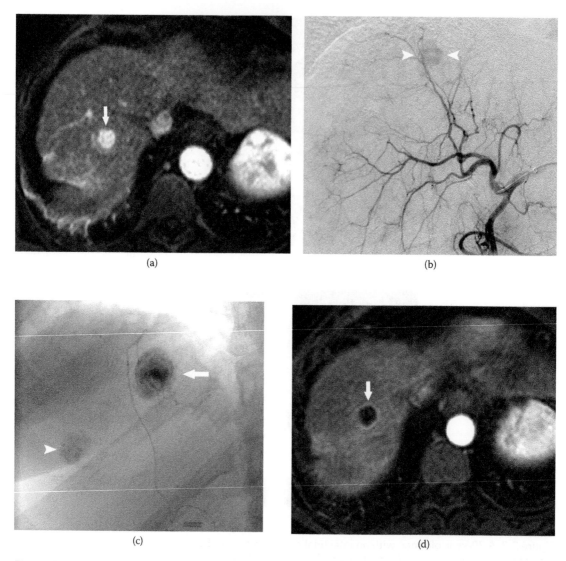

(a)

(b)

(c)

(d)

Figure 2.3 A 56-year-old woman with liver cirrhosis complicated by hepatocellular carcinoma treated with conventional lipiodol-based chemoembolization in preparation for liver transplant. **(a)** Magnetic resonance (MR) image of the liver after intravenous contrast shows intense enhancement in the tumor (arrow). **(b)** Hepatic arteriogram shows enhancement that corresponds to the tumor (arrowheads). **(c)** Chemoembolization during super-selective catheterization of the artery feeding directly into the tumor achieves dense saturation of the tumor with the therapeutic agents (arrow). Another tumor (arrowhead) has residual lipiodol from a previous treatment, a common long-term imaging finding after conventional chemoembolization. **(d)** MR image with intravenous contrast 3 months after treatment shows complete lack of enhancement, which indicates loss of viability (arrow). Pathologic examination of the explanted liver showed complete necrosis of the targeted tumor.

Radioembolization and chemoembolization differ in their respective adverse events. With radioembolization, fatigue is common and postembolization syndrome is uncommon. Radioembolization-induced liver disease (REILD) is an uncommon complication if standard selection criteria are met. Serious complications from nontarget radiation include gastrointestinal ulceration, cholecystitis, pancreatitis, and radiation pneumonitis (Riaz et al., 2009; Molvar and Lewandowski, 2015).

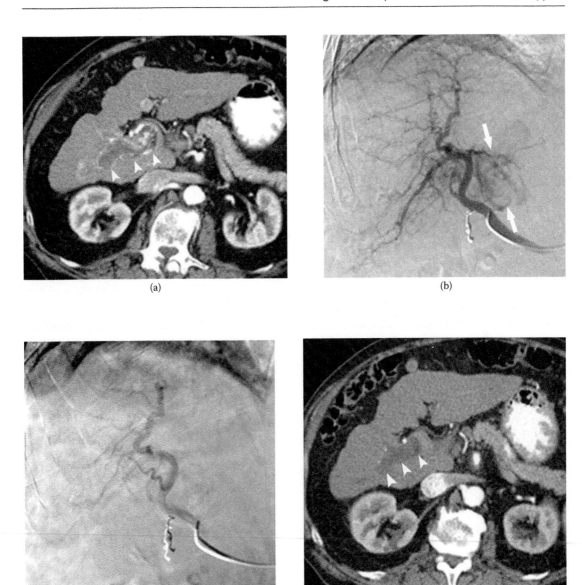

Figure 2.4 A 63-year-old man with a hepatocellular carcinoma with alpha fetoprotein (AFP) tumor marker level of 397 ng/mL. **(a)** CT scan of the cirrhotic liver shows invasion of the portal vein, which intensely enhances with intravenous contrast (arrowheads). **(b)** Hepatic arteriogram shows intense enhancement of the tumor invading the portal vein that matches the CT scan findings (arrows). **(c)** After drug-eluting bead chemoembolization with doxorubicin, the tumor enhancement and overall arterial flow are decreased. **(d)** Two months after treatment, the tumor in the portal vein was smaller, had lost its contrast enhancement (arrowheads), and AFP decreased to 35 ng/mL. Seven months after treatment, the AFP level was 13 ng/mL.

2.4.3.4 External-beam radiation therapy

External-beam radiation has traditionally played a limited role in liver tumor treatment, but recent technological developments have broadened its applicability. Conventional radiation therapy often requires large treatment fields in most but the smallest of tumors. Therefore, too much liver is exposed and can lead to radiation-induced liver disease (RILD),

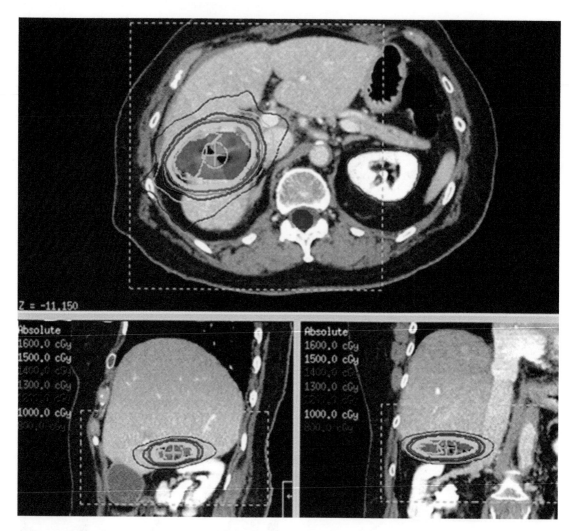

Figure 2.5 Planning radiation dosimetry before stereotactic body radiation therapy for colon cancer that metastasized to the liver. The tightly defined volume exposed to the radiation allows higher dosimetry with greater safety compared with conventional external-beam radiation.

a complication that can be fatal. Three-dimensional conformal radiation therapy (3D-CRT) and stereotactic body radiation therapy (SBRT) are safer. Both methods can deliver high-dose treatment in the tumorcidal range while minimizing the injury to surrounding tissues (Figure 2.5). An advantage of SBRT is that the high-dose treatment can be given in one or a few sessions (Tanguturi et al., 2014).

2.5 HEPATOCELLULAR CANCER

HCC is the most common primary liver tumor and usually occurs alongside cirrhosis (Ferlay et al., 2010). The incidence is on the rise in the United States along with the spread of hepatitis C virus (Davis et al., 2010). Because imaging diagnosis of HCC can be reliably made with high-quality cross-sectional imaging using multiphasic contrast-enhancement patterns, tissue confirmation with percutaneous biopsy is generally reserved for lesions with atypical imaging features (Marrero et al., 2005; Forner et al., 2008; Bruix and Sherman, 2011). There is considerable variation in the epidemiology of HCC based on risk factors present in developing versus developed countries, including hepatitis B and C, alcoholic cirrhosis, nonalcoholic fatty liver disease, and nonalcoholic steatohepatitis. The Barcelona Clinic Liver Cancer (BCLC) classification is the most commonly used staging system in the

Western HCC population and integrates tumor burden, liver function (Child–Pugh classification), and performance (ECOG) status to link prognosis with treatment options. This provides a framework to select patients who receive curative versus palliative treatments (Llovet et al., 1999). Liver transplantation, resection, and ablation are considered curative for very early- (stage 0) and early-stage HCC (stage A). Palliative options include TACE for intermediate-stage HCC (stage B) and radioembolization and sorafenib for advanced-stage HCC (stage C). Best supportive care is utilized in terminal disease (stage D) (Reig et al., 2014).

2.5.1 SURGICAL RESECTION AND LIVER TRANSPLANTATION FOR HCC

Surgical resection remains a curative option for HCC, but wide applicability is limited because of the decreased hepatic reserve found in most cirrhotic livers. Liver transplantation is the definitive therapy for HCC in the setting of cirrhosis. However, the majority of patients never receive transplantation because they do not meet transplantation criteria, because of a lack of access to a transplant center, or because of decreased organ availability (Kim et al., 2005).

2.5.2 TUMOR ABLATION FOR HCC

Image-guided tumor ablation is recommended for patients with very early- and early-stage HCC who are not candidates for surgical resection according to the BCLC criteria. The reference standard for ablation of small HCC is RFA. Studies have demonstrated RFA to be as effective as surgical resection in very early and early HCC (Cho et al., 2010). The evidence available suggests that microwave is at least equivalent to RFA in the treatment of very early- and early-stage HCC (Groeschl et al., 2014; Ziemlewicz et al., 2015; Vogl et al., 2015). RFA and MW can be considered first-line therapy in patients with liver dysfunction who have very early- and early-stage HCC in favorable locations (Wells et al., 2015).

Combination therapy harnesses the synergy between different treatment modalities. With HCC, this most commonly involves the use of particle embolization or TACE with a heat-based ablation. There is no consensus on the order or timing between interventions when using combination therapy. In select situations, combined embolization and ablation appear to improve both technical success and local tumor progression (Takaki et al., 2009; Peng et al., 2010).

2.5.3 CHEMOEMBOLIZATION FOR HCC

The BCLC staging system recommends cTACE for intermediate stage HCC. Conventional TACE may be used as a bridge to transplant or as a palliative therapy for unresectable HCC. Chemoembolization has demonstrated survival benefit over conservative treatment or best supportive care in two well-known randomized controlled trials (RCTs) (Lo et al., 2002; Llovet et al., 2002). In addition, a meta-analysis of 18 RCTs demonstrated a 2-year survival benefit of cTACE over conservative treatment (Camma et al., 2002). The PRECISION V trial was a prospective RCT that compared the efficacy of cTACE to DEB-TACE. DEB-TACE failed to show a response improvement over cTACE at 6 months; however, there was a statistically significant reduction in liver toxicity. Subset analysis of patients with advance disease showed an improved objective response rate with DEB-TACE (Lammer et al., 2010).

2.5.4 RADIOEMBOLIZATION FOR HCC

Radioembolization plays a large role in the treatment of HCC. The safety and efficacy of radioembolization are demonstrated in multiple retrospective studies and cohorts. There are no RCTs comparing radioembolization and other locoregional therapies for HCC. Data available suggest advantages and disadvantages for TACE and radioembolization, but clear superiority has not been shown for either therapy (Moreno-Luna et al., 2013; Minocha et al., 2014).

2.5.5 SYSTEMIC THERAPY FOR HCC

HCC is chemotherapy resistant and systemic chemotherapy is generally not well tolerated in patients with the significant hepatic dysfunction that accompanies cirrhosis. Because survival of patients with advanced HCC is often determined by the degree of hepatic dysfunction rather than the tumor, it is difficult to determine the benefit from chemotherapy. Sorafenib is an orally active multikinase inhibitor acting on the vascular endothelial growth factor receptor (VEGFR). Results from the phase III SHARP (sorafenib hepatocellular carcinoma assessment randomized protocol) trial suggested a survival benefit compared with best supportive care. The multicenter SHARP trial randomly assigned 602 patients with inoperable HCC and Child–Pugh A cirrhosis to sorafenib or placebo. Overall survival was significantly longer in the sorafenib-treated patients (10.7 vs. 7.9 months) along with time to radiologic progression (5.5 vs. 2.8 months). Major adverse effects included diarrhea and hand–foot skin reaction. This study established sorafenib monotherapy as the standard systemic treatment for advanced HCC (Llovet et al., 2008).

2.6 COLORECTAL LIVER METASTASES

Colorectal cancer (CRC) is the second leading cause of cancer deaths in the United States (Siegel et al., 2013). Of patients with CRC, approximately half develop metastases, and the liver is the most common site for these (Yoo et al., 2006). The majority of patients with metastatic disease are unresectable (Muratore et al., 2007). Metastatic colorectal cancer (mCRC) is best treated by a multidisciplinary team of surgeons, oncologists, interventional radiologists, hepatologists, and radiation oncologists.

2.6.1 MEDICAL THERAPY FOR CRC

Systemic chemotherapy is the mainstay of medical therapy. The primary goal is conversion of unresectable disease to resectable disease, when possible. Multiple chemotherapy regimens are currently recommended as first-line therapy, including FOLFOX (folinic acid, fluorouracil, oxaliplatin), FOLFIRI (folinic acid, fluorouracil, irinotecan), CapeOx (capecitabine, oxaliplatin), infusional 5-FU/LV (-fluorouracil, leucovorin) or capecitabine, or FOLFOXIRI (folinic acid, fluorouracil, oxaliplatin, irinotecan) (Benson et al., 2014). Bevacizumab, a monoclonal antibody that blocks VEGF, has been used in conjunction with 5-FU/LV regiments and has shown statistically significant improvement in median survival (Kabbinavar et al., 2005). Multiple reports have

discussed the potential for perioperative complications in patients taking bevacizumab because of its effect on the vasculature (Scappaticci et al., 2005; Gordon et al., 2009). Arterial dissection is a well-known complication that can occur spontaneously and increases the risk and difficulty of transarterial therapies for liver metastases that require catheterization (Aragon-Ching et al., 2008; Brown, 2011; Mantia-Smaldone et al., 2013).

2.6.2 SURGICAL THERAPY FOR CRC

Surgical excision of primary and metastatic disease must be considered in all patients and done when possible because resection provides improved 5-year survival rates compared with nonoperative candidates (Van Cutsem et al., 2006). Surgery is typically performed in conjunction with systemic chemotherapy. The timing of surgical management and administration of chemotherapy can be variable (Benson et al., 2014). PVE and radioembolization of the portion of liver to be resected increase the number of patients eligible for surgery by increasing the volume of the functional liver remnant (Kabbinavar et al., 2005; Abulkhir et al., 2008; Ratti et al., 2010).

2.6.3 CHEMOEMBOLIZATION FOR CRC

Chemoembolization for mCRC is not currently considered first-line therapy. The use of DEB-IRI for treatment of mCRC was reported in 2006 (Aliberti et al., 2006). One study demonstrated improved median survival for patients undergoing treatment with DEB-IRI versus FOLFIRI after failure of first-line agents (Fiorentini et al., 2012). Timing of locoregional therapy is best managed in discussion with other stakeholders in the patient's management.

2.6.4 RADIOEMBOLIZATION

Radioembolization with Yttrium-90 labeled resin beads is approved by the U.S. Food and Drug Administration (FDA) for the treatment of mCRC. Data for the recently completed SIRFLOX trial are available only in abstract form. This study compared Y-90 radioembolization plus FOLFOX chemotherapy versus FOLFOX chemotherapy alone as first-line treatment of nonresectable liver metastases from colorectal cancer. It demonstrated statistically significant improvement in median hepatic progression-free survival and tumor response rates in the group that received radiation (Gibbs et al., 2014).

2.6.5 TUMOR ABLATION FOR CRC

Long-term survival rates of patients carefully selected for treatment with RFA are comparable to rates of resected patients (Raut et al., 2005; Amersi et al., 2006). Local recurrence rates, however, remain higher (Solbiati et al., 1997). Tumor ablation is also frequently used in combination with resection when complete resection is not feasible (Ito et al., 2010).

In general, tumor ablation for mCRC is best used for patients with unresectable liver metastases due to size, location, or comorbidities. RFA has 3-year survival rates of 46% and 5-year survival rates of less than 20%. Ablation is also associated with higher recurrence (1 year: 12%; 5 years: 50%–70%). Lesion size is a predictor of recurrence and overall survival. With lesions less than 2.5 cm, the 5-year survival is 56%. Survival is only 13% with lesions greater than 2.5 cm (Boutros et al., 2010). Location can also affect the rate of complications in patients undergoing ablative therapy (Raman et al., 2004). One study demonstrated no statistically significant difference in survival rates between MW and resection (Shibata et al., 2000). Irreversible electroporation has been used in mCRC but the data are insufficient for outcomes analysis at this time (Scheffer et al., 2015).

2.7 METASTATIC NET

NETs are a heterogenous group of neoplasms that originate from foregut structures (lung and pancreas), midgut (small intestine and part of colon), and hindgut (distal colon). NET can be hormonally functional or nonfunctional depending on the presence of clinical symptoms of hormone/peptide secretion (insulin, glucagon, gastrin, vasoactive intestinal peptide, serotonin). Abdominal carcinoid tumors usually originate in the bowel and present with typical symptoms of flushing, diarrhea, wheezing, or endocardial and valvular heart disease. Pancreatic and bowel NET have a strong tendency for development of liver metastases,

which are responsible for most symptomatology and are the main cause of death.

Gastroenteropancreatic NET can have different degrees of malignant behavior. Pathological criteria of tumor grade are based on markers of cell division, such as the number of mitoses per 10 high-power fields and Ki-67 index, a cell proliferation marker. Low-grade typical carcinoid shows <2 mitoses/10 high-power field (hpf) and <3% Ki-67, intermediate grade atypical carcinoid: 2–20 mitoses/10 hpf or 3–20% Ki-67, and high-grade small- or large-cell neuroendocrine carcinoma >20 mitoses/hpf or >20% Ki-67 (Kunz, 2015).

Numerous factors determine treatment. Management differs if the NET is limited to the primary site, if the NET is metastatic, or if the primary site is unresectable. Liver-dominant metastases, substantial extrahepatic disease, tumor volume, tumor growth rate, origin (pancreatic vs. gastrointestinal), pathologic tumor grade, and whether the tumor is clinically functional are the most critical factors. Pancreatic origin correlates with greater aggressiveness with median overall survival of 24 months compared with 56 months for gastroenteric NET (Yao et al., 2008).

2.7.1 SURGERY FOR NET

When the NET is localized to its organ of origin, surgery is the treatment of choice if the tumor is deemed resectable. When the tumor is metastatic, surgery is still indicated if at least 90% of the disease can be removed and the tumor is of low or intermediate grade, if the tumor location is causing problems, or if removal will reduce hormonal symptoms (Alagusundaramoorthy and Gedaly, 2014).

Surgery does not eliminate the metastatic disease but offers the best achievable tumor control. A retrospective international study compared surgery with transarterial therapy. This was not a randomized study and therefore was subjected to profound patient selection biases (the surgical group had fewer hormonally active tumors and the overall hepatic burden was greater in the transarterial group). The study found that the median and 5-year survival of patients treated with surgery was 123 months and 74% versus 34 months and 30% for transarterial therapy, a highly significant difference. Surgery showed the greatest survival benefit in symptomatic patients with a larger tumor burden, but when the patients were asymptomatic, there was no difference in long-term outcome between surgery and transarterial therapy (Mayo et al., 2011).

In another study of hepatic resection, the overall survival at 5 and 10 years was 63% and 40%, respectively, but most patients had disease progression during follow-up. The authors suggested that aggressive surgery is beneficial for well-differentiated metastatic tumors. Overall survival confirmed large differences between those with well-, moderately-, and poorly differentiated tumors, with median overall survival of approximately 120, 60, and 20 months, respectively. The presence of extrahepatic disease was also a predictor of poor prognosis (Saxena et al., 2011).

Liver transplantation is an option for selected patients with NET liver metastases. When stringent selection criteria are applied, the results can be excellent, with 5-year survival close to 90% and recurrence-free survival close to 80%, but most patients do not qualify (de Herder et al., 2010). A more realistic analysis of the European Liver Transplant Registry reported 5-year overall and disease-free survival rates of 52% and 30%, respectively (Le Treut et al., 2013).

2.7.2 TUMOR ABLATION FOR NET

Tumor ablation can be a stand-alone therapy, an adjunct at the time of surgical resection, for recurrence after previous ablation or in patients who cannot have surgery because of comorbidities and other risk factors. Ablation is most valuable when metastases are few and <5 cm in diameter (ideally <3 cm) and if disease is in a location where adjacent tissues would not be compromised. Intraoperative tumor ablation can address disease that is found beyond the surgical margin and has been found to widen the patient eligibility for surgery and provide additional symptom control (Taner et al., 2013).

Outcomes data for percutaneous ablation of metastatic NET are limited by the absence of randomized trials, the selection biases that are inherent to the assignment of surgical versus percutaneous approaches, and the mixed nature of reports that combine both intraoperative and percutaneous ablation. A recent meta-analysis of 301 patients and 978 tumors confirmed the procedure's safety with a 0.7% mortality rate and 10%

morbidity and 5-year survival rates ranging from 57%–80%. Partial or complete symptom relief was achieved in over 90% of cases (Mohan et al., 2015).

2.7.3 SYSTEMIC THERAPY FOR NET

Systemic therapy for metastatic NET can be directed to target the hormonal/peptide-producing nature of the tumor or the tumor growth. Synthetic somatostatin analogs (SSA) are administered to most patients because most NET overexpress somatostatin receptors and frequently produce hormonal/peptide symptoms. Even in tumors that do not present with clinical hormonal effects, SSA frequently have an antitumor effect with slowing of disease progression and even tumor volume reduction. Octreotide is the most widely prescribed SSA. The PROMID clinical trial of long-acting octreotide showed an antitumor effect, with a median time to progression of 14.3 months for the experimental group, compared with 6 months for placebo (Rinke et al., 2009).

For tumor growth control, various chemotherapeutic agents are used including streptozocin, dacarbazine, 5-fluorouracil, doxorubicin, etoposide, cisplatin, carboplatin, and taxanes. Targeted therapies are valuable, the most useful of which include the tyrosine kinase inhibitors sunitinib and sorafenib, the VEGF monoclonal antibody bevacizumab, and the mammalian target of rapamycin (mTOR) inhibitor everolimus. The development of β-emitting radionuclide-labeled somatostatin analogs 90Y-DOTATOC (DOTA0-D-Phe1-Tyr3-octreotide) and ^{177}Lu-DOTATATE (DOTA-(Tyr^3)-octreotate) offers targeted radiation therapy with a high degree of tissue specificity.

In general, sunitinib or everolimus are first-line options for symptomatic pancreatic, bulky, or progressive NET. For symptomatic, bulky, or progressive gastroenteric carcinoid tumors, the SSA are first-line choices. Where available, the radionuclide-labeled SSA can also be used as a first-line therapy (Castellano et al., 2015).

2.7.4 TRANSARTERIAL THERAPY FOR NET

Transarterial therapies are commonly applied to nonsurgical patients with liver-dominant disease to control tumor growth and symptoms.

Virtually any embolic agent will be effective by inducing ischemia, delivering high-dose chemotherapy, or delivering intra-arterial radioactive beads. The data are insufficient to demonstrate any advantage between inert particles, drug-eluting beads, or conventional lipiodol-based TACE (Brown et al., 1999; Orgera et al., 2015). A study of 100 patients with metastatic NET showed median overall survival of approximately two years (Pitt et al., 2008).

The available trials do not show superiority of any technique, but doxorubicin DEB-TACE has been associated with increased risk for biliary injury and complications (Bhagat et al., 2013). In a retrospective series, the median survival of metastatic NET treated with Y-90 microspheres was 70 months (Kennedy et al., 2008). Direct comparison outcomes data between radioembolization with other transarterial therapies do not exist, and the available data are heavily influenced by selection bias.

2.8 INTRAHEPATIC CHOLANGIOCARCINOMA

ICC is the second most common primary liver malignancy after HCC and originates from bile ducts of small size within the liver. Most ICCs are incurable, respond poorly to therapy, and recur early and often. ICC predominantly spreads in the liver but commonly becomes extrahepatic and produces distant hematogenous metastases (Lafaro et al., 2015).

Surgery can be curative only if the resection is complete. Even if preoperative imaging suggests that a complete resection is possible, at the time of resection many tumors cannot be removed entirely, and the recurrence rate is high. About 30% of cases are candidates for surgery, and the resections are usually extensive and complex, often requiring intervention on extrahepatic bile ducts and vascular structures (Endo et al., 2008). The 5-year overall survival despite surgery is poor, ranging from 14%–40%, and the median disease-free survival is approximately 1 year (Lafaro et al., 2015).

Unresectable patients receive systemic chemotherapy and have a poor prognosis, with variable median survival that averages about 6 months.

Gemcitabine is the most commonly used agent, along with 5-fluorouracil and cisplatin. It is still early to determine whether targeted therapies impact the outcomes of ICC (Lafaro et al., 2015).

External-beam radiation can be used following surgery or as the primary therapy, but studies are small and not controlled. As the primary therapy, the median overall survival can be as high as 13.3 months (Ben-Josef et al., 2005). SBRT can be more effective, but fewer patients qualify for this approach. A study from Mayo Clinic showed an overall survival at 6 and 12 months of 83% and 73%, but this was for a very small, highly selected patient group (Barney et al., 2012). The use of SBRT for ICC is becoming more widely accepted (Tanguturi et al., 2014).

Radioembolization with Y-90 microspheres has shown effectiveness comparable to systemic chemotherapy. A systematic review and pooled analysis of 12 studies and 298 patients yielded a median survival of 15.5 months. An added benefit of radioembolization is the conversion of unresectable to surgically resectable disease (Al-Adra et al., 2015).

ICC treated with TACE has been shown to have complete or partial response in 25% and is associated with improved survival (Hyder et al., 2013). The median overall survival is 9.1 months after TACE, but TACE done with a combination of gemcitabine plus cisplatin may be more effective than single-agent TACE with median survival up to 18.8 months (Gusani et al., 2008).

Tumor ablation is applicable only in a minority of ICCs because of criteria for treatment overlap with surgery. Percutaneous ablation is limited only to those who are not deemed surgical candidates if the size, lesion number, and location are appropriate. Small ICCs can have excellent response to tumor ablation, with a median overall survival of 33 months (Fu et al., 2012).

REFERENCES

Abulkhir, A., Limongelli, P., Healey A.J. et al. (2008). Preoperative portal vein embolization for major liver resection: A meta-analysis. *Ann Surg* 247:49–57.

Ahmed, M. (2014). Image-guided tumor ablation: Standardization of terminology and reporting criteria—A 10-year update. *Radiology* 273:241–260.

Al-Adra, D.P., Gill, R.S., Axford, S.J. et al. (2015). Treatment of unresectable intrahepatic cholangiocarcinoma with yttrium-90 radioembolization: A systematic review and pooled analysis. *Eur J Surg Oncol* 41:120–127.

Alagusundaramoorthy, S.S., Gedaly, R. (2014). Role of surgery and transplantation in the treatment of hepatic metastases from neuroendocrine tumor. *World J Gastroenterol* 20:14348–14358.

Aliberti, C., Tilli, M., Benea, G., Fiorentini, G. (2006). Trans-arterial chemoembolization (TACE) of liver metastases from colorectal cancer using irinotecan-eluting beads: Preliminary results. *Anticancer Res* 26:3793–3795.

Amersi, F.F., McElrath-Garza, A., Ahmad, A. et al. (2006). Long-term survival after radiofrequency ablation of complex unresectable liver tumors. *Arch Surg* 141:581–587; discussion 587–588.

Aragon-Ching, J.B., Ning, Y.M., Dahut, W.L. (2008). Acute aortic dissection in a hypertensive patient with prostate cancer undergoing chemotherapy containing bevacizumab. *Acta Oncol* 47:1600–1601.

Barney, B.M., Olivier, K.R., Miller, R.C., Haddock, M.G. (2012). Clinical outcomes and toxicity using stereotactic body radiotherapy (SBRT) for advanced cholangiocarcinoma. *Radiat Oncol* 7:67.

Ben-Josef, E., Normolle, D., Ensminger, W.D. et al. (2005). Phase II trial of high-dose conformal radiation therapy with concurrent hepatic artery floxuridine for unresectable intrahepatic malignancies. *J Clin Oncol* 23:8739–8747.

Benson AB, 3rd et al. (2014). Colon cancer, version 3.2014. *J Natl Compr Canc Netw* 12:1028–1059.

Bhagat, N., Reyes, D.K., Lin, M. et al. (2013). Phase II study of chemoembolization with drug-eluting beads in patients with hepatic neuroendocrine metastases: High incidence of biliary injury. *Cardiovasc Intervent Radiol* 36:449–459.

Boutros, C., Somasundar, P., Garrean S. et al. (2010). Microwave coagulation therapy for hepatic tumors: Review of the literature and critical analysis. *Surg Oncol* 19:e22–e32.

Brown, D.B. (2011). Hepatic artery dissection in a patient on bevacizumab resulting in pseudoaneurysm formation. *Semin Intervent Radiol* 28:142–146.

Brown, K.T., Koh, B.Y., Brody, L.A. et al. (1999). Particle embolization of hepatic neuroendocrine metastases for control of pain and hormonal symptoms. *J Vasc Interv Radiol* 10:397–403.

Bruix, J., Sherman, M. (2011). Management of hepatocellular carcinoma: An update. *Hepatology* 53:1020–1022.

Camma, C., Schepis, F., Orlando, A. et al. (2002). Transarterial chemoembolization for unresectable hepatocellular carcinoma: Meta-analysis of randomized controlled trials. *Radiology* 224:47–54.

Castellano, D., Grande, E., Valle, J. et al. (2015). Expert consensus for the management of advanced or metastatic pancreatic neuroendocrine and carcinoid tumors. *Cancer Chemother Pharmacol* 75:1099–1114.

Cho, Y.K., Kim, J.K., Kim, W.T., Chung, J.W. (2010). Hepatic resection versus radiofrequency ablation for very early stage hepatocellular carcinoma: A Markov model analysis. *Hepatology* 51:1284–1290.

Davis, G.L., Alter, M.J., El-Serag, H. et al. (2010). Aging of hepatitis C virus (HCV)-infected persons in the United States: A multiple cohort model of HCV prevalence and disease progression. *Gastroenterology* 138:513–521, 521 e511–516.

de Herder, W.W., Mazzaferro, V., Tavecchio, L. et al. (2010) Multidisciplinary approach for the treatment of neuroendocrine tumors. *Tumori* 96:833–846.

Eghtesad, B., Aucejo, F. (2014) Liver transplantation for malignancies. *J Gastrointest Cancer* 45:353–362.

Endo, I., Gonen, M., Yopp, A.C. et al. (2008). Intrahepatic cholangiocarcinoma: Rising frequency, improved survival, and determinants of outcome after resection. *Ann Surg* 248:84–96.

Erinjeri, J.P., Clark, T.W. (2010). Cryoablation: Mechanism of action and devices. *J Vasc Interv Radiol* 21:S187–S191.

Ferlay, J., Shin, H.R., Bray, F. et al. (2010). Estimates of worldwide burden of cancer in 2008: GLOBOCAN 2008. *Int J Cancer* 127:2893–2917.

Fernandez-Ros, N., Silva, N., Bilbao, J.I. et al. (2014). Partial liver volume radioembolization induces hypertrophy in the spared hemiliver and no major signs of portal hypertension. *HPB (Oxford)* 16:243–249.

Fiorentini, G., Aliberti, C., Tilli, M. et al. (2012). Intra-arterial infusion of irinotecan-loaded drug-eluting beads (DEBIRI) versus intravenous therapy (FOLFIRI) for hepatic metastases from colorectal cancer: Final results of a phase III study. *Anticancer Res* 32:1387–1395.

Forner, A., Vilana, R., Ayuso, C. et al. (2008). Diagnosis of hepatic nodules 20 mm or smaller in cirrhosis: Prospective validation of the noninvasive diagnostic criteria for hepatocellular carcinoma. *Hepatology* 47:97–104.

Fu, Y., Yang, W., Wu, W. et al. (2012). Radiofrequency ablation in the management of unresectable intrahepatic cholangiocarcinoma. *J Vasc Interv Radiol* 23:642–649.

Garlipp, B., de Baere, T., Damm, R. et al. (2014). Left-liver hypertrophy after therapeutic right-liver radioembolization is substantial but less than after portal vein embolization. *Hepatology* 59:1864–1873.

Gehl, J. (2003). Electroporation: Theory and methods, perspectives for drug delivery, gene therapy and research. *Acta Physiol Scand* 177:437–447.

Gibbs, P., Gebski, V., Van Buskirk, M. et al. (2014). Selective internal radiation therapy (SIRT) with yttrium-90 resin microspheres plus standard systemic chemotherapy regimen of FOLFOX versus FOLFOX alone as first-line treatment of non-resectable liver metastases from colorectal cancer: The SIRFLOX study. *BMC Cancer* 14:897.

Gordon, C.R., Rojavin, Y., Patel, M. et al. (2009). A review on bevacizumab and surgical wound healing: An important warning to all surgeons. *Ann Plast Surg* 62:707–709.

Groeschl, R.T., Pilgrim, C.H., Hanna, E.M. et al. (2014). Microwave ablation for hepatic malignancies: a multiinstitutional analysis. *Ann Surg* 259:1195–1200.

Guiu, B., Bize, P., Gunthern, D. et al. (2013). Portal vein embolization before right hepatectomy: Improved results using n-butyl-cyanoacrylate compared to microparticles plus coils. *Cardiovasc Intervent Radiol* 36:1306–1312.

Gusani, N.J., Balaa, F.K., Steel, J.L. et al. (2008). Treatment of unresectable cholangiocarcinoma with gemcitabine-based transcatheter arterial chemoembolization (TACE): A single-institution experience. *J Gastrointest Surg* 12:129–137.

Hong, K., Georgiades, C. (2010) Radiofrequency ablation: Mechanism of action and devices. *J Vasc Interv Radiol* 21:S179–S186.

Hyder, O., Marsh, J.W., Salem, R. et al. (2013). Intra-arterial therapy for advanced intrahepatic cholangiocarcinoma: A multi-institutional analysis. *Ann Surg Oncol* 20:3779–3786.

Ito, K., Govindarajan, A., Ito, H., Fong, Y. (2010). Surgical treatment of hepatic colorectal metastasis: Evolving role in the setting of improving systemic therapies and ablative treatments in the 21st century. *Cancer J* 16:103–110.

Kabbinavar, F.F., Hambleton, J., Mass, R.D. et al. (2005). Combined analysis of efficacy: The addition of bevacizumab to fluorouracil/leucovorin improves survival for patients with metastatic colorectal cancer. *J Clin Oncol* 23:3706–3712.

Kennedy, A.S., Dezarn, W.A., McNeillie, P. et al. (2008). Radioembolization for unresectable neuroendocrine hepatic metastases using resin 90Y-microspheres: Early results in 148 patients. *Am J Clin Oncol* 31:271–279.

Kim, R.D., Kim, J.S., Watanabe, G. et al. (2008). Liver regeneration and the atrophy-hypertrophy complex. *Semin Intervent Radiol* 25:92–103.

Kim, W.R., Gores, G.J., Benson, J.T. et al. (2005). Mortality and hospital utilization for hepatocellular carcinoma in the United States. *Gastroenterology* 129:486–493.

Kishi, Y., Abdalla, E.K., Chun, Y.S. et al. (2009). Three hundred and one consecutive extended right hepatectomies: Evaluation of outcome based on systematic liver volumetry. *Ann Surg* 250:540–548.

Komorizono, Y., Oketani, M., Sako, K. et al. (2003). Risk factors for local recurrence of small hepatocellular carcinoma tumors after a single session, single application of percutaneous radiofrequency ablation. *Cancer* 97:1253–1262.

Kunz, P.L. (2015). Carcinoid and neuroendocrine tumors: Building on success. *J Clin Oncol* 33:1855–1863.

Lafaro, K.J., Cosgrove, D., Geschwind, J.F. et al. (2015). Multidisciplinary care of patients with intrahepatic cholangiocarcinoma: Updates in management. *Gastroenterol Res Pract* 2015:860861.

Lammer, J., Malagari, K., Vogl, T. et al. (2010). Prospective randomized study of doxorubicin-eluting-bead embolization in the treatment of hepatocellular carcinoma: Results of the PRECISION V study. *Cardiovasc Intervent Radiol* 33:41–52.

Le Treut, Y.P., Grégoire, E., Klempnauer, J. et al. (2013) Liver transplantation for neuroendocrine tumors in Europe-results and trends in patient selection: A 213-case European liver transplant registry study. *Ann Surg* 257:807–815.

Lencioni, R. (2010). Loco-regional treatment of hepatocellular carcinoma. *Hepatology* 52:762–773.

Lencioni, R., Crocetti, L., Narayanan, G. (2015). Irreversible electroporation in the treatment of Hepatocellular carcinoma. *Tech Vasc Interv Radiol* 18:135–139.

Lewandowski, R.J., Davenport, M.S. (2015). Imaging and image-guided intervention are irrevocably linked. *Radiol Clin North Am* 53:xi.

Liapi, E., Geschwind, J.F. (2010). Intra-arterial therapies for hepatocellular carcinoma: Where do we stand? *Ann Surg Oncol* 17:1234–1246.

Livraghi, T., Solbiati, L., Meloni, M.F. et al. (2003). Treatment of focal liver tumors with percutaneous radio-frequency ablation: Complications encountered in a multicenter study. *Radiology* 226:441–451.

Llovet, J.M., Bru, C., Bruix, J. (1999). Prognosis of hepatocellular carcinoma: The BCLC staging classification. *Semin Liver Dis* 19:329–338.

Llovet, J.M., Real, M.I., Montana, X. et al. (2002). Arterial embolisation or chemoembolisation versus symptomatic treatment in patients with unresectable hepatocellular carcinoma: A randomised controlled trial. *Lancet* 359:1734–1739.

Llovet, J.M., Ricci, S., Mazzaferro, V. et al. (2008). Sorafenib in advanced hepatocellular carcinoma. *N Engl J Med* 359:378–390.

Lo, C.M., Ngan, H., Tso, W.K. et al. (2002). Randomized controlled trial of transarterial lipiodol chemoembolization for unresectable hepatocellular carcinoma. *Hepatology* 35:1164–1171.

Maluccio, M.A., Covey, A.M., Porat, L.B. et al. (2008). Transcatheter arterial embolization with only particles for the treatment of unresectable hepatocellular carcinoma. *J Vasc Interv Radiol* 19:862–869.

Mantia-Smaldone, G.M., Bagley, L.J., Kasner, S.E., Chu, C.S. (2013). Vertebral artery dissection and cerebral infarction in a patient with recurrent ovarian cancer receiving bevacizumab. *Gynecol Oncol Case Rep* 5:37–39.

Marrero, J.A., Hussain, H.K., Nghiem, H.V. et al. (2005). Improving the prediction of hepatocellular carcinoma in cirrhotic patients with an arterially-enhancing liver mass. *Liver Transpl* 11:281–289.

Mayo, S.C. et al. (2011) Surgery versus intra-arterial therapy for neuroendocrine liver metastasis: A multicenter international analysis. *Ann Surg Oncol* 18:3657–3665.

Minocha, J., Salem, R., Lewandowski, R.J. (2014). Transarterial chemoembolization and yittrium-90 for liver cancer and other lesions. *Clin Liver Dis* 18:877–890.

Mohan, H., Nicholson, P., Winter, D.C. et al. (2015). Radiofrequency ablation for neuroendocrine liver metastases: A systematic review. *J Vasc Interv Radiol* 26:935–942.

Molvar, C., Lewandowski, R.J. (2015). Intra-arterial therapies for liver masses: Data distilled. *Radiol Clin North Am* 53:973–984.

Moreno-Luna, L.E., Yang, J.D., Sanchez, W. et al. (2013). Efficacy and safety of transarterial radioembolization versus chemoembolization in patients with hepatocellular carcinoma. *Cardiovasc Intervent Radiol* 36:714–723.

Muratore, A., Zorzi, D., Bouzari, H. et al. (2007). Asymptomatic colorectal cancer with unresectable liver metastases: Immediate colorectal resection or up-front systemic chemotherapy? *Ann Surg Oncol* 14:766–770.

Orgera G, Krokidis M, Cappucci M. et al. (2015). Current status of interventional radiology in the management of gastro-entero-pancreatic neuroendocrine tumours (GEP-NETs). *Cardiovasc Intervent Radiol* 38:13–24.

Peng, Z.W., Chen, M.S., Liang, H.H. et al. (2010). A case-control study comparing percutaneous radiofrequency ablation alone or combined with transcatheter arterial chemoembolization for hepatocellular carcinoma. *Eur J Surg Oncol* 36:257–263.

Pitt, S.C., Knuth, J., Keily, J.M. et al. (2008). Hepatic neuroendocrine metastases: Chemo- or bland embolization? *J Gastrointest Surg* 12:1951–1960.

Raman, S.S., Aziz, D., Chang, X. et al. (2004). Minimizing central bile duct injury during radiofrequency ablation: Use of intraductal chilled saline perfusion–Initial observations from a study in pigs. *Radiology* 232:154–159.

Ratti, F., Soldati, C., Catena, M. et al. (2010). Role of portal vein embolization in liver surgery: Single centre experience in sixty-two patients. *Updates Surg* 62:153–159.

Raut, C.P., Izzo, F., Marra, P. et al. (2005). Significant long-term survival after radiofrequency ablation of unresectable hepatocellular carcinoma in patients with cirrhosis. *Ann Surg Oncol* 12:616–628.

Reig, M., Darnell, A., Forner, A. et al. (2014). Systemic therapy for hepatocellular carcinoma: The issue of treatment stage migration and registration of progression using the BCLC-refined RECIST. *Semin Liver Dis* 34:444–455.

Riaz, A., Lewandowski, R.J., Kulik, L.M. et al. (2009). Complications following radioembolization with yttrium-90 microspheres: A comprehensive literature review. *J Vasc Interv Radiol* 20:1121–1130; quiz 1131.

Rinke, A., Muller, H.H., Schade-Brittinger, C. et al. (2009). Placebo-controlled, double-blind, prospective, randomized study on the effect of octreotide LAR in the control of tumor growth in patients with metastatic neuroendocrine midgut tumors: A report from the PROMID Study Group. *J Clin Oncol* 27:4656–4663.

Saxena, A., Chua, T.C., Sarkar, A. et al. (2011). Progression and survival results after radical hepatic metastasectomy of indolent advanced neuroendocrine neoplasms (NENs) supports an aggressive surgical approach. *Surgery* 149:209–220.

Scappaticci, F.A., Fehrenbacher, L., Cartwright, T. et al. (2005). Surgical wound healing complications in metastatic colorectal cancer patients treated with bevacizumab. *J Surg Oncol* 91:173–180.

Scheffer, H.J., Vroomen, L.G., Nielsen, K. et al. (2015). Colorectal liver metastatic disease: Efficacy of irreversible electroporation—A single-arm phase II clinical trial (COLDFIRE-2 trial). *BMC Cancer* 15:772.

Shibata, T., Niinobu, T., Ogata, N., Takami, M. (2000). Microwave coagulation therapy for multiple hepatic metastases from colorectal carcinoma. *Cancer* 89:276–284.

Shindoh, J., Tzeng, C.W., Aloia, T.A. et al. (2013b). Portal vein embolization improves rate of resection of extensive colorectal liver metastases without worsening survival. *Br J Surg* 100:1777–1783.

Shindoh, J., Vauthey, J.N., Zimmitti, G. et al. (2013a). Analysis of the efficacy of portal vein embolization for patients with extensive liver malignancy and very low future liver remnant volume, including a comparison with the associating liver partition with portal vein ligation for staged hepatectomy approach. *J Am Coll Surg* 217:126–133; discussion 133–124.

Siegel, R., Naishadham, D., Jemal, A. (2013). Cancer statistics, 2013. *CA Cancer J Clin* 63:11–30.

Silk, M., Tahour, D., Srimathveeravalli, G. et al. (2014). The state of irreversible electroporation in interventional oncology. *Semin Intervent Radiol* 31:111–117.

Simon, C.J., Dupuy, D.E., Mayo-Smith, W.W. (2005) Microwave ablation: Principles and applications. *Radiographics* 25(Suppl 1):S69–S83.

Simoneau, E., Aljiffry, M., Salman, A. et al. (2012). Portal vein embolization stimulates tumour growth in patients with colorectal cancer liver metastases. *HPB (Oxford)* 14:461–468.

Solbiati, L., Ierace, T., Goldberg, S.N. et al. (1997). Percutaneous US-guided radio-frequency tissue ablation of liver metastases: Treatment and follow-up in 16 patients. *Radiology* 202:195–203.

Takaki, H., Yamakado, K., Uraki, J. et al. (2009). Radiofrequency ablation combined with chemoembolization for the treatment of hepatocellular carcinomas larger than 5 cm. *J Vasc Interv Radiol* 20:217–224.

Taner, T., Atwell, T.D., Zhang, L. et al. (2013). Adjunctive radiofrequency ablation of metastatic neuroendocrine cancer to the liver complements surgical resection. *HPB (Oxford)* 15:190–195.

Tanguturi, S.K., Wo, J.Y., Zhu, A.X. et al. (2014) Radiation therapy for liver tumors: Ready for inclusion in guidelines? *Oncologist* 19:868–879.

Teo, J.Y., Goh, B.K. (2015). Contra-lateral liver lobe hypertrophy after unilobar Y90 radioembolization: An alternative to portal vein embolization? *World J Gastroenterol* 21:3170–3173.

Teratani, T., Yoshida, H., Shiina, S. et al. (2006). Radiofrequency ablation for hepatocellular carcinoma in so-called high-risk locations. *Hepatology* 43:1101–1108.

Van Cutsem, E., Nordlinger, B., Adam, R. et al. (2006). Towards a pan-European consensus on the treatment of patients with colorectal liver metastases. *Eur J Cancer* 42:2212–2221.

Vogl, T.J., Farshid, P., Naguib, N.N. et al. (2015). Ablation therapy of hepatocellular carcinoma: a comparative study between radiofrequency and microwave ablation. *Abdom Imaging* 40:1829–1837.

Wells, S.A., Hinshaw, J.L., Lubner, M.G. et al. (2015). Liver ablation: Best practice. *Radiol Clin North Am* 53:933–971.

Yao, J.C., Hassan, M., Phan, A. et al. (2008). One hundred years after "carcinoid": Epidemiology of and prognostic factors for neuroendocrine tumors in 35,825 cases in the United States. *J Clin Oncol* 26:3063–3072.

Yoo, P.S., Lopez-Soler, R.I., Longo, W.E., Cha, C.H. (2006). Liver resection for metastatic colorectal cancer in the age of neoadjuvant chemotherapy and bevacizumab. *Clin Colorectal Cancer* 6:202–207.

Ziemlewicz, T.J., Hinshaw, J.L., Lubner, M.G. et al. (2015). Percutaneous microwave ablation of hepatocellular carcinoma with a gas-cooled system: initial clinical results with 107 tumors. *J Vasc Interv Radiol* 26:62–68.

Treatment planning part I: Vascular considerations associated with safety and efficacy in radioembolization

RAY BRADFORD AND J. MARK MCKINNEY

3.1 INTRODUCTION

Yttrium-90 (^{90}Y) radioembolization is delivered via arterial supply to the liver. Hepatic neoplasms receive greater than 80% of their perfusion from the hepatic arteries, while the normal hepatic parenchyma receives the majority of its blood supply from the portal vein. This differential blood supply allows for the arterial delivery of relatively large radiation doses to the tumor with relative sparing of normal liver parenchyma (Welsh et al., 2006). ^{90}Y radioembolization for primary and/or secondary neoplasms of the liver necessitates a thorough understanding of hepatic arterial anatomy to ensure safety and efficacy. The interventionalist must be aware that variations in the arterial supply to the liver are common and that these anatomic variations affect both hepatic lobes.

The arteries perfusing the liver also supply many other important visceral structures including the stomach, duodenum, esophagus, pancreas, gallbladder, and abdominal wall. Consideration must be given to protection against nontarget embolization of these extrahepatic visceral structures. Nontarget embolization is avoided during radioembolization with several techniques, such as adjusting the catheter tip location, employing antireflux catheters, and/or using protective embolic redistribution of arterial flow.

In addition to the classic and variant arterial vascular supply to the liver, the interventionalist must take into account extrahepatic arterial parasitization, which occurs with large or peripheral hypervascular neoplasms. Extrahepatic arteries are parasitized from multiple vascular distributions in proximity to the liver parenchyma. Effective ^{90}Y radioembolization may require catheter-directed therapy to these nonhepatic arterial distributions.

Arterial delivery of ^{90}Y radioembolization requires both training and experience because anatomic treatment decisions are often complex and can be complicated by anatomic variations, extent of disease, and prior treatments. This chapter is not meant to cover every detail of all vascular aspects related to radioembolization, but rather it is meant to provide an overview that can be appreciated and understood by all members of the radioembolization team.

3.2 ARTERIAL ACCESS

Interventionalists have the option to approach the hepatic arterial supply via a femoral or radial artery. Femoral artery access is historically the most common access point for diagnostic and therapeutic arterial interventions. However, with the advances of lower profile catheters and the desire to improve patient satisfaction and comfort with early ambulation, radial artery access has become the favored access point for many interventionalists. Prior to radial artery puncture, patency of the ulnar artery and collateral supply through the palmar arch is confirmed utilizing a Barbeau test (Barbeau et al., 2004). The risk of radial artery spasm and thrombosis is mitigated via intra-arterial infusion of heparin, verapamil, and nitroglycerin (Bishay et al., 2014). A Glidesheath is also utilized to minimize trauma to the radial artery. Approaching the celiac artery from a radial artery approach may have anatomic advantages during celiac and hepatic arterial catheter placement. Potential disadvantages to radial artery access include the need to utilize longer catheter delivery systems and ergonomic challenges for the operator.

3.3 HEPATIC ARTERIAL ANATOMY

Classic hepatic arterial supply arises from the celiac artery with bifurcation of the proper hepatic artery (PHA) into the right and left intrahepatic arterial branches (Figure 3.1). This classic pattern is estimated to be present in greater than 60% of patients (Covey et al., 2002; Lee et al., 2012). Variations of the right hepatic arterial supply include replaced right hepatic artery (RHA) (12%) and accessory

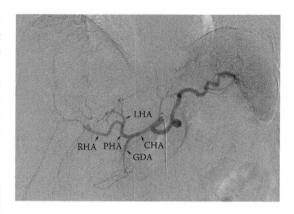

Figure 3.1 Celiac arteriogram demonstrates classic hepatic arterial anatomy. The common hepatic artery (CHA) continues as the proper hepatic artery (PHA) beyond the takeoff of the gastroduodenal artery (GDA). The PHA splits into the right hepatic artery (RHA) and left hepatic artery (LHA).

RHA (6%) from the superior mesenteric artery (Covey et al., 2002; Lee et al., 2012). Variations of the left hepatic artery (LHA) supply include replacement of the LHA (5%) and accessory LHA (15%) from the left gastric artery (LGA) (Covey et al., 2002; Lee et al., 2012) (Figure 3.2). In addition, a middle hepatic artery may present either as an accessory branch of the RHA or a true trifurcation from the PHA (Kerlan and LaBerge, 2006).

Extrahepatic arteries are important for the interventionalist utilizing ^{90}Y radioembolization as locoregional therapy. Extrahepatic arteries are a potential nontarget pathway for ^{90}Y to be diverted from the targeted hepatic parenchyma to nontargeted adjacent viscera or musculoskeletal structures. Specific extrahepatic arteries at risk for nontarget radioembolization typically include the right gastric, gastroduodenal, pancreaticoduodenal, cystic, esophageal, and falciform arteries. Pre-^{90}Y radioembolization planning arteriography is utilized to map and discover potential perihepatic arterial pathways that place the patient at risk for nontarget embolization during ^{90}Y therapy.

The right gastric artery (RGA) frequently arises from LHA, PHA, gastroduodenal artery (GDA), or common hepatic artery (CHA) (VanDamme and Bonte, 1990; Liu et al., 2005). Because both RGA and LGA perfuse the lesser curvature of the stomach, it is important to identify the RGA origin. RGA is characteristically small in caliber and may have a sharply angulated origin

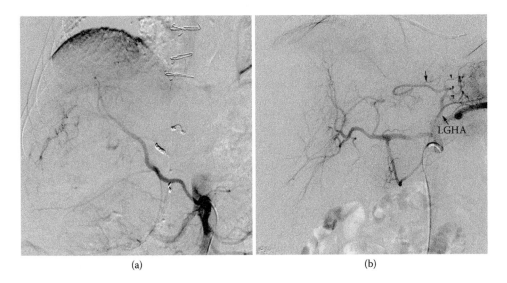

Figure 3.2 Variant hepatic arterial anatomy. **(a)** Replaced right hepatic artery from the SMA. **(b)** Left gastrohepatic artery (LGHA) with branches to gastric fundus (arrowheads) and left hepatic lobe (arrow).

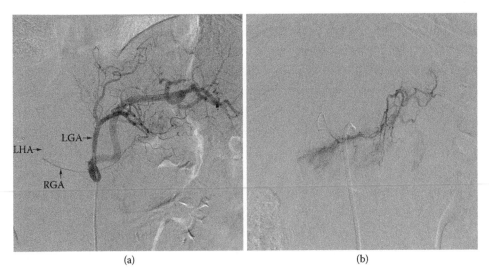

Figure 3.3 Left and right gastric arteries. **(a)** Left gastric arteriogram demonstrates cross filling from the left gastric artery (LGA) to the right gastric artery (RGA) which arises from the left hepatic artery (LHA). **(b)** Direct selective right gastric arteriogram in a different patient.

that makes selective catheterization challenging. When the RGA origin cannot be identified, it is frequently evaluated via a left gastric arteriogram (Figure 3.3).

The gastroduodenal artery is a relatively large artery arising from the CHA and supplies the pancreas, duodenum, and greater curvature of the stomach via the pancreaticoduodenal arcade and gastroepiploic arteries (Figure 3.1). The risk

of nontarget embolization to the gastroduodenal artery must be evaluated due to its continuity with the PHA and the subsequent bifurcation of the PHA into the RHA and LHAs. Depending on treatment intent and catheter tip location for ^{90}Y radioembolization, the gastroduodenal artery may not be at significant risk.

The cystic artery perfuses the gallbladder and typically originates from the proximal RHA.

Clinical symptoms from radioembolic cholecystitis can occur but are usually self-limiting.

Specific attention to the intrahepatic arterial distribution is important for evaluating several extrahepatic arterial perfusion pathways. One of these intrahepatic-to-extrahepatic arterial pathways is the falciform artery, which arises from the left or middle hepatic arteries and perfuses the anterior abdominal wall (Figure 3.4) (Baba et al., 2000; Liu et al., 2005). Other extrahepatic artery

pathways that can complicate radioembolization include the esophageal and gastric branches arising from the LHA (Figure 3.5) and the duodenal branches arising from the central hepatic arteries. With careful analysis, these extrahepatic arterial pathways can be identified and strategies can be developed to protect against nontarget ^{90}Y radioembolization.

Some posttreatment examples of ^{90}Y nontarget embolization through extrahepatic arteries are shown in Chapter 13, associated with clinical sequelae discussed in Chapter 14.

3.4 COMPLICATED HEPATIC ARTERIAL ACCESS

Diffuse or focal vascular disease may be a complicating factor regardless of whether the interventionalist utilizes a femoral or radial arterial approach. Severe peripheral vascular disease with atherosclerotic stenosis of the aorta or iliac arteries may require a contralateral femoral artery approach or radial artery approach. Prior to choice of arterial access, the interventionalist should also be aware of the patient's prior surgical history, which may include aortic, iliac, femoral, or upper extremity surgical grafts.

Celiac artery stenosis from median arcuate ligament syndrome (Figure 3.6) or atherosclerosis

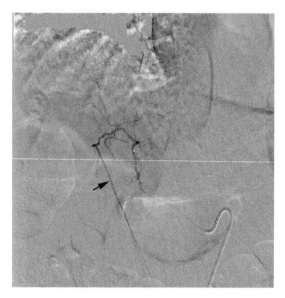

Figure 3.4 Falciform artery (arrow) arises from the left hepatic artery.

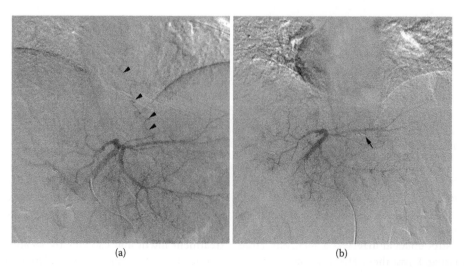

(a)   (b)

Figure 3.5 Esophageal artery. **(a)** An esophageal artery (arrowheads) arises from the left hepatic arteries (arrows). **(b)** Coil embolization (arrow) of the esophageal branches to prevent nontarget radioembolization.

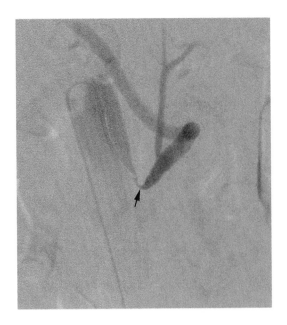

Figure 3.6 Median arcuate ligament narrowing (arrow) the celiac artery.

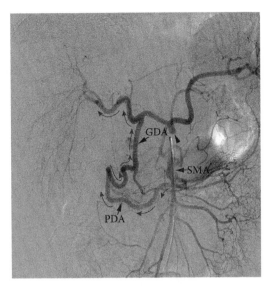

Figure 3.7 Occluded celiac artery (arrowhead). Retrograde microcatheter access (curved arrows) to the hepatic arteries is possible via the superior mesenteric artery (SMA), pancreaticoduodenal arcade (PDA), and gastroduodenal artery (GDA).

may be encountered as a complicating access issue. Stenosis from median arcuate ligament syndrome is often incomplete and allows coaxial microcatheter advancement from an access catheter seated in the narrowed celiac artery. Ultimate relief of median arcuate ligament syndrome is surgical release (Columbo et al., 2015). In the case of atherosclerotic celiac artery stenosis or occlusion, celiac artery access for radioembolization can be achieved via celiac artery stent placement.

In the cases where celiac occlusion is complete and celiac catheter access cannot be achieved, enlarged pancreaticoduodenal arterial collaterals from the superior mesenteric artery provide a retrograde approach to hepatic artery ^{90}Y radioembolization (Figure 3.7). To achieve hepatic artery access via the pancreaticoduodenal collaterals, an access catheter is seated in the superior mesenteric artery and coaxial microcatheter techniques are utilized to advance access serially through the superior mesenteric artery, inferior pancreaticoduodenal trunk, pancreaticoduodenal arcade, gastroduodenal artery, and into the proper and intrahepatic arterial branches. Supraselective access into the hepatic arteries may be limited when taking a circuitous retrograde approach.

3.5 PARASITIZED ARTERIAL PERFUSION

Consideration must be given to extrahepatic arterial pathways that may be recruited and parasitized for perfusion of intrahepatic neoplasms. Characteristics that increase the likelihood of parasitization include peripheral and large tumors and prior hepatic arterial embolization (Abdelmaksoud et al., 2011).

Awareness and identification of parasitized extrahepatic arteries is necessary to completely treat targeted tumor beds. Tumors that receive supplemental arterial blood supply from parasitized extrahepatic arteries are particularly at risk of being undertreated (Abdelmaksoud et al., 2011). Bland, conventional chemoembolization, or drug-eluting bead chemoembolization of parasitized extrahepatic arteries supplying peripherally located hepatic tumors provides therapeutic intent and reestablishment of primary hepatic arterial perfusion through intrahepatic arteries. Extrahepatic arteries most commonly recruited for tumor perfusion include the right inferior phrenic, internal mammary, intercostal, right adrenal, right renal, and greater omental arteries (Figure 3.8) (Abdelmaksoud et al., 2011).

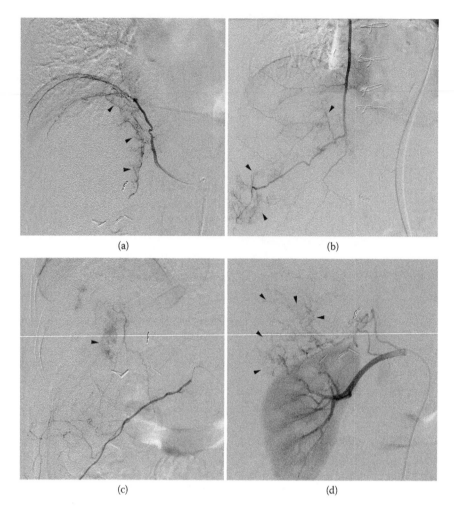

(a) (b)

(c) (d)

Figure 3.8 Parasitized extrahepatic arteries perfusing intrahepatic tumor (arrowheads). **(a)** Right inferior phrenic artery. **(b)** Right internal mammary artery. **(c)** Intercostal artery. **(d)** Right adrenal and renal arteries.

3.6 TECHNIQUES TO PREVENT NONTARGET RADIOEMBOLIZATION

Coil embolotherapy is a well-developed technique utilized by interventionalists to occlude and redirect arterial perfusion. Historically, coil embolotherapy is most frequently utilized in the GDA and RGA during planning arteriography to prevent nontarget embolization.

In the early implementation of radioembolization, the gastroduodenal artery was routinely coil embolized at its origin to prevent nontarget embolization. Gastroduodenal artery coil embolization is performed excluding the GDA as a potential pathway for nontarget embolization. However, additional experience with ^{90}Y radioembolization has shown that gastroduodenal coil embolization is frequently unnecessary and may actually increase the risk of intrahepatic recruitment of duodenal and pancreatic arterial collateral pathways (Hamoui et al., 2013a, 2013b). Avoidance of gastroduodenal coil embolization has been demonstrated to decrease procedure time, contrast volume, and radiation exposure to the patient (Fischman et al., 2014).

The RGA is at a particular risk for nontarget radioembolization due to its proximity to typical radioembolization catheter tip locations. The RGA originates from either the LHA, PHA, GDA, or CHA (Covey et al., 2002; Lee et al., 2012). The RGA is often very small in caliber and has a sharply angulated origin. When technically possible, the RGA is directly accessed via a coaxial microcatheter and its origin is coil embolized. When the RGA cannot be identified or accessed directly, it may be successfully approached via retrograde access from the LGA. An access catheter is seated in the LGA origin, and a coaxial microcatheter is advanced along the communicating artery from the left gastric to the RGA origin where coils are carefully deposited. If ^{90}Y catheter tip delivery does not pose a risk to the RGA, then coil embolization is not necessary (Hamoui et al., 2013a, 2013b).

Coil embolization is also utilized to protect other extrahepatic arterial beds such as the falciform artery. If a falciform artery is identified (Figure 3.4), coil embolization of the falciform artery is performed when technically possible because ^{90}Y radioembolization to the falciform artery can result in a highly localized midabdominal burning sensation for a period of days or weeks (Liu et al., 2005). When the falciform artery cannot be accessed, studies have shown that ice packs to the anterior abdominal wall provide vasoconstriction that reduces nontarget embolization to the terminal falciform arterial branches (Wang et al., 2013).

Coil embolization is also applied when esophageal or gastric branches are identified as originating within the intrahepatic arterial supply. Branches to the esophagus and stomach may originate from the LHA and are embolized to prevent nontarget embolization (Figure 3.5).

Antireflux catheters have been devised to minimize nontarget embolization during radioembolization (Figure 15.2). Catheters are designed to deliver ^{90}Y microspheres in target hepatic arteries ranging from 2 to 6 mm in diameter. Studies have demonstrated increased tumor uptake and decreased nontarget embolization in multiple tumor types (Pasciak et al., 2015). A prospective randomized study of protective embolic coiling versus antireflux catheter delivery demonstrated reduced fluoroscopy time, procedure time, and contrast dose during the planning arteriogram

phase because the need for coil embolotherapy is eliminated or reduced (Fischman et al., 2014).

3.7 TECHNIQUES TO MINIMIZE HEPATOPULMONARY SHUNTING

Hepatopulmonary shunting may lead to nontarget pulmonary embolization and must be recognized during planning and treatment with ^{90}Y radioembolization. Hepatopulmonary shunt fraction is usually evaluated with a test dose of technium-99m microaggregated albumin (^{99m}Tc-MAA) during the arterial planning phase of ^{90}Y treatment, as described in Chapter 4. Arteriovenous shunting into the hepatic veins and portal veins is common with liver tumors (Figure 3.9) (Sugano et al., 1994; Chan et al., 2010). ^{90}Y microspheres can travel to the pulmonary arterial bed via the hepatic and portal veins. Excessive hepatopulmonary shunting with ^{90}Y microspheres may in rare cases lead to radiation pneumonitis. Radiation pneumonitis manifests clinically with nonspecific symptoms of fever, nonproductive cough, and dyspnea.

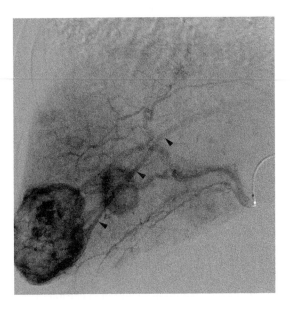

Figure 3.9 Arteriovenous shunting from hypervascular hepatocellular carcinoma leads to early opacification of the draining hepatic vein (arrowheads).

Radiation pneumonitis is radiographically suggested as peribronchial cuffing on chest imaging (Graves et al., 2010) and as a restrictive pattern on pulmonary function testing. Treatment is inhaled and/or systemic corticosteroids (Leung et al., 1995). Ultimately, radiation pneumonitis can lead to debilitating chronic disease.

Because of the potential risks of an elevated pulmonary shunt fraction leading to radiation pneumonitis, manufacturers have released guidelines for ^{90}Y microspheres. The resin ^{90}Y microsphere training manual guidelines by Sirtex Medical (North Sydney, Australia) recommend a lung radiation dose limit of 25Gy per treatment session, not to exceed a 50Gy cumulative dose. For glass ^{90}Y microspheres, the package insert by BTG International (West Conshohocken, Pennsylvania) recommends an upper limit of 16.5 mCi delivered to the lungs. Ho et al. (1997) also recommend restricting the lung radiation absorbed dose to <30 Gy. Dose reductions for an elevated hepatopulmonary shunt fraction have been shown to result in reduced efficacy of ^{90}Y radioembolization therapy (Garin et al., 2015; Lam et al., 2015). However, studies have shown that the risk from an elevated hepatopulmonary shunt fraction is very low. In one series, no patients were found to have radiation pneumonitis with a cumulative lung dose >30 Gy (Salem et al., 2008). Additional considerations associated with hepatopulmonary shunt will be discussed in Chapter 4.

Several non–dose-reducing techniques have been utilized to deal with high hepatopulmonary shunt fraction. The use of systemic sorafenib treatment has been shown to reduce hepatopulmonary shunt fraction by 62%–87% (Theysohn et al., 2012). Transarterial chemoembolization has resulted in reduction of hepatopulmonary shunt fraction by 25%–57% (Rose and Hoh, 2009; Gaba and Vanmiddlesworth, 2012). The use of sorafenib or transarterial chemoembolization may delay ^{90}Y radioembolization; therefore, catheter-based techniques to reduce shunting have been developed for use during ^{90}Y radioembolization rather than reducing the treatment dose. Catheter-based techniques include temporary balloon occlusion of the hepatic veins or portal veins (Bester and Salem, 2007; Murata et al., 2009), embolization of varices, or bland embolization of the hepatic tumor immediately before or following ^{90}Y radioembolization (Ward et al., 2015).

Ward et al. (2015) now recommend that if the expected lung dose is <30 Gy, no shunt mitigation is required. For expected lung dose >30 Gy, catheter-based techniques can be utilized without delay to minimize nontarget radioembolization to the pulmonary arterial bed (Ward et al., 2015).

3.8 COMPLICATIONS

Complications from ^{90}Y radioembolization have been reported in multiple studies. Early complications include fatigue, pain, nausea, emesis, and low-grade fever. This constellation of early symptoms is typically called postembolization syndrome (Riaz et al., 2009). Postembolization syndrome is usually self-limited and gradually resolves over the first 1–2 weeks of treatment.

Late complications of ^{90}Y radioembolization include gastrointestinal ulceration, cholecystitis, pancreatitis, biliary injury, and radiation-induced liver disease (Hamoui and Ryu, 2011), as well as pneumonitis. Gastrointestinal ulceration typically presents weeks after radioembolization as refractory abdominal pain, nausea, and vomiting. Gastrointestinal symptoms may be treated with proton pump inhibitors and sucralfate. Endoscopy may be utilized to confirm the diagnosis of ulceration. Biopsy of the ulcers will often show microspheres in the biopsy specimen.

Radiation-induced liver disease typically occurs 4–8 weeks after radioembolization with elevation of alkaline phosphatase and bilirubin. Radiation-induced liver disease is a clinical diagnosis associated with ascites and jaundice (Sangro et al., 2008; Hamoui and Ryu, 2011). Multiple prior chemotherapy regimens are a risk factor for radiation-induced liver disease. Dosimetric thresholds related to radiation-induced liver disease are discussed in Chapter 5.

Biliary sequelae following ^{90}Y radioembolization are usually clinically inconsequential. As with other liver-directed therapies, biliary complications are seen more commonly with secondary neoplasms than with hepatocellular carcinoma. Potential biliary complications included stricture formation, obstruction, biloma, cholecystitis, hepatic abscess, and serum bilirubin toxicity. Patients are often asymptomatic, even with imaging evidence of biliary complications. Treatment for biliary sequelae is based on clinical presentation and may include antibiotics, percutaneous drainage of fluid collections, biliary decompression, and cholecystectomy (Atassi et al., 2008). Additional discussion of the late complications of radioembolization, including identification using advanced imaging techniques, can be found in Chapters 13 and 14.

3.9 ANGIOGENESIS

Tumor growth and spread is known to be driven by a complex interplay of proangiogenic and antiangiogenic cytokines (Bergers and Benjamin, 2003). Since some patients experience early tumor recurrence following ^{90}Y radioembolization, it is important to consider the role that cytokines may play. Vascular endothelial growth factor (VEGF) levels are known to be associated with suboptimal outcomes in primary and secondary liver neoplasm. In addition, VEGF is associated with hepatocellular carcinoma disease stage, presence of metastasis, vascular invasion, treatment response, and overall survival (Xiong et al., 2004; Sergio et al., 2008). Carpizo et al. (2014) found that VEGF, angiopoietin-2 (Ang-2), platelet-derived growth factor subunit BB (PDGF-BB), and other nonclassic cytokines were temporally associated with ^{90}Y radioembolization. They observed spikes in the cytokine baseline values as sampled following first- and second-stage ^{90}Y radioembolization treatment episodes. This evidence suggests that ^{90}Y radioembolization has the potential to upregulate angiogenic cytokines. When overall survival (OS) is evaluated in association with cytokine release, there is correlation between shortened OS and temporal spikes in VEGF, Ang-2, and PDGF-BB. These cytokines appear to affect OS by promoting angiogenesis. It is plausible that some patients might benefit from antiangiogenic therapy administered before ^{90}Y radioembolization (Carpizo et al., 2014).

3.10 CONCLUSION

Vascular considerations are an important part of ^{90}Y radioembolization planning and therapy. From choosing arterial access to understanding and planning for variations of normal hepatic arterial anatomy, considerable thought must be given to each specific patient's situation. Unexpected complicating factors such as stenotic or occluded celiac access, extrahepatic arterial communications, parasitized arterial perfusion, and hepatopulmonary shunting are frequently encountered. Techniques such as coil embolization, antireflux catheters, and dose modifications allow for safe and efficacious delivery of yttrium-90 to primary and secondary hepatic neoplasms.

REFERENCES

Abdelmaksoud, M.H., Louie, J.D., Kothary, N. et al. (2011). Embolization of parasitized extrahepatic arteries to reestablish intrahepatic arterial supply to tumors before yttrium-90 radioembolization. *J Vasc Interv Radiol* 22:1355–1362.

Atassi, B., Bangash, A.K., Lewandowski, R.J. et al. (2008). Biliary sequelae following radioembolization with yttrium-90 microspheres. *J Vasc Interv Radiol* 19:691–697.

Baba, Y., Miyazono, N., Ueno, K. et al. (2000). Hepatic falciform artery. Angiographic findings in 25 patients. *Acta Radiol* 41:329–333.

Barbeau, G.R., Arsenault, F., Dugas, L. et al. (2004). Evaluation of the ulnopalmar arterial arches with pulse oximetry and plethysmography: Comparison with the Allen's test in 1010 patients. *Am Heart J* 147:489–493.

Bergers, G., Benjamin, L.E. (2003). Tumorigenesis and the angiogenic switch. *Nat Rev Cancer* 3:401–410.

Bester, L., Salem, R. (2007). Reduction of arteriohepatovenous shunting by temporary balloon occlusion in patients undergoing radioembolization. *J Vasc Interv Radiol* 18:1310–1314.

Bishay, V., Patel, R.S., Kim, E. et al. (2014). Transradial approach for hepatic radioembolization: Initial result and technique. *J Vasc Interv Radiol* 25:S88.

Carpizo, D.R., Gensure, R.H., Yu, X. et al. (2014). Pilot study of angiogenic response to yttrium-90 radioembolization with resin microspheres. *J Vasc Interv Radiol* 25:297–306.

Chan, W.S., Poon, W.L., Cho, D.H. et al. (2010). Transcatheter embolisation of intrahepatic arteriovenous shunts in patients with hepatocellular carcinoma. *Hong Kong Med J* 16:48–55.

Columbo, J.A., Trus, T., Nolan, B. et al. (2015). Contemporary management of median arcuate ligament syndrome provides early symptom improvement. *J Vasc Surg* 62:151–156.

Covey, A.M., Brody, L.A., Maluccio, M.A. et al. (2002). Variant hepatic arterial anatomy revisited: Digital subtraction angiography performed in 600 patients. *Radiology* 224:542–547.

Fischman, A.M., Ward, T.J., Patel, R.S. et al. (2014). Prospective, randomized study of coil embolization versus surefire infusion system during yttrium-90 radioembolization with resin microspheres. *J Vasc Interv Radiol* 25:1709–1716.

Gaba, R.C., Vanmiddlesworth, K.A. (2012). Chemoembolic hepatopulmonary shunt reduction to allow safe yttrium-90 radioembolization lobectomy of hepatocellular carcinoma. *Cardiovasc Interv Radiol* 35:1505–1511.

Garin, E., Rolland, Y., Edeline, J. et al. (2015). Personalized dosimetry with intensification using ^{90}Y-loaded glass microsphere radioembolization induces prolonged overall survival in hepatocellular carcinoma patients with portal vein thrombosis. *J Nuclear Med* 56:339–346.

Graves, P.R., Siddiqui, F., Anscher, M.S. et al. (2010). Radiation pulmonary toxicity: From mechanisms to management. *Semin Radiat Oncol* 20:201–207.

Hamoui, N., Minocha, J., Memon, K. et al. (2013b). Prophylactic embolization of the gastroduodenal and right gastric arteries is not routinely necessary before radioembolization with glass microspheres. *J Vasc Interv Radiol* 24:1743–1745.

Hamoui, N., Ryu, R.K. (2011). Hepatic radioembolization complicated by fulminant hepatic failure. *Semin Radiat Oncol* 28:246–251.

Hamoui, N., Salem, R., Lewandowski, R.J. (2013a). Embolization of the gastroduodenal and right gastric arteries prior to radioembolization with glass microspheres: Is it always necessary? Abstract # 413. In *SIR 2013 38th Annual Scientific Meeting*, April 13–18, 2013, New Orleans, LA.

Ho, S., Lau, W.Y., Leung, T.W. et al. (1997). Clinical evaluation of the partition model for estimating radiation doses from yttrium-90 microspheres in the treatment of hepatic cancer. *Eur J Nucl Med* 24:293–298.

Kerlan, R.K.J., LaBerge, J.M. (2006). Liver transplantation: Associated interventions. In Abrams, H.L., Baum, S.B., Pentecost, M.J., (eds.), *Abrams' Angiography: Interventional Radiology*, 2nd Edition. Baltimore, MD: Lippincott Williams & Williams, p. 565.

Lam, M.G., Banerjee, A., Goris, M.L. et al. (2015). Fusion dual-tracer SPECT-based hepatic dosimetry predicts outcome after radioembolization for a wide range of tumour cell types. *Eur J Nucl Med Mol Imaging* 42:1192–1201.

Lee, A.J., Gomes, A.S., Liu, D.M. et al. (2012). The road less traveled: Importance of the lesser branches of the celiac axis in liver embolotherapy. *Radiographics* 32:1121–1132.

Leung, T.W., Lau, W.Y., Ho, S.K. et al. (1995). Radiation pneumonitis after selective internal radiation treatment with intraarterial 90yttrium-microspheres for inoperable hepatic tumors. *Int J Radiat Oncol Biol Phys* 33:919–924.

Liu, D.M., Salem, R., Bui, J.T. et al. (2005). Angiographic considerations in patients undergoing liver-directed therapy. *J Vasc Interv Radiol* 16:911–935.

Murata, S., Tajima, H., Nakazawa, K. et al. (2009). Initial experience of transcatheter arterial chemoembolization during portal vein occlusion for unresectable hepatocellular carcinoma with marked arterioportal shunts. *Eur Radiol* 19:2016–2023.

Pasciak, A.S., McElmurray, J.H., Bourgeois, A.C. et al. (2015). The impact of an antireflux catheter on target volume particulate distribution in liver-directed embolotherapy: A pilot study. *J Vasc Interv Radiol* 26:660–669.

Riaz, A., Lewandowski, R.J., Kulik, L.M. et al. (2009). Complications following radioembolization with yttrium-90 microspheres: A comprehensive literature review. *J Vasc Interv Radiol* 20:1121–1130; quiz 1131.

Rose, S.C., Hoh, C.K. (2009). Hepatopulmonary shunt reduction using chemoembolization to permit yttrium-90 radioembolization. *J Vasc Interv Radiol* 20:849–851.

Salem, R., Parikh, P., Atassi, B. et al. (2008). Incidence of radiation pneumonitis after hepatic intra-arterial radiotherapy with yttrium-90 microspheres assuming uniform lung distribution. *Am J Clin Oncol* 31:431–438.

Sangro, B., Gil-Alzugaray, B., Rodriguez, J. et al. (2008). Liver disease induced by radioembolization of liver tumors: Description and possible risk factors. *Cancer* 112:1538–1546.

Sergio, A., Cristofori, C., Cardin, R. et al. (2008). Transcatheter arterial chemoembolization (TACE) in hepatocellular carcinoma (HCC): The role of angiogenesis and invasiveness. *Am J Gastroenterol* 103:914–921.

Sugano, S., Miyoshi, K., Suzuki, T. et al. (1994). Intrahepatic arteriovenous shunting due to hepatocellular carcinoma and cirrhosis, and its change by transcatheter arterial embolization. *Am J Gastroenterol* 89:184–188.

Theysohn, J.M., Schlaak, J.F., Muller, S. et al. (2012). Selective internal radiation therapy of hepatocellular carcinoma: Potential hepatopulmonary shunt reduction after sorafenib administration. *J Vasc Interv Radiol* 23:949–952.

VanDamme, J.P., Bonte, J. (1990). *Vascular Anatomy in Abdominal Surgery*. New York, NY: Thieme.

Wang, D.S., Louie, J.D., Kothary, N. et al. (2013). Prophylactic topically applied ice to prevent cutaneous complications of nontarget chemoembolization and radioembolization. *J Vasc Interv Radiol* 24:596–600.

Ward, T.J., Tamrazi, A., Lam, M.G. et al. (2015). Management of high hepatopulmonary shunting in patients undergoing hepatic radioembolization. *J Vasc Interv Radiol* 26:1751–1760.

Welsh, J.S., Kennedy, A.S., Thomadsen, B. (2006). Selective internal radiation therapy (SIRT) for liver metastases secondary to colorectal adenocarcinoma. *Int J Radiat Oncol Biol Phys* 66:S62–S73.

Xiong, Z.P., Yang, S.R., Liang, Z.Y. et al. (2004). Association between vascular endothelial growth factor and metastasis after transcatheter arterial chemoembolization in patients with hepatocellular carcinoma. *Hepatobiliary Pancreat Dis Int* 3:386–390.

4

Treatment planning part II: Procedure simulation and prognostication

SHYAM M. SRINIVAS, SANKARAN SHRIKANTHAN, NAICHANG YU, SUSAN D. KOST, RAM GURAJALA, AND KARUNAKARAVEL KARUPPASAMY

4.1 BACKGROUND, PROPERTIES OF ^{99m}TC-MAA, TRENDS TOWARD QUANTIFICATION

4.1.1 BACKGROUND—IMAGING AS A SIMULATION

The entire field of radiation oncology is based on procedure simulation followed by treatment, usually with sealed sources. There are abundant examples in conformal external beam radiation therapy, intensity-modulated radiation therapy (IMRT), stereotactic beam radiotherapy (SBRT), gamma knife, and proton beam therapy. In all these modalities dose calculations are made based on pretreatment simulation, followed by treatment of the patient. This process helps to ensure that the therapeutic index (i.e., treat disease while sparing normal tissue) remains high. With more sophisticated imaging techniques now readily available, simulation is usually based on imaging studies such as computed tomography (CT), magnetic resonance imaging (MRI), positron emission tomography (PET), and/or ultrasound (Pereira et al., 2014).

In nuclear medicine, unsealed source therapy is the predominant treatment modality. There are examples of near-perfect simulation such as distribution of ^{123}I metascan prior to administering ^{131}I for ablation in thyroid cancer or the distribution of ^{99m}Tc-methylene diphosphonate (MDP) prior to administering ^{153}Sm or ^{223}Ra in prostate cancer (Silberstein et al., 2003). More recently with the advent of semiquantitative ^{68}Ga PET imaging, simulations are being performed with ^{68}Ga DOTA-d-Phe(1)-Tyr(3)-octreotide (DOTATOC) for neuroendocrine tumors followed by treatment with a beta emitter such as ^{90}Y or ^{177}Lu DOTATOC (Baum and Kulkarni, 2012). The disconnection between radiation therapy using unsealed internal emitters compared with sealed source internal and external radiation therapy has been and is the lack of quantification. Most radiation oncology treatment plans have isodose curves planned on the images prior to the treatment administration, accurately identifying radiation-absorbed dose to the organs of interest to the nearest Gy. In nuclear medicine, however, dosimetry is vague, since many radiotracers have a multivariable systemic distribution. In the past, the lack of quantitative imaging techniques has led to the omission of this very important question of "how much" of a cytotoxic agent are we going to give.

The ^{90}Y radioembolization community is unique. Its authorized users include radiation oncologists, nuclear medicine physicians, and interventional radiologists with the appropriate training. As a result, there are those who believe in precise quantification and simulation, while others treat by empirical methods. ^{90}Y radioembolization itself is unsealed, but it is not a systemic injection, which allows more control over the process. It is a locoregional therapy delivered through the hepatic arterial vasculature with potential for extrahepatic shunting, as this chapter further explains. While the ^{90}Y microsphere is handled like a radiopharmaceutical, it lacks many of the properties of traditional radiopharmaceuticals. For example, there is no metabolism of radioembolic microspheres as it is an inert permanent implant, resembling a brachytherapy point source.

When glass microspheres (TheraSphere) and resin microspheres (SIR-Spheres) were introduced into clinical usage in the early 2000s, there was no sophisticated mandated simulation technique that would allow for quantification and dosimetry. Instead, the manufacturer's guidance simply asked for documentation of lung shunting using planar scintigraphy of the radiopharmaceutical ^{99m}Tc macroaggregated albumin (MAA). ^{99m}Tc-MAA is used because of the similar particle size to that of the ^{90}Y microspheres, with the idea that this would be a "poor man's" simulation of ^{90}Y radioembolization. As we examine the physical properties of MAA and ^{90}Y microspheres later in the chapter, there will be clarity about why MAA is an imperfect microsphere surrogate.

With the advent of hybrid single-photon emission computed tomography (SPECT)/CT imaging circa 2005, a new opportunity to perform tomographic imaging with the potential for quantification has emerged (Beauregard et al., 2011). It has taken nearly 10 years for the field to mature to realize that there is value in performing such simulations, and the *a priori* information from ^{99m}Tc-MAA SPECT/CT may finally give the type of quantitatively rigorous simulation that is present in the radiation oncology world (Willowson et al., 2009, 2011). So while it is an imperfect surrogate at present, ^{99m}Tc-MAA's potential value is just beginning to be realized. Of course, there are controversies as to whether MAA SPECT/CT does actually predict the distribution of [90Y] microspheres that will also be addressed later in this chapter. For now, it should be clarified that correlation between MAA and [90Y] microspheres has not reached consensus in the literature. While some articles suggest near-perfect agreement (Talanow et al., 2010), others are in disagreement (Wondergem et al., 2013).

4.1.2 [99mTC-MAA] PHYSICAL AND BIOLOGICAL PROPERTIES

Currently, albumin macroaggregates ([99mTc-MAA]) are supplied by Jubilant DraxImage Inc. (Kirkland, Quebec, Canada). The product name is "DRAXIMAGE MAA" and it is a kit containing nonradioactive aggregated human serum albumin (HSA) that can be tagged with [99mTc] pertechnetate (MSDS for Draximage, 2011).

The kit consists of reaction vials that contain the sterile, nonradioactive ingredients necessary to produce [99mTc] albumin-aggregated injection for diagnostic use by intravenous injection. Each 10 mL reaction vial contains 2.5 mg of albumin aggregated, 5.0 mg of albumin human, 0.06 mg stannous chloride, and 1.2 mg of sodium chloride; the contents are in a lyophilized (freeze-dried) form under an atmosphere of nitrogen (Prescribing information for Draximage, 2010).

The aggregated particles are formed by denaturation of human albumin in a heating and aggregation process. Each vial contains 4–8 million particles. By light microscopy, more than 90% of the particles are between 10 and 70 μm, while the typical average size is 20–40 μm; none are greater than 150 μm. No less than 90% of the pertechnetate [99mTc] added to a reaction vial is bound to aggregate at preparation time and [99mTc] remains bound throughout the 6-h lifetime of the preparation (Prescribing information for Draximage, 2010).

[99mTc] decays by isomeric transition with a physical half-life of 6.02 h. Nearly 90% of all disintegrations result in the principal photon emission that is useful for detection and imaging studies, which has a mean energy of 140.5 keV (Prescribing information for Draximage, 2010).

The Food and Drug Administration (FDA) has two currently approved indications and usage for [99mTc-MAA]: the first most common indication is as a lung imaging agent which is to evaluate pulmonary perfusion as part of the "V/Q scan"

or the ventilation–perfusion scan typically used to evaluate for pulmonary embolism (Neumann et al., 1980). The second indication is to aid in the evaluation of the peritoneovenous (LeVeen) shunt patency. This is a rare situation when peritoneal cavity fluid can have systemic circulation entry. The use of [99mTc-MAA] to evaluate for lung shunt fraction and hepatic tumor perfusion with a catheter-based intra-arterial injection is an off-label usage. However, this technique has gained widespread acceptance and is recommended by both microsphere manufacturers for the calculation of the lung shunt (Package Insert for TheraSphere, 2014 and Package Insert for SIR-Sphere, 2014). This "off-label" usage is widely accepted and is also reimbursed by CMS and U.S.-based insurance carriers.

The albumin aggregates are sufficiently fragile for the hepatic tumor capillary micro-occlusion to be temporary. Erosion and fragmentation reduce the particle size, allowing passage of the aggregates through the hepatic capillary bed. The fragments are then accumulated by the reticuloendothelial system. Elimination of the technetium [99mTc] aggregated albumin occurs with a half-life of about 2–3 h (Prescribing information for Draximage, 2010) in lung imaging.

A common occurrence in hepatic tumors, especially hepatocellular carcinoma (HCC), is that of patent intrahepatic arteriovenous shunting. The end result of this shunt is that the [99mTc-MAA] particles will not be mechanically lodged within the tumor vasculature. Instead, they will bypass the tumor and become lodged in the next available capillary bed, which is the pulmonary capillary system. Additionally, there are collateral arterial pathways that can divert the MAA particles to nontargeted capillary beds. Therein lies part of the reason why [99mTc-MAA] is useful prior to radioembolization. It is beneficial to know if the proposed catheter position will expose the patient to either nontarget embolization (NTE) or excessive lung radiation exposure. The [99mTc-MAA] radiotracer, especially when used with tomographic SPECT/CT imaging, will not only quantify lung shunt, but also identify significant deposition in extrahepatic gastrointestinal (GI) tissues. The implications of the result of MAA simulation and subsequent strategies to mitigate potential complications will be discussed in Section 4.4.

For hepatic imaging, the number of [99mTc-MAA] particles per single injection is 200,000–700,000 with the suggested number being approximately 350,000—similar to use in lung perfusion study. Due to breakdown and subsequent disassociation of the MAA particles, free ^{99m}Tc pertechnetate begins to accumulate immediately after injection. Therefore, imaging should be performed as soon as possible after infusion (Prescribing information for Draximage®, 2010). The physical breakdown of the MAA particles in the liver should be longer than in the lung due to lack of motion, leading to a potential for change in relative measured uptake as the time between injection and imaging increases. In addition, the longer the waiting time, the greater the amount of free technetium will be deposited in its expected distribution (salivary and thyroid glands, stomach, and kidneys). Since identification of nontarget embolization via detection of MAA activity in stomach is a major utility of the MAA simulation procedure, free ^{99m}Tc in the stomach can certainly be confounding.

Table 4.1 contrasts the properties of ^{99m}Tc-MAA compared with the two available ^{90}Y radioembolization products. It is quite apparent from Table 4.1 that MAA is close in size to both of the ^{90}Y glass and ^{90}Y resin products; however, there is an order of magnitude between the number of MAA particles and the number of ^{90}Y glass spheres, and two orders of magnitude between the number of MAA particles and the number of ^{90}Y resin spheres. This has caused much controversy as some believe that MAA is not a good surrogate for ^{90}Y while others do believe it is a useful simulation.

Typically, one MAA kit will be combined with 100 mCi (3.7 GBq) of ^{99m}Tc pertechnetate so that multiple doses can be obtained from each kit. An individual dosage can range from 2 to 4 mCi (74–148 MBq) (Package Insert for TheraSphere®, 2014), but most commonly 4 mCi (148 MBq) dosages are used. Occasionally, hepatic arterial anatomy may dictate the requirement of ^{99m}Tc-MAA injection through two distinct catheter positions for demonstration of the entire liver's shunt potential. Alternatively, the interventional radiologist may plan a bilobar radioembolization treatment, which may require infusion at two distinct catheter positions. To attain reasonable surrogacy with MAA, infusion at the same two positions that will be used in subsequent therapy must be performed. In such a situation, two separate 2 mCi (74 MBq) ^{99m}Tc-MAA dosages are often prepared in separate syringes so that they could be injected from the two desired catheter positions.

The effective half-life of ^{99m}Tc-MAA can be calculated from the formula:

$$\frac{1}{t_{\text{effective}}} = (1 / t_{\text{physical}}) + (1 / t_{\text{biological}})$$

where t is the half-life.

With a physical half-life of ^{99m}Tc of 6 h and a biological half-life of approximately 2.5 h in the lungs (Prescribing information for Draximage®, 2010), the above formula suggests an effective half-life of 1.76 h for the lungs. However, this parameter is purely of academic value. The effective half-life of ^{99m}Tc-MAA in the liver should be greater given the longer biologic half-life owing to lack of mechanical breakdown. In fact, the

Table 4.1 ^{99m}Tc-MAA, an imperfect surrogate

	^{99m}Tc-MAA	Y-90 glass	Y-90 resin
Size of particles, mean (range)	20–40 μm (10–70 μm)	20–30 μm (15–35 μm)	30–35 μm (20–60 μm)
No. of particles, mean (range)	350 k (200–700 k)	4 M for 10 GBq vial (1.2–8 M)	30–60 M (10–80 M)

Sources: Prescribing information for Draximage®, Available from: http://www.accessdata.fda.gov/drugsatfda_docs/label/2009/017881s010lbl.pdf, Revised October 2010, Accessed 17 December 2015, 2010; Package Insert for TheraSphere® Yttrium-90 Glass Microspheres, Available from: http://www.therasphere.com/physicians-package-insert/TS_PackageInsert_USA_v12.pdf, Revised August 21, 2014, version 12, Accessed 15 December 2015, 2014; NRC Device Registry for Glass Microspheres, NR-0220-D-131-S, Amended August 10, 2015, 2015; Package Insert for SIR-Sphere® Yttrium-90 Microspheres, Available from: http://www.sirtex.com/media/29845/ssl-us-10.pdf, Revised November 2014, version 10, Accessed 15 December 2015, 2014; Kennedy, A et al., *Int J Radiat Oncol Biol Phys*, 68, 13, 2007.

Note: k, thousand; M, million; μm, microns or micrometers.

biologic half-life in the liver may be closer to 8 h resulting in an effective half-life of 3.4 h for the liver (Grosser et al., 2016). As time progresses following infusion of MAA, the image quality degrades and interpretation is confounded by the 99mTc-MAA breakdown. The presence of even a small amount of free 99mTc may confuse image interpretation. Waiting for several hours would be ill advised, as a significant portion of the 99mTc would be free, rendering the MAA image suboptimal.

Of interest is a recent article (Grosser et al, 2016) that assessed the biodegradation of 99mTc-MAA versus 99mTc HSA. The authors assessed the residual activity in MAA and HSA at three time points, 1, 5, and 24 h postinjection, to assess the lung shunt fraction (LSF). As expected, the LSF calculated using MAA changed by 3.9%, 7.7% and 9.9% at the 3 time points, respectively, representing a bi-exponential decay. Grosser et al. (2016) recommended that MAA shunt calculations not be performed based on images greater than 4 h after injection of MAA and optimally not after 1 h to get an accurate picture of the shunt fraction.

(Amor-Coarasa et al., 2014). If custom kits can be created for 68Ga-MAA synthesis, then it is also possible to increase the number of MAA particles such that it is closer to the number used in the 90Y glass or resin radioembolization. This may reduce the potential for discrepancy in distribution due to the limited embolic effect associated with traditional MAA. The potential for 68Ga-MAA PET/CT is exceptionally exciting as it is possible to run PET/CT scanners with iodinated contrast and a triphasic liver protocol so that structural tumor delineation is paired with the MAA distribution. In addition, PET/MR may be even more useful as liver tumors can be seen with or without gadolinium contrast and diffusion weighted sequences can also provide useful tumor information, although motion artifacts degrage the fusion. There are also newer liver-specific MR contrast agents such as Eovist®/Primavist® that have advantages in imaging hepatic lesions like HCC (Campos et al., 2012). Superimposing the MAA distribution over a detailed anatomical tumor map is a very powerful *a priori* planning tool, especially if the MAA distribution can be quantified.

4.1.3 TRENDS TOWARD QUANTIFICATION

99mTc-MAA is not the only tracer that can be used for simulation in radioembolization. While 99mTc-MAA is imaged using planar scintigraphy, SPECT, or SPECT/CT, the use of PET/CT in pretreatment simulation has sparked some interest. PET is capable of better image quality and more accurate quantification compared with SPECT. In some academic centers, the availability of the 68Ge/68Ga generator has permitted changing the V/Q scan from a planar/SPECT study to that of a PET study. Ament et al. has described the use of a 68Ga tagged aerosol (Galligas) and 68Ga-labeled MAA for the purposes of V/Q scanning (Ament et al., 2013). There are various citations in the literature that describe the radiochemistry required to tag 68Ga to the MAA compound as early as in 1989 (Even and Green, 1989; Mathias and Green, 2008). One recent study describes the use of a lyophilized kit specifically for creation of the PET perfusion agent 68Ga MAA and discusses its potential benefit in radioembolization planning

4.2 IMAGING PROTOCOL

4.2.1 IMAGING PROTOCOL OVERVIEW

99mTc-MAA scintigraphy with angiography for vascular mapping is performed for radioembolization treatment planning and detection of potential complications from extrahepatic deposition of 90Y microspheres. Following infusion of a 99mTc-MAA dosage, planar images of the abdomen and lungs are obtained immediately in anterior and posterior projections to assess the LSF (American College of Radiology–Society of Interventional Radiology [ACR–SIR] practice parameter for radioembolization, 2014). Reasonable image acquisition involves collecting 140 KeV emissions with a planar gamma camera for 2 min using a 20% energy window. A planar dual head camera, utilizing both anterior and posterior heads simultaneously, fulfills the need for geometric mean (GM) calculation. Typically, low-energy high resolution (LEHR) or equivalent collimators are used for this acquisition.

The LSF is calculated using the GM of anterior and posterior projections. Regions of interest (ROIs) are drawn around the whole lung and the whole liver in the anterior and posterior projections and respective counts are obtained. The LSF is then determined using the following formulae:

$$GM = \sqrt{ROI_{Anterior} \times ROI_{Posterior}} \qquad (4.1)$$

$$\text{Lung shunt fraction } (LSF) = \frac{GM_{Lung}}{GM_{Liver} + GM_{Lung}} \qquad (4.2)$$

Example contours are illustrated in Figure 4.1a and b for a patient that has a large (~18 cm) HCC and an LSF of 18%. A coronal hepatic protocol CT scan of this patient is shown in Figure 4.1c for comparison. Total count values from lung fields or the liver field in anterior and posterior views are used to calculate GM and the LSF (Equations 4.1 and 4.2).

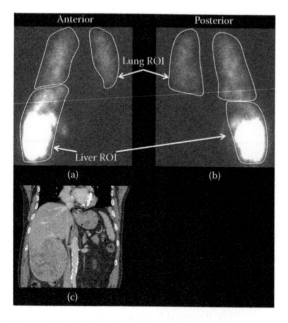

Figure 4.1 Calculation of lung shunt fraction in a patient following injection of 4.0 mCi of ^{99m}Tc-MAA into the right hepatic artery. Contours are drawn to identify counts in the lungs and liver on anterior **(a)** and posterior **(b)** projections. A coronal reformat of a pretreatment hepatic computed tomography scan of the patient's tumor is shown in **(c)**.

4.2.2 TECHNICAL ASPECTS OF MAA INJECTION AND IMAGING

In metastatic liver disease, MAA injection can be performed in the proper hepatic artery, due to the relatively lower incidence of high lung shunting compared with patients with HCC. HCC can be associated with portal vein thrombosis, and an increased prevalence for direct arteriovenous shunting bypassing the capillary bed, increasing the risk of radiation pneumonitis (RP). In the author's experience, significant LSFs (>20%) are seen only in cases of HCC. Among 109 patients who received preradioembolization MAA infusion and imaging over the past year at the author's institution, there were six cases with LSF > 20%, all with HCC. Five of the six also had portal vein thrombosis. These data are representative of the modern worldwide experience with radioembolization.

In patients with bilobar HCC, without gross vascular shunting into the hepatic or portal vein, ^{99m}Tc-MAA can be injected into the proper hepatic artery. However, if shunting is seen, it is often preferable to do a unilobar injection of MAA to assess one lobe at a time. A follow-up MAA scan can be performed separately, prior to treatment of the other lobe. In certain situations of variant arterial anatomy such as replaced right hepatic artery, fractionated dosages of ^{99m}Tc-MAA are injected to cover the entire liver in one sitting if possible, 2–3 mCi of MAA in the replaced right hepatic, and the remaining 2–3 mCi into the left hepatic artery.

As previously mentioned, it is preferable to assess LSF and imaging as soon as possible after the MAA injection. While ^{99m}Tc pertechnetate can confound interpretation of NTE to GI tissue, it can also result in an overestimation of LSF. Compton scattering also artificially increases the LSF and, therefore, the calculated lung dose. The lung dose is important, as a radiation dose >30 Gray (Gy) given to the lungs in one treatment or a cumulative dose >50 Gy in multiple treatments is considered a relative contraindication (Salem and Thurston, 2006a).

Planar scintigraphy is usually employed to calculate LSF. However, Yu et al. (2013) described a new method of calculating the mean lung dose (MLD) for radioembolization of liver cancer based on ^{99m}Tc-MAA SPECT/CT, which can provide a more accurate estimate of radiation risk to the lungs. In addition, SPECT and SPECT/CT also have increased sensitivity of detecting extrahepatic NTE compared with planar imaging. Ahmadzadehfar et al. (2010) reported the sensitivity for detecting extrahepatic deposition increased from 32% with planar imaging and 41% with SPECT alone to 100% with SPECT/CT. The specificity for NTE detection was 98% with planar and SPECT imaging and 93% with SPECT/CT. SPECT/CT imaging is able to detect extrahepatic activity predicting sites at risk for NTE and alter treatment planning in about 29% of cases. Identification of these cases following MAA simulation allows for corrective actions included coiling the responsible vessels, repositioning the infusion catheter, or in some cases cancellation of procedure (Ahmadzadehfar et al., 2010). Examples are shown in Section 4.4.

In addition to pretreatment identification of possible NTE and LSF, the pattern of distribution of MAA activity in the SPECT/CT also helps in the prediction of posttreatment response (Garin et al., 2012). Figure 4.2a shows the ^{18}F-fluorodeoxyglucose (18FDG)-PET/CT of a patient with a tumor in the posterior right hepatic lobe, with the corresponding MAA scan (Figure 4.2b) demonstrating matching

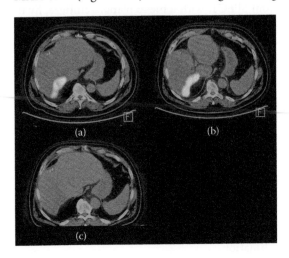

(a)   (b)

(c)

Figure 4.2 The pattern of distribution of macro-aggregated albumin (MAA) activity in the single-photon emission computed tomography (SPECT)/CT helps in predicting posttreatment response. **(a)** Fluorodeoxyglucose (FDG) avid tumor in the posterior right lobe ([18]FDG-PET/CT), with the corresponding MAA scan **(b)** demonstrating matching activity. **(c)** Follow-up [18]FDG-PET/CT scan (about 3 months posttreatment) shows complete response with resolution of the previously present focal FDG uptake on the baseline positron emission tomography (PET) scan.

hepatic distribution. The follow-up [18]FDG-PET/CT scan (Figure 4.2c) about 3 months posttreatment shows complete response with resolution of the previously present focal [18]FDG uptake on the baseline PET scan.

4.3 LUNG DOSIMETRY

4.3.1 CONVENTIONAL TECHNIQUES

Treatment of hepatic cancer with radioembolization relies on administering a high concentration of ^{90}Y microspheres to the tumor. Judicious placement of the injection catheter and relative enhancement of the tumor vasculature make it possible to deliver high doses to the tumor, while controlling the amount of activity to normal liver. Due to the pathological nature of the tumor vasculature, arteriovenous shunting of the injected microspheres may be significant, leading to the deposition of shunted microspheres to the lung and risk of radiation-induced pneumonitis. This risk is controlled through prequalification of patients for the treatment. As previously mentioned, mean lung doses (MLDs) from shunting greater than 30 Gy from a single treatment or cumulative MLD >50 Gy is a contraindication to radioembolization (Riaz et al., 2014).

The LSF can be estimated from pretreatment ^{99m}Tc-MAA scintigraphy, as described in Equations 4.1 and 4.2. Conventionally, the LSF is determined using the partition model (Ho et al., 1996) from the number of counts in the lungs (MAA_{lung}) and liver (MAA_{liver}) designated by regions on a two-dimensional (2D) planar image as previously indicated. Using the ^{99m}Tc-MAA activity distribution as a surrogate for the activity distribution of radioactive microspheres and assuming that all microspheres become trapped in capillaries and decay with rate of the physical half-life only (i.e., there is no biological removal), the medical internal radiation dose (MIRD) formalism can be used to estimate the absorbed dose to the lungs. The MIRD formalism (Loevinger and Berman, 1976) assumes local energy deposition whereby the absorbed dose to a volume of mass M with ^{90}Y activity A is given as follows (Ho et al., 1996):

$$D_{avg}(Gy) = \frac{A\,(GBq) \times 49.98(J \cdot s)}{M\,(kg)} \quad (4.3)$$

Lung dose is determined using the total injected activity A_{total} and the LSF by $A_{lung} = A_{total} \times LSF$ with an assumed mass of 1 kg [International Commission on Radiological Protection (ICRP), 1975]. Patient-specific evaluation of lung mass is not routinely performed for clinical radioembolization procedures owing to the lack of volumetric information provided by planar scintigraphy. This extension for calculating the absorbed dose to the lung is given in Equation 4.4 and can be applied to routine radioembolization treatment scenarios:

$$D_{avg}(Gy) = \frac{A_{total}(GBq) \times LSF \times 49.98(J \cdot s)}{1(kg)} \quad (4.4)$$

While convenient for routine clinical care, estimating lung dose from a 2D projection image has several limitations. Without an anatomical image, lung and liver contours on planar scintigraphy can be subjective. Because of the high uptake of activity in the liver, small variations in the definition of the border between the liver and lung may result in large variations in the measured activity to the lung. In addition, planar scintigraphy is not easily corrected for photon attenuation and scatter, leading to large potential uncertainties in lung and liver MAA activity. Using the GM of LSFs determined from a conjugate pair of anterior and posterior planar images may mitigate the differences in photon attenuation due to depth; however, the GM technique does not correct for scatter and fails to compensate for photon attenuation differences due to varying tissue densities as exhibited in liver and lung.

4.3.2 ADVANCED TECHNIQUES

The use of ^{99m}Tc-MAA SPECT can improve the estimation of lung dose from arteriovenous shunting. Three-dimensional (3D) image reconstruction in conjunction with attenuation and scatter corrections using tissue densities derived from a coregistered anatomical CT image improves the quantification of liver and lung activity uptake for calculation of the LSF. The anatomical CT image also provides information about the lung volume and the ability to perform patient-specific densitovolumetry. However, the time to obtain the necessary planar projections for 3D SPECT reconstruction is lengthy and free breathing during the

image acquisition may result in misregistration of activity around the diaphragm. Liver activity can be blurred into the lung region defined by the CT that was acquired within a few seconds during a specific phase of the respiratory cycle, resulting in an overestimation of the lung activity.

Yu et al. (2013) proposed a method of calculating the MLD based on ^{99m}Tc-MAA SPECT/CT. MLD is approximated by the mean dose in a subregion of the lung (Lung$_{sub}$) that excludes the portion of lung that is within 2 cm of the diaphragm in order to avoid spillover from liver. LSFs to Lung$_{sub}$ were determined from activities in Lung$_{sub}$ and external body within the scan region contoured on the CT. MLD was determined retrospectively for 71 patients by Equation 4.3 using a mass calculated from the volume of Lung$_{sub}$ and an assumed lung density of 0.3 g/cm^3 (Van Dyk et al., 1982). The lung doses calculated from SPECT were compared with those from an LSF estimated from lung and liver regions on planar scintigraphy assuming a total lung mass of 1000 g. In almost all patients, the lung dose from planar scintigraphy exceeded the dose derived from SPECT, with a mean planar scintigraphy to SPECT lung dose ratio of 3.8 ± 4.0. For patients at highest risk for radiation-induced pneumonitis from treatment with estimated lung doses greater than 15 Gy according to planar scintigraphy, the ratio was 2.7 ± 1.1. The authors further studied the effect of ignoring the misregistration due to breathing and failure to make attenuation and scattering corrections during the reconstruction of the SPECT images, concluding that each will result in roughly 50% overestimation of the lung dose. Recognizing the subjectivity and potential inaccuracy of LSF estimation based on planar scintigraphy, the authors recommended using SPECT for more accurate mean lung dose estimation.

A recent study by Kao et al. (2014) further demonstrated the uncertainties using planar images for estimating MLD and the importance of using SPECT/CT for individualized MLD estimation. Kao et al. (2014) compared lung doses calculated from ^{99m}Tc-MAA SPECT/CT to the conventional planar scintigraphy method for a cohort of 30 Southeast Asian patients. The technique for estimating lung dose from 3D SPECT/CT was similar to that used by Yu et al. (2013). LSFs were determined using counts from lung and liver volumes segmented on the CT. A distance of 1.5 cm superior to the domes of the diaphragm was selected to exclude lung counts from misregistration of activity in the liver due to free breathing during image acquisition. MLD was determined using Equation 4.3 with a patient-specific lung parenchyma mass determined by CT densitovolumetry for the lung volume above the exclusion zone. Lung and liver activities were also measured from conventional planar scintigraphy and a statistically significant difference between mean LSFs from planar (5.96 ± 4.59%) and SPECT (7.36 ± 4.96%) was found. However, the mean lung absorbed doses from both methods was not statistically significant. The mean lung mass for their patient cohort was 830 g, not vastly different from the standard lung mass of 1000 g. For patients with particularly smaller lung mass, such as patients postlobectomy, the planar image method may underestimate the MLD.

RP from arteriovenous shunting is a rare complication in treatment of liver cancer using ^{90}Y radioembolization, with an occurrence of less than 1% (Chan et al., 1995; Salem et al., 2008). A contraindication of treatment arises only if lung dose is greater than a single treatment dose of 30 Gy or cumulative dose of 50 Gy (Riaz et al., 2014). The conventional method of determining lung and liver activity from planar scintigraphy of ^{99m}Tc-MAA may result in an overestimation of MLD due to the inaccuracy of delineating regions representing lung and liver volumes and the inability to apply attenuation and scatter corrections to 2D images. Likewise, the necessity of using an assumed standard lung mass of 1000 g for dosimetric calculation from planar scintigraphy may contribute to an over- or underestimation of patient lung dose depending on patient size and lung physiology. Lung dose estimation from 3D ^{99m}Tc-MAA SPECT/CT using patient-specific lung mass offers an improved and more precise assessment of lung dose from shunting. However, even if SPECT/CT is used, interpretation of lung dose from ^{90}Y microsphere activity based on surrogate ^{99m}Tc-MAA scintigraphic imaging is still associated with error. Erosion and fragmentation of the albumin aggregates reduce the particle size resulting in a larger amount of ^{99m}Tc-MAA activity traveling to the lungs compared with ^{90}Y microspheres, which have a rigid structure.

4.4 NONTARGET EMBOLIZATION, RADIATION PENUMONITIS, AND THEIR EFFECTS

Planar ^{99m}Tc-MAA imaging is performed to determine LSF and to exclude gastrointestinal shunting. In particular, ^{99m}Tc-MAA SPECT/CT imaging provides fused functional and anatomic data, adds clarity and confidence in the identification of extra hepatic flow, and can guide target embolization of extrahepatic feeders. However, SPECT/CT should not be solely used to absolutely exclude gastrointestinal shunting. Rather, it should be considered an adjunctive imaging modality of the gastrointestinal tract. Exclusion of gastrointestinal flow should be accomplished by use of the combined information obtained from hepatic angiography, 3D CT angiography, and SPECT imaging.

4.4.1 EXTRAHEPATIC UPTAKE AND CLINICAL STRATEGIES USED TO TREAT

As discussed in Chapter 3, planning angiography is necessary before radioembolization as it provides an overview of normal and variant anatomy. With advanced technologies in C-arm CT and new catheter types, the frequency of prerequisite prophylactic coil embolization to prevent hepaticoeneteric flow has decreased. Nontarget deposition of microspheres can have grave consequences. The most common sites of undesired particle deposition are in the gall bladder, gastrointestinal (GI) system, and pulmonary system.

Radiation cholecystitis results from uptake of radioactive microspheres in the gall bladder. This can be prevented by preemptive identification of the cystic artery and placement of the catheter tip beyond the origin. If blood flow is significant and injection beyond the origin is not possible, then embolization may be considered (Salem and Thurston, 2006b). This is managed by supportive care and if refractory, cholecystectomy should be performed (Atassi, 2008). ^{99m}Tc-MAA simulation provides a sensitive predictor for radiation cholecystitis. Figure 4.3 shows uptake in the gallbladder along with the uptake in the liver and the tumor in the ^{99m}Tc-MAA SPECT/CT. This example illustrates the importance of ⁹⁹ᵐTc-MAA simulation in order to prevent complications such as radiation cholecystitis.

The incidence of GI ulceration is less than 5% if meticulous techniques are used (Mallach et al., 2008; Szyszko et al., 2007). Although pretreatment hepatic angiography should assist in reducing these complications, misdirected microspheres can enter hepaticoenteric flow. These patients may have intense embolic pain during or after the procedure. As shown in Chapter 13, if GI NTE is suspected, confirmation using ⁹⁰Y PET/CT after the microsphere infusion can aid in medical management of these patients. Definitive diagnosis can be made by upper endoscopy. These toxicities can be attributed to unrecognized variants, collateral circulation, and changes in flow dynamics during infusion (Murthy et al., 2007). These patients should be aggressively managed with proton pump inhibitors to prevent more serious complications such as ulceration and/or perforation that may eventually require surgical management.

Figure 4.4 shows gastric uptake in the pylorus identified on ⁹⁹ᵐTc-MAA SPECT/CT that could have potentially resulted in ulceration. Prophylaxis included coiling the culprit gastric artery. Figure 4.5a shows the planning angiogram following routine occlusion of the gastroduodenal artery. The MAA uptake detailed in Figure 4.4 prompted additional occlusion of the right-gastric artery shown in Figure 4.5b. This prophylactic measure allowed for safe administration of ⁹⁰Y radioembolization.

Inadvertent delivery of microspheres can also occur through the vessels supplying the anterior abdominal wall via the falciform artery (Liu et al., 2005) resulting in radiation dermatitis (Leong et al., 2009). These side effects can be avoided by prophylactic embolization during pretreatment angiography (Meyer et al., 2014). At our institute, we use an ice pack/saline bag placed over the abdominal wall that is presumed to cause vasospasm and redirect flow into the intrahepatic circulation (Wang, 2013). Figure 4.6 shows the ⁹⁹ᵐTc-MAA SPECT/CT of a patient with HCC. There is MAA uptake corresponding to the tumor in the right lobe of liver. Additionally, there is a linear focus of activity extending inferiorly into the anterior

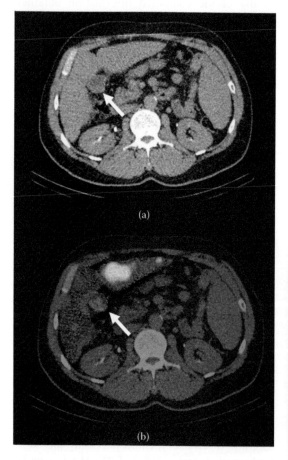

Figure 4.3 Uptake in the gallbladder (arrow) from shunting into the cystic artery, with the risk of radiation-induced cholecystitis. **(a)** Hepatic protocol CT and **(b)** ⁹⁹ᵐTc-MAA SPECT/CT following infusion of 4.0 mCi of ⁹⁹ᵐTc-MAA.

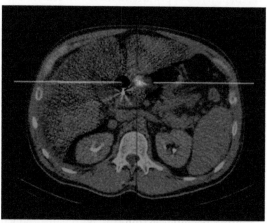

Figure 4.4 ⁹⁹ᵐTc-MAA SPECT/CT image shows potential activity in the pylorus. Additional prophylactic steps must be taken prior to treatment to avoid gastrointestinal (GI) complication.

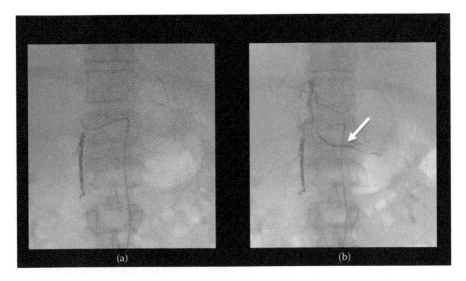

Figure 4.5 Angiography before and after ⁹⁹ᵐTc-MAA SPECT/CT. **(a)** Planning angiogram following routine occlusion of the gastroduodenal artery. **(b)** After identification of nontarget embolization (NTE) to the pylorus on MAA SPECT/CT, coil occlusion of the right-gastric artery was performed.

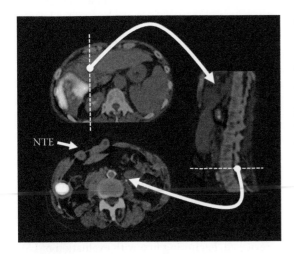

Figure 4.6 Abdominal wall shunting. ⁹⁹ᵐTc-MAA SPECT/CT in a patient with hepatocellular carcinoma. MAA uptake corresponding to the tumor is present in the right lobe of liver. A sagittal reformat shows linear activity extending inferiorly into the anterior abdominal wall in the region of the umbilicus, representing extrahepatic shunting via the falciform artery.

abdominal wall in the region of the umbilicus, representing potential extrahepatic NTE via the falciform artery. Due to pretreatment identification using MAA, the aforementioned ice pack prophylaxis was employed during the treatment of this patient preventing side effects.

4.4.2 EXCESSIVE LUNG SHUNTING AND CLINICAL STRATEGIES USED TO TREAT

Arteriovenous anastomoses or shunts in the liver parenchyma or tumor cause the lung shunting that potentially could result in RP after radioembolization (Wright et al., 2012; Leung et al., 1995). Various strategies for managing excessive LSF are described in the literature. These include cancellation of the procedure in any patient with markedly elevated hepatopulmonary shunting (LSF > 20%) (Leung et al., 1995), reduction in the microsphere dose (10% < LSF < 20%) (Elschot et al., 2011), bland embolization, or chemoembolization of the shunt (Gaba and Vanmiddlesworth, 2012) and balloon occlusion of the hepatic vein while delivering the microspheres could allow safe ⁹⁰Y delivery (Ward et al., 2015). Additional techniques are discussed in other chapters of this book.

4.4.3 RADIATION PNEUMONITIS

RP represents an acute manifestation of radiation-induced lung disease caused by increased hepatopulmonary shunting with an associated increased estimated radiation dose to the lungs. Most of the cases in the literature are a result of external beam irradiation after treatment for lung and breast cancers (Leung et al., 1995), lymphoma, and whole-body irradiation for stem cell transplantation (Camus, 2004). MLD, lung volume receiving a specified dose, and normal tissue complication probability (NTCP) are the three widely studied parameters used to assess the risk for RP (Graves et al., 2010). Rodrigues et al. (2004) highlighted that direct comparisons between studies could not be achieved given the heterogeneity of outcome variables. However, most studies did show an association between dose–volume histogram parameters and RP. It is noted that dosimetric parameters play a lesser role than patient characteristics for the prediction of lung toxicity (Dehing-Oberije et al., 2009). Ramella et al. (2010) determined addition of ipsilateral constraints (i.e., volume of lung receiving 30 Gy) to standard lung dosimetric factors in patients with RP in non–small-cell lung cancer treated with 3D conformal RT and concurrent chemotherapy resulted in a reduction in the incidence of pneumonitis from 14.4% to 6.8%. According to Riaz et al. (2009), RP is a complication that has not been studied optimally. In their report involving 58 patients, none developed RP with cumulative lung doses exceeding 50 Gy (Salem et al., 2008). However, there is a case report (Wright et al., 2012) with RP in a patient with a lung dose of 31.0 ± 13.0 Gy. It is presumed in this case that RP was not predicted using the currently used ^{90}Y dosimetry models that assume uniform distribution in the lungs (Salem et al., 2008), which may explain the findings of Wright et al. (2012).

RP patients generally present with gradual onset of dyspnea, fever, bronchoalveolar lymphocytosis and eosinophilia. This initially presents as a mild restrictive process on pulmonary testing, with ill-defined patchy opacities and ground-glass nodularity in a symmetric (i.e., "bat-wing") pattern with relative peripheral/hilar sparing 1–2 months after therapy. The features can also resemble an organizing or chronic eosinophilic pneumonia. These may resolve or progress toward localized fibrosis, traction bronchiectasis, and focal honeycombing. Late complications include pneumothorax and superinfections (Leung et al., 1995; Camus, 2004; Riaz et al., 2009).

The first line of management of RP is corticosteroids, which may reduce the degree of inflammation (Leung et al., 1995). Rubin and Casarett (1968) demonstrated that when corticosteroids were given after clinical pneumonitis had developed,

an objective response was seen. However, when given prophylactically they failed to prevent RP. Pentoxifylline (a platelet inhibitor with immuno-modulating/anti-inflammatory properties mediated through interleukin-1/tumor necrosis factor) is thought to be helpful in preventing radiotoxicity by inhibiting platelet aggregation and tumor necrosis factor (Ozturk et al., 2004).

In summary, it should be emphasized that the incidence of RP is low. A cautious approach should be of paramount importance when the hepatopulmonary shunt fraction would result in a lung dose exceeding 30 Gy (Murthy et al., 2005).

4.5 CORRELATION BETWEEN ^{99m}TC-MAA AND ^{90}Y RADIOEMBOLIZATION: GENERAL DISCUSSION AND OTHER CONSIDERATIONS

In the current practice of radioembolization, estimating the LSF plays a key role in preventing inadvertent RP. As previously discussed, determination of LSF involves the use of ^{99m}Tc-MAA as a surrogate for radioembolization in a separate planning procedure. The distribution of these particles within the liver, however, is in large part currently ignored in routine clinical practice. Although ^{99m}Tc-MAA is trusted as a surrogate for ^{90}Y microspheres in the measurement of the LSF, its utility in the accurate modeling of hepatic distribution of radioembolization has not been unequivocally demonstrated. In this section, literature reviewing the accuracy of ^{99m}Tc-MAA as a surrogate for radioembolization is reviewed, beginning first with evaluation of LSF with the use of MAA.

4.5.1 LSF EVALUATED USING ^{99m}TC-MAA

It is routine practice to inject ^{99m}Tc-MAA into the hepatic arterial branch 2–4 weeks before the injection of ^{90}Y microspheres. After injection of ^{99m}Tc-MAA, planar scintigraphy is routinely performed and ROIs over the lungs and the liver are used to measure LSF as previously described. In addition, SPECT/CT imaging of the upper abdomen is employed at many sites to visually assess the distribution of particles in the liver and extrahepatic territory to ensure future safe delivery of ^{90}Y radio-active microspheres.

LSF is known to be higher for HCC than for other tumors (8.0% vs. 6.3%; $p = .048$) (Olorunsola et al., 2015). In one study, high LSF (>20%) occurred in 14% of HCC cases but in only 3% of other tumors ($p = .004$) (Gaba et al., 2014). Colorectal cancer (CRC) metastases (median LSF, 10.6%) and HCC (11.7%) are known to have a significantly larger LSF than metastases from breast cancer (7.4%; $p < .005$) (Powerski et al., 2015). Similar results are reported through the literature, and we have noticed consistency at our institution as well in a retrospective review of 39 patients who underwent right lobe ^{99m}Tc-MAA injections (37 lobar, 2 segmental) before radioembolization for HCC at the Cleveland Clinic. Among these patients, the mean ^{99m}Tc-MAA LSF was 5.9% (SD, 0.03%; range, 0.5–12.1%), with no significant difference in LSF among patients with ($n = 7$) or without ($n = 32$) extrahepatic distribution of ^{99m}Tc-MAA.

In a study by Lambert et al. (2010), low-quality whole-body scintigraphy images (defined as visualization of the kidneys when adequate scaling of the whole-body scintigraphy image had to be performed to assess the liver) were correlated with a higher LSF. The authors found that 14% of the 90 studies assessed were considered to be of low quality and suggested that LSF was overestimated in this group.

In the vast majority of patients with primary or secondary hepatic tumors, LSF measured by ^{99m}Tc-MAA is less than 20%. A small fraction of HCC tumors is known to be associated with higher LSF (Refaat and Hassan, 2014). In addition, quality of the scintigraphy images has a significant impact on the measurement of LSF and a low-quality study may result in overestimation of LSF.

4.5.2 EFFECT OF FLOW DYNAMICS AND PARTICLE SIZE

Arterioles feeding liver metastases in humans average 30–40 μm in diameter. When flowing through arteries, particles concentrate in a peri-arteriolar fashion. Hence, the concentration of particles entering a side branch will be lower than the concentration in the main channel. In addition, smaller particles tend to reach the periphery of the liver, whereas larger ones do not. In a study

on rats, the mean tumor to liver arterial perfusion ratio (T:N) was 3:1 for 15- and 32.5-μm spheres but 1:1 for 50-μm microspheres (Van de Wiele et al., 2012). Clearly, the size of the particles plays a key role in their distribution within the liver and subsequent shunting away from the liver.

Two types of microspheres are currently available to perform radioembolization for the treatment of liver cancer: glass-based and resin-based microspheres. Resin ^{90}Y microspheres measure approximately 32.5 ± 2.5 μm in size, whereas the glass-based ^{90}Y microspheres measure approximately 25 ± 5 μm. In addition, the total number of spheres per GBq is approximately 20 million for ^{90}Y resin microspheres and only approximately 400,000 for ^{90}Y glass microspheres (Cremonesi et al., 2014). Thus, glass microspheres have a significantly higher amount of radioactivity per microsphere. As a result, for a similar radiation dose ^{90}Y resin microspheres are expected to be more embolic than ^{90}Y glass microspheres; however, this depends on the prescribed radiation dose and the size of the vascular bed in the liver that is to be treated.

Compared with ^{90}Y microspheres, ^{99m}Tc-MAA particles infused in the planning stage have a wider range of sizes (5–100 μm), with 80%–90% of the particles falling within the range of 10–70 μm (Table 4.1, Zophel et al., 2009). The ^{99m}Tc-MAA particles undergo enzymatic hydrolysis and are phagocytized by reticuloendothelial cells. As opposed to ^{90}Y microspheres, the radioactivity associated with a ^{99m}Tc-MAA dosage does not necessary vary linearly with particle number. For example, doubling the particle size will increase the average radioactivity per particle by a factor of 4. Therefore, a particle with a diameter of 40 μm will contain 16 times more radioactivity than a particle with a diameter of 10 μm (Van de Wiele et al., 2012).

The heterogeneous composition of small (<20 μm) and large (>60 μm) particles with significantly different radiation doses per particle within a dose of ^{99m}Tc-MAA is likely to affect imaged intrahepatic and extrahepatic distribution and have an effect on shunt quantification imaging, since small particles are more likely to pass through the hepatic capillary bed. This effect may combine with the propensity of planar scintigraphy to overestimate LSF due to scatter and attenuation, previously discussed. In a study involving 23 patients with primary and secondary liver malignancies, ^{99m}Tc-MAA scans were found to significantly overestimate LSF when compared with gold standard postradioembolization ^{90}Y PET/CT scans (6% vs. 1.8%; $p < .01$) (Song et al., 2015). In spite of the more heterogeneous composition in a vial, ^{99m}Tc-MAA is still used universally as a simulation surrogate for ^{90}Y radioembolization.

4.5.3 CORRELATION BETWEEN DISTRIBUTION OF ^{99m}TC-MAA AND ABSORBED DOSE

In clinical practice, the distribution of ^{99m}Tc-MAA particles is expected to be similar to the distribution of ^{90}Y microspheres, allowing the particles to serve as a surrogate for the microspheres. However, in several studies, the reliability of ^{99m}Tc-MAA as a surrogate has been questioned. When prescribing the radiation dosage, one should understand the significant differences in the characteristics of ^{99m}Tc-MAA and ^{90}Y resin or glass microspheres, including the size range of the particles/microspheres and the total number of particles/microspheres delivered. Presuming that ^{99m}Tc-MAA and ^{90}Y microspheres are delivered at the same site of infusion in the liver, one can still expect a difference in their distribution due to differences in flow kinetics. The physical properties of the injected agent and blood flow pattern from the tip of the catheter at the moment of infusion will dictate the distribution kinetics of infused particles. This may explain why procedures are rarely cancelled following suboptimal hepatic distribution of ^{99m}Tc-MAA obtained in simulation.

A study cohort of 66 patients with a total of 435 colorectal liver metastases showed that response to ^{90}Y resin microspheres was independent of qualitative grading on the degree of pretreatment ^{99m}Tc-MAA uptake in the tumor. Hence, patients could not be excluded from radioembolization based on ^{99m}Tc-MAA distribution (Ulrich et al., 2013). In response to subsequent questions raised about the possibility of catheter position being an important factor in these results, the authors later reported additional results of a subgroup analysis in which the catheter tip was placed in an identical position for both ^{99m}Tc-MAA and ^{90}Y microspheres (41 of the original 66 patients); there was a similar lack of correlation ($p > .05$) (Amthauer et al., 2014).

However, in a study of 17 patients (14 with HCC, 3 with CRC), Ho et al. (1996) found good correlation between the doses estimated using the partition model based on T:N and intraoperative dosimetry in tumors ($r = 0.862$) and background liver ($r = 0.804$). Ho et al. determined the T:N using ^{99m}Tc-MAA by dividing the average count rates of the tumor by the average count rates of the normal liver. This ratio can be used to estimate the activity of ^{90}Y microspheres that would be partitioned between the tumor and the normal liver compartment. If the activity delivered to the tumor can be estimated, Equation 4.3 can then be employed to determine the amount of ^{90}Y microspheres required to achieve a certain tumoricidal dose or to keep below a tolerance limit of normal liver tissue. Additional details on the utilization of the partition model for hepatic dosimetry are provided in Chapter 5. Unfortunately, this technique is not commonly utilized in routine clinical practice.

As additional conflicting evidence regarding the validity of ^{99m}Tc-MAA as a radioembolization surrogate is discussed below, one should keep in mind that clinical measurement of T:N is likely to vary substantially from patient to patient. This variation is not necessarily indicative of inaccuracy or error and has been shown in large patient studies (Ilhan et al., 2015a). In general, higher values of T:N occur in cases of neuroendocrine tumors, HCC, and cholangiocellular carcinoma, while lower T:N commonly occurs in cases of mammary cancer, CRC, and sarcoma.

4.5.3.1 Data suggesting the validity of ^{99m}Tc-MAA as a radioembolization surrogate

Ilhan et al. (2015b) compared the pattern of uptake in different liver tumors obtained using ^{99m}Tc-MAA SPECT with that of ^{90}Y bremsstrahlung SPECT following radioembolization using ^{90}Y resin microspheres. Among a cohort of 502 patients, 20% had primary hepatic tumors (HCC, 12%; cholangiocellular carcinoma, 8%) and the remaining patients had metastases from several different primary tumors. The ^{99m}Tc-MAA and ^{90}Y bremsstrahlung images were coregistered with contrast-enhanced CT or MR images. Analysis demonstrated that lesions with high uptake on ^{99m}Tc-MAA SPECT also had high uptake of ^{90}Y microspheres. The correlation between ^{99m}Tc-MAA SPECT and ^{90}Y microsphere uptake was significant but weak ($r = 0.26$; $p < .001$) (Ilhan et al., 2015b). Other authors have performed similar analyses comparing ^{99m}Tc-MAA SPECT to posttreatment ^{90}Y bremsstrahlung SPECT with findings suggesting reasonable agreement (Knesaurek et al., 2010).

In Section 4.5.3.2, several reports suggesting poor agreement between ^{99m}Tc-MAA and radioembolization will be reviewed. However, it is important to note that agreement of spatial distribution may not be necessary for ^{99m}Tc-MAA to serve as a valid tool for predictive hepatic dosimetry using the partition model. For example, Kao et al. (2013) compared established ^{90}Y PET/CT to pretreatment ^{99m}Tc-MAA SPECT/CT in 23 patients treated using resin microspheres. Using posttreatment ^{90}Y PET/CT as the gold standard, dosimetry based on MAA SPECT showed good agreement with a median relative error of just 3.8% (max = 13.2%).

4.5.3.2 Data suggesting ^{99m}Tc-MAA is a poor radioembolization surrogate

Several examples from the literature have reported a lack of correlation in the distribution of ^{99m}Tc-MAA and ^{90}Y radiomicrospheres within the liver. In a study to assess the ability of ^{99m}Tc-MAA to predict ^{90}Y distribution in 39 patients treated using ^{90}Y resin microspheres, the predicted amount of ^{90}Y activity in Couinaud liver segments based on ^{99m}Tc-MAA SPECT was compared with the actual amount of ^{90}Y based on ^{90}Y bremsstrahlung SPECT. The absolute mean difference between the estimated and actual ^{90}Y absorbed dose was around 30 Gy, and a difference of more than 30% of the mean activity per milliliter was found in 32% of the 225 segments analyzed (Wondergem et al., 2013).

While data presented by Kao et al. (2013) supported the accuracy of MAA-based tumor predictive dosimetry using ^{90}Y PET/CT as a gold standard, Song et al. (2015) have reported some discordance in a 30 patient cohort. Tumor-absorbed dose estimated using ^{99m}Tc-MAA SPECT/CT was found to be significantly lower than that estimated using ^{90}Y PET/CT (135.4 ± 64.2 Gy vs. 185.0 ± 87.8 Gy; $p < .01$). However, differences in absorbed dose

determined to non-target liver were not statistically different.

For some of the discrepancies reported in the distribution of ^{99m}Tc-MAA particles and ^{90}Y microspheres, differences in the exact location of the catheter tip at the time of delivery of these materials may have played a role. This hypothesis was examined by Jiang et al. (2012) by reviewing the perfusion differences between 81 paired ^{99m}Tc-MAA hepatic SPECT and posttherapy ^{90}Y bremsstrahlung SPECT studies; corresponding angiograms were also reviewed. When the catheter tip was placed in proximity to an arterial bifurcation or a small branch, this seemed to alter microsphere perfusion or trajectory and was found to be associated with mismatch (Jiang et al., 2012).

4.5.3.3 Other limitations of ^{99m}Tc-MAA simulation

Although the anatomical distribution of ^{99m}Tc-MAA particles is considered a surrogate for the distribution of ^{90}Y microspheres, this technique fails to quantify the functional aspect of the liver. Lam et al. (2015) studied the role of intra-arterial injection of ^{99m}Tc-labeled sulfur colloid (SC), which was injected after ^{99m}Tc-MAA-SPECT in the same procedure as a biomarker for functional liver. The authors used the combined information to study voxel-based partitioning and dosimetry for subsequent ^{90}Y radioembolization using resin microspheres in 98 patients and glass microspheres in 24 patients. Through a fusion of ^{99m}Tc-MAA SPECT and ^{99m}Tc-SC images, the liver was divided into four compartments based on uptake (+) and lack of uptake (−) of each tracer: tumor (^{99m}Tc-MAA+, SC−); irradiated functional liver (^{99m}Tc-MAA+, SC+); nonirradiated functional liver (^{99m}Tc-MAA−, SC+); and tumor necrosis, cysts, and major vessels (^{99m}Tc-MAA−, SC−). Independent of the type of microspheres used, HCC had a higher median tumor/median functional liver absorbed dose ratio than other tumor types (median 1.8; $p = .02$). The median tumor absorbed dose was correlated with response in both univariate and multivariate analyses, and the maximum change in toxicity grade from baseline after radioembolization was associated with the absorbed dose in functional liver tissue ($p < .05$). With these results, Lam et al. (2015) demonstrated the potential use of ^{99m}Tc-SC as a tracer that could individualize the tolerance of background liver in each patient, thereby allowing clinicians to adjust the radioembolization dose to maximize tumor response while minimizing the risk of radioembolization-induced liver disease.

4.5.4 SUMMARY OF VALIDITY OF ^{99m}Tc-MAA AS A RADIOEMBOLIZATION SURROGATE

Differences between tracer distributions are likely due to the differences between ^{99m}Tc-MAA particles and ^{90}Y microspheres. The number of ^{90}Y microspheres usually delivered is several orders of magnitude higher than the number of ^{99m}Tc-MAA particles delivered, which likely results in an embolization effect and increases redistribution into background liver, reducing the correlation in measured T:N. Such a difference might be higher with resin microspheres than with glass microspheres due to the vast differences in the number of microspheres within a comparable prescribed radiation dose (Table 4.1). In addition, catheter tip position and changes in tumor vascularity (due to tumor growth or histological changes) between the planning and treatment angiograms might also play a key role in the differences between ^{99m}Tc-MAA particle and ^{90}Y microsphere distribution.

Based on the results of several studies, we can safely conclude that the distribution and tumor uptake of ^{99m}Tc-MAA particles, which are currently used as a surrogate agent, does not consistently demonstrate equivalence with the distribution and uptake of the therapeutic agent (^{90}Y microspheres). The LSF measured by ^{99m}Tc-MAA might be overestimated in some patients and it is therefore possible that some patients may be unnecessarily excluded from radioembolization. Hence, ^{99m}Tc-MAA in its current form is not an ideal surrogate for ^{90}Y radioembolization.

To minimize this discordance, a tighter filtration of ^{99m}Tc-MAA particles to sizes that more closely match the size range of ^{90}Y microspheres might be considered. In addition, efforts to increase the embolic burden of ^{99m}Tc-MAA to more closely approximate ^{90}Y microspheres

may also be considered as a potential method to improve concordance with radioembolization. However, such a practice might result in decreased T:N during treatment due to the pre-embolic effect of MAA if treatment is performed before MAA has been completely cleared. Unfortunately, the time for complete clearance to occur, particularly in the setting of highly variable neoplasm absent of Kupffer cells, is not known. MAA retention is known to be prolonged when there is a reduction in the number of Kupffer cells (Tanaka et al., 1996; Rimola et al., 1984 ; Bilzer et al., 2006).

4.6 ALTERNATIVES TO MAA IN PROCEDURE SIMULATION AND/OR PROGNOSTICATION

4.6.1 USEFULNESS OF PREPROCEDURAL CT/MRI IN PREDICTING LSF

Previous studies have suggested that pretreatment hepatic protocol CT may have a role in prognostication of response to radioembolization in patients with HCC. Tumor hypervascularity compared with background liver and the amount of intratumoral blood flow estimated on CT has been reported to correlate with disease response. In an analysis of CT scans performed before and after ^{90}Y glass microsphere radioembolization in 23 patients with unresectable HCC, prolonged progression-free survival was associated with lower LSF, higher central tumor hypervascularity, and well-defined tumor margins, whereas shorter progression-free survival was associated with abutment of the portal vein by the tumor (Salem et al., 2013).

Morsbach et al. (2013) prospectively evaluated the ability of CT perfusion to predict morphologic response and survival in 38 patients with liver metastases who subsequently underwent ^{90}Y resin microsphere radioembolization; dose was calculated using the BSA method. Five seconds after contrast material injection (50 mL of iopromide), 12 spiral acquisitions covering the liver were obtained in the 4D spiral mode. Arterial perfusion (AP) in target liver lesions was significantly higher in the responders than in the nonresponders (37.5 vs. 11.8 mL/min; $p < .001$). A cutoff AP of 16 mL per 100 mL/min had a sensitivity of 100% and a specificity of 89% for predicting therapy response.

In a single-center retrospective study of 70 patients with HCC, infiltrative morphologic structure, tumor burden greater than 50%, portal vein invasion, and arterioportal shunting were significantly associated with high (>20%) LSF in multivariate analysis (Gaba et al., 2014). Similarly, in a study using pretreatment multiphase CT, strong tumor contrast enhancement was found to be associated with a significantly larger LSF than in tumors with little enhancement (11.7% vs. 8.3%; $p < .001$). In addition, patients with compression (LSF = 13.9%) or tumor thrombosis (15.8%) of a major portal vein branch had a significantly higher LSF than patients with a normal portal vein (8.1%) (both $p < .001$) (Powerski et al., 2015).

Finally, in a multivariate analysis of findings on CT ($n = 134$) or MRI ($n = 18$) among patients with primary and secondary hepatic tumors, early hepatic vein opacification and hepatic vein tumor thrombus or occlusion were associated with a significantly higher LSF. Sensitivity and specificity of early hepatic vein opacification originating from the tumor were 78% and 93%, respectively (positive likelihood ratio, 10.5), for predicting high (>20%) LSF (Olorunsola et al., 2015).

4.6.2 USEFULNESS OF C-ARM CBCT IN ENHANCING SAFETY

With the ability of modern angiographic units to acquire C-arm cone beam CT (CBCT) images, multiplanar evaluation of the tumor, background liver, and extrahepatic enhancement can now be evaluated during planning angiography (Pellerin et al., 2013). This technique has demonstrated increased sensitivity in the detection of extrahepatic enhancement when compared with digital subtraction angiography or ^{99m}Tc-MAA imaging in a small cohort (Louie et al., 2009). A larger study evaluated the utility of pretreatment CBCT to correctly identify the presence of extrahepatic NTE. This effort found that CBCT prior to radioembolization was associated with a negative predictive value for extrahepatic shunting of 95% (van den Hoven et al., 2016). Despite these positive findings, there is no single standard C-arm CBCT protocol in use. In an attempt to optimize a protocol for

identification of extrahepatic shunting and parenchymal enhancement in radioembolization, van den Hoven et al. (2016) conducted a prospective development study. The authors found that the variable contrast and scan delay determined using timing parenchymal enhancement on digital subtraction angiography were more effective than the contrast and scan delay determined with a protocol that used either a fixed 6-s delay and 10-s scan or a 5-s low-dose scan setting applied to reduce breathing artifacts in combination with a variable delay (van den Hoven et al., 2016).

In the future, multispin/multiphase CBCT may be used during planning to identify features associated with safe and favorable clinical outcomes and to aid in modifying required ^{90}Y dose activity. CBCT may also be used to predict high LSF and tumor enhancement characteristics that are unlikely to produce desired outcomes, thus allowing for modification of the treatment plan to other forms of embolization in the same session.

4.6.3 IS RADIOEMBOLIZATION WITHOUT ^{99m}TC-MAA INJECTION FEASIBLE?

In a major proportion of patients who undergo radioembolization, LSF is estimated to be significantly less than 20%. As described in Sections 4.6.1 and 4.6.2, there are several findings on CT/MRI and pretreatment angiography that are associated with a high lung shunt. In addition, CBCT is increasingly used during planning angiography, which has improved our ability to prevent inadvertent extrahepatic uptake in the abdomen, thus reducing our reliance on ^{99m}Tc-MAA SPECT imaging. Hence, obtaining such information before injection of ^{99m}Tc-MAA might help clinicians to predict which patients are likely to have a high LSF.

Among the most recent 39 patients at The Cleveland Clinic to undergo treatment planning using ^{99m}Tc-MAA injection for right lobe HCC, the highest evaluate LSF was 12.1%. At this maximum lung shunt, a radioembolization absorbed dose of 120 Gy to a typical right liver lobe measuring 1000 cc in volume would result in 17.1 Gy to the lungs, well within aforementioned safety limits. Using a predetermined target liver volume, a single-session planning angiogram immediately followed by ^{90}Y microsphere infusion without ^{99m}Tc-MAA infusion might be feasible. However, such a practice would require further validation of methods to prognosticate LSF without simulating using ^{99m}Tc-MAA.

There is already some active effort to condense radioembolization therapy into a single-session treatment. For example, in a recent study, glass microsphere radioembolization was performed on 14 patients by combining the ^{99m}Tc-MAA infusion, LSF assessment, and injection of ^{90}Y microspheres on the same day (Gates et al., 2014). It is possible that such examples are a first step into more aggressive investigation into the possibility of radioembolization treatment on selected patients without scintigraphic evaluation of LSF.

4.6.4 ^{166}HO MICROSPHERES: A POTENTIAL ALTERNATIVE TO ^{99m}TC-MAA AND ^{90}Y RADIOEMBOLIZATION

As discussed in Chapter 1, radioactive Holmium-166 (^{166}Ho) microspheres are seen as a possible alternative to ^{90}Y microspheres. ^{166}Ho microspheres emit high-energy beta-radiation, which is used for therapeutic purposes, as well as gamma-radiation, which allows for direct scintigraphy. In addition, Ho is a highly paramagnetic element that can be visualized effectively on MRI. This allows measurement of radioactivity using diagnostic scintigraphy, which can then be used for accurate therapeutic dosimetry and MRI-based high-resolution imaging of the biodistribution of the tracer.

In a prospective clinical study in 14 patients with unresectable liver metastases treated with ^{166}Ho microsphere radioembolization, the accuracy of lung-absorbed dose estimates by ^{99m}Tc-MAA-based diagnostic imaging was compared with the accuracy of estimates directly measured following radioembolization using ^{166}Ho SPECT/CT. Lung-absorbed doses were significantly overestimated by ^{99m}Tc-MAA imaging (Smits et al., 2012). In addition, as expected, the actual absorbed dose in the lungs was better predicted by diagnostic ^{166}Ho microsphere SPECT/CT imaging (median, 0.02 Gy) than by ^{166}Ho microsphere planar scintigraphy (median, 10.4 Gy; $p < .001$), ^{99m}Tc-MAA SPECT/CT imaging (median, 2.5 Gy; $p < .001$), and ^{99m}Tc-MAA planar scintigraphy (median, 5.5 Gy; $p < .001$) (Elschot et al., 2014). These results call into question the validity of our long-standing practice of relying on ^{99m}Tc-MAA estimates to predict risk of RP.

These data are not mentioned merely to further illustrate potential shortcomings associated with simulation using ^{99m}Tc-MAA, but also a potential benefit of ^{166}Ho radioembolization. Much of this chapter has been devoted to analyzing differences in the physical properties of ^{90}Y radioembolization and ^{99m}Tc-MAA as a source of error. However, since ^{166}Ho radioembolization treatment planning is based on infusion of a scout dose of ^{166}Ho microspheres, the agent used for both simulation and therapy are identical aside from the number of microspheres infused (Prince et al., 2013). Use of ^{166}Ho radioembolization, therefore, has the potential to minimize the mismatches currently reported between the distribution and activity of the diagnostic tracer (^{99m}Tc-MAA) and therapeutic tracer (^{90}Y glass or resin microspheres).

4.7 CONCLUSION

^{99m}Tc-MAA is the standard of care for preradioembolization simulation for the evaluation of LSF to prevent RP. ^{99m}Tc-MAA SPECT/CT can also aide in the prognostication of extrahepatic NTE, allowing for appropriate prophylaxis and avoidance of treatment-related side effects. In addition, pretreatment ^{99m}Tc-MAA SPECT/CT is also used as a treatment planning tool by providing tumor and normal liver predictive dosimetry. Unfortunately, due primarily to physical differences between ^{99m}Tc-MAA and ^{90}Y microspheres, MAA is an imperfect surrogate for radioembolization and the accuracy with which it fulfills any of the aforementioned functions is limited. While alternatives to ^{99m}Tc-MAA are available which may provide superior accuracy, such as ^{166}Ho microspheres, there are many other issues to consider before moving away from an established and effective treatment protocol currently under worldwide clinical use. Instead, the radioembolization treatment team will be well served by carefully understanding and considering the limitations of MAA in the medical management of patients.

REFERENCES

ACR–SIR practice parameter for radioembolization with microsphere brachytherapy device (RMBD) for treatment of liver malignancies: Revised 2014 (resolution 17).

Ahmadzadehfar, H. et al. (2010). The significance of ^{99m}Tc-MAA SPECT/CT liver perfusion imaging in treatment planning for ^{90}Y-microsphere selective internal radiation treatment. *J Nucl Med* 51:1206–1212.

Ament, S. et al. (2013). PET lung ventilation/perfusion imaging using (68)Ga aerosol (Galligas) and (68)Ga-labeled macroaggregated albumin. *Recent Results Cancer Res* 194:395–423.

Amor-Coarasa, A. et al. (2014). Lyophilized kit for the preparation of the PET perfusion agent [68Ga]-MAA. *Int J Mol Imaging* 2014:1–7.

Amthauer, H., Ulrich, G., Grosser, O., Ricke, J. (2014). Reply: Pretreatment dosimetry in HCC radioembolization with (90)Y glass microspheres cannot be invalidated with a bare visual evaluation of (99m) Tc-MAA uptake of colorectal metastases treated with resin microspheres. *J Nucl Med* 55(7):1216–1218.

Atassi, B. et al. (2008). Biliary sequelae following radioembolization with Yttrium-90 microspheres. *J Vasc Interv Radiol* 19(5):691–697.

Baum, R., Kulkarni, H. (2012). THERANOSTICS: From molecular imaging using Ga-68 labeled tracers and PET/CT to personalized radionuclide therapy—The Bad Berka Experience. *Theranostics* 2(5):4375–447.

Beauregard, J. et al. (2011). Quantitative 177Lu SPECT (QSPECT) imaging using a commercially available SPECT/CT system. *Cancer Imaging* 11(1):56–66.

Bilzer, M., Roggel, F., Gerbes, A. (2006). Role of Kupffer cells in host defense and liver disease. *Liver Int* 26(10):1175–1186.

Campos, J., Sirlin, C., Choi, J. (2012). Focal hepatic lesions in Gd-EOB-DTPA enhanced MRI: The atlas. *Insights Imaging* 3(5):451–474.

Camus, P., Kudoh, S., Ebina, M. (2004) Interstitial lung disease associated with drug therapy. *Br J Cancer* 91(suppl 2):S18–S23.

Chan, A. et al. (1995). A prospective randomized study of chemotherapy adjunctive to definitive radiotherapy in advanced nasopharyngeal carcinoma. *Int J Radiat Oncol Biol Phys* 33(3):569–577.

Cremonesi, M. et al. (2014). Radioembolization of hepatic lesions from a radiobiology and dosimetric perspective. *Front Oncol* 4:210.

Dehing-Oberije, C. et al. (2009). The importance of patient characteristics for the prediction of radiation-induced lung toxicity. *Radiother Oncol* 91:421–426.

Elschot, M. et al. (2011). Quantitative evaluation of scintillation camera imaging characteristics of isotopes used in liver radioembolization. *PLoS One* 6(11):e26174.

Elschot, M. et al. (2014). (99m)Tc-MAA overestimates the absorbed dose to the lungs in radioembolization: A quantitative evaluation in patients treated with ^{166}Ho-microspheres. *Eur J Nucl Med Mol Imaging* 41(10):1965–1975.

Even, G., Green, M. (1989). Gallium-68-labeled macroaggregated human serum albumin, 68Ga-MAA. *Int J Rad Appl Instrum B* 16(3):319–121.

Gaba, R., Vanmiddlesworth, K. (2012). Chemoembolic hepatopulmonary shunt reduction to allow safe yttrium-90 radioembolization lobectomy of hepatocellular carcinoma. *Cardiovasc Intervent Radiol* 35 (6):1505–1511.

Gaba, R. et al. (2012). Characteristics of primary and secondary hepatic malignancies associated with hepatopulmonary shunting. *Radiology* 271(2):602–612.

Garin, E. et al. (2012). Dosimetry based on 99mTc-macroaggregated albumin SPECT/ CT accurately predicts tumor response and survival in hepatocellular carcinoma patients treated with 90Y-loaded glass microspheres: Preliminary results. J Nucl Med 53(2):255–263.

Gates, V. et al. (2014). Outpatient single-session yttrium-90 glass microsphere radioembolization. *J Vasc Interv Radiol* 25(2):266–270.

Graves, P., Siddiqui, F., Anscher, M., Movsas, B. (2010). Radiation pulmonary toxicity: From mechanisms to management. *Semin Radiat Oncol* 20:201–207.

Grosser, O. et al. (2016). Pharmacokinetics of ^{99m}Tc-MAA and ^{99m}Tc-HSA microspheres used in pre-radioembolization dosimetry: Influence on the liver-lung shunt. *J Nucl Med* 57(2). Published February 9, 2016 as doi:10.2967/jnumed.115.169987

Ho, S. et al. (1996). Partition model for estimating radiation doses from yttrium-90 microspheres in treating hepatic tumours. *Eur J Nucl Med* 23(8):947–952.

Ilhan, H. et al. (2015a). Systematic evaluation of tumoral [99mTc]-MAA uptake using SPECT and SPECT/CT in 502 patients before 90Y radioembolization. *J Nucl Med* 56(3):333–338.

Ilhan, H. et al. (2015b). Predictive value of [99mTc]-MAA SPECT for 90Y-labeled resin microsphere distribution in radioembolization of primary and secondary hepatic tumors. *J Nucl Med* 56(11):1654–1660.

International Commission on Radiological Protection. (1975). *Reference Man: Anatomical Physiological and Metabolic Characteristics. ICRP Publication 23.* Oxford: Pergamon Press.

Jiang, M. et al. (2012). Segmental perfusion differences on paired Tc-99m macroaggregated albumin (MAA) hepatic perfusion imaging and yttrium-90 (Y-90) Bremsstrahlung imaging studies in SIR-sphere radioembolization: Associations with angiography. *J Nucl Med Radiat Ther* 3:122.

Kao, Y. et al. (2014). Personalized predictive lung dosimetry by technetium-99m macroaggregated albumin SPECT/CT for yttrium-90 radioembolization. *EJNMMI Res* 4(1):33.

Kao, Y. et al. (2013). Post-radioembolization yttrium-90 PET/CT—part 2: Dose–response and tumor predictive dosimetry for resin microspheres. *EJNMMI Res* 3(57):1–12.

Kennedy, A. et al. (2007). Recommendations for radioembolization of hepatic malignancies using yttrium-90 microsphere brachytherapy: A consensus panel report from The Radioembolization Brachytherapy Oncology Consortium. *Int J Radiat Oncol Biol Phys* 68:13–23.

Knesaurek, K. et al. (2010). Quantitative comparison of yttrium-90 (90Y)-microspheres and technetium-99m (99mTc)-macroaggregated albumin SPECT images for planning 90Y therapy of liver cancer. *Tech Cancer Res Treatm* 9(3):253–261.

Lam, M. et al. (2015). Fusion dual-tracer SPECT-based hepatic dosimetry predicts outcome after radioembolization for a wide range of tumour cell types. *Eur J Nucl Med Mol Imaging* 42(8):1192–1201.

Lambert, B. et al. (2010). 99mTc-labelled macroaggregated albumin (MAA) scintigraphy for planning treatment with 90Y microspheres. *Eur J Nucl Med Mol Imaging* 37(12):2328–2333.

Leong, Q. et al. (2009). Radiation dermatitis following radioembolization for hepatocellular carcinoma: A case for prophylactic embolization of a patent falciform artery. *J Vasc Interv Radiol* 20:833–836.

Leung, T., Lau, W., Ho, S. (1995). Radiation pneumonitis after selective internal radiation treatment with intraarterial 90yttrium-microspheres for inoperable hepatic tumors. *Int J Radiat Oncol Biol Phys* 33:919–924.

Liu, D. et al. (2005). Angiographic considerations in patients undergoing liver-directed therapy. *J Vasc Interv Radiol* 16:911–935.

Loevinger, R., Berman, M. (1976). *A Revised Schema for Absorbed-Dose Calculations for Biologically Distributed Radionuclides. MIRD Pamphlet 1 Biologic.* New York, NY: Society of Nuclear Medicine.

Louie, J. et al. (2009). Incorporating cone-beam CT into the treatment planning for yttrium-90 radioembolization. *J Vasc Interv Radiol* 20(5):606–613.

Mallach, S. et al. (2008). An uncommon cause of gastro-duodenal ulceration. *World J Gastroenterol* 14:2593–2595.

Material Safety Data Sheet for DRAXIMAGE®-MAA. (2011). Available from: http://www.draximage.com/data/PDF/51_en.pdf. Jubilant DraxImage Inc., revised July 2011, revision no. 4. Accessed 15 December 2015.

Mathias, C., Green, M. (2008). A convenient route to [Ga-68]Ga-MAA for use as a particulate PET perfusion tracer. *Appl Radiat Isotopes* 66(12):1910–1912.

Meyer, C. et al. (2014). Feasibility of temporary protective embolization of normal liver tissue using degradable starch microspheres during radioembolization of liver tumours. *Eur J Nucl Med Mol Imaging* 41:231–237.

Morsbach, F. et al. (2013). Computed tomographic perfusion imaging for the prediction of response and survival to transarterial radioembolization of liver metastases. *Invest Radiol* 48(11):787–794.

Murthy, R. et al. (2005). Yttrium-90 microsphere therapy for hepatic malignancy: Devices, indications, technical considerations, and potential complications. *Radiographics* 25(Suppl 1):S41–S55.

Murthy, R. et al. (2007). Gastrointestinal complications associated with hepatic arterial yttrium-90 microsphere therapy. *J Vasc Interv Radiol.* 18:553–561.

Neumann, R., Sostman, H., Gottschalk, A. (1980). Current status of ventilation-perfusion imaging. *Semin Nucl Med* 10(3):198–217.

NRC Device Registry for Glass Microspheres. (2015). NR-0220-D-131-S, Amended August 10, 2015.

Olorunsola, O. et al. (2015). Imaging predictors of elevated lung shunt fraction in patients being considered for yttrium-90 radioembolization. *J Vasc Interv Radiol* 26(10):1472–1478.

Ozturk, B., Egehan, I., Atavci, S., Kitapci, M. (2004). Pentoxifylline in prevention of radiation-induced lung toxicity in patients with breast and lung cancer: A double-blind randomized trial. *Int J Radiat Oncol Biol Phys* 58:213–219.

Package Insert for TheraSphere® Yttrium-90 Glass Microspheres. (2014). Available from: http://www.therasphere.com/physicians-package-insert/TS_PackageInsert_USA_v12.pdf. Revised August 21, 2014, version 12. Accessed 15 December 2015.

Package Insert for SIR-Sphere® Yttrium-90 Microspheres. (2014). Available from: http://www.sirtex.com/media/29845/ssl-us-10.pdf. Revised Nov 2014, version 10. Accessed 15 December 2015.

Pellerin, O. et al. (2013). Can C-arm cone-beam CT detect a micro-embolic effect after TheraSphere radioembolization of neuroendocrine and carcinoid liver metastasis? *Cancer Biother Radiopharm* 28(6):459–465.

Pereira, G., Traughber, M., Muzic, R. (2014). The role of imaging in radiation therapy planning: Past, present, and future. *BioMed Res Int* 2014:1–9.

Powerski, M. et al. (2015). Hepatopulmonary shunting in patients with primary and secondary liver tumors scheduled for radioembolization. *Eur J Radiol* 84(2):201–207.

Prince, J. et al. (2013). Holmium-166 microspheres for image-guided radioembolization: No need for patient isolation after treatment [abstract]. *J Vasc Interv Radiol* 24(4):S157–S158.

Ramella, S. et al. (2010) Adding ipsilateral V20 and V30 to conventional dosimetric constraints predicts radiation pneumonitis in stage IIIA-B NSCLC treated with combined-modality therapy. *Int J Radiat Oncol Biol Phys* 76:110–115.

Refaat, R., Hassan, M. (2014). The relationship between the percentage of lung shunting on Tc-99m macroaggregated albumin (Tc-99m MAA) scan and the grade of hepatocellular carcinoma vascularity. *Egypt J Radiol Nucl Med* 45(2):333–342.

Riaz, A. et al. (2009). Complications following radioembolization with yttrium-90 microspheres: A comprehensive literature review. *J Vasc Interv Radiol* 20:1121–1130.

Riaz, A., Rafia, A., Riad, S. (2014). Side effects of yttrium-90 radioembolization. *Frontiers Oncol.* 4. (198):1–11

Rimola, A. et al. (1984). Reticuloendothelial system phagocytic activity in cirrhosis and its relation to bacterial infections and prognosis. *Hepatology* 4(1):53–58.

Rodrigues, G. et al. (2004). Prediction of radiation pneumonitis by dose–volume histogram parameters in lung cancer—A systematic review. *Radiother Oncol* 71:127–138.

Rubin, P., Casarett, G. (1968). *Clinical Radiation Pathology.* Philadelphia, PA: W.B. Saunders.

Salem, M. et al. (2013). Radiographic parameters in predicting outcome of patients with hepatocellular carcinoma treated with yttrium-90 microsphere radioembolization. *ISRN Oncol* 2013:538376.

Salem, R. et al. (2008). Incidence of radiation pneumonitis after hepatic intra-arterial radiotherapy with yttrium-90 microspheres assuming uniform lung distribution. *Am J Clin Oncol* 31(5):431–438.

Salem, R., Thurston, K. (2006a). Radioembolization with 90-yttrium microspheres: A state-of-the-art brachytherapy treatment for primary and secondary liver malignancies: Part 1—Technical and methodologic considerations. *J Vasc Interv Radiol* 17:1251–1278.

Salem, R., Thurston, K. (2006b). Radioembolization with 90-yttrium microspheres: A state-of-the-art brachytherapy treatment for primary and secondary liver malignancies: Part 2: Special topics. *J Vasc Interv Radiol* 17:1425–1439.

Silberstein, E. et al. (2003). Society of nuclear medicine procedure guideline for palliative treatment of painful bone metastases version 3.0. Approved January 25, 2003. Available from: http://interactive.snm.org/docs/pg_ch25_0403.pdf. Accessed 31 January 2016.

Smits, M. et al. (2012). Holmium-166 radioembolisation in patients with unresectable, chemorefractory liver metastases (HEPAR trial): A phase 1, dose-escalation study. *Lancet Oncol* 13(10):1025–1034.

Song, Y. et al. (2015). PET/CT-based dosimetry in 90Y-microsphere selective internal radiation therapy: Single cohort comparison with pretreatment planning on (99m)Tc-MAA imaging and correlation with treatment efficacy. *Medicine (Baltimore)* 94(23):e945.

Szyszko, T. et al. (2007). Management and prevention of adverse effects related to treatment of liver tumours with 90Y microspheres. *Nucl Med Commun* 28:21–24.

Talanow, R. et al. (1996). Ability of pre-therapy Tc99m MAA SPECT/CT to predict the distribution of Y90 radiomicrosphere therapy as defined by Bremsstrahlung SPECT/CT. *Abstract and poster presented at RSNA annual meeting in 2010.*

Tanaka, M. et al. (1996). Pathomorphological study of Kupffer cells in hepatocellular carcinoma and hyperplastic nodular lesions in the liver. *Hepatology* 24(4):807–812.

Ulrich, G. et al. (2013). Predictive value of intratumoral 99mTc-macroaggregated albumin uptake in patients with colorectal liver metastases scheduled for radioembolization with 90Y-microspheres. *J Nucl Med* 54(4):516–522.

Van de Wiele, C. et al. (2012). SIRT of liver metastases: Physiological and pathophysiological considerations. *Eur J Nucl Med Mol Imaging* 39(10):1646–1655.

Van den Hoven, A. et al. (2016). Use of C-arm cone beam CT during hepatic radioembolization: Protocol optimization for extrahepatic shunting and parenchymal enhancement. *Cardiovasc Intervent Radiol* 39(1):64–73.

Van Dyk, J., Keane, T., Rider, W. (1982). Lung density as measured by computerized tomography: Implications for radiotherapy. *Int J Radiat Oncol Biol Phys* 8:1363–1372.

Wang, D. et al. (2013). Prophylactic topically applied ice to prevent cutaneous complications of nontarget chemoembolization and radioembolization. *J Vasc Interv Radiol* 24(4): 596–600.

Ward, T. et al. (2015). Management of high hepatopulmonary shunting in patients undergoing hepatic radioembolization. *J Vasc Interv Radiol* 26(12):1751–1760.

Willowson, K., Bailey, D., Baldock, C. (2009). Quantitative analysis of [Tc-99m]-MAA lung uptake in the clinical work-up for the treatment of liver tumours with [Y-90]-SirSpheres. *J Nucl Med* 50 (suppl 2):1438.

Willowson, K., Bailey, D., Baldock, C. (2011). Quantifying lung shunting during planning for radio-embolization. *Phys Med Biol* 56(13):N145.

Wondergem, M. et al. (2013). 99mTc-macroaggregated albumin poorly predicts the intrahepatic distribution of 90Y resin microspheres in hepatic radioembolization. *J Nucl Med* 54(8):1294–1301.

Wright, C. et al. (2012). Radiation pneumonitis following yttrium-90 radioembolization: Case report and literature review. *J Vasc Interv Radiol.* 23:669–674.

Yu, N. et al. (2013). Lung dose calculation with SPECT/CT for 90Yttrium radioembolization of liver cancer. *Int J Radiat Oncol Biol Phys* 85(3):834–859.

Zophel, K., Bacher-Stier, C., Pinkert, J., Kropp, J. (2009). Ventilation/perfusion lung scintigraphy: What is still needed? A review considering technetium-99m-labeled macroaggregates of albumin. *Ann Nucl Med* 23(1):1–16.

5

Treatment planning part III: Dosimetric considerations in radioembolization with glass and resin microspheres

ALEXANDER S. PASCIAK, AUSTIN C. BOURGEOIS, AND YONG C. BRADLEY

5.1 BACKGROUND

As mentioned in Chapter 1, the most widely used radioembolization products worldwide are SIRTeX SIR-Spheres (SIRTeX Technology Pty, Lane Cove, Australia) and BTG Theraspheres (BTG International Ltd., London, United Kingdom), often referred to as resin and glass microspheres, respectively, in reference to their material composition.

However, differences between the two products extend beyond microsphere composition and include the method in which yttrium-90 (^{90}Y) is bound to each microsphere, the presence of other radioactive contaminants, and, most importantly, the activity of ^{90}Y per sphere. Treatment planning for the two products differs in some ways due to the activity per microsphere, as discussed at a microdosimetric level in Chapter 9. However, the primary differences in treatment planning methodologies discussed later in this chapter may be related more to the use of legacy planning models, which have shown utility in tens of thousands of treatments.

Some of the differences between the two primary radioembolization products were introduced in Chapter 1 but have also been restated in Table 5.1.

It should be noted that for resin microspheres, the projected lower-limit activity per microsphere listed in Table 5.1 is based on the shelf-life of the product as determined by the manufacturer. At the end of the 15-day shelf-life of the glass product (Giammarile et al., 2011), the low-end activity per microsphere is approximately 50 Bq rather than the commonly referenced lower limit of 300 Bq. After 15 days, the total activity in the largest 20 GBq vial of glass microspheres would be far less than 1 GBq, which is insufficient for an effective treatment. The use of glass microspheres approaching their expiration date is consequentially impractical in many situations.

The focus of this chapter is on dosimetric treatment planning for liver cancer using both of these radioembolization products, highlighting their similarities and differences. This content will serve as a guide for nuclear medicine physicians, medical physicists, and radiation oncologists participating and assisting with radioembolization treatment planning.

5.2 ABSORBED DOSE, EFFECTIVE DOSE, AND DOSAGE

Ariel and Padula (1978a, 1978b) reported the first cases of the clinical use of ^{90}Y microspheres in the intra-arterial treatment of colorectal metastases to the liver. Ariel combined intra-arterial infusion of resin ^{90}Y microspheres with chemotherapy in the form of 5-fluorouracil. Ariel's patients received relatively large dosages of ^{90}Y, ranging from 100 to 150 mCi (3.7–5.5 GBq) that lead to a robust response in their cohort of 65 patients. This method involved ^{90}Y microsphere delivery using both percutaneous injection and intra-arterial approaches via a surgically placed hepatic arterial catheter. Since this initial experience, intra-arterial ^{90}Y microsphere therapy has significantly evolved in both treatment and manufacturing processes.

Despite the now widespread use of radioembolization in the treatment of a growing list of malignancies, there is often a great deal of confusion among individuals involved in these procedures on the subject of "dose." This is at least in part due to the variable training and backgrounds of the individuals involved in these treatments. The word "dose" is a term that is somewhat ambiguous and is commonly used to describe numerous differing

Table 5.1 SIRTex SIR-Spheres and BTG TheraSpheres: an overview of the properties of ^{90}Y radioembolization products

	SIRTex SIR-Spheres	BTG TheraSpheres
Composition	Resin with ^{90}Y bound to the surface	Glass permanently impregnated with ^{90}Y
Size (μm)	32.5 ± 5	20–30
Density (g/cc)	1.6	3.3
Number of spheres per vial	40–80 × 10^6	1.2–8 × 10^6
Activity per sphere at calibration	40–70 Bq	2500 Bq
Long-lived contaminants?	No	Yes
Time postinfusion when 90% of absorbed dose is delivered	~9 days	~9 days
Shelf-life from calibration	24 hours (Giammarile et al., 2011)	15 days (Giammarile et al., 2011)
Activity per sphere at treatment	50–38.6[a] Bq	2500–300 Bq (Kritzinger et al., 2013)
Activity per vial at calibration	3 GBq	3, 5, 7, 10, 15, 20 GBq

Source: Salem, R. and Thurston, K. G., J. Vasc. Interv. Radiol., 17, 2006.

[a] Based on decay at expiration/shelf-life of the product (Giammarile et al., 2011).

concepts in radioembolization, including absorbed dose, effective dose to the patient or staff, and the dosage delivered to the patient. To prevent confusion, these terms need clear definition in the context of radioembolization.

5.2.1 ABSORBED DOSE

The *absorbed dose* is a physical dose quantity (D) representing the mean energy imparted to matter per unit mass by ionizing radiation. The SI (International System of Units) unit for the absorbed dose is the gray (Gy). In radioembolization, as in radiation therapy, the absorbed dose to the tumor and to uninvolved liver tissues is of primary concern. The goal of radioembolization therapy is to obtain a sufficiently high absorbed dose to the tumor to produce a therapeutic effect while limiting the absorbed dose to normal liver tissue, preventing clinically significant toxicity. Of course, limiting the absorbed dose to extrahepatic tissues is also an important concern. This is particularly important in the lung and gastrointestinal (GI) tract, where cases of radiation pneumonitis and GI tract ulceration have been reported (Carretero et al., 2007; South et al., 2008a, 2008b; Naymagon et al., 2010). It is in the calculation of dose to tumor and nontarget tissue that the term *absorbed dose* should be used in the setting of radioembolization.

5.2.2 DOSAGE

In contrast to "dose" or "absorbed dose," the term *dosage* is used to quantify the size or frequency of a dose of a medicine or drug. Ariel's first radioembolization patients (Ariel and Padula, 1978a, 1978b) were treated with large dosages ranging from 100 to 150 mCi (3.7–5.5 GBq). These dosages were larger than those used in modern lobar therapies according to the treatment planning models suggested by both major microsphere manufacturers. Because radionuclides are quantified in terms of activity, rather than weight or international unit (IU), treatment dosages of ^{90}Y radioembolization are quantified in mCi or GBq. The dosage of ^{90}Y radioembolization has a weak correlation to the absorbed dose that will be received by the patient's tumor, normal hepatic parenchyma, or extrahepatic tissues. Factors such as liver and tumor volume, infusion technique, lung shunt fraction (SF), presence of patent gastroenteric collaterals,

catheter position, and tumor-to-normal uptake ratio (T:N) form the complex relationship relating dosage to absorbed dose in a particular tissue.

5.2.3 EFFECTIVE DOSE

Effective dose is a term that is sometimes used in the context of radioembolization treatment planning. Unfortunately, the term is somewhat confusing in the context of radioembolization as the term "effective" does not refer to effectiveness of treatment for the patient. However, this term is important for staff members since its primary use is for radiation protection and not for radiation therapy. The effective dose (E) considers the neoplastic radiation sensitivity of each organ irradiated and combines them into a single term that can be used to estimate cancer risk, that is, the risk of stochastic effects. For example, effective dose (E) is often used to quantify the increase in cancer risk later in life when a patient receives a diagnostic computed tomography (CT) scan. It is usually not appropriate to apply this term to a patient receiving radiation therapy. However, the effective dose can be used to quantify the radiation exposure received by staff members involved in ^{90}Y radioembolization procedures, such as nuclear medicine technologists preparing the dose and the interventional radiologists delivering it. This concept is discussed more in Chapter 7.

5.3 ONE MICROSPHERE OR MILLIONS OF MICROSPHERES?

By convention, absorbed dose is used in the setting of radioembolization to quantify the average effects of ionizing radiation within a particular volume of tissue. In contrast, *microdosimetry* refers to the analysis of variable absorbed doses to the tissue immediately adjacent to a single source on a microscopic level. Commonly accepted methods calculating average absorbed dose, some of which are discussed later in this chapter, assume uniform distribution of microspheres. In reality, the distribution of ^{90}Y microspheres is never uniform and the absorbed dose varies drastically on a microscopic scale. Microdosimetry sheds light on the complex dose–response relationship in radioembolization and helps to explain major dose–response differences between radioembolization and external beam radiation therapy (EBRT).

To illustrate microdosimetry in radioembolization, consider how radiation dose varies in tissue in the area surrounding a single ^{90}Y microsphere. Figure 5.1 is a three-dimensional representation of the dose profile surrounding a single ^{90}Y microsphere as it decays with analysis performed at the submillimeter scale. Note the extremely heterogeneous nature of the dose distribution in tissues surrounding this microsphere. Nearby tissues and cells will receive large doses of radiation (>100 Gy) while tissues just a few millimeters away will receive nonlethal doses. Figure 5.1 was generated using a Monte Carlo radiation transport code (MCNPX; McKinney et al., 2006) with a resolution of 50 μm. Similar data are available in the literature in the form of published dose-point kernels (DPK) for varied voxel sizes (Strigari et al., 2006; Lanconelli et al., 2012) based on calculations using established codes such as MCNPX (McKinney et al., 2006) and FLUKA (Ballarini et al., 2007).

It is important to keep in mind that radioembolization therapy involves deposition of millions of individual sources, each capable of widely variable dose deposition to the immediately adjacent tissue. The absorbed dose is therefore very heterogeneous when viewed on a microscopic scale. In other words, there is potential for variable local dose deposition at different points within the tumor. However, it is possible to quantify the microscopic dose profile from infusion of many microspheres if one assumes that ^{90}Y microspheres will deposit randomly with equal probability in all parts of the tumor. Figure 5.2 shows a 4 cm mass demonstrated on a magnetic resonance imaging (MRI) with a line subtending the long diameter of the mass. Assuming a randomized, uniform filling pattern of the tumor with ^{90}Y microspheres, the resulting absorbed dose profile along the line in Figure 5.2 is shown in Figure 5.3. Note that the dose profile varies substantially with the number of microspheres per unit volume in the tumor. Using microspheres with a higher specific activity, that is, 2500 Bq per microsphere versus 50 Bq per microsphere (see Table 5.1), will result in a substantially lower

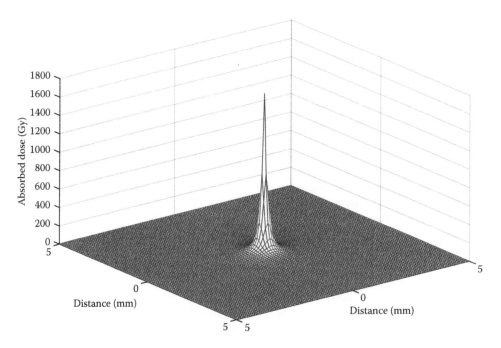

Figure 5.1 Three-dimensional representation of the dose profile surrounding a single ^{90}Y microsphere as it decays with analysis performed at the submillimeter scale. Note the extremely heterogeneous nature of the dose distribution in tissues surrounding this microsphere. Nearby tissues and cells will receive large doses of radiation (hundreds of Gy) while tissues just a few millimeters away will receive nonlethal doses.

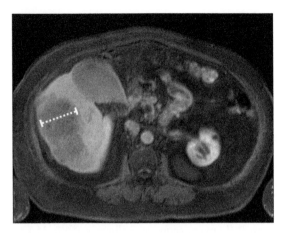

Figure 5.2 Hepatic protocol MRI demonstrating a 4 cm lesion with a line subtending the short diameter of the mass.

number of microspheres per unit volume in tissue and, therefore, larger variations in the local absorbed dose at a microscopic level. The microdosimetry of radioembolization is discussed in much greater detail in Chapter 9.

5.4 DISTRIBUTION IN TISSUE

The simulated absorbed dose profile in Figure 5.3 relies on the assumption that microspheres are randomly deposited in tumor with equal probability, that is, a microsphere has both the same chance of landing in the center of the tumor and in the periphery. This is rarely true (Fox et al., 1991; Campbell et al., 2000, 2001; Kennedy et al., 2004). In a pivotal paper

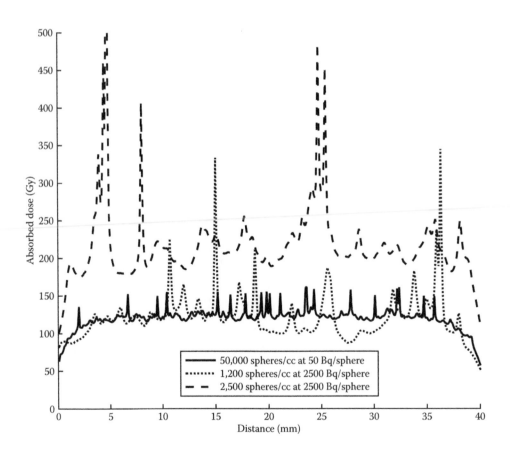

Figure 5.3 Dose profile showing microdosimetric absorbed dose along the profile in Figure 5.2 for three different average microsphere concentrations. Dose was determined assuming a randomized, uniform filling pattern of the tumor with ^{90}Y microspheres. Dosimetry was performed with 0.05 mm resolution.

by Campbell et al. (2000, 2001), light microscopy was used to evaluate the distribution of ^{90}Y microspheres directly from microscopic analysis of tissue samples taken from a patient postradioembolization. This patient received 3 GBq of 32-μm resin microspheres to treat an 8-cm metastatic liver tumor. Campbell's analysis noted that the microspheres were highly concentrated in the periphery of the tumor, creating very large absorbed doses ranging from 200 to 600 Gy. However, the average absorbed dose in the center of the tumor was only 6.8 Gy. Dose distribution was equally inhomogeneous in uninvolved hepatic tissue with an average dose of 8.9 Gy with approximately 1% of normal liver receiving more than 30 Gy.

The dose heterogeneity determined by microscopic analysis of Campbell's biopsy samples has been confirmed on a larger scale with a similar analysis of whole livers (Kennedy et al., 2004) after radioembolization. Large tumor dose heterogeneity ranging between 100 and 3000 Gy was found in patients with both primary and metastatic liver cancer treated with glass and resin microspheres, respectively (Kennedy et al., 2004). Owing to the different specific microsphere activity in the glass and resin products, livers treated with resin radioembolization had more microspheres per cluster within the vessels (Kennedy et al., 2004).

Evaluation of intratumoral dose heterogeneity has also been performed noninvasively, using various imaging modalities and methods. Three-dimensional dose distributions can be inferred from ^{99m}Tc macroaggregated albumin (^{99m}Tc-MAA) single-photon emission computed tomography (SPECT)/CT as also demonstrated by Kennedy et al. (2011). In addition, the dose distribution can be determined directly from posttreatment [90]Y

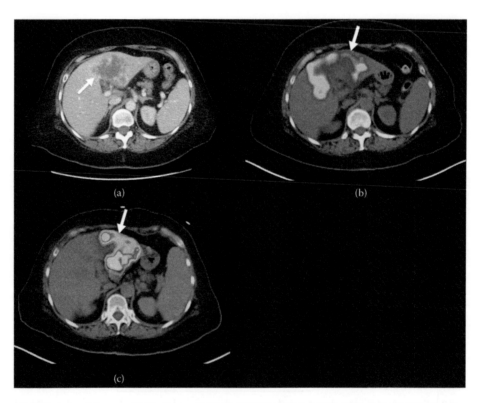

Figure 5.4 **(a)** Pretreatment contrast-enhanced CT of a patient with cholangiocarcinoma. Left lobe tumor demonstrates hypodense necrotic core with enhancing peripheral areas of active tumor **(b).** Correlating pretreatment [18]FDG-PET/CT shows hypermetabolic activity along the peripheral tumor margin. The necrotic, low attenuating tumor component manifests relatively little FDG avidity. **(c)** [90]Y PET/CT imaging following embolization of this patient's left lobe with 821 MBq (22.2 mCi) of resin microspheres. The low absorbed dose in the center of the necrotic region is plainly visible on posttreatment ^{90}Y PET/CT.

bremsstrahlung SPECT or ^{90}Y positron emission tomography/computed tomography (PET/CT), as described in detail in Chapters 10 and 11. Figure 5.4a illustrates the pretreatment contrast-enhanced CT in a patient with cholangiocarcinoma. Figure 5.4b demonstrates the correlating 2-deoxy-2-(^{18}F) fluoro-D-glucose (FDG)-PET/CT of the same patient, with hypermetabolic activity along the peripheral tumor margin. The necrotic, low attenuating tumor component manifests relatively little FDG avidity. Figure 5.4c shows ^{90}Y PET/CT imaging following embolization of this patient's left lobe with 821 MBq (22.2 mCi) of resin microspheres. The low-absorbed dose in the center of the necrotic region is plainly visible on posttreatment ^{90}Y PET/CT. Quantification of the PET data in Figure 5.4c using methods described in Chapters 11 and 12 yields a maximum tumor dose of 440 Gy with a minimum dose of just 4.6 Gy at the necrotic center of the tumor.

5.5 CALCULATING THE ABSORBED DOSE

Yttrium-90 is one of a handful of radionuclides that is considered to be a pure β-emitter, emitting no gamma rays at any appreciable yield following decay. Although several rare decay pathways of ^{90}Y are discussed in the following chapters, ^{90}Y is considered a pure β-emitter for all practical purposes related to dosimetry. 90Y releases a higher energy electron than other pure β-emitters that have been used in internal radionuclide therapies, with a maximum and average energy of 2.28 and 0.935 MeV, respectively. The range of the maximum energy ^{90}Y β-particle is 11 mm in tissue, while its average energy β particle has a range slightly less than 4 mm. The penetration depth of the high-energy ^{90}Y β-particle is a key component of this radionuclide's success in radioembolization, allowing for high dose deposition into the tissues between embolized capillaries. However, when the absorbed dose of ^{90}Y therapy is considered for a tumor or nontarget site, an important simplifying assumption can be made: β-radiation released from microspheres within a given organ will be fully absorbed by that organ. This assumption is easy to justify based on the average 4-mm ^{90}Y β range in tissue. However, this does not account for secondary radiation, which may extend beyond the 11 mm range of a ^{90}Y β-particle. As discussed in Chapter

1, high-energy ^{90}Y β-radiation will create secondary X-rays in the form of bremsstrahlung and characteristic emissions. These photons will penetrate through the liver and contribute to dose in extrahepatic tissues and even to individuals in close proximity to the patient. However, as discussed by Stabin et al. (1994), this will have little effect on the absorbed dose in the organ treated with radioembolization. A second assumption that becomes important is the permanence of ^{90}Y radioembolization—neither glass nor resin microspheres are biodegradable, and once infused, both products form a permanent implant. Combining these two assumptions allows for easy calculation of *average absorbed dose* to an organ of interest on a macroscopic scale.

This calculation is derived in Equations 5.1 through 5.3:

$$E_{avg} = \int_0^\infty E\varphi(E)dE = 0.935 \, \text{MeV}$$
$$= 1.498 \cdot 10^{-13} \, \text{J} \tag{5.1}$$

$$E_{tot}(J) = A_0 E_{avg} \int_0^\infty e^{-\lambda t} \, dt = \frac{A_0}{\lambda} \cdot \left(1.498 \cdot 10^{-13}\right)$$
$$= A_0 \, (\text{GBq}) \cdot 49.86 \, (\text{J} \cdot \text{s}) \tag{5.2}$$

$$D_{avg}(\text{Gy}) = \frac{A_0 \, (\text{GBq}) \cdot 49.98 \, (\text{J} \cdot \text{s})}{M_{liver} \, (\text{kg})} \tag{5.3}$$

where E_{avg} is the average energy released per decay of ^{90}Y based on the probability density function (E) for emission (Eckerman et al., 1994), and λ is the Y^{90} decay constant based on a half-life of 64.24 hours. A_0 is the ^{90}Y activity present in the organ or organ segment of interest in GBq, and E_{tot} is the total energy released by A_0 from the time that it is infused until it has fully decayed. The absorbed dose (D) is expressed in Gy and can be obtained by dividing E_{tot} by the mass of the treated liver, M_{liver}. Alternatively, the volume of the tissue can be obtained from tomographic imaging and multiplied by established tissue densities (International Commission on Radiation Units and Measurements [ICRU], 1992). It must be reiterated that the absorbed dose calculation in Equation 5.3 is only valid for ^{90}Y radioembolization and only representative of *average absorbed dose* in an organ. Equation 5.3 cannot be used for absorbed dose calculation for radioembolization using radio-isotopes other than ^{90}Y such as ^{166}Ho, which decays with the emission of a prompt gamma ray and a β-particle, thus violating the assumption that all energy emitted during decay is absorbed by the organ of interest.

5.6 TUMOR AND NORMAL LIVER ENDPOINTS

Careful consideration of tumor and nontarget tissue endpoints is paramount to hepatic treatment planning for radioembolization. In addition to nontarget hepatic dose deposition, radiation exposure to extra-hepatic tissues, including the lungs and tissues in the GI tract, are of concern. Consideration of lung dose, from arteriovenous shunting as well as managing patients with extrahepatic nontarget embolization, is covered in detail elsewhere in this book. As such, this section will emphasize dosimetry endpoints in tumor and uninvolved liver tissue.

5.6.1 NORMAL LIVER ENDPOINTS

Radiation-induced liver disease (RILD) is a term that is commonly referenced in the literature as ^{90}Y radioembolization has gained traction as a standard-of-care treatment for primary and metastatic liver cancer. However, the meaning and context of RILD are often ambiguous and inconsistent. We define RILD in patients following radioembolization as clinically significant manifestations of hepatic insufficiency such as weight gain, ascites, anicteric hepatomegaly, and/or pain in the right upper quadrant within 90 days of treatment (Guha and Kavanagh, 2011). In patients without underlying cirrhosis, these symptoms may be accompanied by serologic evidence of hepatic toxicity, manifested by rising alkaline phosphatase and glutamyl transpeptidase (Sangro et al., 2008). It is important to note that RILD represents a spectrum of clinical and serologic manifestation of acute hepatic injury. RILD is often self-limited, although in its most severe form can result in hepatic vascular endothelial damage leading to a fatal condition resembling veno-occlusive disease (VOD). On the other end of the spectrum, it should be emphasized that Grades 1 and 2 mild liver toxicity is very common following radioembolization (Goin et al., 2005; Gulec et al., 2007; Kennedy et al., 2009) and does not fit into the common definition of RILD.

Normal liver tissue has a relatively low tolerance to radiation (Dawson and Guha, 2008). Data from EBRT suggest that the threshold for RILD following whole-liver irradiation is between 30 and 40 Gy (Emami et al., 1991; Lau et al., 1994; Cremonesi et al., 2008). Specific cases have demonstrated that liver failure can occur from VOD in the setting of EBRT at doses as low as 35 Gy (Sempoux et al., 1997). It is important to consider that the toxicity thresholds from EBRT can only be applied to radioembolization in a very limited fashion. In fact, the liver can tolerate higher absorbed doses from radioembolization than it can from fractionated EBRT. This phenomenon may be explained by differences between the dose rate of fractionated EBRT and radioembolization, extent of nontumor involved liver, as well as the degree of hypoxia created by capillary occlusion in radioembolization. In addition, the heterogeneity of absorbed dose at a microscopic scale also contributes to the decreased hepatic toxicity per gray of radioembolization compared with EBRT due to microscopic sparing of normal tissue. The importance of microscopic dosimetry is discussed in detail in Chapter 9.

The dose threshold for hepatic toxicity is variable among patients and reflects hepatic functional reserve, liver volume, concomitant diseases and medications, tumor volume, and other patient-specific factors. The most widely supported upper absorbed dose toxicity limit to normal liver from radioembolization is 70–80 Gy (Fox et al., 1991; Lau et al., 1994; Campbell et al., 2001; Salem and Thurston, 2006). Because underlying cirrhosis decreases the tolerance of the liver to radiation (Dawson and Guha, 2008), 70 Gy should be considered the maximum in cirrhotic patients. However, there have been published cases where the absorbed dose to normal liver has safely exceeded these thresholds, approaching 100 Gy (Gulec et al., 2007). Further, treatment planning models for the glass radioembolization product support average doses of up to 150 Gy in treated tissue (Lewandowski et al., 2005; Salem and Thurston, 2006).

Cirrhosis and other underlying chronic liver disease such as viral hepatitis are not the only conditions that could decrease the tolerance of the liver to radiation. It has been shown that previous chemotherapy can increase the occurrence of VOD following radioembolization (Sangro et al., 2008) by contributing to intrahepatic vascular endothelial damage. Previous research has demonstrated that patients with metastatic liver disease receive a higher average dose to nontarget liver even when the same dosage of radioembolization is delivered (Sangro et al., 2008). Therefore, one must carefully consider prior chemotherapy in the treatment planning process of patients undergoing radioembolization for metastatic disease, just as the severity of cirrhosis affects treatment planning in hepatocellular carcinoma (HCC) patients.

5.6.2 TUMOR ENDPOINTS

Tumoricidal endpoints in theory depend on many radiobiologic factors, including tumor type, absorbed dose, heterogeneity of absorbed dose, and prior radiological or chemotherapeutic treatments. Because the number of infused microspheres and the activity per microsphere also vary widely between glass and resin microspheres, dose–response data of one product cannot necessarily be applied to the other.

The largest cohort of published data describing tumoricidal endpoints is for hepatocellular carcinoma. When treating HCC, 120 Gy should be considered to be a reasonable minimum tumor target dose (Yoo et al., 1989; Lau et al., 1994; Ho et al., 1997; Kennedy et al., 2004; Strigari et al., 2010). Efforts to achieve this absorbed dose in the tumor can be made using the methods described in the following sections, provided the maximum tolerable dose to normal liver and extrahepatic tissues is not exceeded.

Because of the histologic and biologic variability of metastatic liver tumors, dose–response data are less certain. However, worldwide clinical

trials are currently underway to use advanced quantitative imaging to identify dose–response thresholds for metastatic disease. While published data in the literature is widely varying, it is likely that neuroendocrine metastases to the liver or neuroendocrine tumor (NET) are likely to show a strong response to radioembolization at a lower absorbed dose than HCC. On the other hand, metastatic colorectal cancer (mCRC) may require an absorbed dose equal to or higher than HCC (Gulec et al., 2007) to achieve the desired therapeutic response.

5.7 TUMOR-TO-NORMAL UPTAKE RATIO

The tumor-to-normal uptake ratio (T:N) is an important quantity in radioembolization with glass and resin ^{90}Y microspheres as well as ^{166}Ho radioembolization. This quantity has a substantial role in treatment planning and can impact the tumor-absorbed dose, liver dose, and the toxicity or efficacy of the treatment. T:N is defined in Equation 5.4:

$$T{:}N = \frac{A_{90Y,\,Tumor}\,/\,V_{tumor}}{A_{90Y,\,Normal}\,/\,V_{normal}} \qquad (5.4)$$

where $A_{90Y,Tumor}$ is the ^{90}Y activity (MBq) deposited in the tumor and $A_{90Y,Normal}$ is the activity deposited in uninvolved liver tissue. V_{normal} and V_{tumor} are the respective volumes of each. Table 5.2 lists a range of measured T:N for different tumor types from several publications in the literature.

While Table 5.2 shows just a few examples from the literature, one thing is immediately clear: T:N

Table 5.2 T:N from published sources

Disease	T:N median/ mean	T:N range	Number of patients or tumors	Measurement method	Reference
HCC	11.5:1	7:1–16:1	2	Pathologic	Kennedy et al. (2004)
HCC	7.0:1	3.9:1–9.2:1	5	^{99m}Tc-MAA	Gulec et al. (2007)
HCC	3.5:1	1:10–13.5:1	27	^{99m}Tc-MAA	Lau et al. (1994)
mCRC	6.8:1	2.9:1–15.4:1	15	^{99m}Tc-MAA	Gulec et al. (2007)
mCRC	—	2:5–2:1	2	Pathologic	Kennedy et al. (2004)
NET	5.9:1	3.5:1–11.1:1	20	^{99m}Tc-MAA	Gulec et al. (2007)

varies significantly even in cases of the same tumor type. Tumor size, percentage infiltration, presence of centralized necrosis, and the technique used to measure T:N all contribute to these variations.

5.7.1 USING MAA AS AN ESTIMATION TOOL

Before reviewing how T:N can be used in radioembolization treatment planning, it is important to understand methods that can be used to quantify it. The most widely used established method for pretreatment *estimation* of T:N is MAA SPECT/CT. This technique is convenient since it is simply an extrapolation of the data acquired in the pretreatment lung-shunt study. This method assumes the validity of ^{99m}Tc-MAA as a valid surrogate for ^{90}Y microspheres. There are many reasons why this may be untrue, several of which are outlined in Table 5.3.

Many authors have evaluated the validity of MAA as a radioembolization surrogate, with no clear consensus (Knesaurek et al., 2010; Kao et al., 2012; Lam and Smits, 2013; Lam et al., 2013; Wondergem et al., 2013; Garin et al., 2014; Lam and Sze, 2014). The position of the catheter tip during both infusion of MAA and radioembolization is among the most critical factors to the prognostic utility of T:N measurements made from ^{99m}Tc-MAA SPECT/CT. Positioning of the catheter tip becomes especially critical when it is near a bifurcation or when it is positioned in a tortuous vessel (Jiang et al., 2012; Wondergem et al., 2013). However, in spite of variable correlation between MAA and ^{90}Y microspheres in the literature, the

Table 5.3 Potential sources of error when using MAA as a microsphere surrogate

Issue	Comments	Impact	Amelioration
Specific gravity differences between macroaggregated albumin (MAA) and microspheres	Significant difference in the case of glass microspheres	Moderate	N/A
Particle size and shape	MAA particles generally have a wider size range compared to either resin or glass microspheres	Moderate	N/A
Free ^{99m}Tc	May not significantly affect T:N measurements but could obscure detection of gastric nontarget embolization (NTE)	Moderate	Prophylactic oral administration of 500 mg perchlorate (Ahmadzadehfar et al., 2010); use MAA right after preparation
Embolic differences	Stasis may be reached in some tumors treated with resin microspheres but not MAA	High	Understand the conditions under which MAA is likely to act as a good surrogate based on tumor type and size
Catheter position and centering	Catheter positioning differences between MAA and ^{90}Y infusions can drastically effect deposition	High	Confirm catheter positioning

majority of authors agree that MAA is an excellent option for treatment planning and predictive dosimetry. Substantial additional discussion related to the use of MAA as a radioembolization surrogate is presented in Chapter 4.

5.7.2 OTHER TOOLS?

Cone-beam CT (CBCT) is a promising method of defining three-dimensional vascular anatomy and elucidating sources of nontarget embolization (NTE). Louie et al. (2009) evaluated the utility of CBCT as a tool to augment ^{99}Tc-MAA for the identification of both extrahepatic NTE and poor T:N. In this study, CBCT was used to identify areas of extrahepatic localization in 22 out of 42 patients using CBCT. Perhaps most significantly, CBCT identified incomplete tumor visualization in 8 out of 42 patients, indicating a secondary vascular supply to the tumor (Louie et al., 2009). Cases such as these are often a result of so-called "parasitization" of vascularity, in which the angiogenic factors produced by hypervascular tumors recruit flow from otherwise insignificant vessels. This phenomenon is common following one or more chemoembolization treatments, which preferentially depend on small vessel embolization to achieve treatment efficacy. If identified during pretreatment angiography, collateral vessels may be prophylactically embolized, thus redirecting more vascular flow to the treated vessel and increasing the likelihood of technical success. In extreme cases, extrahepatic collateral vessels may provide the majority of a tumor's blood supply, and standard delivery of ^{90}Y microspheres via the hepatic arteries can result in a T:N less than 1. In situations such as these, CBCT can be a powerful tool in treatment planning and delivery.

CBCT has also been investigated as a potential tool in quantifying T:N. A study by Jones and Mahvash (2012) used gelatin phantoms injected with different concentrations of iodinated contrast to show that CBCT is able to quantify *relative* contrast enhancement. Since T:N is a ratio, absolute quantification is not necessary for its calculation. In other words, relative densities in tumor and normal liver in pre- and postcontrast CBCT can be effectively used to estimate T:N. In this manner, CBCT can be used as an excellent way to augment ^{99m}Tc-MAA scintigraphy in the estimation of T:N for use in partition model treatment planning. One must be careful to differentiate this method from those that allow absolute quantification, such as ^{90}Y PET/CT. In ^{90}Y PET/CT, following image acquisition, a region-of-interest drawn in the tumor will report a value that is directly representative of the activity of ^{90}Y in that region (Pasciak et al., 2014). These distinctions are explained in detail in Chapter 11.

5.8 TREATMENT PLANNING STRATEGIES

5.8.1 THE EMPIRIC MODEL FOR RESIN MICROSPHERE TREATMENT PLANNING

The simplest model for radioembolization treatment planning with resin microspheres is the empiric model. The empiric model can be used to determine treatment dosages for whole-liver radioembolization based purely on the percent hepatic tumor involvement. This model has fallen out of favor and its clinical use has been largely replaced by preferred methods such as the body surface area (BSA) methods and partition models, discussed in the following sections. However, it should be noted that the empiric model is still included on the SIRTeX SIR-Sphere package insert (SIRTeX Technology Pty, Lane Cove, NSW, Australia). In addition, the empiric model was the treatment strategy used during the initial SIRTeX SIR-Sphere clinical trials.

In the largest single cohort of retrospective data of patients who underwent radioembolization using resin microspheres, 28 expired with a cause of death attributable to complications related to liver toxicity. Of these deaths, 21 were from a single center that used the empiric model exclusively for treatment planning (Kennedy et al., 2009). This does not suggest that the empiric model is unsafe; however; it is imperative that the model be used correctly.

When using the empiric model, it is critical to understand that its definition in Table 5.4 is for whole-liver radioembolization. For lobar or segmental therapy, one must modify the recommended dosage (A_{emp}) in Table 5.4 according to Equation 5.5:

$$A(\text{GBq}) = A_{emp}(\text{GBq}) \cdot \left[\frac{V_{treated}}{V_{wholeliver}}\right] \quad (5.5)$$

Table 5.4 Description of the empiric model used for ^{90}Y radioembolization with resin microspheres

The percentage of tumor involvement in the liver	Recommended dosage, A_{emp} (GBq)
More than 50%	3.0 GBq
25%–50%	2.5 GBq
Less than 25%	2.0 GBq

where A_{emp} is the recommended dosage from the empiric model as defined in Table 5.4. $V_{treated}$ and $V_{whole\ liver}$ are the volumes of the portion of the liver to be treated and entire liver volume, respectively.

5.8.2 THE BSA MODEL FOR RESIN MICROSPHERE TREATMENT PLANNING

The most widely used treatment planning technique for radioembolization with resin microspheres is the BSA method. This method may have some familiarity to many clinicians since BSA-based calculations are widely used in medicine for determining dosages for medications such as chemotherapy. Treatment dosage using the BSA model is strongly dependent on the patient's height and weight and moderately dependent on the percent tumor infiltration. Equations 5.6 and 5.7 describe the BSA treatment planning method:

$$\text{BSA}\,(\text{m}^2) = 0.2025 \cdot \text{height}^{0.725}\,(\text{m})$$
$$\cdot \text{weight}^{0.425}(\text{kg}) \quad (5.6)$$

$$A_{BSA}\,(\text{GBq}) = \text{BSA} - 0.2 + \left[\frac{V_{tumor}}{V_{tumor} + V_{normal}}\right] \quad (5.7)$$

V_{tumor} and V_{normal} are the respective volumes of tumor and uninvolved liver tissue in the portion of the liver to be treated. The BSA model assumes a relationship between the physical size of the patient and ability to tolerate increasing dosage. The concept that larger patients (not necessarily with larger livers) are more tolerant to increased dosages of ^{90}Y has been shown in the literature (Sangro et al., 2008). The BSA model was also found to have a lower risk of liver toxicity than the empiric model in the aforementioned cohort of 680

patients treated with resin microspheres (Kennedy et al., 2009).

When using the BSA model, it is critical to understand that its definition in Equations 5.6 and 5.7 is for whole-liver radioembolization. For lobar or segmental therapy, one must modify the recommended dosage, A_{BSA}, in Equation 5.7 according to Equation 5.8:

$$A\,(\text{GBq}) = A_{BSA}\,(\text{GBq}) \cdot \left[\frac{V_{treated}}{V_{whole\ liver}}\right] \quad (5.8)$$

where A_{BSA} is the recommended dosage from the BSA model as defined in Equations 5.6 and 5.7. $V_{treated}$ and $V_{whole\ liver}$ are the volumes of the portion of the liver to be treated and entire liver volume, respectively.

5.8.3 GLASS MICROSPHERE TREATMENT PLANNING

Standard treatment planning for radioembolization using glass microspheres is a relatively straightforward process, giving treating physicians significant latitude to consider patient-specific factors. The foundational principle is based on Equation 5.3, which describes the average dose in a tissue volume as a function of ^{90}Y dosage. Equation 5.3 can be rewritten to solve for the treatment dosage, A_o, as shown in Equation 5.9:

$$A_o\,(\text{GBq}) = \frac{D_{avg}\,(\text{Gy}) \cdot M_{liver}\,(\text{kg})}{49.98(\text{J}\cdot\text{s})} \quad (5.9)$$

where D_{avg} is the target-absorbed dose in the *portion of the liver* to be treated and M_{liver} is the mass of treated liver tissue. Liver mass is extrapolated from volumetric analysis performed on pretreatment CT data for planning lobar therapy. In the setting of radiation segmentectomy, segmental volume may be calculated by preprocedural cross-sectional imaging or from CBCT during the pretreatment mapping procedure, discussed in additional detail in Chapter 6. Once hepatic volume is obtained, a conversion factor of 1.05 kg/L is used to convert hepatic tissue volume to mass (ICRU, 1992).

The average dose endpoint (D_{avg}) used in treatment planning is recommended to be in the range between 80 and 150 Gy as specified in the

Therasphere package insert (BTG International Ltd., London, UK). The endpoint used should be selected based on tumor burden, health of uninvolved liver tissue, previous therapy, and other clinical factors that may affect toxicity and/or response.

Finally, it should be mentioned that Equation 5.9 does not account for lung shunting or residual activity remaining in the delivery system, both of which will decrease the dosage of radioactivity deposited in the liver, A_o, and therefore D_{avg}.

5.8.4 THE PARTITION MODEL

The partition model for radioembolization (Ho et al., 1996, 1997), also referred to as the medical internal radiation dose (MIRD) model (Gulec et al., 2006), is a three-compartment model that takes into account patient-specific data to tailor the desired dosimetric endpoints in the tumor, normal liver, and lungs. Unlike the previously discussed models that are specific to resin or glass microspheres, the partition model can potentially be used for either product. However, one must always keep in mind that tumor and normal liver dose–response relationships are likely to vary between resin and glass products due to variable-specific activity, sphere number/density, and sphere composition.

The lung SF is an important input into the partition model that is used along with T:N, tumor volume (V_{tumor}), and nontarget liver volume (V_{normal}) to determine the fractional uptake (FU) into a compartment. Equations 5.10 and 5.11 are the equations for the calculation of FU in the uninvolved liver and tumor:

$$U_{normal} = (1-SF)\left[\frac{V_{normal}}{T:N \cdot V_{tumor} + V_{normal}}\right] \quad (5.10)$$

$$U_{tumor} = (1-SF)\left[\frac{T:N \cdot V_{tumor}}{T:N \cdot V_{tumor} + V_{normal}}\right] \quad (5.11)$$

The FU can be used to determine the average absorbed dose in both tumor and normal liver compartments according to Equations 5.12 and 5.13:

$$D_{normal}(Gy) = \frac{A_0(GBq) \cdot 49.98 \cdot FU_{normal}}{M_{normal}(kg)} \quad (5.12)$$

$$D_{tumor}(Gy) = \frac{A_0(GBq) \cdot 49.98 \cdot FU_{tumor}}{M_{tumor}(kg)} \quad (5.13)$$

where A_o is the activity of ^{90}Y to be infused into the liver. Thus, Equation 5.13 can be rearranged to derive the prescribed ^{90}Y dosage in Equation 5.14:

$$A_0(GBq) = \frac{D_{tumor}(Gy) \cdot M_{tumor}(kg)}{49.98 \cdot FU_{tumor}} \quad (5.14)$$

The partition model equation uses patient-specific tumor and liver volumes, along with predetermined T:N from pretreatment ^{99m}Tc-MAA SPECT/CT and/or CBCT. Thus, this method represents the most tailored treatment planning algorithm, allowing for accurate estimation of absorbed dose to tumor, nontarget liver tissue, and lungs. Prior research has shown that treatment planning with the partition model based on ^{99m}Tc-MAA SPECT/CT can improve clinical outcomes (Gulec et al., 2006).

As discussed in a previous section, when treating an HCC patient with radioembolization, one may wish to set D_{tumor} to a minimum of 120 Gy. Solving for A_o, however, is not the only necessary step in the treatment plan. A_0 must be back-substituted into Equation 5.11 to ensure that the absorbed dose to normal liver tissue is below suggested limits. In addition, the dosimetric techniques described in chapter 4 should also be applied to ensure an acceptable lung dose.

It is critical to note that the partition model can be effectively used for any radioembolization product including resin or glass ^{90}Y microspheres and ^{166}Ho radioembolization. However, tumor and uninvolved liver efficacy and safety endpoints are likely to differ between radioembolization products.

5.8.5 COMPARING TREATMENT PLANNING MODELS

In the previous sections, four treatment planning models have been introduced for radioembolization using resin microspheres, glass microspheres, or both. However, it is difficult to understand simply by examining the equations how the treatment activity prescribed using these models differs in patients of varying size, tumor burden, and T:N. Figure 5.5a compares the recommended dosage of resin microspheres based on the empiric, BSA, and partition

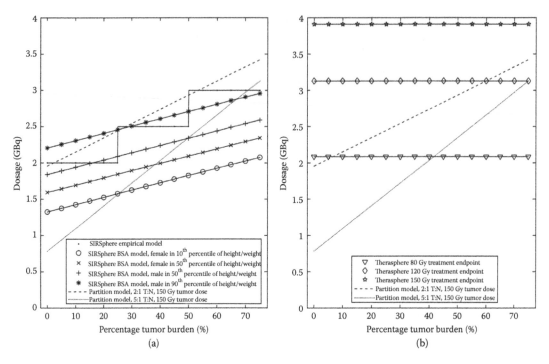

Figure 5.5 Prescribed dosage recommendation for **(a)** resin microspheres and **(b)** glass microspheres as a function of patient size, tumor burden, and T:N. Data are presented for a 1300 mL liver. Patient size data are from published U.S. census demographic information 2007–2008.

models as a function of patient size, tumor burden, and T:N. One can immediately see that the empiric model yields a dosage that is higher than most of the other models, and it does this regardless of circumstance (i.e., varying patient size or liver size). On the other hand, patient size (height + weight) has a large impact on the BSA treatment planning model. A female in the 10th percentile of height and weight will be treated with about half the dosage as a male in the 90th percentile of height and weight. The partition model with a T:N of 5:1 yields the lowest dosage at a low percentage tumor infiltration but rises steeply as the tumor burden increases. Use of the partition model with a low T:N (2:1 or less) can result in a large dosage and the potential for high absorbed doses to uninvolved liver. Figure 5.5b describes the dosage recommended for glass treatment planning at 80, 120, and 150 Gy absorbed-dose endpoints. Comparing Figure 5.5a with 5.5b, it is quickly apparent that larger dosages are used for radioembolization with glass microspheres than resin microspheres. However, this does not suggest that one product carries less toxicity or more efficacy

than the other since toxicity and efficacy endpoints differ between the two products. Chapter 9 will explain this phenomenon at a microscopic level.

5.9 CONTOURING LIVER AND TUMOR VOLUMES

Each of the treatment planning methodologies presented in the previous section require measurement of tissue volume through some form of volumetry based on CT, MR, or hybrid (e.g., PET/CT) tomographic imaging. The empiric and BSA models are reliant on measurement of the percent tumor infiltration and also the percentage of treated liver tissue relative to total liver volume. Treatment planning for glass microspheres depends strongly on the total volume of liver tissue being treated, while the partition model relies on both volume of tumor and uninvolved liver tissue. These volumes can be determined using automatic, semiautomatic, or manual techniques as described in Sections 5.9.1 and 5.9.2.

5.9.1 AUTOMATIC AND SEMIAUTOMATIC SEGMENTATION

Automatic and semiautomatic organ and tumor segmentation have been widely used in areas of health physics, radiation therapy, radiology, surgery, and other facets of medicine for years with different levels of accuracy and required human input (van Ginneken and Haar Romeny, 2000; Ghanei et al., 2001; Marroquín et al., 2002; Caon, 2004; François et al., 2004; Buie et al., 2007; Fripp et al., 2007; Klein et al., 2008; Metzger et al., 2013; Kockelkorn et al., 2014). Techniques such as thresholding and region growing (El-Baz et al., 2011) can be quite effective in high-contrast anatomical areas that are clearly demarcated. These techniques have even been used successfully in the semiautomatic segmentation of the liver based on CT imaging (Foruzan et al., 2009).

In the context of radioembolization treatment planning, available segmentation tools will dictate the degree of automation that can be achieved in the clinical workflow. However, some automation can be achieved in almost every clinical environment without specialized tools. For example, consider patients with metastatic liver disease who have received a pretreatment [18]FDG-PET/CT exam. In many of these patients, manually contouring multiple liver lesions to determine the percent tumor involvement can be a time-consuming task. However, determining the tumor burden is a necessary component of treatment planning for radioembolization using resin microspheres. Automated 3D thresholding tools available on basic radiology reading stations can be used in conjunction with [18]FDG-PET/CT to delineate hypermetabolic tumor boundaries. The threshold can be adjusted on a per-patient basis or set quantitatively as a function of maximum standardized uptake value (SUV_{max}). This technique is useful for many patients with FDG-avid metastatic liver tumors treated with radioembolization. An example of thresholding at 40% of whole-liver SUV_{max} is shown in Figure 5.8b.

If automated segmentation tools are available, atlas-based segmentation (Ghanei et al., 2001; Bondiau et al., 2005; Zhang et al., 2006; Reed et al., 2009; Linguraru et al., 2010; El-Baz et al., 2011; Park et al., 2014) is a method that has the potential for more accurate organ segmentation and may require less user input compared to alternatives such as thresholding and region growing. The degree of automation of atlas-based segmentation is greater, in part, because it is reliant on a standard atlas of reference segmented organ contours. These contours are predefined and validated by radiologists or anatomists. The atlas is then matched to the dataset of interest using deformable registration. To improve the accuracy of the process and account for variant anatomy, multiatlas-based segmentation is often employed. In this technique, many deformable registrations are performed with atlases that feature varying anatomy and those producing the best match are combined or fused together to produce a final optimal segmentation. Several fusion methods are employed with the simplest a majority vote fusion to more complex techniques that incorporate statistical models into the fusion process (Warfield et al., 2004).

In the context of radioembolization treatment planning, automated delineation of lobar or segmental liver volumes can significantly speed the planning process. While atlas-based segmentation usually requires some postsegmentation manual correction (Figure 5.6), this task can be accomplished much faster than manual segmentation alone. Small inaccuracies in the measured liver lobe volume, segment volume, or tumor volume should be considered in the context of the sensitivity of the treatment planning model to volume changes discussed in Section 5.8.5 (Figure 5.5).

5.9.2 MANUAL SEGMENTATION

At institutions where radioembolization treatment planning is not assisted by a radiation oncology department, sophisticated thresholding and segmentation tools may be unavailable. In these cases, manual segmentation may be the only option. The authors of this chapter have found that DICOM viewers that feature spline-based regions of interest (ROIs) combined with interpolative volumetry can significantly reduce the time required for manual liver and tumor delineation since ROIs need not be drawn on every slice. The three-dimensional liver volume rendered in Figure 5.7 was generated with contours on only six axial CT slices using the free DICOM viewer OsiriX 5.8.2 (Osirix Foundation, Geneva, Switzerland).

Once again, errors in either tumor or lobe delineation resulting from manual contouring with or without interslice interpolation should be weighed against the sensitivity of the treatment planning model as discussed in a Section 5.8.5 (Figure 5.5).

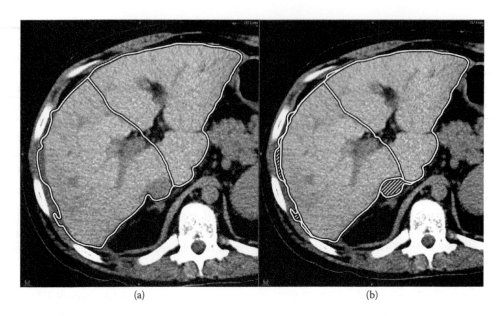

(a)   (b)

Figure 5.6 **(a)** Results of fully automated atlas-based segmentation of the left and right liver lobes using the MIM 6 (MIM Software Inc., Cleveland, OH) software package. **(b)** Atlas-based segmentation followed by manual correction (shaded areas).

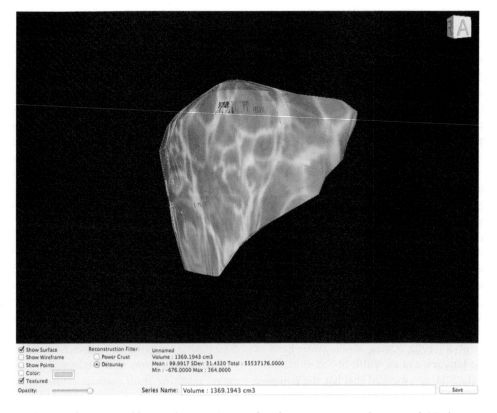

Figure 5.7 Three-dimensional liver volume generated with contours on only six axial CT slices using the free DICOM viewer OsiriX 5.8.2 (Osirix Foundation, Geneva, Switzerland).

5.10 TREATMENT PLANNING CALCULATIONS

This section functions as a guide that will provide instructive examples of radioembolization treatment planning in both lobar and segmental therapies. Different treatment planning models are considered for both resin and glass microspheres.

5.10.1 LOBAR THERAPY

As a surgeon aims to resect a margin of normal parenchyma surrounding a liver tumor, radioembolization is often purposefully nonselective, so as to treat microscopic metastatic disease. Patients with liver metastases are usually noncirrhotic and have a greater functional hepatic reserve, and are therefore commonly treated with microsphere infusion to an entire hepatic lobe. Lobar therapy is also desirable in the setting of liver metastases due to the propensity for multifocal disease, making superselective infusion impractical. It should be noted that this methodology applies to HCC patients with multifocal disease in the absence of clinical and serologic indicators of severe underlying liver disease. However, in the setting of underlying liver impairment, minimizing nontarget embolization to nontumor hepatic tissue is preferred, so as to preserve hepatic function (Salem and Thurston, 2006).

5.10.1.1 Example case

A 61-year-old female presents with lung cancer and four metastatic lesions to the left hepatic lobe. There is no serologic or radiologic evidence of cirrhosis or biliary obstruction. Figure 5.8a shows a hepatic protocol CT of one centrally necrotic metastatic lesion in the left hepatic lobe, with corresponding hypermetabolic activity by [18]FDG-PET/CT (Figure 5.8b). In all treatment planning scenarios (1–4 below) a left lobar therapy is assumed using conventional anatomic boundaries. Each lobe was contoured in three dimensions to determine relative volumes of the left lobe, right lobe, and total liver (Figure 5.8c).

The volume of the left lobe tumors can be contoured based on the contrast-enhanced CT or [18]FDG-PET/CT (Figure 5.8a and b). In this particular case, the tumor is not avidly enhancing

on the portal venous phase CT and manual segmentation of the CT image set would be limited to corresponding hypodensity visualized in Figure 5.8a. The [18]FDG-PET/CT identifies the tumor as centrally necrotic and highly FDG-avid along its periphery. Due to these differences, manual contouring from contrast-enhanced CT and semiautomated threshold-based contouring based on PET/CT yielded left lobe tumor volumes of 67 and 122 mL, respectively. These volumes include out-of-slice lesions not visualized in Figure 5.8.

5.10.1.2 Scenario 1: Empiric model for resin microspheres

Based on these measurements, the percentage tumor infiltration in the liver is as follows in Equations 5.15 and 5.16 for the tumor volumetry performed on CT and PET/CT, respectively:

$$I_{\text{tumor,CT}} = \frac{V_{\text{tumor}}}{V_{\text{tumor}} + V_{\text{normal}}} = \frac{67 \text{ mL}}{424 \text{ mL}} \times 100\%$$

$$= 15.8\% \tag{5.15}$$

$$I_{\text{tumor,PET}} = \frac{V_{\text{tumor}}}{V_{\text{tumor}} + V_{\text{normal}}} = \frac{122 \text{ mL}}{424 \text{ mL}} \times 100\%$$

$$= 28.7\% \tag{5.16}$$

It is important to note in Equations 5.15 and 5.16 that the percent infiltration was calculated only for the lobe to be treated, that is, using the volume of tumor and volume of normal liver *within the left lobe only*. As denoted in Table 5.4, the recommended whole-liver treatment dosage (A_{emp}) for this level of infiltration is 2.0 and 2.5 GBq, for 15.8% and 28.7% tumor infiltration, respectively. Recall that the empiric model is defined for whole-liver therapy and the dosage must be scaled for the treated lobe. Equation 5.17 shows this calculation for the 2.5 GBq prescription:

$$A(\text{GBq}) = A_{\text{emp}}(\text{GBq}) \cdot \left[\frac{V_{\text{treated}}}{V_{\text{wholeliver}}}\right] = 2.5$$

$$\text{GBq} \cdot \left[\frac{424 \text{ mL}}{986 \text{ mL} + 424 \text{ mL}}\right] = 0.75 \text{ GBq} \tag{5.17}$$

The final step illustrated in Equation 5.17 is critical. This can be appreciated using Equation 5.3 to

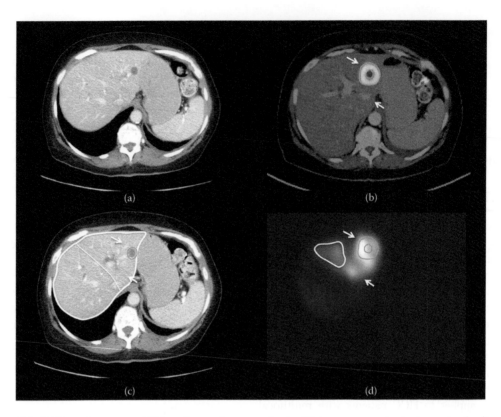

Figure 5.8 (a) Hepatic protocol CT performed on a patient with lung cancer metastases to the liver. A centrally necrotic metastatic lesion is present in the lateral left hepatic lobe **(a)** and corresponding hypermetabolic activity by [18]FDG-PET/CT is shown in **(b).** Automatic contouring of the lesion was performed in **(b)** based on a threshold of 40% of SUV_{max} (blue line). **(c)** The hepatic protocol CT was used to contour both the left and right lobe volume (white line) as well as the tumor based (blue line) on anatomical boundaries. **(d)** Pretreatment ^{99m}Tc-MAA SPECT with tumor boundary (blue line) and normal left lobe tissue region (white line).

calculate the average absorbed dose to the left liver lobe if the step in Equation 5.17 is not performed. Applying a density of 1.05 kg/L to this patients' 424 mL left lobe, we obtain a mass of 0.445 kg. A dosage of 2.5 GBq, according to Equation 5.3, would yield an average absorbed dose of 280 Gy to the lobe. This level of absorbed dose would likely result in severe toxicity in the uninvolved tissues of that lobe.

5.10.1.3 Scenario 2: Resin microspheres using the BSA model

Now let us change gears and consider the BSA treatment planning model for this scenario. This female patient has a weight of 64 kg (141 lb.) and a height of 1.62 m (5 feet, 4 inches). Therefore, her body surface area can be calculated according to Equation 5.6:

$$\text{BSA } (\text{m}^2) = 0.2025 \times \text{height}^{0.725}(\text{m}) \cdot$$
$$\text{weight}^{0.425} (\text{kg})$$
$$= 0.2025 \times 1.62^{0.725}(\text{m}) \cdot 64^{0.425}(\text{kg})$$
$$= 1.68 \, \text{m}^2 \tag{5.18}$$

One must keep in mind that the units of height (m) and weight (kg) are critical as shown in Equation 5.18:

$$A_{\text{bsa}} (\text{GBq}) = \text{BSA} - 0.2 + \left[\frac{V_{\text{tumor}}}{V_{\text{tumor}} + V_{\text{normal}}} \right]$$
$$= 1.68 - 0.2 + \frac{122 \, \text{mL}}{424 \, \text{mL}}$$
$$= 1.767 \, \text{GBq} \tag{5.19}$$

In Equation 5.19, the BSA dosage (A_{BSA}) has been calculated based on the tumor volumes identified using ^{18}FDG-PET/CT. A_{BSA} would be slightly lower had the tumor volumes from contrast-enhanced CT been used.

Once again, the BSA model is defined for whole-liver therapy and the dosage (A_{BSA}) must be scaled for the treated lobe. This process is shown in Equation 5.20. Note that the final recommended treatment activity using the BSA method (0.53 GBq) is less than when the empiric method is used (0.75 GBq):

$$A(GBq) = A_{BSA}(GBq) \cdot \left[\frac{V_{treated}}{V_{wholeliver}}\right]$$

$$= 1.767 \, GBq \cdot \left[\frac{424 \, mL}{986 \, mL + 424 \, mL}\right]$$

$$= 0.53 \, GBq \qquad (5.20)$$

5.10.1.4 Scenario 3: Glass microspheres using the recommended treatment planning method

Treatment planning for the left lobe therapy of the same patient using glass microspheres is considerably more straightforward. The percent tumor burden is not necessary and the solution is dependent only on the total mass of tissue in the left lobe: 0.445 kg. In fact, the most difficult part is selecting an absorbed dose endpoint within the recommended 80–150 Gy range.

For the purposes of this example, selecting a conservative 100 Gy absorbed dose treatment endpoint, we obtain

$$A_0(GBq) = \frac{D_{avg}(Gy) \cdot M_{liver}(kg)}{49.98 \, (J \cdot s)}$$

$$= \frac{100 \, Gy \cdot 0.445 \, kg}{49.98}$$

$$= 0.89 \, GBq \qquad (5.21)$$

Note that the absorbed dose endpoint shall be selected based on the many clinical and serologic factors discussed in the case introduction. However, in this patient without chronic liver disease or biliary obstruction, any endpoint in 100–120 Gy would likely be appropriate.

5.10.1.5 Scenario 4: Resin microspheres using the partition model

Lobar therapies will result in microsphere deposition in normal liver parenchyma and microsphere deposition in the tumor according to the T:N ratio. The partition model accounts for measured T:N as well as normal liver and tumor volumes to design a patient-specific treatment plan. Figure 5.8d describes the measurement process of T:N following ^{99m}Tc-MAA SPECT/CT imaging. A left lobar infusion of 150 MBq of ^{99m}Tc-MAA was performed 2 weeks before treatment with the catheter tip positioned carefully at the expected treatment location. Tumor contours from the metabolically active region on ^{18}FDG-PET/CT (Figure 5.8b) have been copied onto the ^{99m}Tc-MAA SPECT image set (Figure 5.8d). Areas of uninvolved liver are also shown. T:N was calculated in this case according to Equation 5.4 to be 2.8:1. Based on the pretherapy MAA mapping procedure, a lung SF for this patient was measured to be 5%. With this information, we can calculate the FU in tumor and uninvolved liver.

To compute the FU, we must determine the volume of uninvolved parenchyma in the left lobe (V_{normal}), which was not previously included in Table 5.5. If we assume tumor volumes based on ^{18}FDG-PET/CT, then $V_{normal} = 424$–122 mL = 302 mL. The corresponding mass of uninvolved left lobe parenchyma is 0.317 kg:

$$FU_{normal} = (1 - SF)\left[\frac{V_{normal}}{T{:}N \cdot V_{tumor} + V_{normal}}\right]$$

$$= (1 - 0.05)\left[\frac{302 \, mL}{2.8 \cdot 122 \, mL + 302 \, mL}\right]$$

$$= 0.45 \qquad (5.22)$$

$$FU_{tumor} = (1 - SF)\left[\frac{\frac{T}{N} \cdot V_{tumor}}{T{:}N \cdot V_{tumor} + V_{normal}}\right]$$

$$= (1 - 0.05)\left[\frac{2.8 \cdot 122 \, mL}{2.8 \cdot 122 \, mL + 302 \, mL}\right]$$

$$= 0.50 \qquad (5.23)$$

Table 5.5 Volumetric measurements of the patient presented in Figure 5.8

Structure	Volume (mL)	Mass (kg)[a]
Total liver	1,410	1.480
Total right lobe[b]	986	1.035
Total left lobe[b]	424	0.445
Left lobe tumor, manually contoured from CT	67	0.070
Left lobe tumor, automatically measured from ^{18}FDG-PET/CT with threshold = $0.4 \cdot$ SUVmax	122	0.128
Right lobe tumor	0 mL	0 kg

Note: The patient has a weight of 64 kg (141 lb.) and a height of 1.62 m (5 feet, 4 inches).
[a] Mass calculated by multiplying volume (L) by the density of liver tissue, 1.05 (kg/L) (ICRU, *Photon, Electron, Proton and Neutron Interaction Data for Body Tissues,* International Commission on Radiation Units and Measurements, 1992).
[b] Includes contributions from uninvolved tissue and tumor tissue.

We can now determine the activity required to reach a tumor-absorbed dose endpoint as shown in Equation 5.24. Given that this patient has metastatic liver cancer with no underlying chronic liver disease, a tumor-absorbed dose of 150 Gy for radioembolization using resin microspheres is appropriate:

$$A_0\,(\text{GBq}) = \frac{D_{\text{tumor}}\,(\text{Gy}) \cdot M_{\text{tumor}}\,(\text{kg})}{49.98 \cdot \text{FU}_{\text{tumor}}}$$
$$= \frac{150\,(\text{Gy}) \cdot 0.128\,(\text{kg})}{49.98 \cdot 0.50}$$
$$= 0.77\ \text{GBq} \qquad (5.24)$$

Assuming accuracy of T:N measurement, a dosage of 0.77 GBq will result in a 150 Gy average tumor dose. However, this dosage must still be back substituted into Equation 5.12 to ensure that normal liver toxicity thresholds are not exceeded. This is illustrated in

$$D_{\text{normal}}\,(\text{Gy}) = \frac{A_0\,(\text{GBq}) \cdot 49.98 \cdot \text{FU}_{\text{normal}}}{M_{\text{normal}}\,(\text{kg})}$$
$$= \frac{0.77\,(\text{GBq}) \cdot 49.98 \cdot 0.45}{0.317\,(\text{kg})}$$
$$= 55\ \text{Gy} \qquad (5.25)$$

As previously discussed, the maximum tolerable dose to normal, noncirrhotic, liver tissue from radioembolization is approximately 80 Gy. A treatment dosage of 0.77 GBq in this case is likely to be safe, particularly in the setting of a left-lobe

therapy due to the small fraction of liver tissue irradiated. This particular point is carefully detailed in Chapter 8.

5.10.1.6 Other considerations

Each case above represented an example of treatment planning for a single-session therapy. Fractionation of radioembolization treatments is a recommended practice, which may decrease toxicity to normal liver (Salem and Thurston, 2006; Cremonesi et al., 2008). Multicycle therapy will be discussed in more detail in Chapter 8.

5.10.2 SELECTIVE THERAPY

Segmental and selective therapies will result in microsphere deposition in tumor and normal tissues within the segment according to T:N and should yield little or no deposition outside of that segment except in cases of reflux. While outside the scope of this chapter, vascular planning for segmental therapy has been discussed elsewhere (Salem and Thurston, 2006; Riaz et al., 2011) and in Chapter 3. Instead, this section will focus primarily on dosimetric aspects of selective therapy. In all cases of lobar therapy reviewed in the previous section, treatment planning was always dependent on the volume of the treated lobe. In selective and segmental therapy, one must instead identify the volume of tissue perfused by the treatment that may consist of only tumor or tumor and normal liver tissue within one or more segments.

As shown in Figure 5.9, the patient has a 22 cm solitary liver metastasis from parotid adenoid cystic carcinoma (ACC). The patient's total liver volume is 3500 mL, with 2200 mL attributable to tumor. Contrast-enhanced CT and pretreatment angiography suggest that the tumor is well perfused and solid, with no substantial areas of necrosis. This patient was referred for radioembolization using resin microspheres. Unfortunately, this case represents a clinical and technical challenge owing to the large tumor size. Targeting of this patient's entire tumor would result in a subtumoricidal response, even if the entire 3 GBq resin microsphere dosage was delivered. Therefore, the tumor was treated in multiple stages. Figure 5.10a shows the pretreatment angiography with the catheter positioned in the right hepatic artery and two dominant branches supplying the tumor. Figure 5.10b shows a selective segmental right hepatic arteriogram prior to ^{99m}Tc-MAA injection. The ^{99m}Tc-MAA SPECT/CT in Figure 5.11 shows the volume of tumor perfused to be 48% of the total tumor volume. Since MAA deposited almost exclusively in the tumor, Equation 5.3 could be used directly to determine the dosage necessary to reach a 120 Gy endpoint in 48% of the tumor volume:

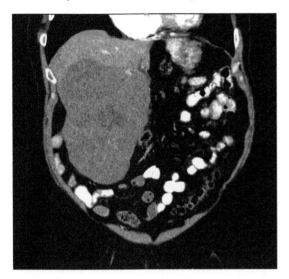

Figure 5.9 Contrast-enhanced CT of a patient with a 22 cm solitary liver metastasis from parotid ACC. Contrast-enhanced CT and pretreatment angiography suggest that the tumor is well perfused and solid, with no substantial areas of necrosis.

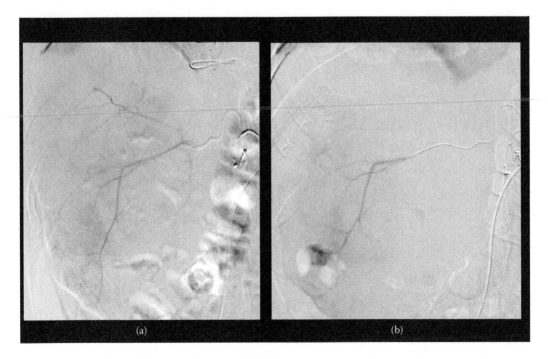

Figure 5.10 (a) Pretreatment angiography with the catheter positioned in the right hepatic artery and two dominant branches supplying the tumor. (b) Selective segmental right hepatic arteriogram prior to ^{99m}Tc-MAA injection.

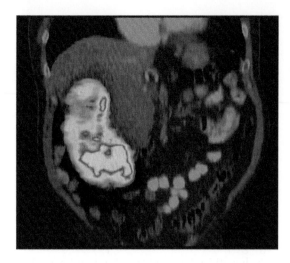

Figure 5.11 ^{99m}Tc-MAA SPECT/CT showing the volume of tumor perfused at the catheter location shown in Figure 5.10b. Perfused volume of MAA was measured to be 48% of the total tumor volume.

$$A_0(\mathrm{GBq}) = \frac{D_{\mathrm{tumor}}(\mathrm{Gy}) \cdot M_{\mathrm{tumor}}(\mathrm{kg})}{49.98}$$

$$= \frac{120\,(\mathrm{Gy}) \cdot 0.48 \cdot 2.2\,(\mathrm{L}) \cdot 1.05\,(\mathrm{kg/L})}{49.98}$$

$$= 2.66\,\mathrm{GBq} \tag{5.26}$$

Large tumor size is certainly not the only reason to treat in stages. A segmental approach to the treatment of multifocal disease is also an excellent option, which can reduce toxicity in normal liver (Riaz et al., 2011). Figures 5.12a and 5.12b show a hepatic protocol MRI and ^{111}In Octreotide SPECT of a patient with NET metastases. Multiple metastatic lesions are present throughout the right lobe. Given the absence of underlying chronic liver disease and the typically high efficacy of radioembolization in treating NET, a selective multistep approach was employed. A lateral right hepatic artery branch was selected (Figure 5.12c and d)

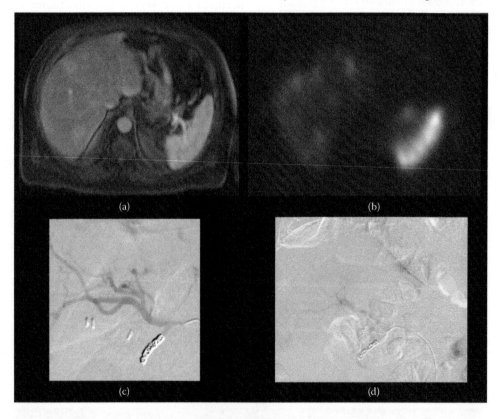

Figure 5.12 **(a)** Hepatic protocol MRI and **(b)** ^{111}In octreotide SPECT performed on a patient with neuroendocrine metastases to the liver. Multiple metastatic lesions are present throughout the right lobe. **(c)** Angiogram showing right hepatic arterial vasculature. **(d)** Selective microcatheter positioning in lateral right hepatic arterial branch that, based on pretreatment ^{99m}Tc-MAA SPECT, perfused the posterior half of the right lobe volume.

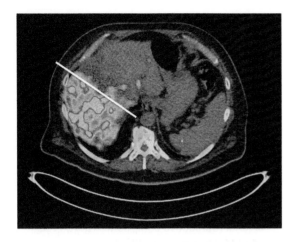

Figure 5.13 Posttreatment ^{90}Y PET/CT verifying that the radioembolization dosage was delivered to the posterior half of the right hepatic lobe.

that, based on pretreatment ^{99m}Tc-MAA SPECT, perfused approximately 50% of the right lobe volume. Treatment planning was performed according to the BSA model for radioembolization using resin microspheres. The dosage recommended by the BSA model was reduced by 50% due to the estimated perfused volume of the treatment territory. Following delivery of 0.81 GBq of ^{90}Y, a ^{90}Y PET/CT scan was performed verifying limited microsphere deposition outside of the treatment zone (Figure 5.13). One month later, this patient underwent selective radioembolization of the remaining untreated lesions in the anterior right lobe.

5.11 CONCLUSIONS

Treatment planning for radioembolization is not only a multidisciplinary process requiring input from different medical specialists but also a process that requires significant experience. Differences between patients, even among those with the same disease type, can vastly effect the treatment decision and planning process. These differences necessitate a patient-specific approach to treatment planning. This chapter is but a brief overview of the factors that must be considered by the interventional radiologists, medical physicists, nuclear medicine physicians, and/or radiation oncologists involved in the therapy. In addition to the instruction provided in this chapter, the most important suggestion that the authors wish to convey is the importance of

a team-based approach to treatment planning. A periodic meeting to discuss upcoming cases consisting of the treating interventional radiologist, medical physicist, nuclear medicine physician, and/or radiation oncologist will help to elucidate optimal treatment plans through multidisciplinary discussion. While such meetings can be difficult to schedule in a busy clinical setting, their importance cannot be overemphasized and will undoubtedly lead to improved patient outcomes.

REFERENCES

Ahmadzadehfar, H., Sabet, A., Biermann, K. et al. (2010). The significance of 99mTc-MAA SPECT/CT liver perfusion imaging in treatment planning for 90Y-microsphere selective internal radiation treatment. *J Nucl Med* 51:1206–1212.

Ariel, I.M., Padula, G. (1978a). Treatment of symptomatic metastatic cancer to the liver from primary colon and rectal cancer by the intra-arterial administration of chemotherapy and radioactive isotopes. *Prog Clin Cancer* 7:247–254.

Ariel, I.M., Padula, G. (1978b). Treatment of symptomatic metastatic cancer to the liver from primary colon and rectal cancer by the intra-arterial administration of chemotherapy and radioactive isotopes. *J Surg Oncol* 10:327–336.

Ballarini et al. (2007). The physics of the FLUKA code: Recent developments. *Adv Space Res* 40:1–11.

Bondiau, P.Y., Malandain, G., Chanalet, S. (2005). Atlas-based automatic segmentation of MR images: Validation study on the brainstem in radiotherapy context. *Int J Radiat Oncol Biol Phys* 61:289–298.

Buie, H.R., Campbell, G.M., Klinck, R.J., MacNeil, J.A. (2007). Automatic segmentation of cortical and trabecular compartments based on a dual threshold technique for in vivo micro-CT bone analysis. *Bone* 41:505–515. Epub 2007 Jul 18.

Campbell, A.M., Bailey, I.H., Burton, M.A. (2000). Analysis of the distribution of intra-arterial microspheres in human liver following hepatic yttrium-90 microsphere therapy. *Phys Med Biol* 45:1023–1033.

Campbell, A.M., Bailey, I.H., Burton, M.A. (2001). Tumour dosimetry in human liver following hepatic yttrium-90 microsphere therapy. *Phys Med Biol* 46:487–498.

Caon, M. (2004). Voxel-based computational models of real human anatomy: A review. *Radiat Environ Biophys* 42:229–235. Epub 2004 Jan 17.

Carretero, C., Munoz-Navas, M., Betes, M. et al. (2007). Gastroduodenal injury after radioembolization of hepatic tumors. *Am J Gastroenterol* 102:1216–1220.

Cremonesi, M., Ferrari, M., Bartolomei, M. et al. (2008). Radioembolisation with 90Y-microspheres: Dosimetric and radiobiological investigation for multi-cycle treatment. *Eur J Nucl Med Mol Imaging* 35:2088–2096.

Dawson, L.A., Guha, C. (2008). Hepatocellular carcinoma: Radiation therapy. *Cancer J* 14:111–116. doi: 10.1097/PPO.0b013e31816a0e80.

Eckerman, K.F., Westfall, R.J., Ryman, J.C., Cristy, M. (1994). Availability of nuclear decay data in electronic form, including beta spectra not previously published. *Health Phys* 67:338–345.

El-Baz, A.S., U, R.A., Mirmehdi, M., Suri, J.S. (2011). *Multi-Modality State-of-the-Art Medical Image Segmentation and Registration Methodologies*. New York, NY: Springer Science and Business Media.

Emami, B., Lyman, J., Brown, A. et al. (1991). Tolerance of normal tissue to therapeutic irradiation. *Int J Radiat Oncol Biol Phys* 21:109–122.

Foruzan, A.H., Zoroofi, R.A., Hori, M., Sato, Y. (2009). Liver segmentation by intensity analysis and anatomical information in multi-slice CT images. *Int J Comput Assist Radiol Surg* 4:287–297.

Fox, R.A., Klemp, P.F., Egan, G. et al. (1991). Dose distribution following selective internal radiation therapy. *Int J Radiat Oncol Biol Phys* 21:463–467.

François, C.J., Fieno, D.S., Shors, S.M., Finn, J.P. (2004). Left ventricular mass: Manual and automatic segmentation of true FISP and FLASH cine MR images in dogs and pigs. *Radiology* 230:389–395. Epub 2003 Dec 29.

Fripp, J., Crozier, S., Warfield, S.K., Ourselin, S. (2007). Automatic segmentation of articular cartilage in magnetic resonance images of the knee. *Med Image Comput Comput Assist Interv* 10:186–194.

Garin, E., Boucher, E., Rolland, Y. (2014). 99mTc-MAA-Based dosimetry for liver cancer treated using 90Y-loaded microspheres: Known proof of effectiveness. *J Nucl Med* 55:1391–1392.

Ghanei, A., Soltanian-Zadeh, H., Ratkewicz, A. (2001). A three-dimensional deformable model for segmentation of human prostate from ultrasound images. *Med Phys* 28:2147–2153.

Giammarile, F., Bodei, L., Chiesa, C. et al. (2011). EANM procedure guideline for the treatment of liver cancer and liver metastases with intra-arterial radioactive compounds. *Eur J Nucl Med Mol Imaging* 38:1393–1406.

Goin, J.E., Salem, R., Carr, B.I. et al. (2005). Treatment of unresectable hepatocellular carcinoma with intrahepatic yttrium 90 microspheres: Factors associated with liver toxicities. *J Vasc Intervent Radiol* 16:205–213.

Guha, C., Kavanagh, B.D. (2011). Hepatic radiation toxicity: Avoidance and amelioration. *Semin Radiat Oncol* 21:256–263.

Gulec, S.A., Mesoloras, G., Dezarn, W.A. et al. (2007). Safety and efficacy of Y-90 microsphere treatment in patients with primary and metastatic liver cancer: The tumor selectivity of the treatment as a function of tumor to liver flow ratio. *J Transl Med* 5:15.

Gulec, S.A.S., Mesoloras, G.G., Stabin, M.M. (2006). Dosimetric techniques in 90Y-microsphere therapy of liver cancer: The MIRD equations for dose calculations. *J Nucl Med* 47:1209–1211.

Ho, S., Lau, W.Y., Leung, T.W. et al. (1996). Partition model for estimating radiation doses from yttrium-90 microspheres in treating hepatic tumours. *Eur J Nucl Med* 23:947–952.

Ho, S.S., Lau, W.Y.W., Leung, T.W.T. et al. (1997). Clinical evaluation of the partition model for estimating radiation doses from yttrium-90 microspheres in the treatment of hepatic cancer. *Eur J Nucl Med* 24:293–298.

ICRU. (1992). *Photon, Electron, Proton and Neutron Interaction Data for Body Tissues*. Bethesda, MD: International Commission on Radiological Units and Measurements.

Jiang, M., Fischman, A., Nowakowski, F.S. et al. (2012). Segmental perfusion differences on paired Tc-99m macroaggregated albumin (MAA) hepatic perfusion imaging and yttrium-90 (Y-90) bremsstrahlung imaging studies in SIR-sphere radioembolization: Associations with angiography. *J Nucl Med Radiat Ther* 3 :122. doi:10.4172/2155–9619.1000122.

Jones, A.K., Mahvash, A. (2012). Evaluation of the potential utility of flat panel CT for quantifying relative contrast enhancement. *Med Phys* 39:4149–4154.

Kao, Y.H., Hock Tan, A.E., Burgmans, M.C. et al. (2012). Image-guided personalized predictive dosimetry by artery-specific SPECT/CT partition modeling for safe and effective 90Y radioembolization. *J Nucl Med* 53:559–566.

Kennedy, A., Dezarn, W., Weiss, A. (2011). Patient specific 3D image-based radiation dose estimates for 90Y microsphere hepatic radioembolization in metastatic tumors. *J Nucl Med Radiat Ther* 2:1–8.

Kennedy, A.S. et al. (2009). Treatment parameters and outcome in 680 treatments of internal radiation with resin 90Y-microspheres for unresectable hepatic tumors. *Int J Radiat Oncol Biol Phys* 74:1494–1500.

Kennedy, A.S.A., Nutting, C.C., Coldwell, D.D. et al. (2004). Pathologic response and microdosimetry of 90Y microspheres in man: Review of four explanted whole livers. *Int J Radiat Oncol Biol Phys* 60:12.

Klein, S., van der Heide, U.A., Lips, I.M., van Vulpen, M. (2008). Automatic segmentation of the prostate in 3D MR images by atlas matching using localized mutual information. *Med Phys* 35:1407–1417.

Knesaurek, K., Machac, J., Muzinic, M. et al. (2010). Quantitative comparison of yttrium-90 (90Y)-microspheres and technetium-99m (99mTc)-macroaggregated albumin SPECT images for planning 90Y therapy of liver cancer. *Technol Cancer Res Treatm* 9:253–261.

Kockelkorn, T.T.J.P., Schaefer-Prokop, C.M., Bozovic, G. et al. (2014). Interactive lung segmentation in abnormal human and animal chest CT scans. *Med Phys* 41:081915–081915.

Kritzinger, J., Klass, D., Ho, S. et al. (2013). Hepatic embolotherapy in interventional oncology: Technology, techniques, and applications. *Clin Radiol* 68:1–15.

Lam, M.G.E.H., Goris, M.L., Iagaru, A.H. et al. (2013). Prognostic utility of 90Y radioembolization dosimetry based on fusion 99mTc-macroaggregated albumin-99mTc-sulfur colloid SPECT. *J Nucl Med* 54:2055–2061.

Lam, M.G.E.H., Smits, M.L.J. (2013). Value of 99mTc-macroaggregated albumin SPECT for radioembolization treatment planning. *J Nucl Med* 54:1681–1682.

Lam, M.G.E.H., Sze, D.Y. (2014). Reply: 99mTc-MAA-based dosimetry for liver cancer treated using 90y-loaded microspheres: Known proof of effectiveness. *J Nucl Med* 55:1392–1393.

Lanconelli, N., Pacilio, M., Meo Lo, S. et al. (2012). A free database of radionuclide voxel S values for the dosimetry of nonuniform activity distributions. *Phys Med Biol* 57:517–533.

Lau, W.Y., Leung, W.T., Ho, S. et al. (1994). Treatment of inoperable hepatocellular carcinoma with intrahepatic arterial yttrium-90 microspheres: A phase I and II study. *Br J Cancer* 70:994–999.

Lewandowski, R.J., Thurston, K.G., Goin, J.E. et al. (2005). 90Y microsphere (TheraSphere) treatment for unresectable colorectal cancer metastases of the liver: Response to treatment at targeted doses of 135–150 Gy as measured by [18F]fluorodeoxyglucose positron emission tomography and computed tomographic imaging. *J Vasc Interv Radiol* 16:1641–1651.

Linguraru, M.G., Sandberg, J.K., Li, Z. et al. (2010). Automated segmentation and quantification of liver and spleen from CT images using normalized probabilistic atlases and enhancement estimation. *Med Phys* 37:771–783.

Louie, J.D., Kothary, N., Kuo, W.T. et al. (2009). Incorporating cone-beam CT into the treatment planning for yttrium-90 radioembolization. *J Vasc Interv Radiol* 20:606–613.

Marroquín, J.L., Vemuri, B.C., Botello, S. et al. (2002). An accurate and efficient Bayesian method for automatic segmentation of brain MRI. *IEEE Trans Med Imaging* 21:934–945.

McKinney, G.W., Durkee, J.W., Hendricks, J.S. (2006). *MCNPX overview*. Los Alamos, NM: Los Alamos National Laboratory.

Metzger, M.C., Bittermann, G., Dannenberg, L. et al. (2013). Design and development of a virtual anatomic atlas of the human skull for automatic segmentation in computer-assisted surgery, preoperative planning, and navigation. *Int J Comput Assist Radiol Surg* 8:691–702.

Naymagon, S., Warner, R.R.P., Patel, K. et al. (2010). Gastroduodenal ulceration associated with radioembolization for the treatment of hepatic tumors: An institutional experience and review of the literature. *Dig Dis Sci* 55:2450–2458.

Park, S.H., Gao, Y., Shi, Y., Shen, D. (2014). Interactive prostate segmentation using atlas-guided semi-supervised learning and adaptive feature selection. *Med Phys* 41:111715.

Pasciak, A.S., Bourgeois, A.C., McKinney, J.M. et al. (2014). Radioembolization and the dynamic role of (90)Y PET/CT. *Front Oncol* 4:38.

Reed, V.K., Woodward, W.A., Zhang, L. et al. (2009). Automatic segmentation of whole breast using atlas approach and deformable image registration. *Int J Radiat Oncol Biol Phys* 73:1493–1500.

Riaz, A., Gates, V.L., Atassi, B. et al. (2011). Radiation segmentectomy: A novel approach to increase safety and efficacy of radio-embolization. *Int J Radiat Oncol Biol Phys* 79:163–171.

Salem, R., Thurston, K.G. (2006). Radioembolization with 90Yttrium micro-spheres: A state-of-the-art brachytherapy treatment for primary and secondary liver malignancies. Part 1: Technical and methodologic considerations. *J Vasc Interv Radiol* 17:1251–1278.

Sangro, B., Gil-Alzugaray, B., Rodriguez, J. et al. (2008). Liver disease induced by radioembolization of liver tumors: Description and possible risk factors. *Cancer* 112:1538–1546.

Sempoux, C., Horsmans, Y., Geubel, A. et al. (1997). Severe radiation-induced liver disease following localized radiation therapy for biliopancreatic carcinoma: Activation of hepatic stellate cells as an early event. *Hepatology* 26:128–134.

South, C.D., Meyer, M.M., Meis, G. et al. (2008a). W1672 yttrium-90 microsphere Induced gastrointestinal tract ulceration. *Gastroenterology* 134:A-904.

South, C.D., Meyer, M.M., Meis, G. et al. (2008b). Yttrium-90 microsphere induced gastrointestinal tract ulceration. *World J Surg Oncol* 6:93.

Stabin, M.G., Eckerman, K.F., Ryman, J.C. et al. (1994). Bremsstrahlung radiation dose in yttrium-90 therapy applications. *J Nucl Med* 35:1377–1380.

Strigari, L., Menghi, E., D'Andrea, M., Benassi, M. (2006). Monte Carlo dose voxel kernel calculations of beta-emitting and Auger-emitting radionuclides for internal dosimetry: A comparison between EGSnrcMP and EGS4. *Med Phys* 33:3383.

Strigari, L., Sciuto, R., Rea, S. et al. (2010). Efficacy and toxicity related to treatment of hepatocellular carcinoma with 90Y-SIR spheres: Radiobiologic considerations. Journal of nuclear medicine: Official publication. *Soc Nucl Med* 51:1377–1385.

van Ginneken, B., Haar Romeny ter, B.M. (2000). Automatic segmentation of lung fields in chest radiographs. *Med Phys* 27:2445.

Warfield, S.K., Zou, K.H., Wells, W.M. (2004). Simultaneous truth and performance level estimation (STAPLE): An algorithm for the validation of image segmentation. *IEEE Trans Med Imaging* 23:903–921.

Wondergem, M., Smits, M.L.J., Elschot, M. et al. (2013). 99mTc-macroaggregated albumin poorly predicts the intrahepatic distribution of 90Y resin microspheres in hepatic radioembolization. *J Nucl Med* 54:1294–1301.

Yoo, H.S., Park, C.H., Suh, J.H. et al. (1989). Radioiodinated fatty acid esters in the management of hepatocellular carcinoma: Preliminary findings. *Cancer Chemother Pharmacol* 23(Suppl):S54–S58.

Zhang, L.Z.L., Hoffman, E.A., Reinhardt, J.M. (2006). Atlas-driven lung lobe segmentation in volumetric X-ray CT images. *IEEE Trans Med Imaging* 25:1–16.

Radioembolization in segmentectomy, lobectomy, and future liver remnant hypertrophy

EDWARD KIM, JOSEPH TITANO, AND SAFET LEKPERIC

6.1 INTRODUCTION

As the role of ^{90}Y transarterial radioembolization has evolved, specific treatment paradigms have led to the development of radioembolization applications analogous to surgical liver interventions. The first such application is "radiation lobectomy," in which radioembolization is performed to treat an unresectable right lobe liver lesion with the threefold intention of treating the tumor, allowing a biological test of time to select for less aggressive lesions and causing left lobe hypertrophy as a means of enabling right lobe surgical resection. The second analogous concept is "radiation segmentectomy," in which a large dose of radiation is delivered to a small volume of liver thereby imparting a highly tumoricidal dose to the perfused target, while sparing adjacent and nontarget liver parenchyma. These concepts have broadened the armament physicians have available to treat liver tumors by allowing for treatment of lesions with complex anatomic locations and by expanding the patient population eligible for interventions. The radiation biology, physics, nuclear medicine, and interventional radiology concepts related to these treatment entities are discussed in this chapter.

6.2 TREATMENT PLANNING AND DELIVERY

As described in Chapter 4, once a patient is selected for radioembolization therapy, treatment planning begins with the completion of a technetium-99m macroaggregated albumin (^{99m}Tc-MAA) study. The purpose of this study, again, is multifaceted. First, it is undertaken to visualize the vascular supply to the liver and to the lesion of interest. This allows for the potential embolization of vessels that have the potential to divert the glass or resin ^{90}Y microspheres away from the intended target thereby causing adverse events. In addition, it allows for calculation of the lung shunt fraction (Uliel et al., 2012).

Safe and effective delivery of radioembolization therapy—whether for lobar or segmental infusion—requires careful evaluation in the interventional suite as well as the patient. Detailed discussion of radiation safety will be provided in Chapter 7. Briefly, the interventional suite should be outfitted with radiation detection equipment such that a thin-window Geiger–Müller counter that is able to detect radiation levels under 0.1 mR/hour is available to detect the contamination of personnel, garbage, and the interventional suit equipment. In addition, an ionization chamber able to detect radiation doses of 1 mrem/hour should be available to localize the dose delivery site and measure activity remaining in the dose vial. A large drape should be prepared in close proximity to the fluoroscopy table so that potential leaks are contained immediately. An acrylic desiccator is also required in order to house the dose vile, tubing, and catheter following radioembolization (Salem and Thurston, 2006c).

Following standard prepping and draping, arterial access is gained—typically via the common femoral artery. A 4-French or 5-French catheter system is generally utilized to navigate the aorta and to cannulate the celiac axis or the superior mesenteric artery. Radioembolization is then generally performed through a coaxial 0.0325-inch system within the target vessel. During infusion, care must be taken to avoid stasis and reflux of the ^{90}Y microspheres in order to avoid nontarget embolization, which could result in gastrointestinal ulceration, lung parenchymal injury, or pancreatitis among other complications (Salem and Thurston, 2006c).

Following radioembolization completion and catheter removal, postprocedure imaging with single-photon emission computed tomography (SPECT)/computed tomography (CT) or positron emission tomography (PET)/CT/magnetic resonance imaging (MRI) may be obtained to evaluate for nontarget deposition of ^{90}Y microspheres or to assess the distribution of microspheres within the liver and tumor, respectively. Recently, ^{90}Y PET/CT has gained increasing popularity compared with ^{90}Y-bremsstrahlung SPECT/CT on account of its greater dosimetry accuracy (Braat et al., 2015). While further study is still required, delivered-dose calculation on posttreatment imaging may predict tumor response and thereby allow for early planning of repeat interventions (Braat et al., 2015). More detailed discussion of these modalities will be provided in Chapters 10 and 11.

6.3 RADIATION LOBECTOMY AND FUTURE LIVER REMNANT HYPERTROPHY

6.3.1 DEFINITION AND TREATMENT RATIONALE

Radiation lobectomy, as its name implies, entails lobar infusion of ^{90}Y microspheres. Future liver remnant (FLR) hypertrophy specifically refers to right lobar delivery of ^{90}Y microspheres in patients with right-sided tumors who would be candidates for resection if the FLR were adequate. Adequate volumes for the FLR have been cited between 20% and 40% of total liver volume, with a larger remnant recommended for cirrhotic patients (Kubota et al., 1997; Zorzi et al., 2007; Shindoh et al., 2013; Vouche et al., 2013). The intention behind radiation lobectomy is threefold: (1) treat the right-sided tumor, (2) simultaneously induce left liver lobe hypertrophy such that an adequate FLR is achieved allowing the patient to proceed with potentially curative procedures such as surgical resection, and (3) allow for a test of time to identify less aggressive tumors in the hopes of limiting recurrence rates postresection (Gaba et al., 2009; Inarrairaegui et al., 2012; Salem et al., 2013).

Portal vein embolization (PVE) is an alternative to radiation lobectomy that is also employed with the intention of inducing lobar hypertrophy in order to produce an adequate FLR. While PVE is perhaps even more effective than radioembolization (Azoulay et al., 2000; Pamecha et al., 2009; Garlipp et al., 2014) at increasing the FLR volume, there are several advantages offered by radiation lobectomy. First, PVE does not directly treat the liver tumor; while it interrupts portal flow to the lesion, the arterial supply from which tumors draw the majority of their nutrition remains intact. This means that lesions remain unchecked while awaiting the FLR hypertrophy following PVE. Radiation lobectomy is also a microembolic therapy causing radiation-induced atrophy of the target lobe allowing for a delayed diversion of portal blood flow from the right lobe to the left lobe, which may allow for a greater accommodation of increased blood flow by the FLR (Jakobs et al., 2008; Gaba et al., 2009; Vouche et al., 2013). This pattern of hypertrophy allows for a test-of-time through which tumor response to therapy may be assessed and those patients with positive tumor biology may then be selected for further curative interventions (Vouche et al., 2013). Finally, the microembolic nature of radiation lobectomy allows for expansion of the treatment population to patients with portal vein thrombus (PVT). While resection of tumors associated with PVT is rare in the United States, these resections are performed frequently in Asian hospital centers (Vouche et al., 2013).

The combination of transarterial chemoembolization (TACE) and PVE has been offered to account for the lack of tumor control offered by PVE alone. While one might expect that embolization of the portal and arterial vessels supplying the same region of liver parenchyma may be associated with hepatic injury, increased toxicity is only transient (Aoki et al., 2004). In addition to safety, sequential TACE and PVE have been shown to increase the rate of the FLR hypertrophy, improve recurrence-free survival, and increase overall survival in patients with hepatocellular carcinoma (HCC) (Ogata et al., 2006; Yoo et al., 2011). However, the FLR hypertrophy achieved by the combination of TACE and PVE is equivalent to or arguably inferior to that of PVE alone (Teo et al., 2015).

6.3.2 RADIATION BIOLOGY AND RADIATION LOBECTOMY

Before examining the details of liver parenchymal change following radiation lobectomy, a general discussion of liver tissue response to insult—and specifically radiation-induced insult—is warranted. A more robust discussion of radiation biology can be found in Chapters 8 and 9. Following parenchymal injury, hepatic stellate cells migrate to the affected region and begin to produce extracellular matrix leading to fibrosis (Clement et al., 1986; Jakobs et al., 2008). As shown in studies primarily focused on external beam radiation, once the radiation dose applied to liver tissue exceeds 30–40 Gy additional pathological and morphological changes consistent with veno-occlusive disease (VOD) are seen including sinusoidal congestion, hemorrhage, atrophy, and necrosis (Fajardo and Colby, 1980; Jakobs et al., 2008). The implication, then, is that radioembolization may induce a degree of portal hypertension (Jakobs et al., 2008; Gaba et al., 2009). The restorative mechanisms associated with hepatic parenchymal insult are also initiated following radioembolization. Hepatocyte proliferation is the end result of multiple inputs directed by cytokines, growth factors, signaling pathway cascades, and transcription factor activation (Gaba et al., 2009).

While local veno-occlusive changes follow radioembolization, radiation lobectomy does not necessarily induce portal hypertension globally. In a study by Jakobs et al. (2008), in which volumetric changes following radioembolization were explored, a subgroup analysis of those patients who underwent unilateral treatment of the right liver lobe showed a significant increase in left liver lobe volume, a significant decrease in right liver lobe volume, increased left portal vein diameter, and no change in splenic volume. Splenic volumes were significantly increased in patients who underwent bilobar radioembolization (Jakobs et al., 2008). Together, these findings suggest that portal venous flow is diverted into the left-sided portal system—as evidenced by the increased diameter of the left portal vein—without the development of secondary signs of portal hypertension such as an increase in splenic volume following right lobar radioembolization.

Recently, Fernandez-Ros et al. (2015) described in detail the mechanisms of the biological response to radioembolization specifically. They describe oxidative stress as the driver behind endothelial cell injury as well as activation of coagulation and proinflammatory pathways. While it remains unclear if the elevation of proinflammatory markers is a primary effect of radioembolization or follows secondarily from VOD, the role of coagulation in VOD (as shown in the pathologically similar VOD following bone marrow transplantation) has been better established (Fernandez-Ros et al., 2015).

At a cellular level, the changes that have been described following portal vein ligation and partial hepatic resection have shown that hepatocytes proliferate initially followed by nonparenchymal cells (Michalopoulos and DeFrances, 1997; Fernandez-Ros et al., 2015). In addition to the redistribution of portal blood flow, mitogens including hepatocyte growth factor (HGF), fibroblast growth factor type 19 (FGF-19), interleukin 6 (IL-6), and insulin are also increased following these procedures. Specifically following radioembolization, significant increases in HGF and FGF-19 are noted (Fernandez-Ros et al., 2015). Both of these factors drive transcription factor activation and initiate hepatocyte regeneration. In addition, IL-6 and tumor necrosis factor-alpha (TNF-α) also drive hepatocyte replication by initiating the transition of these cells from G0 to G1 (Fernandez-Ros et al., 2015). Sustained increases in TNF-α and IL-6 are observed following treatment and likely contribute to the FLR hypertrophy (Fernandez-Ros et al., 2015). In contrast to TACE following which transient increases in IL-6 and HGH have been documented, these factors were noted to be elevated months after radioembolization—a timeline commensurate with that of the FLR hypertrophy (Yamazaki et al., 1996; Kim et al., 2013; Fernandez-Ros et al., 2015). This contrast with TACE also provides a basis for the suggestion that IL-6 and HGH sustained elevations are more likely the result of radiation effects than embolic effects of therapy (Fernandez-Ros et al., 2015).

Radiation lobectomy, then, harnesses multiple processes leading to both atrophy and hypertrophy contributing to the therapeutic aim of increasing the volume of the FLR. The treated right lobe undergoes the development of sinusoidal congestion, atrophy, and necrosis while the tumor itself is also treated. The veno-occlusive changes that follow radioembolization then lead to a slow redirection of portal vein blood flow to the untreated left liver lobe. Simultaneously then, signaling pathways are initiated and proliferative mediators are recruited while portal venous flow is directed toward the FLR (Gaba et al., 2009; Vouche et al., 2013).

6.3.3 PATIENT SELECTION

Generally, radioembolization is indicated for patients with unresectable HCC, cholangiocarcinoma, or with metastatic liver lesions. In 1999, the U.S. Food and Drug Administration (USFDA) issued a humanitarian device exemption for glass microspheres as neoadjuvant therapy prior to surgery or transplantation in patients with unresectable HCC. Similarly, in 2002, the USFDA approved the use of ^{90}Y resin microspheres for the treatment of unresectable metastatic liver tumors from primary colorectal cancer with adjuvant intrahepatic artery chemotherapy.

Against this background, radiation lobectomy is a specific application of radioembolization in patients with hepatic tumor lesions that would be eligible for definitive therapy with hepatic lobar resection if the remaining FLR were adequate (Gaba et al., 2009; Siddiqi and Devlin, 2009; Vouche et al., 2013). Radiation lobectomy allows for the treatment of right hepatic lobe tumor burden while inducing the FLR hypertrophy through the redirection of portal blood flow and by the production of growth factors and cytokines. In addition, the time interval required to allow for future remnant hypertrophy mandates a period prior to surgical intervention that allows for aggressive tumors to declare themselves on follow-up imaging (Vouche et al., 2013).

6.3.4 LOBECTOMY DOSIMETRY

Generally, dose calculation for radiation lobectomy and the FLR hypertrophy is performed by completing calculations for lobar therapy as discussed in Chapter 5. Briefly, under assumptions of uniform dose distribution and complete ^{90}Y decay as elaborated upon in the previous chapter, the administered activity is calculated using Equation 5.9 where A_o is the treatment activity, D_{avg} is the desired average absorbed dose, and M_{liver} is the mass of the liver *to be treated*:

$$A_o(GBq) = \frac{D_{avg}(Gy) \cdot M_{liver}(kg)}{49.98 \, (J \cdot s)}$$

Again, the lobar mass is utilized as the mass of liver intended to undergo treatment in this dose calculation paradigm. The mass of the liver to be treated, M_{liver}, is obtained by measuring the target liver volume and converting the volume measure to a calculated mass value 1.05 kg/L.

Following treatment, the actual dose delivered may be determined utilizing Equation 6.1:

$$D_{delivered}(Gy) = \frac{A_o(GBq)(1-SF)(1-R) \cdot 49.98 \, (J \cdot s)}{M_{liver}(kg)}$$

(6.1)

where SF is the lung shunt fraction and R is the percentage of dose remaining within the vial at the completion of treatment. Currently, there is no universal dosing pattern for radiation lobectomy and the FLR hypertrophy. The dosing information in the major studies of radiation lobectomy is provided in Table 6.1. In the largest study of resin microspheres utilized for radiation lobectomy and the FLR hypertrophy, Fernandez-Ros et al. (2014) found no correlation between dose and volume changes and concluded that—while it is currently unknown increased dosing could enhance the FLR hypertrophy—hypertrophy does indeed occur at therapeutic doses. Similarly, in the largest study of glass microspheres used for radiation lobectomy and the FLR hypertrophy, Vouche et al. (2013) reported a median dose of 112 Gy (range: 74–215 Gy) delivered to the treatment site and included dose ≤100 Gy and dose >120 Gy as variables in their multivariate analysis of %FLR hypertrophy (defined below), with neither of the conditions meeting statistical significance.

6.3.5 RADIATION LOBECTOMY AND THE FUTURE LIVER REMNANT HYPERTROPHY OUTCOME DATA

6.3.5.1 Imaging response

In the assessment of the FLR hypertrophy, volumetric measures of liver parenchyma are obviously essential. In the largest cohort of glass microsphere radiation lobectomy cases, Vouche et al. (2013) completed computer-assisted volumetric assessment of the liver on either gadolinium-enhanced magnetic resonance imaging (SHARP or VIBE sequences) or computed tomography. Boundaries for the right lobe and the left lobe were delineated by the left hepatic vein in the upper lobe and a line drawn from the inferior vena cava to the insertion of the falciform ligament in the lower lobes. The portal triad, gallbladder, and inferior vena cava were excluded from volumetric analysis. According to this system, the right liver lobe volume consists of the combined measures of segments 1, 4, 5, 6, 7, and 8 while the left liver lobe volume consists of the combined measures of segments 2 and 3.

While usage of the term FLR has been defined in several ways, we have selected to define it as the percentage of the FLR volume as a ratio of total liver volume (Vouche et al., 2013). Integral to the discussion of liver volume changes following radiation lobectomy are the following equations:

$$FLR\% = \frac{LLPV}{TLPV} \cdot 100\%$$

(6.2)

%FLR Hypertrophy

$$= \frac{FLR_{post-Y90} - FLR_{pre-Y90}}{FLR_{pre-Y90}} \cdot 100\%,$$

(6.3)

where LLPV is defined as the volume of segments 2 and 3 less the total volume of tumor within the left lobe, and TLPV is defined as the total liver volume less the volume of total tumor burden within the liver (Vouche et al., 2013).

6.3.5.2 Liver volume changes, subsequent therapies, and survival outcomes

Several studies have evaluated the efficacy of radiation lobectomy in producing the FLR hypertrophy. The findings of these studies are provided in Table 6.1. Jakobs et al. (2008) evaluated volumetric changes in the liver following lobar treatment with ^{90}Y in a cohort predominantly of colorectal metastasis cases; a subanalysis of the those patients who received unilateral right lobar treatment revealed significantly decreased right lobe volume with subsequent left lobe hypertrophy and no significant increase in spleen volume (Jakobs et al., 2008). These findings contributed to the notion that radiation lobectomy allowed for a gradual, well-compensated diversion of portal venous flow from

Table 6.1 Radiation lobectomy and the FLR hypertrophy selected studies

Study	Number of patients	Patient age (years)	Tumor path	Micro-spheres	Single or multiple treatments	Activity/dose information	Time to volume measure	FLR hypertrophy	Imaging response summary	Overall survival	Future resection or transplantation
Ahmadza-dehfar et al. (2013)	24	Median 63 (range 44–78)	CRC 15, breast 5, pancreatic 2, gastric 1, unknown primary 1	Resin	Staged lobar treatments 4–6 weeks apart for 17 patients with bilobar disease; single treatment for 7 patients with unilobar disease	Mean activity 1.67 GBq, median activity 1.75 GBq (range 0.40–3.90 Gbq), dose NR	Mean 44 days, median 36 days	Mean 47%, median 34%	RECIST, r disease group right lobe CR 2, PR 14, PD 1; bilobar disease group left lobe PR 1, SD 4, PD 12; unilobar group right lobe PR 4, SD 2, PD 1	NR	One patient with unilobar CRC metastases went on to right hepatectomy
Edeline et al. (2013)	34	NR	HCC	Glass 30, Resin 4	Single	Median dose to treated segment 122.1 Gy (90.4–210.5 Gy)	3 months	Mean 29% at 3 months Mean 42% at all available time points	mRECIST CR 29.6%, PR 33.3%, SD 29.6%, PD 7.4%	Median 13.5 months	NR
Fernandez-Ros et al. (2014)	83	Median 66 (IQR 53–79)	HCC 52, CRC 13, IHC 4, Other 14	Resin	Single	NR	4–26+ weeks	Mean approximately 45%	NR	NR	NR
Garlipp et al. (2014)	26	Mean 59.2 (standard deviation 11.1)	CRC 18, Breast 8, Other 6	Resin	Single	Median activity 1.2 GBq (range 0.8–1.7 GBq) Dose NR	Median 46 days (27–79 days)	Mean 29% (standard deviation 22.9%), median 25.3%	RECIST CR 1, PR 19, SD 5, PD 1	NR	NR
Teo et al. (2014)	17	Median 72 (42–78)	HCC	Resin	Single	NR	Median 5 months (range 2–12 months), mean 5.7 months	Mean 34.2% (standard deviation 34.9%; range 19.0–106.5%)	RECIST, CR 2, PR 5, SD 6, PD 4	NR	One patient underwent surgical right lobectomy
Theysohn et al. (2014)	45	Mean 71.9 (range 55–90)	HCC	Glass	Single	Mean dose to right lobe 112 Gy (range 100–160 Gy)	1–12 months	Mean 50.46% at 6 months, mean 56.49% at 12 months	NR	NR	NR
Vouche et al. (2013)	83	Median 68 (range 36–89)	HCC 67, IHC 8, CRC 8	Glass	Single	Median dose to treatment site 112 Gy (range 74–215 Gy)	1–9+ months	Median maximal FLR 26%, median 45% at > 9 months (5–186)	NR	Median survival BCLC B and C patients was 34.4 and 9.6 months, respectively	Five patients underwent surgical right lobectomy; 6 patients underwent OLT

Note: CR, complete response; CRC, colorectal cancer; IQR, interquartile range; NR, not reported; PD, progressive disease; PR, partial response; SD, stable disease.

the right lobe to the FLR without inducing global portal hypertension.

A study by Gaba et al. (2009) provided similar results in a cohort of HCC and cholangiocarcinoma cases with statistically significant increases and decreases in left lobe and right lobe volumes, respectively, measured at an average of 18 months posttreatment (Gaba et al., 2009). Similarly, Vouche et al. (2013) demonstrated statistically significant lobar volume changes and showed a linear, time-dependent hypertrophy of the FLR. In their cohort of 83 patients, 5 went on to lobar resection and 6 underwent liver transplantation. Vouche et al. (2013) also showed that volumetric changes are apparent as early as 1 month posttreatment with maximum FLR hypertrophy achieved at approximately 9 months posttreatment. Interestingly, the presence of portal vein thrombosis was a significant predictor of the FLR hypertrophy >40% with the implication that existing portal vein thrombus might act as a naturally occurring portal vein embolization with even earlier diversion of portal flow to the FLR (Vouche et al., 2013).

In addition to portal vein thrombus predicting increased FLR hypertrophy, a recent study by Teo et al. (2014) showed that HCC patients with underlying hepatitis B may achieve greater FLR hypertrophy than their counterparts with hepatitis C or alcoholic liver disease. This finding, combined with a trend observed by Fernandez-Ros et al. (2014) toward reduced the FLR hypertrophy, implies that cirrhosis may somewhat limit the benefits of lobectomy in generating FLR hypertrophy.

Survival data for studies focused on radiation lobectomy and FLR hypertrophy specifically are reported at up to a median of 36.6 months and are in line with concurrently published prospective and retrospective cohorts (Gaba et al., 2009; Vouche et al., 2013).

6.4 RADIATION SEGMENTECTOMY

6.4.1 DEFINITION AND TREATMENT RATIONALE

Radiation segmentectomy is defined as radioembolization of two or fewer hepatic segments—as delineated by the Couinaud system—during a single treatment session (Rhee et al., 2005; Riaz et al., 2011). The term segmentectomy was utilized in reference to the resultant atrophy of the treated segments seen at follow-up imaging, which is analogous to segmental surgical hepatic resection.

Several factors created the clinical need for radiation segmentectomy. First, small tumors (those ≤3 cm) are generally considered for curative therapies including transplantation, surgical resection, and ablation (Llovet et al., 1999). If, however, a lesion is not amenable to curative intervention on account of anatomic considerations (e.g., adjacent to the dome of the liver and diaphragm or in close proximity to large vessels), comorbidities, or inadequate functional liver reserve, radiation segmentectomy remains a viable treatment option for these patients. Second, several authors have shown that increased radiation dose is associated with improved tumor response (Ben-Josef et al., 2005; Riaz et al., 2011; Vouche et al., 2014), and radiation segmentectomy allows for greater activity delivery directly to a target lesion. Further, it has also been theorized that lower radiation dose applied to normal hepatic parenchyma minimizes injury to the normal tissue allowing for greater physiologic regeneration of normal parenchyma (Riaz et al., 2011).

In direct comparison with ablative procedures, there are several advantages and disadvantages to radiation segmentectomy. Advantages of radiation segmentectomy include the obviation of percutaneous needle and probe placement with the associated theoretical risk of tract seeding and the ability to target high-risk ablation lesions (Riaz et al., 2011; Vouche et al., 2014). Disadvantages of segmentectomy relative to ablative procedures potentially include cost and radiation exposure, although ablation probes are often placed with CT guidance making this a relative disadvantage (Vouche et al., 2014).

In addition to its complimentary role with other ablative therapies, radiation segmentectomy also has several advantages compared with external-beam radiation therapy. These advantages are mainly linked to anatomic and practical considerations regarding treatment planning and delivery. Lesions within the caudate lobe and the dome of the liver put adjacent structures such as the lung parenchyma and porta hepatis ducts and vessels at increased risk (Riaz et al., 2011). In addition, while dose fractionation has shown benefits in targeting radiosensitive as well as resistant malignant cells, this therapeutic approach requires multiple treatment sessions (Riaz et al., 2011). A final practical consideration is that respiratory motion potentially puts lung parenchyma at risk during external-beam radiation delivery in a manner that is avoided with transarterial delivery of radiation with the maximum tissue penetration of 11 mm associated with ^{90}Y (Salem and Thurston, 2006a; Riaz et al., 2011).

6.4.2 PATIENT SELECTION

Based on the definition of radiation segmentectomy, a lesion must be isolatable within only two segments of the liver supplied by the hepatic arterial vessel selected for delivery of ^{90}Y microspheres. As discussed above, segmentectomy is complimentary to ablation in that suboptimal lesions for ablation may be treated by radiation segmentectomy. The most often cited reason that a lesion is deferred for ablation is that a lesion is located at the dome of the liver in close proximity to the diaphragm and lung tissue (Riaz et al., 2011). Additional anatomic considerations leading to the choice of radiation segmentectomy over ablation include proximity to vessels and biliary structures, caudate lobe location, and proximity to small bowel, large bowel, the gallbladder, or the heart (Vouche et al., 2014). Recent publications have also demonstrated the safety and efficacy of radiation segmentectomy in patients with moderate hepatic dysfunction and advanced disease including portal vein invasion (Padia et al., 2014).

6.4.3 SEGMENTECTOMY DOSIMETRY

Generally, dose calculation for radiation segmentectomy is performed by completing calculations intended for treatment of the entire lobe in which the lesion is located; however, intra-arterial injection of the lobar dose is performed from a segmental vessel supplying one or two segments as described previously (Rhee et al., 2005; Vouche et al., 2014). Equation 5.9 may be utilized in order to calculate the activity to be delivered.

Following treatment, the actual dose delivered may be determined utilizing Equation 6.1. However, accurate calculation of activity delivered

to a tumor is complicated by the physiology of blood flow to liver tumors and to normal hepatic parenchyma. First, the formulae applied for dose calculations often assume uniform distribution of microspheres within the treated volume of liver. However, it has been shown through a number of modalities that blood flow is preferentially diverted toward tumor compared with normal liver parenchyma (Lau et al., 1994; Ho et al., 1996; Campbell et al., 2000; Sarfaraz et al., 2003; Riaz et al., 2011). Intuitively, this matches an essential tenant of transarterial liver tumor therapy that liver tumors draw a majority of their blood supply from the hepatic arterial system while normal parenchyma receives a majority of its blood supply from the portal venous system.

Attempts to account for the nonuniform distribution of blood flow—and therefore of ^{90}Y microspheres—have been made previously. Riaz et al. (2011) identified the problems associated with the assumption of uniformity in microsphere distribution and sought to account for these issues by incorporating a subjectively determined ratio of tumor hypervascularity relative to adjacent normal liver tissue following a review of angiography and cross-sectional imaging studies. Although not an ideal means of quantifying the asymmetric distribution of blood flow to tumor relative to surrounding normal tissue, this method demonstrated that such differences in calculation lead to more than doubling of the median calculated dose delivered to tumor—from 521 Gy (95% CI: 404–645 Gy) to 1214 Gy (95% CI: 961–1546 Gy) in their cohort of 84 patients (Riaz et al., 2011).

6.4.4 RADIATION SEGMENTECTOMY OUTCOME DATA

6.4.4.1 Radiation segmentectomy imaging response

Imaging response in the studies focused on the methodology of radiation segmentectomy is summarized in Table 6.2. Riaz et al. (2011) presented imaging response in accordance with World Health Organization (WHO) and European Association for the Study of the Liver (EASL) guidelines. EASL response was reported in 81%

Table 6.2 Radiation segmentectomy selected studies

Study	Number of patients	Patient age (years)	Tumor path	Micro-spheres	Dose	Follow-up time	Imaging response summary	Overall survival
Padia et al. (2014)	20	Median 61 (range 54–76)	HCC	Glass	Median dose to segment 254 Gy (range 105–1055 Gy), median dose to tumor 536 Gy (range 203–1618 Gy)	Median 275 days (range 32 –677 days)	Time to EASL response: 33 days (range 5–133 days). EASL: CR 19, SD 1	90% at median follow-up of 275 days (range 32–677 days)
Rhee et al. (2005)	14	Mean 62 (range 41–78)	HCC	Glass	Median dose to segment 348 Gy (range 105–857 Gy)	Median 185 days (range 35–600 days)	NR	NR
Riaz et al. (2011)	84	Median 68 (43–90)	HCC	Glass	Median dose to segment 521 Gy (range 404–645 Gy)	NR	TTP 13.6 months (95% CI, 9.3–18.7 months); EASL: response in 81% of patients; median time to response 1.2 months (95% CI, 1.1–1.4 months); WHO: response in 59% of patients; median time to response 7.2 months (95% CI, 4.2–8.5 months)	Median overall survival 26.9 months (95% CI, 20.5–30.2 months)
Vouche et al. (2014)	102	Median 64 (IQR, 58–74)	HCC	Glass	Median dose to segment 242 Gy (IQR, 173–369 Gy)	Median 27.1 months	mRECIST: CR 47%, PR 39%, SD 12%, PD 1%	Median overall survival uncensored 53.4 months; median overall survival censored for transplantation 34.5 months

Note: CR, complete response; IQR, interquartile range; NR, not reported; PD, progressive disease; PR, partial response; SD, stable disease; TTP, time to progression; WHO, World Health Organization.

of patients with median time-to-disease response 1.2 months, and WHO response was reported in 59% of patients with median time-to-disease response of 7.2 months. The largest series published to date by Vouche et al. (2014) demonstrated complete response in 47% of patients, partial response in 39% of patients, and stable disease in 12% of patients according to modified Response Evaluation Criteria for Solid Tumors (mRECIST). Median time-to-disease progression in this series was 33.1 months with new intrahepatic lesions responsible for disease progression in a majority of cases (Vouche et al., 2014). For comparison, median time-to-disease progression has been reported between 7.9 and 33.3 months in prospective studies and large cohort retrospective studies (Lewandowski et al., 2009; Salem et al., 2010; Sangro et al., 2011).

6.4.4.2 Radiation segmentectomy survival outcomes

Median overall survival for patients undergoing radiation segmentectomy has been reported at 13.6–53.4 months (Riaz et al., 2011; Vouche et al., 2014). Censored for transplantation, the overall survival in the 34.5 months in the Vouche et al. (2014) study compared with the uncensored median overall survival of 53.4 months. These overall survival rates are comparable with or exceed overall survival reported in other large cohorts of patients undergoing radioembolization (Sangro et al., 2011). The variability in the reported segmentectomy median overall survival is likely secondary to the duration of the largest studies of radiation segmentectomy—5 and 8 years, respectively (Riaz et al., 2011; Vouche et al., 2014). In addition, sorafenib was approved late in 2007 and was therefore unavailable for approximately half of the Riaz et al. (2011) study period, while it was available for the majority of the Vouche et al. (2014) study duration (Llovet et al., 2008).

Because radiation segmentectomy is ideally suited as a complimentary therapy to ablation in instances wherein ablation options are limited, comparison to overall survival in ablation studies is warranted. Vouche et al. (2014) argue that overall survival as well as local control of tumor lesions in cases of radiation segmentectomy does not differ dramatically from overall survival achieved by ablation when stratified by measure of baseline liver dysfunction (e.g., Childs–Pugh score)—given the inherent limitations in comparing locoregional therapy survival with that of ablation (Lencioni et al., 2005; Chen et al., 2006; Livraghi et al., 2008; Pompili et al., 2013; Vouche et al., 2014). Further, Vouche et al. (2014) postulate that radiation segmentectomy might offer similar survival outcomes as those seen in Barcelona Clinic Liver Cancer (BCLC) A patients. While controversies exist regarding the limitations of percutaneous ablation, radiation segmentectomy remains a viable option for complex lesions with the potential for complete pathological necrosis of the target lesion (Salem et al., 2015; Seror et al., 2015).

6.4.4.3 Radiation segmentectomy dose and response relationship

As the goal of radiation segmentectomy is to achieve complete pathological necrosis (CPN) equivalent to that seen in analogous ablation procedures, several authors have focused on the radiation dose necessary to achieve such an outcome. By utilizing ^{99m}Tc-MAA SPECT/CT as a surrogate in dosimetry calculations, Garin et al. (2012) established a threshold dose of 205 Gy as predictive of EASL imaging response, progression-free survival, and overall survival. Specific to radiation segmentectomy, in a study by Vouche et al. (2014), 33 of the 102 patients studied went on to receive liver transplantation; explant analysis revealed that a dose >190 Gy was associated with CPN. Their findings suggest that a threshold radiation dose exists to achieve CPN. In their study of highly selective radiation segmentectomy, Padia et al. (2014) reported a median dose of 255 Gy (range 105–1055 Gy) with a complete EASL response noted in 19 of 20 patients in their cohort. A summary of dose, imaging response, and survival outcomes for several radiation segmentectomy studies is included in Table 6.2.

6.5 RADIOEMBOLIZATION TOXICITIES AND COMPLICATIONS

The most commonly encountered clinical adverse effect following radioembolization is fatigue (Salem and Thurston, 2006a). In addition, patients may experience fever and chills up to several days following treatment—likely secondary to the effects of radiation on normal hepatic parenchyma and subsequent release of endogenous pyrogens (Murthy et al., 2005; Salem et al., 2005; Salem and Thurston, 2006c). Additional clinical toxicities described include abdominal pain, nausea, vomiting, anorexia, diarrhea, and weight loss (Riaz et al., 2011; Vouche et al., 2014);. While these adverse effects are often self-limited, endoscopic evaluation should be considered if symptoms are persistent as they may reflect the development of gastrointestinal ulceration (Salem and Thurston, 2006c).

Laboratory toxicities following radioembolization most commonly include abnormalities in bilirubin, albumin, AST, ALT, alkaline phosphatase, platelet levels, lymphocyte counts, and international normalized ratio (INR). Care must be taken in the interpretation of laboratory toxicities as it is often difficult to separate progressive liver disease secondary to underlying disease versus treatment toxicity. In a majority of cases, laboratory abnormalities are of limited clinical significance and occasionally reflect preexisting poor liver function and cirrhosis (Rhee et al., 2005; Vouche et al., 2014).

Radioembolization-induced liver disease (REILD) is analogous to the radiation-induced liver disease (RILD) described in the external beam radiation therapy literature and characterized by jaundice, fatigue, and the development of ascites typically 1–2 months posttreatment in the absence of tumor progression or bile duct obstruction (Lawrence et al., 1995; Sangro et al., 2008). REILD represents a form of VOD as does RILD; the two entities differ in that RILD presentation is described as an "anicteric ascites" with a proportional elevation of liver enzymes in contrast to REILD, which is characterized by markedly increased bilirubin (Sangro et al., 2008). Risk factors for the development of REILD have been identified as prior treatment with chemotherapy, relative young age, elevated baseline bilirubin, cirrhosis at baseline, whole-liver treatment (versus lobar therapy), and the ratio of activity administered relative to the volume of treated liver (Sangro et al., 2008; Riaz et al., 2009a; Fernandez-Ros et al., 2015). Additional discussion related to the side effects of radioembolization can be found in other chapters in this book. Chapter 14 focuses on identifying and managing clinical sequelae using posttreatment imaging.

6.5.1 RADIATION SEGMENTECTOMY TOXICITIES

Of particular interest to radiation segmentectomy is bilirubin toxicity as the concentration of a lobar dose within one or two segments imposes the theoretical risk of increased biliary complications (Riaz et al., 2011). REILD is less of a concern in radiation segmentectomy compared with treatments involving greater volumes of liver as the target liver parenchyma volume is reduced without limiting tumor treatment efficacy (Sangro et al., 2008). However, the increased radiation dose delivered to a smaller volume of liver parenchyma raises concern for injury to the biliary system. Rhee et al. (2015) reported a statistically significant increase in bilirubin following segmental infusion of a lobar radiation dose but deemed this change clinically insignificant as only one of the 14 patients included in their study had a change in Childs–Pugh class following radioembolization (Rhee et al., 2005). Riaz et al. (2011) reported 5% of patients within their cohort of 84 patients undergoing radiation segmentectomy developed small bilomas postprocedure as well as 1% of patients developing biliary stricture within the treated segment (Riaz et al., 2011). Vouche et al. (2014) did not report biloma or biliary stricture development in their multicenter cohort of 102 patients. Padia et al. (2014) also did not report biloma or biliary stricture in their 20 patient cohort. For comparison, in a study of 327 patients who underwent standard lobar treatment protocols, Atassi et al. (2008) found 17 cases of biliary necrosis, 8 cases of biliary strictures, and 3 cases of biloma.

6.6 POSTTREATMENT PATIENT MANAGEMENT

6.6.1 PATIENT CARE IN THE IMMEDIATE POSTPROCEDURE SETTING

Radioembolization is generally performed on an outpatient basis. Following treatment, patients recover for 2–6 hours prior to discharge home. Practice patterns vary between institutions, but patients may be sent home with gastrointestinal ulcer prophylaxis such as proton pump inhibitors or with steroid tapers for treatment of fatigue. All patients require instruction regarding radiation safety precautions as their body surface readings may reach 1 mrem/hour following radioembolization (Salem and Thurston, 2006c). Further, patients treated with resin should be advised that urinary excretion of radioactivity is possible at trace levels (25–50 kBq per liter of urine per GBq of dose), which may be safely disposed of with standard precautionary methods. While the toxicity profile of radioembolization is limited, patients should again be reminded of potential treatment adverse effects that they may experience following discharge.

6.6.2 FOLLOW-UP EVALUATION

Patients are typically seen in an outpatient setting 2 weeks following treatment with radioembolization. At the time of this assessment, providers should focus their history and examination on the preservation of the patient's performance status in addition to monitoring for adverse effects, tumor lysis syndrome, or toxicities of nontarget organs such as the lungs and the gastrointestinal tract. Laboratory testing may be obtained at this point and is likely to show lymphopenia, transient increases in aminotransferase levels, and transient increases in tumor markers (Salem and Thurston, 2006c). Follow-up imaging is obtained as described below.

6.6.3 IMAGING RESPONSE

Generally, follow-up imaging with contrast-enhanced CT or MRI is obtained at 1–3 months posttreatment and then at 3- to 6-month intervals for lesion surveillance and future treatment planning regardless of whether radioembolization is performed in segmental or lobar treatment. Optimal imaging response is expected to be seen approximately 3–6 months following therapy with median time to response 6.6 months for change in size and 1.2 months for evidence of necrosis (Sangro et al., 2006; Riaz et al., 2009b; Salem et al., 2010, 2013).

6.7 CLINICAL CASE EXAMPLES

6.7.1 SAMPLE RADIATION LOBECTOMY AND THE FUTURE LIVER REMNANT HYPERTROPHY CASE

A 68-year-old male with a past medical history significant for hepatitis C cirrhosis was screened for liver cancer using gadolinium contrast-enhanced magnetic resonance imaging that revealed a right hepatic lobe, segment 8 HCC (Figure 6.1). The patient was scheduled for right radioembolization lobectomy following a tumor board conference attended by transplant surgeons, hepatologists, medical oncologists, and interventional radiologists. Resin microspheres were utilized in this

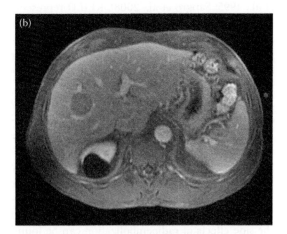

Figure 6.1 Magnetic resonance T1-weighted gadolinium-enhanced arterial phase (a) and portal venous phase (b) sequences demonstrating a right liver lobe mass measuring >3 cm with early arterial enhancement and portal venous washout consistent with hepatocellular carcinoma.

patient's radioembolization lobectomy procedure. The activity calculations are provided following the body surface area method presented in Chapter 5—specifically Equations 5.6 and 5.7:

$$BSA(m^2) = 0.2025 \cdot height^{0.725}(m) \cdot weight^{0.425}(kg)$$

$$BSA(m^2) = 0.2025 \cdot 1.88^{0.725} \cdot (87.1)^{0.425}$$

$$BSA(m^2) = 0.2025 \cdot 1.58 \cdot 6.68$$

$$BSA = 2.13\,m^2$$

$$A_{bsa}(GBq) = BSA - 0.2 + \frac{V_{tumor}}{V_{tumor} + V_{normal}}$$

$$A_{bsa}(GBq) = 2.13 - 0.2 + \frac{38.8\,mL}{38.8\,mL + 1381.2\,mL}$$

$$A_{bsa}(GBq) = 2.13 - 0.2 + 0.03$$

$$A_{bsa} = 1.96\,GBq$$

If glass microspheres had been utilized in this radiation lobectomy case, a sample of the hypothetical activity calculation is provided utilizing Equation 5.9 following the conversion of the lobar volume to mass (conversion factor of 1.05 kg/L):

$$A_o(GBq) = \frac{D_{avg}(Gy) \cdot M_{liver}(kg)}{49.98\,(J \cdot s)}$$

$$A_o(GBq) = \frac{120\,Gy \cdot 1.42\,kg}{49.98\,(J \cdot s)}$$

$$A_o = 3.42\,GBq$$

Following successful radioembolization of the right lobe, subsequent right hepatic lobe atrophy and compensatory left hepatic lobe

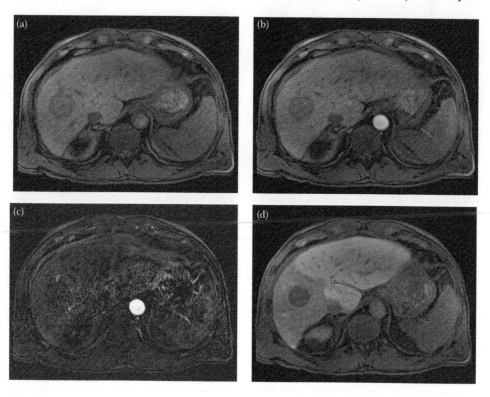

Figure 6.2 Magnetic resonance T1-weighted precontrast (a), T1-weighted gadolinium-enhanced arterial phase (b), and subtracted arterial phase (c) sequences demonstrating a right liver lobe treatment cavity with increased T1 signal intensity on the precontrast sequence and lack of arterial enhancement as evidenced by the subtracted sequence. Baseline increased T1 signal intensity is likely due to hemorrhagic changes within the lesion following RE. Magnetic resonance T1-weighted gadolinium-enhanced 20-minute delayed sequence (d) demonstrates a right lobe treatment cavity, decreased right lobar contrast retention suggesting decreased hepatocyte function and atrophy, and left lobar contrast retention within normal limits.

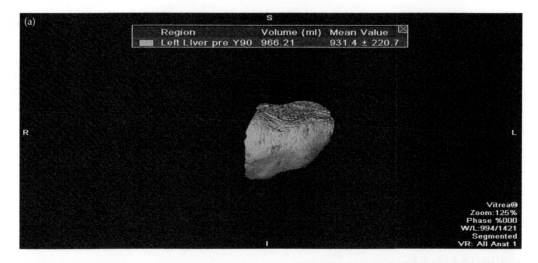

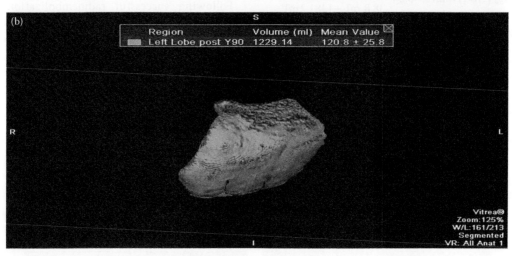

Figure 6.3 Volumetric rendering using Vitrea software (Vital Imaging, Inc., Minnetonka, Minnesota) of the left lobe at the time of diagnosis **(a)** and 3 months following ^{90}Y radioembolization lobectomy **(b)** demonstrating interval hypertrophy of the future liver remnant.

hypertrophy were noted. At 3-month follow-up, complete mRECIST response was noted in the right lobe HCC (Figure 6.2). Left liver lobe volumes prior to and 3 months following radioembolization lobectomy are provided (Figure 6.3). The patient underwent surgical right lobe hepatectomy 5 months following radioembolization lobectomy. As such, this case represents successful treatment of the right hepatic tumor along with left lobe hypertrophy in preparation for surgical right lobectomy. After three years of

right lobe hepatectomy, the patient remains free of recurrence.

6.7.2 SAMPLE RADIATION SEGMENTECTOMY CASE

A 75-year-old male with a past medical history significant for nonalcoholic hepatic steatosis was screened for liver cancer using contrast-enhanced magnetic resonance imaging revealing a right hepatic dome HCC (Figure 6.4). While

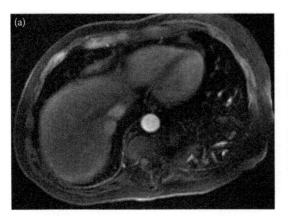
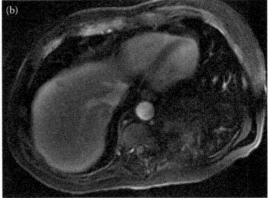

Figure 6.4 Magnetic resonance T1-weighted gadolinium-enhanced arterial phase **(a)** and portal venous phase **(b)** sequences demonstrating a hepatic dome mass measuring 2.5 cm with early arterial enhancement and portal venous washout consistent with hepatocellular carcinoma. Note the location of the lesion, which is adjacent to the inferior vena cava, the right hepatic vein, and the diaphragm.

the size of the lesion (2.5 cm in greatest axial diameter) is amenable to percutaneous ablation, the location of the lesion adjacent to the inferior vena cava, the right hepatic vein, and the diaphragm limit the role of percutaneous ablation. The patient was listed for liver transplantation, received sorafenib therapy, and scheduled for radioembolization segmentectomy following a tumor board conference consisting of transplant surgeons, hepatologists, medical oncologists, and interventional radiologists. The volume of the right hepatic lobe was found to be 789.9 mL; utilizing the conversion factor of 1.05 kg/L, the mass of the right lobe was calculated at 0.83 kg. The activity calculations utilizing the MIRD model described in Chapter 5 are provided—specifically utilizing Equation 5.9:

$$A_o\,(\text{GBq}) = \frac{D_{avg}\,(\text{Gy})\cdot M_{liver}\,(\text{kg})}{49.98\,(\text{J}\cdot\text{s})}$$

$$A_o\,(\text{GBq}) = \frac{120\ \text{Gy}\cdot 0.83\,\text{kg}}{49.98\,(\text{J}\cdot\text{s})}$$

$$A_o = 2.00\ \text{GBq}$$

Given the segmental administration of the dose calculated for lobar infusion, we can calculate the dose delivered using Equation 6.1 with a 5% lung shunt fraction, minimal dose remaining within the vial (1%) at completion of radioembolization, and a segmental volume of 89.1 mL (0.09 kg utilizing the 1.05 kg/L conversion):

$$D_{delivered}\,(\text{Gy}) = \frac{A_o\,(\text{GBq})(1-\text{SF})(1-R)\cdot 49.98\,(\text{J}\cdot\text{s})}{M_{liver}\,(\text{kg})}$$

$$D_{delivered}\,(\text{Gy}) = \frac{2\ \text{GBq}\cdot(1-0.05)\cdot(1-0.01)\cdot 49.98\ (\text{J}\cdot\text{s})}{0.09\ \text{Kg}}$$

$$D_{delivered}\,(\text{Gy}) = \frac{2\ \text{GBq}\cdot 0.95\cdot 0.99\cdot 49.98\ (\text{J}\cdot\text{s})}{0.09\ \text{Kg}}$$

$$D_{delivered} = 1042\ \text{Gy}$$

At follow-up imaging, the patient demonstrated mRECIST complete response with progressive decrease in size of both the treatment cavity and the treated segment (Figure 6.5). The patient has remained tumor free with continued HCC surveillance 3 years after radiation segmentectomy.

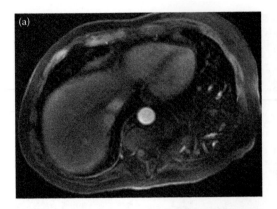
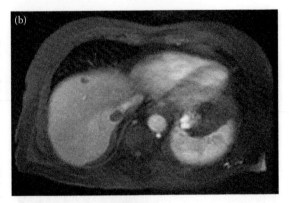
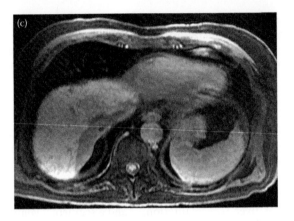

Figure 6.5 Magnetic resonance T1-weighted gadolinium-enhanced arterial phase **(a)** at the time of diagnosis. Magnetic resonance T1-weighted gadolinium-enhanced 20-minute delayed sequences 3 months after radioembolization **(b)** and 2 years after radioembolization **(c)** demonstrate progressive decrease in size of both the treatment cavity and the treated segment, which demonstrates decreased contrast retention suggesting decreased hepatocyte function and atrophy.

6.8 CONCLUSIONS

The preceding discussion and sample cases highlight the range of radioembolization applications available to physicians for patients with varying disease severity. From BCLC A patients with anatomically challenging lesions otherwise amenable to ablation to patients with metastatic disease deemed unresectable at the time of diagnosis, radioembolization offers solutions in a variety of challenging clinical situations. Understanding the underlying radiation and tumor biology, dosimetry considerations, and patient selection criteria are integral to proper application of the breadth of radioembolization techniques.

REFERENCES

Ahmadzadehfar, H., Meyer, C., Ezziddin, S. et al. (2013). Hepatic volume changes induced by radioembolization with 90Y resin microspheres. A single-centre study. *Eur J Nucl Med Mol Imaging* 40:80–90.

Aoki, T., Imamura, H., Hasegawa, K. et al. (2004). Sequential preoperative arterial and portal venous embolizations in patients with hepatocellular carcinoma. *Arch Surg* 139:766–774.

Atassi, B., Bangash, A.K., Lewandowski, R.J. et al. (2008). Biliary sequelae following radioembolization with Yttrium-90 microspheres. *J Vasc Interven Radiol* 19:691–697.

Azoulay, D., Castaing, D., Smail, A. et al. (2000). Resection of nonresectable liver metastases from colorectal cancer after percutaneous portal vein embolization. *Ann Surg* 231:480–486.

Ben-Josef, E., Normolle, D., Ensminger, W.D. et al. (2005). Phase II trial of high-dose conformal radiation therapy with concurrent hepatic artery floxuridine for unresectable intrahepatic malignancies. *J Clin Oncol Official J Am Soc Clin Oncol* 23:8739–8747.

Braat, A.J., Smits, M.L., Braat, M.N. et al. (2015). ^{90}Y Hepatic radioembolization: An update on current practice and recent developments. *J Nucl Med Official Publ Soc Nucl Med* 56:1079–1087.

Campbell, A.M., Bailey, I.H., Burton, M.A. (2000). Analysis of the distribution of intra-arterial microspheres in human liver following hepatic yttrium-90 microsphere therapy. *Phys Med Biol* 45:1023–1033.

Chen, M.H., Wei, Y., Yan, K. et al. (2006). Treatment strategy to optimize radiofrequency ablation for liver malignancies. *J Vasc Interven Radiol* 17:671–683.

Clement, B., Grimaud, J.A., Campion, J.P. et al. (1986). Cell types involved in collagen and fibronectin production in normal and fibrotic human liver. *Hepatology* 6:225–234.

Edeline, J., Lenoir, L., Boudjema, K. et al. (2013). Volumetric changes after (90)y radioembolization for hepatocellular carcinoma in cirrhosis: An option to portal vein embolization in a preoperative setting? *Ann Surg Oncol* 20:2518–2525.

Fajardo, L.F., Colby, T.V. (1980). Pathogenesis of veno-occlusive liver disease after radiation. *Arch Pathol Lab Med* 104:584–588.

Fernandez-Ros, N., Inarrairaegui, M., Paramo, J.A. et al. (2015). Radioembolization of hepatocellular carcinoma activates liver regeneration, induces inflammation and endothelial stress and activates coagulation. *Liver Int Official J Int Assoc Study Liver* 35:1590–1596.

Fernandez-Ros, N., Silva, N., Bilbao, J.I. et al. (2014). Partial liver volume radioembolization induces hypertrophy in the spared hemiliver and no major signs of portal hypertension. *Official J Int Hepato Pancreato Biliary Assoc* 16:243–249.

Gaba, R.C., Lewandowski, R.J., Kulik, L.M. et al. (2009). Radiation lobectomy: preliminary findings of hepatic volumetric response to lobar yttrium-90 radioembolization. *Ann Surg Oncol* 16:1587–1596.

Garin, E., Lenoir, L., Rolland, Y. et al. (2012). Dosimetry based on 99mTc-macroaggregated albumin SPECT/CT accurately predicts tumor response and survival in hepatocellular carcinoma patients treated with 90Y-loaded glass microspheres: Preliminary results. *J Nucl Med Official Publ Soc Nucl Med* 53:255–263.

Garlipp, B., de Baere, T., Damm, R. et al. (2014). Left-liver hypertrophy after therapeutic right-liver radioembolization is substantial but less than after portal vein embolization. *Hepatology* 59:1864–1873.

Ho, S., Lau, W.Y., Leung, T.W. et al. (1996). Partition model for estimating radiation doses from yttrium-90 microspheres in treating hepatic tumours. *Eur J Nucl Med* 23:947–952.

Inarrairaegui, M., Pardo, F., Bilbao, J.I. et al. (2012). Response to radioembolization with yttrium-90 resin microspheres may allow surgical treatment with curative intent and prolonged survival in previously unresectable hepatocellular carcinoma. *Eur J Surg Oncol J Eur Soc Surg Oncol Br Assoc Surg Oncol* 38:594–601.

Jakobs, T.F., Saleem, S., Atassi, B. et al. (2008). Fibrosis, portal hypertension, and hepatic volume changes induced by intra-arterial radiotherapy with 90yttrium microspheres. *Dig Dis Sci* 53:2556–2563.

Kim, M.J., Jang, J.W., Oh, B.S. et al. (2013). Change in inflammatory cytokine profiles after transarterial chemotherapy in patients with hepatocellular carcinoma. *Cytokine* 64:516–522.

Kubota, K., Makuuchi, M., Kusaka, K. et al. (1997). Measurement of liver volume and hepatic functional reserve as a guide to decision-making in resectional surgery for hepatic tumors. *Hepatology* 26:1176–1181.

Lau, W.Y., Leung, T.W., Ho, S. et al. (1994). Diagnostic pharmaco-scintigraphy with hepatic intra-arterial technetium-99m macroaggregated albumin in the determination of tumour to non-tumour uptake ratio in hepatocellular carcinoma. *Br J Radiol* 67:136–139.

Lawrence, T.S., Robertson, J.M., Anscher, M.S. et al. (1995). Hepatic toxicity resulting from cancer treatment. *Int J Radiat Oncol Biol Phys* 31:1237–1248.

Lencioni, R., Della Pina, C., Bartolozzi, C. (2005). Percutaneous image-guided radiofrequency ablation in the therapeutic management of hepatocellular carcinoma. *Abdominal Imaging* 30:401–408.

Lewandowski, R.J., Kulik, L.M., Riaz, A. et al. (2009). A comparative analysis of transarterial downstaging for hepatocellular carcinoma: Chemoembolization versus radioembolization. *Am J Transpl Official J Am Soc Transpl Am Soc Transpl Surg* 9:1920–1928.

Livraghi, T., Meloni, F., Di Stasi, M. et al. (2008). Sustained complete response and complications rates after radiofrequency ablation of very early hepatocellular carcinoma in cirrhosis: Is resection still the treatment of choice? *Hepatology* 47:82–89.

Llovet, J.M., Bru, C., Bruix, J. (1999). Prognosis of hepatocellular carcinoma: The BCLC staging classification. *Semin Liver Dis* 19:329–338.

Llovet, J.M., Ricci, S., Mazzaferro, V., Hilgard, P. (2008). Sorafenib in advanced hepatocellular carcinoma. *N Engl J Med* 359:378–390.

Michalopoulos, G.K., DeFrances, M.C. (1997). Liver regeneration. *Science* 276:60–66.

Murthy, R., Nunez, R., Szklaruk, J. et al. (2005). Yttrium-90 microsphere therapy for hepatic malignancy: Devices, indications, technical considerations, and potential complications. *Radiogr A Rev Publ Radiol Soc N Am* 25 (Suppl 1):S41–S55.

Ogata, S. Belghiti, J., Farges, O. et al. (2006). Sequential arterial and portal vein embolizations before right hepatectomy in patients with cirrhosis and hepatocellular carcinoma. *Br J Surg* 93:1091–1098.

Padia, S.A., Kwan, S.W., Roudsari, B. et al. (2014). Superselective yttrium-90 radioembolization for hepatocellular carcinoma yields high response rates with minimal toxicity. *J Vasc Interven Radiol* 25:1067–1073.

Pamecha, V., Glantzounis, G., Davies, N. et al. (2009). Long-term survival and disease recurrence following portal vein embolisation prior to major hepatectomy for colorectal metastases. *Ann Surg Oncol* 16:1202–1207.

Pompili, M., Saviano, A., de Matthaeis, N. et al. (2013). Long-term effectiveness of resection and radiofrequency ablation for single hepatocellular carcinoma </=3 cm. Results of a multicenter Italian survey. *J Hepatol* 59:89–97.

Rhee, T.K., Omary, R.A., Gates, V. et al. (2005). The effect of catheter-directed CT angiography on yttrium-90 radioembolization treatment of hepatocellular carcinoma. *J Vasc Interven Radiol* 16:1085–1091.

Riaz, A., Gates, V.L., Atassi, B. et al. (2011). Radiation segmentectomy: A novel approach to increase safety and efficacy of radioembolization. *Int J Radiat Oncol Biol Phys* 79:163–171.

Riaz, A., Kulik, L., Lewandowski, R.J. et al. (2009b). Radiologic-pathologic correlation of hepatocellular carcinoma treated with internal radiation using yttrium-90 microspheres. *Hepatology* 49:1185–1193.

Riaz, A., Lewandowski, R.J., Kulik, L.M. et al. (2009a). Complications following radioembolization with yttrium-90 microspheres: A comprehensive literature review. *J Vasc Interven Radiol* 20:1121–1130; quiz 1131.

Salem, R., Mazzaferro, V., Sangro, B. (2013). Yttrium 90 radioembolization for the treatment of hepatocellular carcinoma: Biological lessons, current challenges, and clinical perspectives. *Hepatology* 58:2188–2197.

Salem, R., Thurston, K.G. (2006a). Radioembolization with yttrium-90 microspheres: a state-of-the-art brachytherapy treatment for primary and secondary liver malignancies: Part 3: Comprehensive literature review and future direction. *J Vasc Interven Radiol* 17:1571–1593.

Salem, R., Thurston, K.G. (2006b). Radioembolization with 90yttrium microspheres: a state-of-the-art brachytherapy treatment for primary and secondary liver malignancies. Part 2: Special topics. *J Vasc Interven Radiol* 17:1425–1439.

Salem, R., Thurston, K.G. (2006c). Radioembolization with 90Yttrium microspheres: A state-of-the-art brachytherapy treatment for primary and secondary liver malignancies. Part 1: Technical and methodologic considerations. *J Vasc Interven Radiol* 17:1251–1278.

Salem, R., Lewandowski, R.J., Atassi, B. et al. (2005). Treatment of unresectable hepatocellular carcinoma with use of 90Y microspheres (TheraSphere): Safety, tumor response, and survival. *J Vasc Interven Radiol* 16:1627–1639.

Salem, R., Lewandowski, R.J., Mulcahy, M.F. et al. (2010). Radioembolization for hepatocellular carcinoma using Yttrium-90 microspheres: A comprehensive report of long-term outcomes. *Gastroenterology* 138:52–64.

Salem, R., Vouche, M., Habib, A. et al. (2015). Reply: to PMID 24691943. *Hepatology* 61:407.

Sangro, B., Bilbao, J.I., Boan, J. et al. (2006). Radioembolization using 90Y-resin microspheres for patients with advanced hepatocellular carcinoma. *Int J Radiat Oncol Biol Phys* 66:792–800.

Sangro, B., Gil-Alzugaray, B., Rodriguez, J. et al. (2008). Liver disease induced by radioembolization of liver tumors: Description and possible risk factors. *Cancer* 112:1538–1546.

Sangro, B., Salem, R., Kennedy, A. et al. (2011). Radioembolization for hepatocellular carcinoma: A review of the evidence and treatment recommendations. *Am J Clin Oncol* 34:422–431.

Sarfaraz, M., Kennedy, A.S., Cao, Z.J. et al. (2003). Physical aspects of yttrium-90 microsphere therapy for nonresectable hepatic tumors. *Med Phys* 30:199–203.

Seror, O., Nault, J.C., Nahon, P. et al. (2015). Is segmental transarterial yttrium 90 radiation a curative option for solitary hepatocellular carcinoma ≤5 cm? *Hepatology* 61:406–407.

Shindoh, J., Tzeng, C.W., Aloia, T.A. et al. (2013). Optimal future liver remnant in patients treated with extensive preoperative chemotherapy for colorectal liver metastases. *Ann Surg Oncol* 20:2493–2500.

Siddiqi, N.H., Devlin, P.M. (2009). Radiation lobectomy—A minimally invasive treatment model for liver cancer: Case report. *J Vasc Interven Radiol* 20:664–669.

Teo, J.Y., Allen, J.C., Jr., Ng, D.C. et al. (2015). A systematic review of contralateral liver lobe hypertrophy after unilobar selective internal radiation therapy with Y90. *HPB (Oxford)*. 2015 Oct 16. doi: 10.1111/hpb.12490 [Epub ahead of print].

Teo, J.Y., Goh, B.K., Cheah, F.K. et al. (2014). Underlying liver disease influences volumetric changes in the spared hemiliver after selective internal radiation therapy with 90Y in patients with hepatocellular carcinoma. *J Dig Dis* 15:444–450.

Theysohn, J.M., Ertle, J., Muller, S. et al. (2014). Hepatic volume changes after lobar selective internal radiation therapy (SIRT) of hepatocellular carcinoma. *Clin Radiol* 69:172–178.

Uliel, L., Royal, H.D., Darcy, M.D. et al. (2012). From the angio suite to the gamma-camera: Vascular mapping and 99mTc-MAA hepatic perfusion imaging before liver radioembolization—A comprehensive pictorial review. *J Nucl Med Official Publ Soc Nucl Med* 53:1736–1747.

Vouche, M., Habib, A., Ward, T.J. et al. (2014). Unresectable solitary hepatocellular carcinoma not amenable to radiofrequency ablation: Multicenter radiology-pathology correlation and survival of radiation segmentectomy. *Hepatology* 60:192–201.

Vouche, M., Lewandowski, R.J., Atassi, R. et al. (2013). Radiation lobectomy: Time-dependent analysis of future liver remnant volume in unresectable liver cancer as a bridge to resection. *J Hepatol* 59:1029–1036.

Yamazaki, H., Oi, H., Matsumoto, K. et al. (1996). Biphasic changes in serum hepatocyte growth factor after transarterial chemoembolization therapy for hepato-cellular carcinoma. *Cytokine* 8:178–182.

Yip, D., Allen, R., Ashton, C., Jain, S. (2004). Radiation-induced ulceration of the stomach secondary to hepatic embolization with radioactive yttrium microspheres in the treatment of metastatic colon cancer. *J Gastroenterol Hepatol* 19:347–349.

Yoo, H., Kim, J.H., Ko, G.Y. et al. (2011). Sequential transcatheter arterial chemoembolization and portal vein embolization versus portal vein embolization only before major hepatectomy for patients with hepatocellular carcinoma. *Ann Surg Oncol* 18:1251–1257.

Zorzi, D., Laurent, A., Pawlik, T.M. et al. (2007). Chemotherapy-associated hepatotoxicity and surgery for colorectal liver metastases. *Br J Surg* 94:274–286.

PART 3

Treating Patients with Radioembolization

7

Radiation safety concerns associated with preparing the dosage, treating and releasing the patient, and managing radioactive waste

WILLIAM D. ERWIN

7.1 INTRODUCTION

Radioembolization, being a form of internal administration of radioactive material (RAM) in humans, necessarily has associated with it a number of patient, personnel, and general public-related radiation safety concerns. Among these are patient absorbed dose prescription or radioactivity prescription (also referred to as dosage throughout this chapter) and associated calculations; radioactivity (heretofore abbreviated to activity) delivery to and preparation in the radiopharmacy (or hot lab); transport to interventional radiology (IR) and

135

infusion into the patient; handling of the radioactive waste and contamination from the procedure; internal radiation exposure of the patient; and external exposure from both the patient and the radioactive waste (Salem and Thurston, 2006; Gulec and Siegel, 2007; Dezarn et al., 2011). These various radiation safety aspects of radioembolization are the subject of this chapter.

The two radiolabeled microspheres products approved by the Food and Drug Administration (FDA) for liver-directed radioembolization, TheraSphere® (BTG International Ltd., Canada) and SIR-Spheres® (Sirtex Medical Limited, Lane Cove, NSW, Australia), have both similarities and differences from the perspective of radiation safety (Dezarn et al., 2011). On the one hand, both employ the same radionuclide for the β radiotherapy (yttrium-90, ^{90}Y), as well as a similar range of particle sizes (20–60 μm and 20–30 μm spheres, respectively). The radiopharmaceutical ^{99m}Tc-radiolabeled macroaggregated albumin (MAA) is employed for both as a surrogate for pretreatment microcatheter placement imaging, lung shunt quantification, and treatment planning absorbed dose or activity calculations. Finally, the administration kits of both have the same basic components: a sealed v-vial containing the microspheres housed in an acrylic β shield; an acrylic infusion delivery box in which the v-vial-in-shield is placed that serves as additional β radiation shielding as well as a RAM containment vessel; and needles and associated infusion lines (tubing) for connection to and flow between the injection syringe, v-vial, and the prepositioned microcatheter. On the other hand, TheraSphere consists of 3.29 g/mL specific gravity glass spheres with naturally occurring ^{89}Y embedded in the material that are irradiated in a nuclear reactor to produce ^{90}Y via the ^{89}Y(n,γ) ^{90}Y reaction (with a number of radionuclidic impurities being produced in the process), while SIR-Spheres consists of lower density (1.6 g/mL) resin-based spheres with ^{90}Y extracted from a strontium-90 (^{90}Sr)/^{90}Y generator and chemically bound to the surface, which results in no radionuclidic impurities other than a trace amount of ^{90}Sr due to "breakthrough" in the extraction process. Finally, a higher amount of activity, in general, is administered with TheraSphere.

Yttrium-90 microspheres are recognized in the United States by the Nuclear Regulatory Commission (NRC) and associated Agreement States as permanent implant manual brachytherapy sources, as approval for both radioembolization products as a device rather than a radiopharmaceutical was sought and obtained from the FDA. However, the NRC has developed a Microsphere Brachytherapy Sources and Devices licensing guidance document (NRC, 2012) relating how its "Medical Use of Byproduct Material" (NRC, 2013) regulations apply to radioembolization due to its unique nature (the infusion of millions of individual sources in solution directly into the hepatic arterial bloodstream). The NRC has chosen to categorize radioembolization under 10 (Code of Federal Regulations CFR) 35 Subpart K—Other Medical Uses of Byproduct Material or Radiation From Byproduct Material (10 CFR 35.1000) as opposed to Subpart F—Manual Brachytherapy; and the guidelines describe how licensees comply with certain aspects of the 10 CFR 35 general requirement Subparts A, B, C, L, and M. The guidelines specifically address authorized user (AU) training and experience; the written directive; ^{90}Y microspheres source leak tests (explicitly stated as not applicable) and inventory; patient release; labeling; medical event reporting; and waste disposal issues related to radionuclidic impurities in the microspheres. Certain important statements in the NRC guidelines regarding the various radiation safety aspects of radioembolization covered in this chapter will be mentioned in the corresponding sections that follow.

7.2 THE WRITTEN DIRECTIVE

Internal administration of a therapeutic amount of RAM requires, by regulation, what is called a written directive, defined by the NRC as "an authorized user's (AU) written order for the administration of byproduct material or radiation from byproduct material to a specific patient or a human research subject" (NRC, 2013). The information required to be present in a written directive is specific to the category under which a particular RAM therapy falls, and there are currently six categories defined in the NRC regulations. Radioembolization, due to its unique nature, has characteristics that resemble those in both "unsealed byproduct material other than sodium iodide I-131" and "all other brachytherapy" categories. As a consequence, the

NRC has established special case written directive requirements for ^{90}Y microspheres liver-directed therapy (NRC, 2012). First, the prescription can be in terms of either radiation absorbed dose (rad or Gy) or activity (mCi or GBq). The documented written directive must include (in addition to the obvious patient identification, date, and AU signature) the following information:

Prior to treatment:

1. Treatment site (anatomical target, e.g., whole liver)
2. Radionuclide and physical form (i.e., "^{90}Y microspheres")
3. Prescribed absorbed dose or activity
4. Manufacturer of the particular ^{90}Y microspheres device used
5. The statement "or dose delivered at stasis" or "or activity delivered at stasis" (if appropriate, e.g., for SIR-Spheres)
6. Maximum acceptable absorbed dose or activity in the lungs and any other possible sites (e.g., gastrointestinal tract) due to shunting

After treatment (within 24 hours):

1. Absorbed dose or activity delivered to the treatment site
2. Absorbed dose or activity delivered to nontarget sites (e.g., lungs)
3. If either stasis or a patient condition emerging during the treatment occurred and prevented the full absorbed dose or activity from being delivered (≥80% of prescribed, per NRC regulations), a statement to that effect must be recorded

(If multiple sites, e.g., right and left lobes of the liver, are treated separately, then a separate written directive is required for each.)

7.2.1 AUTHORIZED USER

AU status for the medical use of RAMs is restricted to those physicians (or dentists or podiatrists) who meet certain minimum didactic, laboratory, and clinical training, as well as board certification and licensure requirements (NRC, 2015). These requirements are specific to each diagnostic and therapeutic use category and correspond to specific physician specialties (e.g., "AU Eligible" Radiologist, Nuclear Medicine Physician,

Radiation Oncologist). The NRC recognizes three types of AU for ^{90}Y microspheres liver-directed therapy (NRC, 2012):

1. A physician board certified in radiation oncology, qualified and identified as an AU on a licensee's RAM license for manual brachytherapy
2. A physician board certified in nuclear medicine, qualified and identified as an AU on a licensee's RAM license for all therapeutic uses of unsealed RAM
3. An IR physician meeting a specific set of requirements related to ^{90}Y microsphere's liver-directed therapy:
   a. 80 hours of didactic and laboratory training in the radiation physics, instrumentation, protection, activity measurement (assay), and biology aspects of RAM (including ^{90}Y)
   b. ^{90}Y microspheres AU supervised work experience:
      i. RAM ordering, receiving, surveying, and unpacking
      ii. Activity assay instrument quality control procedures
      iii. Absorbed dose and activity aspects of radioembolization treatment planning
      iv. ^{90}Y microspheres activity calculation, assay, and preparation
      v. Medical event prevention administrative controls
      vi. RAM spill containment and decontamination procedures, including those specific to microspheres
      vii. Radioembolization patient case history follow-up and review

In addition to the aforementioned requirements, each candidate AU must undergo documented training on the operation, safety, and clinical applications of each ^{90}Y microspheres delivery system for which AU status is sought and successfully complete at least three hands-on treatments under the direct supervision of either the manufacturer or a current AU of the delivery system. In the case of training by the manufacturer, at least three directly supervised *in vitro* simulated treatments should be performed first. Finally, the AU is responsible for ensuring that all personnel to whom the AU delegates ^{90}Y microsphere preparation, measurement, absorbed dose calculations, and administration are adequately trained (e.g., AU ineligible interventional radiologists to whom the AU has delegated performing the actual infusion of the microspheres).

7.2.2 YTTRIUM-90 MICROSPHERES ABSORBED DOSE-RELATED PROPERTIES

Before presenting the specifics of the methods employed for determining the absorbed dose or activity required for SIR-Spheres and TheraSphere radioembolization, it is important to understand the underlying radiation absorbed dose-related properties of ^{90}Y microspheres upon which the prescriptions of the two products are currently based.

Yttrium-90 is considered a pure beta (β) emitter, whereby (1) 100% of its decay occurs via β$^-$ emission to a stable daughter as opposed to a β-emitting radionuclide that also has significant alternative electron (e$^-$) capture and/or positron (e$^+$) branching fractions (e.g., copper-64). (2) Its secondary Auger and internal conversion (IC) e$^-$, gamma (γ), and characteristic daughter x-rays emissions are negligible (Eckerman and Endo, 2008; Dezarn et al., 2011). The dominant (99.9885%) β$^-$ emission has average/endpoint energies of 0.935 MeV/2.280 MeV, decaying to ground state zirconium-90 (^{90}Zr). The remaining decays (0.0115%) are lower-energy (0.186 MeV/0.519 MeV) β$^-$ emissions resulting in a 1.760 MeV excited state of ^{90}Zr, which immediately transitions to its ground state via either IC (dominant), a two γ-emission or e$^+$–e$^-$ pair production (Ford, 1955; Johnson, 1955; Selwyn et al., 2007). (A third, even lower-energy emission occurs, but its yield is so small that it can be ignored.) As a consequence, essentially all of the ^{90}Y decay energy emitted (excluding that carried away by antineutrinos) is e$^-$ kinetic energy. The ranges in tissue for the average (0.935 MeV) and maximum (2.280 MeV) energy electrons emitted by ^{90}Y are 4.0 and 11.3 mm, respectively (Cole, 1969; ICRU, 1984); and 50% and 90% of the absorbed dose from a point source of ^{90}Y are deposited within 2.3 and 5.2 mm, respectively (Berger, 1976). Thus, 100% of the energy emitted can be assumed to be deposited locally at a macroscopic level of an organ or tumor (Pasciak and Erwin, 2009; Dezarn et al., 2011). (Some energy is not absorbed locally, as

secondary bremsstrahlung x-ray and γ radiations do occur. However, their yields are quite small, and thus 100% local deposition is a reasonable approximation.)

The mean energy emitted per ^{90}Y decay is 1.498E–13 Gy-kg/Bq-s (Eckerman and Endo, 2008); and the two microspheres products are biocompatible but not biodegradable, and therefore are permanent implants once they are trapped at the capillary level (in either tumors, the normal liver or the lungs). Thus, no ^{90}Y is cleared biologically, and all activity decays in place with an effective half-life equal to the physical half-life of ^{90}Y (64.24 hours). The cumulative energy deposited in tissue per GBq of ^{90}Y microspheres in that tissue is

$$49.98 \text{ Gy} \cdot \text{kg} / \text{GBq} = \frac{\begin{array}{l} 1.498E - 13 \text{ Gy} \cdot \text{kg} / \text{Bq} \cdot \text{s} \\ \times 1E9 \text{ Bq} / \text{GBq} \\ \times 3.6E3 \text{ s} / \text{h} \times 64.24 \text{ h} \end{array}}{\ln(2)} \quad (7.1)$$

where 64.24 hours/ln(2) is the integral of $e^{-\ln(2)t/64.24\,h}$ from $t = 0$ to infinity (Dezarn et al., 2011) (49.98 is often rounded off to 50). This greatly simplifies the calculation of absorbed dose at the level of tumors, normal liver, and lungs and is exploited by the current methods of prescribing SIR-Spheres and TheraSphere.

7.2.3 SIR-SPHERES PRESCRIPTION

Three methods of prescribing SIR-Spheres have been developed and recommended by the manufacturer; two are activity-based empiric (so-called "basic" and "body surface area" or "BSA") methods, and the third is an absorbed dose-based partition model (Sirtex Medical Limited, 2003, 2010; Salem and Thurston, 2006; Dezarn et al., 2011; Lam et al., 2014; Braat et al., 2015). All three will be described, although the basic empiric method is now considered a legacy method, having been essentially superseded by the BSA method.

SIR-Spheres was originally developed as a whole-liver treatment for hepatic metastases from colorectal cancer with adjuvant chemotherapy, both delivered via the hepatic artery; and the 3 GBq maximum administered activity and 0.20 lung shunt fraction (SF) limits were established empirically early on (Kennedy et al., 2007). These limits correspond to maximum tolerated liver and lung absorbed doses of 80 and 30 Gy, respectively. The 3 GBq limit was derived from a worst-case uniform whole-liver uptake and 80 Gy absorbed dose with no lung shunting, the energy deposited per GBq, and a 1.91 kg reference adult liver mass:

$$3\,\text{GBq} = \frac{80\ \text{Gy} \times 1.91\,\text{kg}}{50\,\text{Gy} \cdot \text{kg} / \text{GBq}} \qquad (7.2)$$

The maximum SF (0.20) was derived from a 30 Gy lung absorbed dose limit, the energy deposited per GBq, a 1 kg reference adult male lung mass, and the maximum 3 GBq:

$$0.20 = \frac{30\,\text{Gy} \times 1\,\text{kg}}{50\,\text{Gy} \cdot \text{kg} / \text{GBq} \times 3\,\text{GBq}} \qquad (7.3)$$

7.2.3.1 Empiric methods

The basic empiric method starts with the 3 GBq activity limit and scales the activity downward to a more conservative level according to both the estimated fractional tumor involvement (TI) in the liver and SF as shown in Tables 7.1 and 7.2.

The BSA method Chapter 5, Section 5.8.2 was later developed after unacceptable levels of clinical and laboratory toxicities were found when using the basic method. The BSA formula for calculation of the base amount of activity to administer (Equations 5.6 and 5.7) was developed empirically; and although the basis upon which the specific formula parameters were derived is nowhere reported in the literature, it has nonetheless effectively

Table 7.1 SIR-Spheres basic empiric method base prescribed activity level

% Tumor	GBq
>50	3
25≤% tumor≤50	2.5
<25	2

Table 7.2 SIR-Spheres basic and BSA empiric method prescribed activity level reduction

% Lung shunt	GBq % reduction
<10	0
10 ≤ % LS < 15	20
15 ≤ % LS ≤ 20	40
>20	No treatment

supplanted the basic method as the empiric method of choice. (The reduction in the base amount of activity to administer as a function of % SF is identical to that used for the basic method.)

The basic formula for the BSA method (Equation 5.7) assumes a whole-liver treatment. However, lobar (either separate left and right or just one or the other) and segmental SIR-Spheres treatments have become commonplace over time. The total activity for lobar or segmental treatments is typically scaled according to their estimated percentage of total liver volume (Equation 5.8). The standard (nominal normal) right/left lobar volume split is 70%/30%, but the split is typically determined on a patient-by-patient basis as diseased liver anatomy often diverges from normal and % TI between lobes and segments to be treated is often variable.

The pretreatment part of the written directive for the BSA method should include the lung shunt (% or fraction), the prescribed activity for each treatment target anatomy (whole liver, lobe, or segment) A(GBq) (Equation 5.8), and the activity to be delivered to the lungs for each treatment target:

$$A\,(\text{GBq}) \times \text{SF} \qquad (7.4)$$

as well as the overall cumulative activity to be delivered to the lungs for all treatments. The posttreatment part should include the net total and lung activities delivered for each treatment:

$$A\,(\text{GBq}) \times (1 - W) \qquad (7.5)$$

$$A(\text{GBq}) \times \text{SF} \times (1 - W) \qquad (7.6)$$

as well as net overall cumulative activity delivered to the lungs. Here, A(GBq) is the preinfusion dose calibrator assay of the treatment target prescribed activity (see Section 7.3); W is the infusion waste activity fraction computed as the ratio of the infusion waste activity and pretreatment A(GBq) survey meter readings (see Section 7.3 and 7.7), and all measurements are decay corrected to the time of infusion.

7.2.3.2 Partition model method

The third manufacturer-recommended method of prescribing SIR-Spheres is a more robust, patient-specific, and absorbed dose-based method called the "partition model" (Ho et al., 1996; Sirtex Medical Limited, 2003; Dezarn et al., 2011; Braat et al., 2015), whereby the total administered activity A_0(GBq) is assumed to be apportioned among three partitions—tumor, normal liver, and lung—according to relative blood flow and SF Chapter 5, Section 5.8.4). The activity to administer is back calculated from prescribed target tumor (D_{tumor}(Gy)), normal liver (D_{normal}(Gy)), and lung (D_{lung}(Gy)) maximum tolerated absorbed doses and estimates of tumor-to-normal liver blood flow ratio (T:N), target tumor (M_{tumor}(kg)), and normal liver (M_{normal}(kg)) masses, and optionally lung mass, M_{lung}(kg), although 1 kg is typically assumed (the 0.8 kg reference adult female lung mass could be used for female patients) as an estimate of patient-specific lung mass that is difficult to obtain. The absorbed dose-based constraints on A_0(GBq) are then derived using the model parameters (Equations 5.10 through 5.14):

$$A_{lung}(GBq) \leq \frac{D_{lung}(Gy) \times M_{lung}(kg)}{(49.98 \text{ Gy} \cdot \text{kg} / \text{GBq} \times \text{SF})} \quad (7.7)$$

$$A_{normal}(GBq) \leq \frac{D_{normal}(Gy) \times \left[\begin{array}{c}(T:N \times M_{tumor}(kg)) \\ +M_{normal}(kg)\end{array}\right]}{\left[49.98 \text{ Gy} \cdot \text{kg} / \text{GBq} \times (1-\text{SF})\right]} \quad (7.8)$$

The prescribed activity A_0(GBq) is constrained by upper limits on D_{normal}(Gy) and D_{lung}(Gy), and thus would be constrained to be the lesser of A_{normal}(GBq) or A_{lung}(GBq). If A_0(GBq) is constrained, then D_{tumor}(Gy) and either D_{lung}(Gy) or D_{normal}(Gy) would be recalculated based on the constrained value for A_0(GBq) and the model dose equations. The manufacturer recommends lung and normal liver absorbed dose limits of 25 and 80 Gy (70 Gy if cirrhotic), respectively (Sirtex Medical Limited, 2003; Braat et al., 2015), although a 30 Gy limit for lung and lower values for normal liver, e.g., Emami 30 Gy whole-liver external beam tolerance dose (TD) 5/5 limit (Emami et al., 1991), is commonly employed. The partition model equations are only valid for solitary tumors or multiple tumors with identical T:N values, within the treatment target.

The pretreatment part of the written directive for the partition model method should include the lung shunt (% or fraction), the tumor, normal liver, and lung masses and prescribed absorbed doses, and prescribed total activity A_0(GBq). The posttreatment part should include the net total activity infused and adjusted tumor, normal liver, and lung absorbed doses.

7.2.4 THERASPHERE PRESCRIPTION

The current manufacturer-recommended method of prescribing TheraSphere consists of calculating the infused activity required for the whole liver or separate lobar or segmental treatment targets (A_0(GBq)) based on a treatment target absorbed dose (80–150 Gy is considered therapeutic) assuming a uniform microsphere distribution (D_{avg}(Gy)) and lung 30 Gy (D_{lung}(Gy)) single treatment session and 50 Gy cumulative (repeat treatment) absorbed dose limits (Chapter 5, Section 5.8.3):

$$A_0(GBq) = \frac{D_{avg}(Gy) \times M_{liver}(kg)}{49.98 \text{ Gy} \cdot \text{kg} / \text{GBq} \times (1-\text{SF})}$$
$$\leq \frac{D_{lung}(Gy) \times M_{lung}(kg)}{49.98 \text{ Gy} \cdot \text{kg} / \text{GBq}} \quad (7.9)$$

The 1 kg reference adult male lung mass for M_{lung} (kg) is assumed, unless the reference mass for female patients (0.8 kg) or a patient-specific lung mass estimate is employed (Simon, 2000; Salem and Thurston, 2006; BTG International Ltd., 2010, 2014; Dezarn et al., 2011; Busse et al., 2013; Braat et al., 2015).

The pretreatment part of the written directive for TheraSphere should include the lung shunt (% or fraction); the mass, prescribed activity, absorbed dose for each treatment target and cumulative absorbed dose to the lung. The posttreatment part should include the net total, target, and lung activities infused and adjusted treatment target and lung absorbed doses (Equation 7.10).

7.2.5 ALTERNATIVE RADIOEMBOLIZATION PRESCRIPTION METHODS

The use of other more sophisticated prescription methods, such as artery-specific partition modeling (Kao et al., 2012) or three-dimensional absorbed dose calculation (absorbed dose point kernel, Monte Carlo, or deterministic) and treatment-planning techniques analogous to those used for external beam radiation therapy or brachytherapy (Dieudonné et al., 2011; Petitguillaume et al., 2014), is not precluded for radioembolization. The accuracy of such methods should be validated prior to clinical use, and the NRC states in the ^{90}Y microspheres licensing guidance document that such alternative methods should be submitted (presumably to the NRC or Agreement State regulatory body, although that is not stated explicitly).

7.2.6 REPORTABLE MEDICAL EVENT

The NRC requires that the absorbed dose or activity delivered to each treatment target be either within 20% or within the therapeutic range of that prescribed, unless the reason for falling outside those limits was due to intervention by the patient (NRC, 2013). If the infusion was terminated early due to stasis or other emergent patient condition, it is necessary to include a statement to that effect in the posttreatment written directive (NRC, 2012). The ±20% limit of prescribed activity (empiric method) would be applicable to SIR-Spheres. The ±20% limit of prescribed absorbed dose would be applicable to TheraSphere, SIR-Spheres if the partition model was used, and either product if an alternative absorbed dose-based method of treatment planning was employed. (One could argue that within the prescribed range is applicable for TheraSphere, as a delivered absorbed dose anywhere between 80 and 150 Gy is considered therpeutic [BTG International Ltd., 2014]). If the delivered absorbed dose or activity falls outside these limits and it was not due to the patient, but either a failure of the device, improper technique (e.g., use of a microcatheter with too small a bore size, incorrect microcatheter placement or device assembly), or administering the incorrect activity (e.g., accidently switching separate left and right lobe treatment dosage vials), then it may have to be reported to the NRC or Agreement State regulatory body. The NRC has established conditions under which such a misadministration is considered reportable, as well as timeliness of the reporting following discovery (notification of the referring physician and patient within 24 hours, phone call within 24 hours, and written report within 15 days to the NRC or Agreement State),

and what the written notification must contain (names of licensee and prescribing physician, brief description of event and why it occurred, effect on patient, actions taken or planned to prevent recurrence, and certification that the individual or responsible relative or guardian was notified or why not). The conditions applicable to radioembolization for a reportable medical event are (NRC, 2012):

1. A dose that differs from the prescribed absorbed dose or that which would have resulted from the prescribed dosage by more than
   a. 0.05 Sv (5 rem) effective dose equivalent
   b. 0.5 Sv (50 rem) equivalent dose to an organ or tissue
   c. 0.5 Sv (50 rem) shallow dose equivalent to the skin

   and

   i. The total absorbed dose delivered differs from the prescribed absorbed dose by >±20%
   ii. The total dosage delivered differs from the prescribed dosage by 20% or more or falls outside the prescribed dosage range

2. A shallow dose equivalent to the skin, an equivalent dose to an organ or a tissue other than the treatment site that exceeds
   a. by 0.5 Sv (50 rem) to an organ or a tissue
   b. 50% or more of that expected from the administration defined in the written directive

(excluding, for permanent implants, seeds that were implanted in the correct site but migrated outside the treatment site).

Calculating the estimated difference between delivered and prescribed absorbed dose for a possible reportable medical event where either too much or too little activity was inadvertently delivered to the intended target tissue (normal liver, lungs, and/or tumor) is relatively straightforward. The target prescribed activity, A(GBq); the target mass, M [kg, e.g., computed tomography (CT) or magnetic resonance (MR) volume based]; the estimated fraction of prescribed activity actually delivered to the target, F (can be either less than or greater than 1.0); and the constant 49.98

Gy-kg/GBq can be combined to compute the estimated absorbed dose (Gy) actually delivered to the target tissue:

$$D = \frac{49.98 \times A \times F}{M} \qquad (7.10)$$

and the difference from the prescribed absorbed dose then calculated using D.

Estimating the absorbed dose to an unintended target, whether hepatic or extrahepatic, is more complicated, as a posttherapy ^{90}Y bremsstrahlung single-photon emission computed tomography (SPECT)/CT or positron emission tomography (PET)/CT scan from which accurate estimates of both absolute activity within and mass of the unintended target can be obtained, must be relied upon, and standardized methods of generating such quantitative scans do not yet exist.

7.3 DOSAGE DELIVERY TO AND PREPARATION IN THE HOT LAB

The final form of each radioembolization dosage ready to be administered to the patient is similar for both products, that is, radiolabeled microspheres in solution (sterile water or saline) in an infusion v-vial contained within an acrylic shield. However, the process by which each dosage is dispensed into that final form currently differs greatly between the two. Thus, the specifics of dosage delivery and preparation of each radioembolization product will be discussed separately.

Licensee verification of the activity of dosages of either radioembolization product requires the calibration of all dose calibrators that will be used for this purpose. Dose calibrator manufacturers have established standard settings for each of their dose calibrators for assaying sources of radionuclides commonly used in nuclear medicine and that emit an abundance of characteristic gamma and/or x-ray photons (Carey et al., 2012). For some of those radionuclides, minor setting adjustments or correction factors based on source form factor (e.g., syringe versus vial) may be necessary due to a slight variation in the radiation output of abundant low-energy photon emissions and a result of location of the source within the dose calibrator well

[self-attenuation within the source; attenuation by the well insert, vial/syringe dipper, or liner (e.g., copper) employed; or a combination thereof]. The external radiation from a source of a given activity of a pure β-emitting radionuclide such as ^{90}Y, on the other hand, is highly variable. This is due to the predominance of secondary bremsstrahlung x-ray radiation, the amount of which is highly dependent upon source material composition and geometry, as well as intervening material. As a consequence, a separate calibration is needed for each ^{90}Y source form factor encountered.

Calibration of a dose calibrator for assay of dosages of either radioembolization product is currently accomplished by requiring the manufacturer to ship a certificate of analysis with at least the first three dosages, indicating the calibrated (as opposed to nominal or estimated) activity of the particular dosage with a reasonably small uncertainty, along with the date and time of the activity calibration (Dezarn et al., 2011). The calibrated activity of each dosage is decay corrected to the time of measurement with each dose calibrator providing an expected reading and the dose calibrator setting adjusted until the expected reading is achieved. Repeating this process for at least three dosages allows for calculation of an average setting in an effort to improve the overall accuracy of subsequent dosage activity assays. Periodic recalibration (e.g., annually) would be prudent to maintain equivalence of activity assay with the manufacturer. (Optimal and consistent placement of the radioembolization dosage within the dose calibrator will yield the best calibration result. A vial/syringe dipper such as the Biodex Atomlab model 086-242 allows such placement of both radioembolization product dosage vials. In addition, the shipping vial containing SIR-Spheres must first be gently shaken back and forth to suspend the microspheres uniformly in solution. Tipping the vial upside down should be avoided during the process of suspension, as that may cause some of the microspheres to be trapped around the periphery of the vial's septum.)

Calibration of dose calibrators for SIR-Spheres dosage assay is currently not traceable to the U.S. National Institutes of Standards and Technology (NIST), although the Australian Nuclear Science and Technology Organization (ANSTO, Australian equivalent of NIST) and the Australian Radiopharmaceuticals and Industrials have made activity measurements of SIR-Spheres, and (as of 4 years ago) the manufacturer was still using an ANSTO-traceable calibrated ion chamber for activity measurement (Dezarn et al., 2011). The manufacturer of TheraSphere participates in the NIST Radioactivity Measurement Assurance Program, where settings for commercial Capintec dose calibrators have been established; a secondary measurement standard for routine calibration is maintained; and the manufacturer's dose calibrator measurement of activity is periodically verified (Dezarn et al., 2011). However, no standards organization-traceable dose calibrator standards or accredited laboratory calibration methods exist for either products at this time, although methods for developing NIST-traceable activity standards and calibrations for SIR-Spheres on an institution-by-institution basis have been published (Mo et al., 2005; Selwyn et al., 2007, 2008). Until such activity standards and methods are available for all institutions to exploit for periodic dose calibrator calibration (including SIR-Spheres in variable solution volumes in the shipping vial, as exemplified in Figure 7.1), each institution must rely on cross-calibration with the manufacturer. As a result, there will most likely be a larger uncertainty in the measurement of radioembolization dosages compared with, e.g., that for the ^{90}Y radioimmunoconjugate Zevalin[*], for which both a NIST-traceable transfer standard (Figure 7.2) and a calibration procedure exist (Thieme et al., 2004). (One could attempt a self-calibration for assaying radioembolization form factor dosages by first calibrating the dose calibrator with the Zevalin transfer standard and assaying a source of ^{90}Y chloride in the Zevalin form factor, calibrating for microspheres suspended in solution in the shipping vial [SIR-Spheres] and v-vial within its acrylic shield [both products] by dispensing a known amount of the calibrated ^{90}Y chloride activity plus fluid resulting in a volume employed for the particular radioembolization product, and adjusting the dose calibrator setting until the correct reading is obtained. Uncertainties associated with such a calibration include that for the transfer standard itself [95% confidence interval = ±4.7%]; activity loss during the dispensing that is not accounted for; and the fact that chloride solution does not mimic microspheres exactly, which have a density substantially greater than that of water and thus for which a larger amount of secondary bremsstrahlung radiation for the same activity will be produced, and which precipitate to the bottom of the vial as opposed to remaining uniformly suspended in solution.)

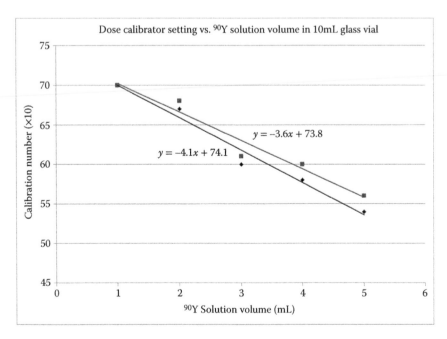

Figure 7.1 Settings for two Capintec CRC-15R dose calibrators for assaying ^{90}Y in solution in a 10 mL glass vial as a function of solution volume. Employing the 5 mL of volume setting to assay 1 mL of solution results in an overestimation of activity on the order of 10%.

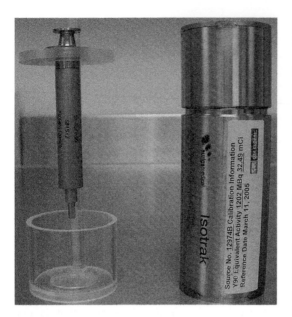

Figure 7.2 NIST-traceable ^{90}Sr/^{90}Y transfer standard for calibration of dose calibrators for assaying ^{90}Y Zevalin ($2\sigma = 4.7\%$) that simulates 4 mL of ^{90}Y Zevalin in a 10 mL Becton–Dickinson plastic syringe. The standard is accompanied by a table of correction factors for other solution volumes that vary linearly from 0.983 at 9 mL to 1.017 at 1 mL.

7.3.1 SIR-SPHERES DOSAGE DISPENSING AND PREPARATION

SIR-Spheres are shipped from the manufacturer to the hot lab on the day of treatment, as a nominal amount of activity (3 GBq ± 10%) in 5 mL of sterile water contained within a 10 cc glass vial inside a lead pot, regardless of dosage (Figure 7.3a). The activity in the shipped vial is calibrated for 18:00 hours United States Eastern Time (U.S. ET) on the day of treatment, and the product has a shelf-life of 24 hours postcalibration (Sirtex Medical Limited, 2003, 2010; Dezarn et al., 2011). (One may request the delivery of the product for treatment one day prior to the calibration date, if the total prescribed activity is substantially greater than 3 GBq based on a treatment-planning technique that differs from those recommended by the manufacturer.) An administration kit, infusion v-vial, and acrylic shield for each dosage to be dispensed are shipped along with the activity but the licensee is responsible for dispensing each of the one or more

prescribed SIR-Spheres treatment dosages from the shipping vial to an infusion v-vial (Figure 7.3b).

Preparation and dispensing of each SIR-Spheres treatment dosage is a multistep process (Dezarn et al., 2011). The first step consists of assaying the initial activity in the shipping vial. The microspheres must be uniformly suspended in solution (in the same fashion as that for initial dose calibrator calibration), and the vial is assayed immediately in the dose calibrator using the previously established dose calibrator setting. This assay allows the activity concentration in the shipping vial to be calculated (by division by 5 cc). The second step consists of the calculation of the volume to withdraw for the first dosage by dividing the required activity (prescribed activity decay-corrected from the anticipated time of infusion back to the time of dispensing) by the activity concentration, and then, transferring that volume from the shipping vial into the infusion v-vial via a 5 cc syringe. (A 3 cc syringe may be preferable for volumes less than 2 cc, for more accurate volume withdrawal.) The third step consists of assaying the activity remaining in the shipping vial and calculating the activity dispensed as the difference between the initial and remaining activities. If the difference is too high, then too much activity was withdrawn and some must be transferred back to the shipping vial, and if too low then additional activity needs to be transferred to the infusion v-vial. (It is worth noting here that dose calibrator cross-calibration with the manufacturer is only for the initial activity in 5 mL, whereas it can be demonstrated that a 5 mL volume dose calibrator setting can cause an overestimate on the order of 10% in a 1 mL residual activity assay. On the other hand, no standardized methodology exists yet for calibrating the dose calibrator for assaying the shipping vial containing microspheres in a variable volume of solution.) Steps 2 and 3 must be performed quickly, while the microspheres are kept in suspension, and are repeated for each additional treatment dosage to be dispensed. In addition, the shipping vial, v-vial, and syringe should all be appropriately shielded during the process to minimize radiation exposure (especially hand β absorbed dose) to the person performing the dispensing. (It may be possible to wait to transfer the activity from the syringe to the v-vial until it has been determined that the correct amount of activity has been withdrawn from the shipping vial into the syringe. However,

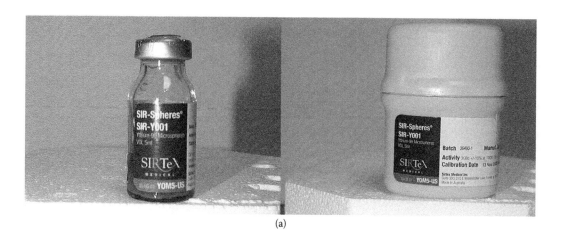

(a)

(b)

Figure 7.3 SIR-Spheres manufacturer's dose vials. **(a)** A ten milliliters glass shipping vial containing 3 GBq ± 10% at 18:00 U.S. ET and associated lead pot and **(b)** patient dose delivery v-vial containing dispensed target volume activity and associated acrylic shield.

caution must be observed with such an approach to ensure the procedure is performed quickly enough to avoid the aggregation of microspheres at the bottom of the syringe, which could impede the transfer of the entire contents of the syringe to the v-vial.) A detailed step-by-step procedure for dispensing SIR-Spheres dosages is provided in the manufacturer's package insert (Sirtex Medical Limited, 2010). An example dispensing worksheet is shown in Figure 7.4.

7.3.2 THERASPHERE DOSAGE DISPENSING AND PREPARATION

The activities of all TheraSphere dosages are calibrated for Sunday at 12:00 U.S. ET and each dosage has a treatment shelf-life of 12 days postcalibration (BTG International Ltd., 2010, 2014; Dezarn et al., 2011). Originally, the manufacturer only allowed the ordering of dosages with calibrated

SIR-SPHERES THERAPY DOSE DISPENSING/ASSAY WORKSHEET

Patient name :_____   **MRN:**_____   **Date:**_____

Time:_____

Prescribed Dose: _____ **mCi**   **Actual Dose Administered:** _____ **mCi**

Supplies needed to prepare dose: (enter value calculated at bottom of worksheet)

Y-90 SIR-Spheres dose (5 ml in 10 ml glass vial in lead (Pb) pig)
V-vial in acrylic shield
2 venting needles
B-D 5 ml syringe (if more than 2 ml to be drawn) OR 3 ml syringe (if 2 ml or less to be drawn) and Pb pig
10 ml sterile water
Alcohol prep pads

Procedure

1. Calculate concentration:
A. Remove Pb pig cap, gently shake vial in pig 5 to 6 times to uniformly suspend spheres in the solution
B. Set pig on counter behind Pb drawing station shield and assay dose in the PET bench calibrator (S/N 151034)
C. Set dose calibrator Cal. # for number of ml in vial (___ x 10 from Table below) and obtain actual activity (mCi) and place it back in the pig. (Note: record the initial stable reading, as it will then drift due to settling of the spheres!)
D. Divide activity by the number of ml in vial to calculate mCi per ml

Original activity in vial = _____ **mCi ÷** _____ **ml =** _____ **mCi/ml Concentration**

2. Calculate volume required for the prescribed dose:
(Divide the prescribed dose by the mCi per ml calculated above)

Prescribed dose _____ **mCi ÷** _____ **mCi/ml Concentration =** _____ **ml required**

3. Draw the calculated volume from the dose vial:
A. Open up the V-vial by flipping up the metal tab, place the vial in an acrylic shield, wipe with an alcohol prep pad and insert a venting needle
B. Open the dose vial by flipping up the metal tab, wipe the top of the vial with an alcohol prep pad and gently shake vial to suspend the spheres before drawing the calculated volume
C. Draw the calculated volume using the 3 ml (if <= 2 ml needed) or 5 ml syringe (if > 2 ml needed) without any air. (Note: Place the syringe into its Pb pig, but do NOT dispense into v-vial at this point!)

4. Measure residual to estimate activity drawn:
Using the same motion, remix residual in the dose vial and assay, using the residual ml Cal. # in the table below

_____ **ml –** _____ **ml drawn into syringe =** _____ **ml remaining (round to nearest 0.2 ml)**

Vial Vol (ml) Cal. #				
0.2 72 x 10	1 69 x 10	2 65 x 10	3 61 x 10	4 57 x 10
0.4 71 x 10	1.2 68 x 10	2.2 64 x 10	3.2 60 x 10	4.2 56 x 10
0.6 71 x 10	1.4 67 x 10	2.4 63 x 10	3.4 60 x 10	4.4 55 x 10
0.8 70 x 10	1.6 67 x 10	2.6 62 x 10	3.6 58 x 10	4.6 54 x 10
	1.8 66 x 10	2.8 62 x 10	3.8 58 x 10	4.8 53 x 10

Residual activity = _____ **mCi**   5 U5 (53x10)

Original _____ **mCi – Residual** _____ **mCi in dose vial =** _____ **mCi in syringe**

If estimated mCi is not within +/-10% of prescribed dose, empty or add activity from/to syringe into/from dose vial!
After the estimated activity in the syringe is satisfactory, inject the syringe contents into the V-vial and qs the V-vial to 3/4th full with sterile water using the same syringe and needle used to draw the dose.
Remove the venting needle, replace the shield cap, and place the dose in the Sir-Spheres acrylic drawing station.

Figure 7.4 Example SIR-Spheres patient dose dispensing worksheet. The amount of activity required for each SIR-Spheres patient dose must be transferred from the manufacturer's shipping vial to the delivery v-vial based upon the computed activity concentration at the time of dispensing.

activities of 3, 5, 7, 10, 15, or 20 GBq. Dosages may now be ordered with activities between 3 and 20 GBq in 0.5 GBq increments. Also originally, each dosage had a treatment shelf-life of only 5 days postcalibration. However, the manufacturer later increased that up to 12 days to allow what it calls extended shelf-life treatment, whereby an enhancement of the embolic effect for the same activity delivered to the same treatment volume may be achieved. The enhancement is achieved by ordering the dosage at a much higher calibrated activity 1 week earlier and having it decay longer until it reaches the treatment activity level resulting in an infusion of a much larger number of spheres for the same infused activity.

Preparation of TheraSphere dosages is much simpler than that for SIR-Spheres (BTG International Ltd., 2010; Dezarn et al., 2011). During the planning phase, the treatment is scheduled for a particular day of the week and the activity required on the date of calibration (Sunday of that week or the week before) that will decay to the prescribed activity on the day of treatment is

computed and ordered. The manufacturer provides a so-called "treatment window illustrator" electronic spreadsheet, where treatment target volume, desired absorbed dose, variance from U.S. ET, % lung shunt, and anticipated % waste are entered; and tables of estimated treatment target absorbed doses are computed for specific times on each day of the first or second week postcalibration for each allowed noon Sunday U.S. ET calibrated activity (Figure 7.5). The calibrated activity for which those calculated absorbed doses are within the range of the prescribed absorbed dose on the planned day of treatment (i.e., within the treatment time window)

| Patient Name: | XYZ (enter data) | | Patient ID: | ### (enter data) | Target Tissue: | X Lobe (enter data) |

| Target Volume (cc): | 1000.0 | | Target Liver Mass (kg): | 1.030 |

| Desired Dose (Gy): | 120 |

| Time Zone Variance (h): | 1 | (see Time Zones tab for details) Places in this Time Zone: Dallas Texas / Mexico city Mexico |

| Lung Shunt Fraction (% LSF): | 5.00% |

| Anticipated Residual Waste (%): | 1.00% | Optional estimated value |

| Previous Dose to the Lungs (Gy): | 0 |

| Required Activity at Administration (GBq): | 2.63 | This value is corrected for LSF and Residual Waste if values are entered above. |

| Calculated Dose to Lungs (Gy): | 6.51 | Dose Limit to the Lungs per treatment (Gy): | 30 | See Package Insert or Instructions for Use |
| Lung Dose within recommended limit for treatment |
| Cumulative Dose to Lungs (Gy): | 6.51 | Cumulative Dose Limit to the Lungs (Gy): | 50 |
| Lung Dose within recommended cumulative limit for treatment |

Use the following tables to select a dose size where the Desired Dose (above) is at a suitable treatment time.

Dose Size Selected (GBq): 10 GBq Optional field for Medical Professional to document treatment dose selected

Date & Time for Administration: Thursday, Week 1 Optional field for Medical Professional to document treatment window selected

Tables below show the dose to perfused target tissue, accounting for target mass, time zone variance, lung shunt fraction and residual waste.

Dose Delivered (Gy) for: 3 GBq dose size — Week 2 treatment

Time	Sunday	Monday	Tuesday	Wednesday	Thursday	Friday	Saturday	Sunday	Monday	Tuesday	Wednesday	Thursday	Friday
8:00 AM	Calibration Day @ 12:00 Eastern Time	109	84	65	50	39	30	23	18	14	11	8	6
12:00 PM		105	81	62	48	37	29	22	17	13	10	8	6
4:00 PM		100	77	60	46	35	27	21	16	13	10	7	6
8:00 PM		96	74	57	44	34	26	20	16	12	9	7	6

Dose Delivered (Gy) for: 5 GBq dose size — Week 2 treatment

Time	Sunday	Monday	Tuesday	Wednesday	Thursday	Friday	Saturday	Sunday	Monday	Tuesday	Wednesday	Thursday	Friday
8:00 AM	Calibration Day @ 12:00 Eastern Time	182	140	108	84	64	50	38	30	23	18	14	10
12:00 PM		174	134	104	80	62	48	37	28	22	17	13	10
4:00 PM		167	129	99	77	59	46	35	27	21	16	12	10
8:00 PM		160	123	95	73	57	44	34	26	20	15	12	9

Dose Delivered (Gy) for: 7 GBq dose size — Week 2 treatment

Time	Sunday	Monday	Tuesday	Wednesday	Thursday	Friday	Saturday	Sunday	Monday	Tuesday	Wednesday	Thursday	Friday
8:00 AM	Calibration Day @ 12:00 Eastern Time	255	196	152	117	90	70	54	41	32	25	19	15
12:00 PM		244	188	145	112	86	67	51	40	31	24	18	14
4:00 PM		234	180	139	107	83	64	49	38	29	23	17	13
8:00 PM		224	173	133	103	79	61	47	36	28	22	17	13

Dose Delivered (Gy) for: 10 GBq dose size — Week 2 treatment

Time	Sunday	Monday	Tuesday	Wednesday	Thursday	Friday	Saturday	Sunday	Monday	Tuesday	Wednesday	Thursday	Friday
8:00 AM	Calibration Day @ 12:00 Eastern Time	364	281	216	167	129	99	77	59	46	35	27	21
12:00 PM		348	269	207	160	123	95	73	57	44	34	26	20
4:00 PM		334	257	199	153	118	91	70	54	42	32	25	19
8:00 PM		320	246	190	147	113	87	67	52	40	31	24	18

Dose Delivered (Gy) for: 15 GBq dose size — Week 2 treatment

Time	Sunday	Monday	Tuesday	Wednesday	Thursday	Friday	Saturday	Sunday	Monday	Tuesday	Wednesday	Thursday	Friday

Figure 7.5 Top portion of the TheraSphere manufacturer-provided treatment window illustrator spreadsheet used for determining how much activity calibrated at Sunday noon U.S. ET to order given a specified target absorbed dose and scheduled day of treatment postcalibration.

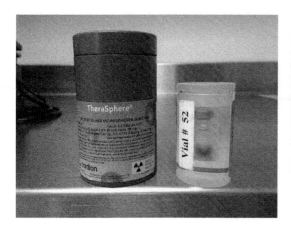

Figure 7.6 TheraSphere patient dose delivery v-vial in its acrylic shield and associated lead pot.

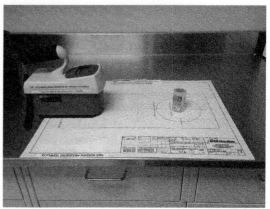

Figure 7.7 Setup for pretreatment ion chamber survey meter exposure rate reading of a TheraSphere dose delivery v-vial, using the fixed-geometry template provided by the manufacturer (30 cm v-vial center-to-ion chamber center distance).

is the one that is ordered. Each ordered dosage is then delivered prior to or on the day of treatment, with the microspheres already in 0.6 mL of sterile water in the infusion v-vial, sealed in its acrylic shield, and ready for administration to the patient (Figure 7.6). Thus, the only dosage "preparation" required is activity assay (and exposure rate measurement, which is described in the next section).

7.3.3 EXPOSURE RATE MEASUREMENT OF DOSAGE VIALS

A measurement of the exposure rate or dose equivalent rate (e.g., mR/h, μSv/h) from each dosage vial in its acrylic shield using a calibrated ion chamber survey meter and a fixed geometry is required for both radioembolization products (Figure 7.7). This measurement is necessary in order to estimate the infusion waste activity fraction and subsequently net activity administered as the only way to "assay" the waste is by survey meter measurement (see Section 7.7.1), the measurement of which is then compared with that of the dosage vial to compute the estimated fraction. Furthermore, this measurement must be made correctly as it is the only basis upon which a possible misadministration (net administered activity or absorbed dose outside of ±20% of prescribed) can be detected. An incorrect survey meter measurement of the dosage vial results in a corresponding incorrect estimated net administered activity (and by way of direct proportionality, net absorbed dose) and could result in either a missed or false-positive misadministration.

A linear regression analysis of exposure rate versus dosage vial activity for the first 20 or so measurements may be useful (Figure 7.8) to derive a dosage vial exposure rate constant (e.g., mR/mCi-h, μSv/GBq-h) that can then be used for comparing future dosage vial measurements against an expected value (exposure rate constant times assayed activity) to assess whether or not the measurement was made correctly (Erwin, 2012; Gress and Erwin, 2015). A geometric template for survey meter measurement of both the dosage vial and radioactive waste container has been developed for one of the radioembolization products (Figure 7.7).

7.3.4 DOSAGE CART PREPARATION

After each treatment dosage has been prepared, assayed, and had its exposure rate measured, the components of the device apparatus for each treatment dosage are placed on a separate cart for transport from the hot lab to the IR procedure room. The top surface should be sterilized and draped with fresh chux to absorb any spilled or leaked microspheres. The SIR-Spheres infusion system consists of the delivery box containing the dosage v-vial in its acrylic shield placed in the retaining ring and both the acrylic shield cap and delivery box lid on and the dosage delivery set (infusion lines plus needles) in its sterile package (Figure 7.9a). The TheraSphere infusion system consists of the delivery box containing the

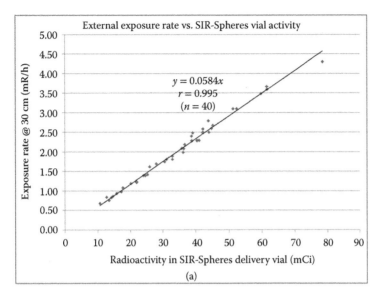

(a)

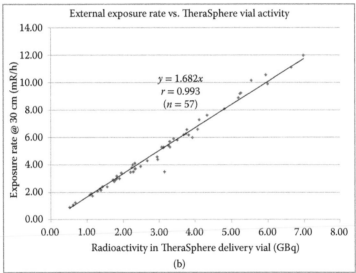

(b)

Figure 7.8 Graphs of fixed-geometry ion chamber survey meter exposure rate reading of ^{90}Y microspheres dose delivery v-vial in acrylic shield as a function v-vial activity for **(a)** SIR-Spheres and **(b)** TheraSphere. An exposure rate constant (e.g., mR/mCi-h) can be computed as either the slope of a linear regression of measured exposure rate versus activity or the statistical average ratio of exposure rate to activity, and then used to compare the actual reading of each subsequent vial with an expected reading to ensure a proper reading for the later estimation of fraction of total v-vial activity delivered to the patient.

dosage v-vial in its lead pot placed inside in the retaining ring and both the lids (delivery box, lead pot) on a solid-state electronic dosimeter attached to its holder on the outside of the delivery box and the administration kit (infusion lines, needle plunger assembly, and pressure-relief valve and vial) in its sterile package (Figure 7.9b). Duplicate labels should be affixed to both the acrylic shield (SIR-Spheres) or lead pot (TheraSphere) and the delivery box, containing

1. Patient name and medical record number
2. Procedure (e.g., radioembolization)
3. Radionuclide (^{90}Y)
4. Product (SIR-Spheres or TheraSphere)
5. Activity (mCi or GBq)

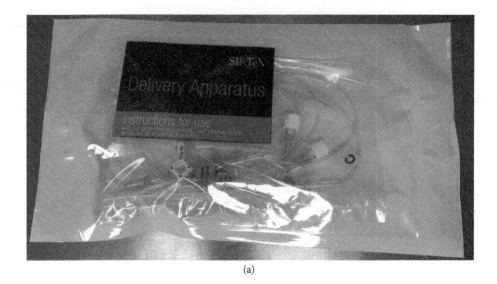

(a)

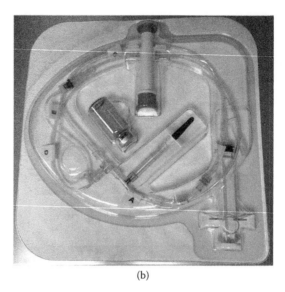

(b)

Figure 7.9 Yttrium-90 microspheres dose administration kits. **(a)** SIR-Spheres (infusion lines and needles) and **(b)** TheraSpheres (infusion lines, needle plunger assembly, infusion syringe, and pressure-relief valve and vial).

6. Anatomical treatment target (e.g., R Lobe)
7. Dates and times of assay (and by whom), anticipated infusion, and expiration

This will help to ensure the correct dosage is administered to the correct anatomical target in the correct patient. Fresh, sterile gloves should be worn while assembling and transporting the cart; and once assembled, the cart may need to be covered with plastic if being transported to a sterile IR suite. Additional items to place on the

lower shelves of the cart are a (preferably pancake) probe Geiger–Mueller (G–M) survey meter; a calibrated ion chamber survey meter (if monitoring of exposure or dose equivalent rate from the patient will be performed); a β detector survey meter (for TheraSphere); a 2-L Nalgene waste container (jar), and an associated acrylic β shield for storing the residual activity in the infusion system (one per treatment dosage); and one or more biological waste trash bags (Figures 7.10 and 7.11). (Labeling each waste jar prior to transport with "^{90}Y,"

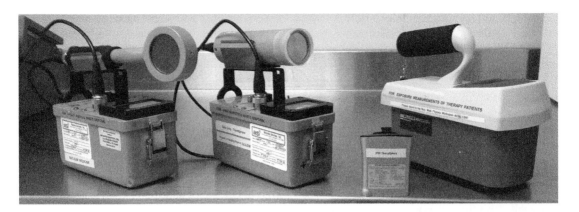

Figure 7.10 Radiation detectors employed during ⁹⁰Y microspheres radioembolization. From left to right: pancake probe G–M survey meter (protective cover removed for detecting β particles); β probe survey meter with plastic attenuator (for measuring count rate from TheraSphere microspheres at the output line–microcatheter junction); electronic personal dosimeter (for measuring the external bremsstrahlung dose rate from the TheraSphere vial); and ion chamber survey meter (for measuring external bremsstrahlung exposure rate from the patient's liver containing microspheres).

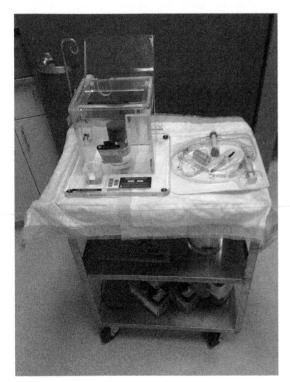

Figure 7.11 TheraSphere treatment delivery cart. Top shelf: dose v-vial in acrylic shield and lead pot, inside the infusion delivery box, and dose administration kit. Middle shelf: biohazard bag for collecting potentially radioactive garments and surface coverings, and a 2-L Nalgene waste jar inside its shield. Bottom shelf: β scintillator, pancake G–M, and ion chamber survey meters.

"SIR-Spheres," or "TheraSphere," date, and anatomical treatment target is advisable, so that only estimated activity and time of day need to be added after postprocedure assay.) Each treatment dosage can be kept behind lead shielding and placed in the delivery box only after IR informs the hot lab that the IR physician is almost ready to infuse in order to reduce the exposure to hot lab personnel.

7.4 RADIATION SAFETY DURING TREATMENT IN THE IR SUITE

7.4.1 IMMEDIATE PREADMINISTRATION

The treatment dosage should be requested and the dosage cart assembled and transported to the IR suite, shortly before the IR physician has completed the embolization of any nontarget vessels into which microspheres could potentially be diverted and positioned the microcatheter in the desired target treatment location in the vasculature. A just-in-time delivery of treatment dosages, as opposed to transporting them to the IR suite well in advance of the radioembolization procedure, has the advantages of both reducing exposure to IR personnel who are unaccustomed to being in the presence of sources of radioactivity and minimizing the possibility of radioactive sources being left unattended in a potentially unsecured area. If the IR suite is within a defined sterile zone, a sterile plastic cover should be placed over the dosage cart once it is assembled for transport and kept in place until the cart is ready to be wheeled into the procedure room.

The first step in a radioembolization is a time-out prior to the dosage cart being wheeled into the procedure room to ensure the correct dosage will be administered to the correct treatment target in the correct patient. Once that has been verified, the cart is wheeled into the procedure room and into position for infusion (as close as possible to the fluoroscopy table, with the delivery box oriented such that its output side is on the patient side of the cart) by the NM technologist or physicist. Towels or absorbent pads should then be placed between the dosage cart and patient entry point with overlap to confine any spill or leakage of microspheres along the path between the output side of the delivery box and femoral artery entry point of the microcatheter. Next, the input and output lines plus needles are removed from their package and primed with sterile water or saline according to the manufacturer's instructions to ensure that there is no air in the infusion lines. Finally, the infusion system is assembled, again according to the manufacturer's instructions. Developing a checklist with time-outs at various steps in the process is highly recommended (Figure 7.12) to help ensure that all assembly steps are correctly followed (Salem and Thurston, 2006). Radioembolization is typically performed infrequently and is a unique form of brachytherapy that differs substantially from routine NM and IR procedures. Furthermore, there will be turnover of NM and IR personnel participating in this procedure over time. These factors increase the likelihood of mistakes that could be detrimental to either the patient or personnel if all the required steps in the process are not adhered to strictly (i.e., misadministrations or contaminations) and justify the use of a checklist even if certain personnel believe that they have mastered the procedure.

Since radioembolization involves RAM and is performed under fluoroscopic guidance, all personnel in the room during the procedure must wear appropriate radiation protection garments (lead aprons and collars, protective eyewear) and whole-body dosimeters (and ring dosimeters for anyone who might receive measurable hand exposure from fluoroscopy x-ray or microsphere β-radiation). In addition, disposable surgical shoe covers, gowns, and caps must be worn by any personnel who could come in direct contact with microspheres (most likely the physician administering the dosage, but also potentially anyone directly assisting with the infusion). The fluoroscopy system should be covered with disposable clear plastic and disposable absorbent material should be placed over the patient and attached to the floor, with a minimum of 6 ft × 6 ft area of floor coverage recommended (Salem and Thurston, 2006), where spilled or leaked microspheres have a high likelihood of landing.

7.4.2 DURING ADMINISTRATION

Once assembly of the infusion system has been completed and verified, the physician begins to administer the microspheres to the patient. The system should be visually monitored at the start of infusion, and then periodically throughout, to ensure that the system remains "closed" (i.e., there is no leakage of fluid, especially around the septum of the v-vial and at the output line–microcatheter junction). Leakage from the septum of the SIR-Spheres v-vial can be visualized directly as the v-vial is only shielded with transparent acrylic, whereas the TheraSphere v-vial in its acrylic shield is kept within a lead pot and the needle plunger is attached to the acrylic shielding so that what is monitored instead is leakage from the seal around the plunger connection to the acrylic shield. An ion chamber survey meter may be held externally above the patient's liver (at approximately 6 in.) for real-time verification that microspheres are being delivered to the liver. (The exposure rate or dose equivalent rate reading should increase as the infusion progresses. Note, however, that this form of verification is limited, as such meters tend to be omnidirectional and are thus unable to localize the direction of the source of radiation with high accuracy and precision.)

SIR-Spheres dosage administration consists of sequences of syringe-pulsed infusions of microspheres (0.25–0.5 mL per pulse, at a rate of not more than 5 mL per minute) using sterile water, with periodic pauses for contrast fluoroscopy to ensure that flow to the target is being maintained and the location of the microcatheter has not changed (Sirtex Medical Limited, 2003; Salem and Thurston, 2006). A control knob on the delivery box allows the physician to toggle the input to the

THERASPHERE ADMINISTRATION

☐ Prepare Nalgene waste container
☐ Get beta meter ready for background measurement
☐ Have paper and pen ready to record administration time

IR Physicians to do the following:

☐ IR physician may choose to verify priming
☐ Time out verification of correct dose, correct target volume, correct patient
☐ If TheraSphere dose has been stagnant for several minutes during patient prep, IR physician should repeat solution perturbation
☐ Fully extend stainless steel arm
☐ Remove lead pot lid and place it upside down on a non-sterile surface (IR physician may ask you to do this)
☐ Use hemostat to remove purple seal from top of dose vial acrylic shield and discard it in Nalgene waste container
☐ Use sterile adhesive strip to remove dose vial acrylic shield plug
☐ Discard plug and sterile adhesive strip in Nalgene waste container
☐ Use hemostat to swab dose vial septum with alcohol swab
☐ Discard used alcohol swab in Nalgene waste container
☐ Close clamp on outlet tubing, labeled E

Please observe the following steps intently:

☐ Insert needle injector assembly into acrylic dose vial shield
☐ Press on GREEN CAP to lock and seal needle injector assembly in place
　o YOU WILL HEAR OR FEEL A "SNAP"
☐ IR PHYSICIAN MUST LIFT UP ON NEEDLE INJECTOR ASSEMBLY TO VERIFY PROPER LOCK AND SEAL OF NEEDLE INJECTOR ASSEMBLY TO DOSE VIAL
☐ Place inlet tubing through slot B in acrylic box
☐ Place outlet tubing through slot D in acrylic box
☐ Loop tubing around the side and place the fitting into the holder at location labeled C
☐ Push YELLOW TABS on needle injector assembly to puncture dose vial septum and lock needles in place
　o YOU WILL HEAR OR FEEL A "SNAP"
☐ Assure that acrylic box lid is properly placed with sloped shield toward patient
☐ Assure that tubing is not pinched or kinked, except for outlet tubing clamp
☐ Move cart close to patient; patient bed should be in lowest position
☐ Place sterile towel under steel extension arm holder E, and under holder at location C
☐ Place sterile towel across gap between acrylic box and patient
☐ IR physician must flush infusion catheter to verify positioning and patency of catheter
☐ Disconnect outlet tubing labeled E from priming tubing at holder C
☐ Firmly connect outlet tubing E to catheter
☐ Place catheter connection into slotted holder E at end of extended arm
☐ Outlet tubing E must be above holder with infusion catheter hanging vertically below
☐ RELEASE PINCHED CLAMP FROM OUTLET TUBING AND MASSAGE OR ROLL TUBING WITH FINGERS TO MINIMIZE KINK IN TUBING

THERASPHERE ADMINISTRATION, cont'd

NMT to do the following:

☐ Record RADOS dosimeter initial reading for dose vial; monitor reading throughout administration
☐ Using beta meter, measure background signal of TheraSphere administration tubing on patient (hot) side of acrylic box
☐ Record start time of administration

IR Physicians to do the following:

☐ Infuse TheraSpheres using steady pressure on syringe plunger
☐ Infuse continuously until 20 mL syringe is empty
☐ Observe outlet line and catheter for proper function; if problems are observed, stop infusion, inform team and take corrective action
☐ Refill syringe for subsequent flushes by pulling back on syringe plunger
☐ Minimum of three total flushes (60 mL) is recommended
☐ If infusion pressure exceeds 30 psi, excess fluid will drip into 20 mL vent vial
　o If this occurs, reduce pressure until there is no fluid venting into 20 mL vial
☐ Continue flushes until desired RADOS dosimeter reading (0.0 mR/hr) is achieved
☐ Cut inlet tubing at indicated location
☐ Remove acrylic box lid
☐ Remove infusion catheter from patient and lift catheter out of extended holder E
☐ Leaving catheter connected to outlet tubing, use hemostat to place tubing, catheter, and dose vial in Nalgene waste container
☐ Remove gloves and place them in Nalgene waste container

NMT to do the following:

☐ After collecting IR physician's gloves, survey all IR staff that operated TheraSphere Administration Set for radioactive contamination using GM pancake survey meter
☐ If IR physician gown is contaminated, or if physician insists, place gown in red biohazard bag

Figure 7.12 Example TheraSphere dose administration checklist. Checklists for (1) items required; (2) documentation required; (3) preadministration activity measurements; (4) cart preparation; (5) postadministration; (6) residual measurement; and (7) radiology information system (RIS) records and documentation are also contained in the document from which the checklist shown was obtained.

three-way stopcock on the output side between the output line of the v-vial and the contrast/sterile water line. Stasis is a common occurrence with SIR-Spheres due to complete embolization of the target as a consequence of the large number of microspheres involved. A temporary loss of antegrade flow may also occur due to vascular spasms in reaction to the infusion of the microspheres. Infusion must not continue if stasis has occurred or until antegrade flow has been restored as inadvertent delivery of microspheres to nontarget tissues via arteries upstream from the position of the microcatheter may occur, causing a misadministration (and possibly a reportable medical event), as well as possibly causing radiation-related normal tissue complications (e.g., ulcerations) requiring medical intervention. The infusion proceeds until either the entire dosage has been administered, stasis has been reached, or early termination occurs due to an emergent patient condition (e.g., an unacceptable level of abdominal pain). An entire SIR-Spheres dosage has essentially been infused when the fluid in the v-vial has transitioned from its initial sandy-yellow translucent color to being almost clear. An air-filled syringe can be used to flush the remaining fluid from the v-vial. A single SIR-Spheres dosage administration may take up to 20 minutes due to the slow nature of the infusion process.

Administration of a TheraSphere dosage occurs much more rapidly than that for SIR-Spheres (Salem and Thurston, 2006; BTG International Ltd., 2010, 2014). Twenty milliliters of fluid is drawn from a ≥100-mL saline bag attached to the infusion system and into the integrated 20-mL syringe and then flushed through the v-vial using steady pressure (≤30 psi). (A relief valve that diverts fluid into a vented 20-mL vial is integrated into the input line to the v-vial, should the pressure exceed 30 psi.) One-way valves in the lines between the saline bag and syringe and syringe and v-vial are incorporated to prevent reverse flow of fluid. The TheraSphere system is designed to infuse the vast majority of the microspheres into the patient with the initial 20 mL of flush and infuse nearly all of the remaining microspheres via a minimum of two additional 20 mL of flushes. Two methods are employed for verification that the TheraSphere infusion has been completed (Salem and Thurston, 2006; BTG International Ltd., 2010, 2014). First, the electronic dosimeter attached to the input side

of the acrylic delivery box measures the external dose equivalent rate from the lead pot containing the dosage vial. An initial reading just before the start of infusion is recorded, and a second recorded when the physician has stopped the infusion. The second reading should be zero or a very small fraction (<0.05) of the initial reading under normal circumstances. (The readings must be recorded only when the fluoroscopy x-ray beam is off and the dosimeter reading has stabilized. The scattered x-ray radiation will cause the dosimeter reading to increase substantially and a 5- to 10-second delay after the x-ray beam is turned off before recording the reading is required due to the slow response time of the dosimeter.) The second method of verification is the counts per minute (cpm) reading of a β detector (thin-wafer plastic scintillator insensitive to gamma and x-ray radiation) survey meter in close proximity to, and directed at, the infusion system output line/microcatheter junction after the initial 20 mL of flush. The initial high cpm reading should stabilize at a much lower reading after two or more follow-on 20 mL flushes. (Depending on the sensitivity of the β detector, a plastic attenuator over the entrance window may be required to achieve counts per minute (cpm) readings within the lower range settings of the survey meter.) The reading will not decrease to zero as some of the microspheres, albeit a small fraction, will inevitably become attached to or trapped in the v-vial, output line, microcatheter, and, in particular, the output line–microcatheter junction. Periodic gentle tapping on the output line–microcatheter fitting with a hemostat during flushing can free up some of the microspheres that have become trapped at that point.

7.4.3 IMMEDIATE POSTADMINISTRATION

After the administration of each treatment dosage has either been completed or terminated early due to stasis or emergent patient condition, the radioactive items from the infusion must be segregated and placed in the Nalgene radioactive waste container for later measurement of the estimated fraction of the total activity that was not delivered to the patient for that particular treatment (see Section 7.7.1). Since the input and output lines, needles (and needle plunger for TheraSphere), dosage vial in its acrylic shield, and microcatheter have all been interconnected, they are to be treated as a single radioactive item at the end of infusion (as opposed to being disassembled). The SIR-Spheres input lines should be recapped after removing the attached syringes, and the TheraSphere input line is cut immediately distal to the pressure-relief valve. The dosage vial in its shield followed by the infusion lines should then be placed in the waste container first. The infusion microcatheter is then slowly withdrawn from the patient. Coiling up the microcatheter on the towel or absorbent pad underneath the base catheter as it is being removed is recommended, and as soon as the tip of the microcatheter appears, it is prudent to clamp it with a hemostat (although covering it with gauze has also been suggested). (The same hemostat could be used for cleaning the dosage vial septum, tapping the TheraSphere output line–microcatheter junction, and clamping the microcatheter.) Blood leaking from the base catheter should be allowed to drip onto the same towel or absorbent pad before wrapping it around the microcatheter and hemostat and placing all of them in the waste container. (If a syringe is used to withdraw blood from the base catheter after removal of the microcatheter, it should be considered radioactive and placed in the waste container.) Other items to be assumed radioactive are the removed acrylic shield plug, the hemostat and alcohol swab(s) used to clean the dosage vial septum, and any gloves that may have come into contact with microspheres (all to be placed in the waste container), as well as the acrylic delivery box. Items on the input side of the infusion system that would not come into contact with microspheres under normal circumstances, and can thus be considered nonradioactive, are used sterile water, saline, and contrast syringes; towels and/or absorbent chux; packaging materials; and the TheraSphere saline bag, pressure-relief vial, and infusion line up to where it was cut (as well as the scissors used to cut the line). The chux draping the surface of the dosage cart should be surveyed with the G–M meter, and if found to be radioactive, placed in the waste container. (The entrance window of the G–M detector must be exposed for surveying surfaces for ^{90}Y microspheres. Otherwise, most of the β particles will be absorbed by the protective cover, resulting in either underestimating or missing contaminations.)

Personnel who were in the procedure room during the radioembolization(s) and who could possibly have been contaminated with microspheres must be surveyed head-to-toe with the G–M survey meter before leaving the room (Dezarn et al., 2011). If any radioactivity is detected on protective coverings (cap, gown, gloves, shoe covers), they must be removed and placed in the designated decay-in-storage biohazard bag. Afterward, a microspheres contamination survey of the room should be performed, paying particular attention to the coverings over the patient, fluoroscopy system, and floor used during the procedure, and anything containing fluid or blood (Dezarn et al., 2011). (This survey may have to be performed when the patient is no longer in the procedure room, if the bremsstrahlung radiation emanating from the patient would mask detection of low levels of β radiation from microsphere contamination.) Alternatively, the coverings used could be carefully folded inward together to confine any microspheres and placed in a receptacle that could be later surveyed and sequestered for decay-in-storage if found radioactive. If any surfaces in the room are found to be radioactive, the institution's Radiation Safety Officer (RSO) should be contacted for guidance and oversight regarding attempts at removal and/or cover up of the source(s) of radiation. If radiation exposure to personnel is deemed excessive and removal is not possible, then acrylic (≥0.5 inch) should be placed over the source(s) of radiation to absorb essentially all of the β particles from ^{90}Y until decay to background or a level considered safe (Dezarn et al., 2011). (Secondary bremsstrahlung radiation will be generated in the acrylic but it should constitute a negligible fraction of the total energy of the incoming β radiation.)

7.5 PATIENT RELEASE

Radioembolized patients will be a source of radiation exposure above background for an extended period of time postinfusion. The duration will be related to both the total infused activity and half-life of ^{90}Y as the internalized activity decreases by physical decay only (i.e., there is no biological clearance). Thus, the external radiation exposure of others is of concern, in particular, those individuals for whom the exposure would be unexpected, whether they are members of the general public (when the patient is released from the licensee's control) or hospital personnel categorized as non-radiation workers (if the patient needs to be admitted following treatment).

7.5.1 TO THE PUBLIC

The NRC allows the release of a radioactive patient from radiation confinement as long as the estimated cumulative (total) radiation exposure to any other individual does not exceed 5 mSv effective dose equivalent (EDE). However, if the estimated maximum EDE exceeds 1 mSv, then verbal and written instructions must be provided to the patient and a record of them maintained by the licensee in order to keep the exposure as low as reasonably achievable (ALARA) (NRC, 2013; Siegel, 2004). The only exposure others will normally receive from a patient to whom permanent radioactive implants have been administered is external (as opposed to a radiopharmaceutical such as iodine-131 sodium iodide, which must take into account potential internal exposure due to intake of activity that is excreted).

Radioembolization is considered a type of permanent implant. The equation governing the estimation of the total EDE (TEDE) received by the most exposed person for the case of permanent implants is (NRC, 2008)

$$TEDE\,(mSv) = \frac{0.346 \times \Gamma \times Q_0 \times T_p \times E}{r^2} \quad (7.11)$$

where $0.346 = 0.01$ mSv/mrem $\times$ 24 hours per day/ln(2); Γ is the radionuclide exposure rate constant (mR-m^2/mCi-h); Q_0 is the amount of activity remaining in the patient at a proposed time of release (mCi); T_p is the radionuclide half-life (d); E is an occupancy factor (0.25 for T_p greater than 1 day); r is the distance (m) from the patient; and a Γ-to-EDE rate (mrem/mR) conversion factor of unity is assumed. As stated earlier, ^{90}Y is a pure beta emitter. Thus, the only external radiation from the patient would be that due to secondary bremsstrahlung x-rays produced that are not absorbed by the patient. A published value for the bremsstrahlung exposure rate constant for ^{90}Y based on uniform distribution in a reference adult model is 5.64×10^{-4} mR-m^2/mCi-h (Zanzonico et al., 1999) and values of $1.84 \pm 0.53 \times 10^{-3}$ and 2.82×10^{-3} mR-m^2/mCi-h have been obtained from measurements at 1 m from radioembolization patients (Figure 7.13) (Gulec and Siegel, 2007; Erwin et al., 2012). Assuming the worst case, the highest of the three Γ values for ^{90}Y, the amount of infused activity at which release instructions would be required (TEDE = 1 mSv), would be

$$57\ GBq = \frac{0.037\ GBq\,/\,mCi \times 1\ mSv}{0.346 \times 2.83E - 3 \times 2.67\ d \times 0.25} \quad (7.12)$$

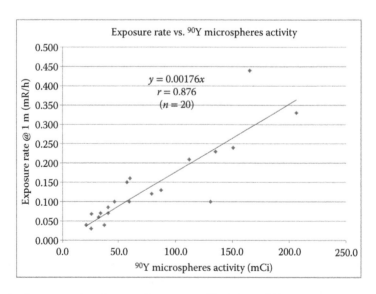

Figure 7.13 Graph of measured posttreatment external bremsstrahlung exposure rate versus net administered activity for 20 ^{90}Y microspheres patients (9 SIR-Spheres, 11 TheraSpheres). The slope of the linear regression or the statistical average ratio of exposure rate to activity represents a measured ^{90}Y microspheres patient bremsstrahlung exposure rate constant.

and the activity above which a period of radiation confinement before release is required would be five times that amount. The theoretical upper limit on the infused activity for a single radioembolization treatment is currently 20 GBq and the typical infused activity is roughly an order of magnitude below that resulting in 1 mSv TEDE. Thus, all radioembolization patients may currently be released immediately without the documentation of verbal and written instructions. However, it is prudent to provide all such patients with documentation indicating that they are radioactive and for how long, especially if they intend to travel soon enough after the treatment to potentially trigger radiation detectors at an airport or border crossing. Trace amounts of unbound ^{90}Y from SIR-Spheres have been detected in urine (25–50 kBq/liter per GBq) within the first 24 hours after implant (Sirtex Medical Limited, 2003; Lambert et al., 2011). Although not considered a significant source of radiation exposure, all patients could be asked to flush twice after urination and male patients asked to sit during urination for 1 day after treatment (Dezarn et al., 2011).

7.5.2 HOSPITAL ADMISSION

As the above ^{90}Y radioembolization patient release calculation demonstrates, if a patient must be admitted after treatment, there is currently no requirement to place them in radiation confinement under the care of personnel trained in radiation safety and allowed to receive exposures above that to members of the public (including nonradiation hospital workers). However, one may choose to do so if the patient is admitted soon after treatment and will require intensive care, whereby personnel will spend a substantial amount of time at close proximity to the patient and it is deemed possible that they will receive substantially higher than general public limit exposures. A standardized document could be developed and provided with each patient to notify the hospital staff caring for the patient that universal precautions should be observed while caring for a ^{90}Y microspheres patient while he or she is still radioactive (Figure 7.14). Such a document is especially useful when patients are admitted to a regular room where personnel are not trained as radiation workers. If a SIR-Spheres patient requires urinary catheterization requiring the changing of collection bags, then it would be prudent for the personnel handling them to be gloved and empty the bags into the patient's toilet followed by two flushes until the radiation level in the bag reaches background. If any radioembolized patient requires abdominal drainage, then an assessment of the radiation level in the drainage bag should be made, and if radioactive then medical intervention may be indicated, as under normal circumstances there should be no radioactivity in the abdominal fluid.

Y-90 Microsphere Treatment

This patient has received a Y-90 microsphere treatment. Y-90 microspheres are radioactive and emit beta radiation. Once implanted in the patient, the microspheres do not present a radiological risk to care givers, family members, or others who come into contact with the patient. Therefore, special isolation measures, such as those used for most radioactive patients, are not required to protect care givers from the radiation present in patients who have received a Y -90 microsphere treatment. Universal precautions should be followed when providing care for this patient. If you have any questions or concerns, please contact the Radiation Safety Officer.

Figure 7.14 Example notification of ^{90}Y microspheres patient, should he or she need to be admitted to the hospital while still considered radioactive. Such patients can be released to the general public immediately without radiation precaution instructions. However, if they need inpatient care, prudence suggests that hospital staff caring for the patient be made aware that he or she is radioactive and what precautions, if any, are necessary.

7.6 SIGNIFICANT POSTTREATMENT EVENTS

7.6.1 SURGERY: LIVER RESECTION OR TRANSPLANT, OR OTHER

One of the clinical indications of TheraSphere is radiotherapy neoadjuvant to liver resection or transplantation. Therefore, a percentage of patients treated with TheraSphere will proceed to surgery. If surgery is planned for a radioembolized patient, then how long before resection or transplant can occur from a radiation safety standpoint must be a consideration. A surface dose equivalent rate of 20 μSv/h above the liver is a generally accepted threshold below which radiation safety precautions are not required for surgery postradioembolization (Salem and Thurston, 2006). A posttreatment surface dose equivalent rate measurement combined with the physical half-life of ^{90}Y can be used to predict when surgery can be performed without concern regarding the radiation exposure to personnel (in particular, the surgeon). The explanted liver tissue should be placed in a leakproof container containing formaldehyde and refrigerated and located behind lead shielding for radioactive decay-in-storage if the dose equivalent rate at the surface of the container is greater than 50 μSv/h (Salem and Thurston, 2006). Surgery due to an emergent condition while the liver is still considered radioactive should not be prohibited as the welfare of the patient outweighs an infrequent and not excessive personnel radiation exposure. Under all circumstances, the institution's RSO should be consulted to provide guidance related to the surgery itself (e.g., the use of lead surgical gloves and other radioprotective apparel and body and extremity dosimeters); handling of the explanted liver tissue (by both surgery and pathology) and any other items that may be radioactive as a result (e.g., blood, surgical garments and instruments, and towels and other surface-covering materials); surveys of all personnel and areas where RAM contaminations may have occurred; labeling of all radioactive items; and appropriate signage in the surgical suite and other areas where RAM is present (Salem and Thurston, 2006; Dezarn et al., 2011).

7.6.2 AUTOPSY, BURIAL, OR CREMATION

Individuals currently treated with TheraSphere or SIR-Spheres tend to be late-stage cancer patients. Thus, there exists a possibility of a patient expiring posttreatment while still radioactive. As explained above (see Section 7.5), radioembolization patients may be released to the general public without radiation precaution instructions as the most exposed person is estimated to receive an exposure about an order of magnitude lower than the limit for a member of general public. The exposure to others, including embalmers and funeral workers, during body preparation, visitation, funeral, and/or burial following the death of a radioembolization patient is anticipated to be even lower than that to the most exposed person under normal circumstances (all these individuals being exposed over a much shorter time period overall), and thus should not be of concern (Dezarn et al., 2011). Furthermore, no special precautions are required during embalming using standard methods. Microspheres are permanent implants, and thus there will be no radioactivity in the aspirated blood (aside from possible trace amounts of ^{90}Y dissociated from SIR-Spheres, which has been detected in urine and which suggests it was present in blood and extracted by the kidneys).

If an autopsy is to be performed, then depending upon how long after infusion of the microspheres the autopsy is scheduled, it may be prudent for the pathologist to explant the liver and have it relocated for radioactive decay-in-storage before proceeding with the autopsy on the remainder of the body to minimize unintended exposure. According to the recommendations found in International Commission on Radiological Protection publication 94 (Harding et al., 2004), routine autopsy procedures may be followed if a corpse contains less than 0.45 GBq of sealed ^{90}Y; but if still considered a radiation hazard and the radioactivity is confined to a specific organ, then that organ should be explanted and stored for radioactive decay.

Cremation of a corpse containing radioactive microspheres presents a special case radiation hazard, with the possibility of not only external exposure but also internal exposure due to the inhalation of RAM in the residual ashes or crematorium effluent (NCRP, 2006; Nelson et al., 2008; Dezarn et al., 2011). The radiation hazard is complicated by the presence of long-lived impurities in the microspheres (see Section 7.7, "Radioactive Waste"). The limit on the amount of ^{90}Y radioactivity in a body to be cremated varies country by country, with values ranging from 0.047 GBq in the United States to 1 GBq in Australia (Harding et al., 2004). Local regulations regarding limits on radioactivity in the effluent from the cremation process may also apply (Nelson et al., 2008). The ideal scenario is to have the liver explanted (and possibly the lungs as well, in the case of substantial shunting of microspheres) and stored for the radioactive decay prior to presenting the body for cremation (Dezarn et al., 2011). If the liver is not explanted (nor are radioactive lungs), then the body will have to be stored for a period of time before cremation, if the on-board activity at the time of death exceeds the local regulatory limit.

General recommendations regarding handling of radioactive patients post-mortem can be found in National Council on Radiation Protection Report Nos. 155 and 161 (NCRP, 2006, 2010), in addition to ICRP publication 94. The institution's RSO should be consulted for guidance specific to radioembolized patients, regarding radiation safety precautions related to the corpse, excised liver (and possibly lungs), and potentially radioactive blood or urine, and to ensure compliance with local regulations. Finally, depending upon whether the death occurs inside or outside of the institution where the patient underwent the procedure, either a physician or an RSO involved or a family member of the deceased should inform the morgue, funeral home, and/or crematorium that the decedent underwent a radioembolization procedure and when. Those entities can then consult with the treating institution's AU or RSO to assess whether or not a radiation hazard is presented at the various stages post-mortem (NCRP, 2006).

7.7 RADIOACTIVE WASTE

7.7.1 RESIDUAL ACTIVITY ASSAY

As mentioned in Section 7.4.3, the residual activity for each radioembolization treatment dosage will reside not only in the dosage vial but can also reside in the infusion needles and tubing, the microcatheter, the towels or absorbent pads under the base and microcatheters (and any syringe into which radioactive blood was withdrawn from the base catheter), the acrylic shield plug, the alcohol swab used to clean the dosage vial septum, hemostat(s) used to wipe the septum and clamp the tip of the microcatheter postinfusion, and the acrylic delivery box (and elsewhere if a contamination or spill occurs) (Salem and Thurston, 2006; Dezarn et al., 2011). Therefore, a direct measurement of the residual activity with a dose calibrator is not possible. Instead, an estimate must be derived from calibrated ion chamber exposure rate or dose equivalent rate readings of the bremsstrahlung radiation from the dosage vial prior to administration and the waste container (inside its acrylic shield) afterward, using a fixed measurement geometry (Figure 7.15) (Salem and Thurston, 2006; BTG International Ltd., 2010). The dosage vial is essentially a radially symmetric radiation source, so only a single reading from any direction is necessary. However, the residual activity within the waste jar is not isotropic; therefore, an average of readings for multiple rotations of the acrylic shield in front of the survey meter (typically four, at 0°, 90°, 180°, and 270° of rotation)

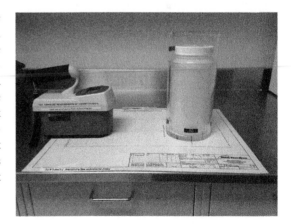

Figure 7.15 Setup for TheraSphere posttreatment ion chamber survey meter exposure rate reading of residual activity, using the fixed-geometry template provided by the manufacturer (30-cm waste jar center-to-ion chamber center distance). The average of four readings, with the waste jar plus shield rotated 90° between each, is compared with that from the v-vial in shield prior to treatment, to estimate the fraction of total v-vial activity delivered to the patient.

is computed. (A background reading is also obtained at the time of, and subtracted from, the dosage vial and waste container readings.) The two measured rates (R) are decay corrected to the time of infusion and the estimated residual activity is then calculated:

$$\text{residual } A_{\text{residual}}(GBq) = A(GBq) \times W \quad (7.13)$$

where initial $A(GBq)$ is the preadministration vial activity calibrator assay of the treatment dosage vial (decay corrected to the time of infusion), and W is the ratio of measured rates ($R_{\text{waste}}/R_{\text{vial}}$). A spreadsheet that automates the calculation of residual activity, as well as the estimates of net activity administered and absorbed (and percent of prescribed) dose, is illustrated in Figure 7.16.

7.7.2 EXAMPLE PROCEDURE MEASUREMENTS AND CALCULATIONS

The measurements and calculations surrounding the radioembolization procedure itself that are used to estimate the net activity administered and absorbed dose delivered, as well as assess whether or not a reportable medical event has occurred, are illustrated in this section. Two examples are provided, both based on actual cases. The first example demonstrates a case where the outcome of the infusion was as expected. The second example illustrates a misadministration that occurred as a result of a failure of the device, resulting in a reportable medical event.

Example 7.1: Successful infusion
A patient whose entire right lobe of the liver was replaced by a pancreatic neuroendocrine cancer metastasis was treated with SIR-Spheres based on a target absorbed dose prescription.
Target M_{liver}(kg): 2.98 kg (1.03 kg/l × 2.893 L estimated using CT)
Prescribed D_{tumor}(Gy): 70 Gy
Lung shunt fraction (SF): 0.0577
Required $A(GBq)$ for 70 Gy:

$$4.43 \text{ GBq} = \frac{70 \text{ Gy} \times 2.98 \text{ kg}}{(49.98 \text{ Gy} \cdot \text{kg} / \text{GBq} \times [1 - 0.0577])}$$

Dispensed $A(GBq)$ assay: 4.21 GBq (decayed to time of infusion)
Dosage vial R: 6.63 mR/h (decayed to time of infusion)

Average residual waste R: 0.0852 mR/h (decayed to time of infusion)
Waste ratio W: 0.013
Net activity administered:
$$4.16 \text{ GBq} = 4.21 \times (1 - 0.013)$$
Percent of D_{tumor}(Gy) delivered:

$$93.9 \% = 100 \times \frac{4.16 \text{ GBq}}{4.43 \text{ GBq}} \quad (D_{\text{tumor}}(Gy) \propto A(GBq))$$

The estimated absorbed dose delivered to the target was well within the regulatory limit of $\pm 20\%$ of the prescribed absorbed dose.

Example 7.2: Reportable medical event
A patient with hepatocellular carcinoma and bi-lobar disease was treated with two dosages of TheraSphere, one for each lobe of the liver. The right lobe infusion was successful, resulting in the delivery of an estimated 96% of the prescribed 115 Gy absorbed dose. However, the left lobe infusion resulted in a reportable medical event due to a breach of the seal around the plunger assembly vial interface.
Target M_{liver}(kg): 0.755 kg (1.03 kg/L × 0.733 L estimated using CT)
Prescribed D_{tumor}(Gy): 80 Gy
Lung shunt fraction (SF): 0.1035
Required $A(GBq)$ for 80 Gy:

$$1.35 \text{ GBq} = \frac{80 \text{ Gy} \times 0.755 \text{ kg}}{(49.98 \text{ Gy} \cdot \text{kg} / \text{GBq} \times [1 - 0.1035])}$$

Dispensed $A(GBq)$ assay: 1.32 GBq (decayed to time of infusion)
Dosage vial R: 2.19 mR/h (decayed to time of infusion)
Average residual waste R: 0.82 mR/h (decayed to time of infusion)
Waste ratio W: 0.374
Net activity administered:
$$0.83 \text{ GBq} = 1.32 \times (1 - 0.374)$$
Net absorbed dose delivered:

$$49 \text{ Gy} = \frac{49.98 \text{ Gy} - \text{kg} / \text{GBq} \times 0.83 \text{ GBq} \times (1 - 0.1035)}{0.755 \text{ kg}}$$

Percent of D_{tumor}(Gy) delivered: $61 \% = \dfrac{49 \text{ Gy}}{80 \text{ Gy}}$

(|Delivered – Prescribed|): 31 Sv (1 Gy = 1 Sv for electrons and photons)
The estimated absorbed dose delivered to the target was less than 80% of the prescribed

Date:		**Target Tissue:**	right lobe

TheraSphere Therapy Calculations

Patient Name:	
Patient ID:	

Calibration Date:		mm/dd/yyyy (calibrated at 12:00 EST)	
Lot # and Vial #:		Adminstration Kit Lot #:	
mg of microspheres:	244.00	Adminstration Kit Expiration Date:	

ACTIVITY ASSAY

Assay:	70.8	mCi	2.62 GBq)
Military Time of Day:	9:13	hh:mm	

PRE-ADMINISTRATION EXPOSURE RATE MEASUREMENTS (mR/h at 30 cm)

Military Time of Day:	9:15	hh:mm	
Background:	0.00	mR/h	

Activity, V-vial in Acrylic Shield

Vial Reading:	4.35	mR/h	Expected mR/h (1.692 x GBq)
Net Vial - Background:	4.35	mR/h	4.43 ± 10% (~95% CI)

RADOS RAD 60R DOSIMETER V-VIAL READING   Acceptable Range:

Pre-Implant:	7.40	mR/h	Low:	3.99 mR/hr
Post-Implant:	0.00	mR/h	High:	4.88 mR/hr
Estimated V-Vial Residual:	0.0%			

DOSE ADMINISTRATION

Military Time of Day:	11:31	hh:mm

POST-ADMINISTRATION EXPOSURE RATE MEASUREMENTS (mR/h at 30 cm)

Military Time of Day:	11:45	hh:mm
Background:	0.00	mR/h

V-vial Residual + Waste in Nalgene Waste Container

0 Degrees:	0.15	mR/h
90 Degrees:	0.11	mR/h
180 Degrees:	0.11	mR/h
270 Degrees:	0.13	mR/h
Average - Background:	0.13	mR/h

POST-ADMINISTRATION CALCULATIONS

Total Activity @ Admin. Time:	69.1	mCi	2.56 GBq)
Decay Corrected mR/h PRE:	4.24		
Decay Corrected mR/h POST:	0.13		
Estimated Total Residual:	3.0%		
Residual Activity:	2.04	mCi (@ Admin. Time)	
NET Activity Administered:	67.0	mCi	2.48 GBq)
NET Percent Administered:	97.0%	%	
Estimated Target Tissue Dose:	91.8	Gy	
Percent of Prescribed Dose:	96.6%	%	
Estimated Lung Dose:	6.9	Gy	

Figure 7.16 Example TheraSphere day-of-therapy spreadsheet that includes a comparison of measured and expected exposure rates from the delivery v-vial in acrylic shield, as well as calculations of net activity and absorbed dose delivered to the patient and comparison to the prescribed dose.

absorbed dose, and the difference between the estimated absorbed dose delivered and the prescribed absorbed dose exceeded the threshold for reporting a misadministration (0.5 Sv equivalent dose to an organ or tissue). Thus, reporting this medical event to the NRC or Agreement State regulatory authority was required.

7.7.3 STORAGE FOR DECAY AND DISPOSAL

The radioactive waste from each radioembolization procedure requiring a period of decay-in-storage prior to disposal typically consists of (1) one or more waste jars containing the residual infusion system activity (sealed and removed from the reusable acrylic shield); (2) any biohazard bags containing microspheres-contaminated items such as gloves, garments, surface coverings, and absorbent media employed in the removal of microspheres from contaminated surfaces; and (3) any unused treatment doses due to procedure cancellation or an emergent patient condition. Each of these sources of radioactive waste must be appropriately labeled. The label for each should include

1. Radionuclide (^{90}Y)
2. Product (SIR-Sphere or TheraSphere)
3. Date and time
4. Activity (mCi or GBq) for each residual activity waste jar (Figure 7.17) and unused dosage

The ^{90}Y in SIR-Spheres is eluted from a ^{90}Sr/^{90}Y generator and thus there would ideally be no radionuclidic impurities in the radioactive waste. However, trace amounts of ^{90}Sr have been detected in SIR-Spheres waste due to "breakthrough" during the elution process, with amounts of ^{90}Sr on the order of 3 Bq per GBq of ^{90}Y at the time of manufacturer assay reported (Metyko et al., 2014). As a consequence, SIR-Spheres waste will remain slightly radioactive with both ^{90}Sr and the ^{90}Y it is generating (both decaying with the 28.79 year half-life of ^{90}Sr) after all of the initial ^{90}Y activity has essentially decayed away (Figure 7.18). On the other hand, the negligible amount of ^{90}Sr impurity suggests that the duration of time over which the radiation level external to the source container is distinguishable from background before it may be disposed of as regular, biohazard waste may not

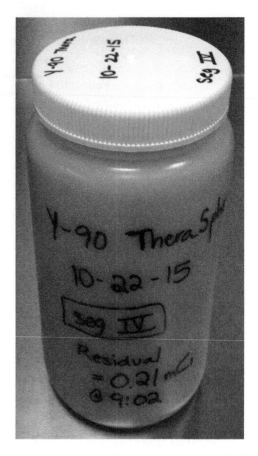

Figure 7.17 Example ^{90}Y microspheres residual waste jar, labeled for decay-in-storage and later disposal.

be significantly affected. (Of course, the effect will depend on the geometry and self-shielding of the container and its contents, and the distribution of microspheres within, in addition to the amount of ^{90}Sr impurity.)

Yttrium-90 is activated in TheraSphere microspheres by neutron bombardment. The microspheres are manufactured with ^{89}Y as a constituent of the glass matrix, and ^{90}Y is produced via the irradiation of the microspheres in a nuclear reactor (^{89}Y(n,γ)^{90}Y). TheraSphere microspheres are known to contain a number of gamma-emitting radionuclidic impurities as a result of neutron activation of other elements within the glass matrix (NRC, 2007; Ostrowski et al., 2007; Nelson et al., 2008; Metyko et al., 2012). The impurities that have been detected have much longer half-lives than that of ^{90}Y. The two most prominent impurities are ^{88}Y (106.6-day half-life) and ^{91}Y (58.5-day half-life) due to the

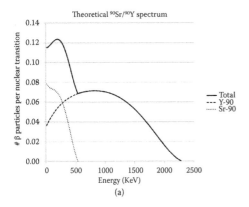

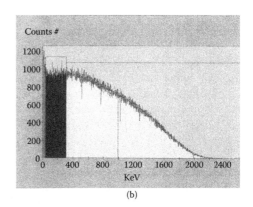

Figure 7.18 Strontium-90 impurity in SIR-Spheres. **(a)** Theoretical β energy spectrum for ⁹⁰Sr/⁹⁰Y. **(b)** Liquid scintillation counter β energy spectrum of an aliquot from an unused SIR-Spheres v-vial, long after the original ⁹⁰Y radioactivity had decayed to the level of background, demonstrating the production of ⁹⁰Y from the decay of on-board ⁹⁰Sr.

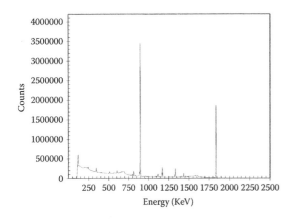

Figure 7.19 Radionuclidic impurities in TheraSphere. A high-purity germanium detector, high-resolution γ-ray emission spectrum from an unused TheraSphere v-vial, long after the original ⁹⁰Y radioactivity had decayed to the level of background. A number of γ-emitting impurities were present, most notably ⁸⁸Y (898 keV and 1.836 MeV).

reactions ^{89}Y(n,2n)^{88}Y and ^{89}Y(2n,γ) ^{91}Y, respectively (Figure 7.19). Both produce high-energy gamma emissions as a result of radioactive decay and two from ^{88}Y are high yield (898 keV/93.4%, 1.836 MeV/99.4%). Depending upon an institution's RAM license and local regulations and whether or not the radioactive impurities in the procedure waste are detectable above background, the radioactive waste from TheraSphere procedures may have to be shipped to an outside entity for decay-in-storage and disposal.

7.8 CONCLUSIONS

Radioembolization is a unique therapeutic application of RAM and as a consequence presents correspondingly unique challenges from a radiation safety perspective. The intent of this chapter is to arm the radioembolization team (interventional radiologists, medical physicists, nuclear medicine physicians, and/or radiation oncologists) with a thorough understanding of the dosage preparation, patient treatment and release, and radioactive waste management radiation concerns specific to both SIR-Spheres and TheraSphere. All members of the interdisciplinary radioembolization team should be keenly aware of and adhere strictly to the various regulatory and good radiation safety practice aspects associated with this treatment modality. Doing so will help ensure that a licensee establishes and maintains a radioembolization program that is both safe and effective for patients, while simultaneously safeguarding personnel and members of the public against excessive radiation exposure.

REFERENCES

Berger, M.J. (1976). *MIRD Pamphlet 7: Distribution of Absorbed Dose Around Point Sources of Electrons and Beta Particles in Water and Other Media*. Reston, VA: The Society of Nuclear Medicine and Molecular Imaging.

Braat, A.J.A.T. et al. (2015). ^{90}Y hepatic radio-embolization: An update on current practice and recent developments. *J Nucl Med* 56:1079–1087.

BTG International Ltd. (2010). *TheraSphere® Reference Manual [Accessed February 24, 2011]*.

BTG International Ltd. (2014). *Package Insert – TheraSphere® Yttrium-90 Glass Microspheres*. Available from: http://www.therasphere.com/physicians-package-insert/TS_PackageInsert_USA_v12.pdf [Accessed August, 2015].

Busse, N., Erwin, W., Pan, T. (2013). Evaluation of a semiautomated lung mass calculation technique for internal dosimetry applications. *Med Phys*, 40:122503.

Carey, J.E. et al. (2012). AAPM report no. 181: The selection, use, calibration, and quality assurance of radionuclide calibrators used in nuclear medicine. American Association of Physicists in Medicine, One Physics Ellipse, College Park, MD, 20740-3846.

Cole, A. (1969). Absorption of 20-eV to 50,000 eV electron beams in air and plastic. *Radiat Res* 38:7–33.

Dezarn, W.A. et al. (2011). Recommendations of the American Association of Physicists in Medicine on dosimetry, imaging, and quality assurance procedures for ^{90}Y microsphere brachytherapy in the treatment of hepatic malignancies. *Med Phys* 38:4824–4845. Available from: http://www.aapm.org/pubs/reports/RPT_144.pdf [Accessed October 17, 2011].

Dieudonné, A. et al. (2011). Clinical feasibility of fast 3-dimentional dosimetry of the liver for treatment planning of hepatocellular carcinoma. *J Nucl Med* 52:1930–1937.

Eckerman, K.F., Endo, A. (2008). *MIRD: Radionuclide Data and Decay Schemes*, Second Edition. Reston, VA: The Society of Nuclear Medicine and Molecular Imaging.

Emami,B. et al. (1991). Tolerance of normal tissue to therapeutic irradiation. *Int J Radiat Oncol Biol Phys* 21:109–122.

Erwin, W. et al. (2012). A measured bremsstrahlung dose rate constant for yttrium-90 microsphere liver directed brachytherapy. *J Nucl Med* 53(Suppl 1):2230.

Erwin, W. (2012). Accuracy of yttrium-90 microspheres delivery vial exposure rate measurement: Effect on estimated percent of activity delivered. *J Nucl Med* 53(Suppl 1):2244.

Ford, K. (1955). Predicted 0+ level of Zr^{90}. *Phys Rev* 98:1516.

Gress, D., Erwin, W. (2015). SIR-Spheres dose vial activity assay cross-calibration with an outside radiopharmacy. *Med Phys* 42:3498.

Gulec, S.A., Siegel, J.A. (2007). Posttherapy radiation safety considerations in radiomicrosphere treatment with ^{90}Y-microspheres. *J Nucl Med* 48:2080–2086.

Ho, S. et al. (1996). Partition model for estimating radiation doses from yttrium-90 microspheres in treating hepatic tumours. *Eur J Nucl Med* 23:947–952.

Harding, L.K. et al. (2004). ICRP publication 94: Release of patients after therapy with unsealed radionuclides. *Ann ICRP* 34:1–80.

ICRU. (1984). *Report 37: Stopping Power for Electrons and Positrons*. Bethesda, MD: International Commission on Radiation Units and Measurements.

Johnson, O., Johnson, R., Langer, L. (1955). Evidence for a 0+ first excited state in Zr^{90}. *Phys Rev* 98:1517–1518.

Kao, Y.H. et al. (2012). Image-guided personalized predictive dosimetry by artery-specific SPECT/CT partition modeling for safe and effective ^{90}Y radioembolization. *J Nucl Med* 53:559–566.

Kennedy, A. et al. (2007). Recommendations for radioembolization of hepatic malignancies using yttrium-90 microsphere brachytherapy: A consensus panel report from the radioembolization brachytherapy oncology consortium. *Int J Radiat Oncol Biol Phys* 68:13–23.

Lam, M.G.E.H. et al. (2014). Limitations of body surface area-based activity calculation for radioembolization of hepatic metastases in colorectal cancer. *J Vasc Interv Radiol* 25:1085–1093.

Lambert, B.L. et al. (2011). Urinary excretion of yttrium-90 following intraarterial microsphere treatment for liver tumours. *J Nucl Med* 52(suppl 1):1744.

Metyko, J., Erwin, W., Poston, J. Jr, Jimenez, S. (2014). ^{90}Sr content in ^{90}Y-labeled SIR-Spheres and Zevalin. *Health Phys* 107:S177–S180.

Metyko, J. et al. (2012). Long-lived impurities of ^{90}Y-labeled microspheres, TheraSphere and SIR-Spheres, and the impact on patient dose and waste management. *Health Phys* 103(5S):S204–S208.

Mo, L. et al. (2005). Development of activity standard for ^{90}Y microspheres. *Appl Radiat Isot* 63:193–199.

NCRP. (2006). *Report 155: Management of Radionuclide Therapy Patients*. Bethesda, MD: National Council on Radiation Protection and Measurements.

NCRP. (2010). *Report 161: Management of Persons Contaminated with Radionuclides: Handbook*. Bethesda, MD: National Council on Radiation Protection and Measurements.

Nelson, K., Vause, P.E., Jr., Koropova, P. (2008). Post-mortem considerations of yttrium-90 (^{90}Y) microsphere therapy procedures. *Health Phys* 95(Suppl 5):S156–S161.

NRC. (2007). *Yttrium-90 TheraSphere and SIR-Spheres Impurities: NRC Information Notice 2007–10*. U.S. Nuclear Regulatory Commission. Washington, DC. Available from: http://www.nrc.gov/reading-rm/doc-collections/gen-comm/info-notices/2007/in200710.pdf [Accessed October 7, 2011].

NRC. (2008). *Consolidated Guidance About Material Licenses: NUREG-1556 Vol. 9 Rev. 2*. U.S. Nuclear Regulatory Commission. Washington, DC: Office of Federal and State Materials and Environmental Management Programs. Available from: http://pbadupws.nrc.gov/docs/ML0734/ML073400289.pdf [Accessed March 21, 2011].

NRC. (2012). *Microsphere Brachytherapy Sources and Devices: Licensing Guidance – TheraSphere® and SIR-Spheres® Yttrium-90 ZMicrospheres*. Washington, DC: U.S. Nuclear Regulatory Commission. Available from: http://pbadupws.nrc.gov/docs/ML1217/ML12179A353.pdf [Accessed January 14, 2013].

NRC. (2013). *10 Code of Federal Regulations Part 35*. Washington, DC: U.S. Nuclear Regulatory Commission. Available from: http://www.nrc.gov/reading-rm/doc-collections/cfr/part035/ [Accessed April 25, 2016].

NRC. (2015). *Specialty Board(s) Certification Recognized by NRC under 10 CFR Part 35*. U.S. Nuclear Regulatory Commission. Washington, DC. Available from: http://www.nrc.gov/materials/miau/med-use-toolkit/spec-board-cert.html [Accessed March 31, 2016].

Ostrowski, R. et al. (2007). Radionuclidic purity assessment of yttrium-90 (Y-90) microspheres. *J Nucl Med* 48(Suppl 2):2206.

Pasciak, A.S., Erwin, W.D. (2009). Effect of voxel size and computation method on Tc-99m MAA SPECT/CT-based dose estimation for Y-90 microsphere therapy. *IEEE Trans Med Imag* 28:1754–1758.

Petitguillaume, A. et al. (2014). Three-dimensional personalized Monte Carlo dosimetry in ^{90}Y resin microspheres therapy of hepatic metastases: Nontumoral liver and lungs radiation protection considerations and treatment planning optimization. *J Nucl Med* 55:405–413.

Salem, R., Thurston, K.G. (2006). Radioembolization with ^{90}Yttrium microspheres: A state-of-the-art brachytherapy treatment for primary and secondary liver malignancies. *J Vasc Interv Radiol* 17:1251–1278.

Siegel, J.A. (2004). *Guide for Diagnostic Nuclear Medicine and Radiopharmaceutical Therapy*. Reston, VA: The Society of Nuclear Medicine and Molecular Imaging.

Selwyn, R.G. et al. (2007). A new internal pair production branching ratio of ^{90}Y: the development of a non-destructive assay for ^{90}Y and ^{90}Sr. *Appl Radiat Isot* 65:318–327.

Selwyn, R. et al. (2008). Technical note: The calibration of ^{90}Y-labeled SIR-Spheres® using a nondestructive spectroscopic assay. *Med Phys* 35:1278–1279.

Simon, B.A. (2000). Non-invasive imaging of regional lung function using x-ray computed tomography. *J Clin Monit Comput* 16:433–442.

Sirtex Medical Limited. (2003). *SIR-Spheres® Training Program – Physicians and Institutions* [Accessed June 30, 2010].

Sirtex Medical Limited. (2010). *Package Insert – SIR-Spheres® Microspheres (Yttrium-90 Microspheres)*. Available from: http://www.sirtex.com/media/29845/ssl-us-10.pdf [Accessed August, 2015].

Thieme, K., Beinlich, U., Fritz, E. (2004). Transfer standard for beta decay radionuclides in radiotherapy. *Appl Radiat Isot* 60:519–522.

Zanzonico, P.B., Binkert, B.L., Goldsmith, S.J. (1999). Bremsstrahlung radiation exposure from pure β-ray emitters. *J Nucl Med* 40:1024–1028.

The radiation biology
of radioembolization

MARTA CREMONESI, FRANCESCA BOTTA, MAHILA FERRARI, LIDIA
STRIGARI, GUIDO BONOMO, FRANCO ORSI, AND ROBERTO ORECCHIA

8.1 INTRODUCTION

In recent years, the use of dosimetry to support radionuclide therapy has gained importance as documented by the increased number of articles addressing dosimetry. The latest literature reports some remarkable correlations between absorbed dose delivered, response, and toxicity, which have advanced the understanding of radiobiological effects. In a recent review published by Strigari et al. (2011), dose–effect relationships were collected and these indicated that dosimetry-based personalized treatments would improve outcomes and increase survival and open the way toward predictivity and personalization of therapy. Available evidence covers nearly all widely used nuclear medicine therapies, including

the treatment of differentiated thyroid cancer and benign thyroid disease with ^{131}I, neuroblastomas with ^{131}I-mIBG (metaiodobenzylguanidine), neuroendocrine tumors with ^{177}Lu and ^{90}Y radiopeptides, bone pain palliation with ^{153}Sm-ethylene diamine tetramethylene phosphonate (^{153}Sm-EDTM), and also radioembolization of primary and secondary liver cancer with ^{90}Y-microspheres.

In particular for radioembolization, significant dose–effect correlations have been provided in the literature, including a study by Strigari et al. (2010), which was the first to describe a model to interpret toxicity and tumor response for hepatocellular carcinoma (HCC) treated with ^{90}Y-labeled resin spheres; studies by Garin et al. (2012, 2015, 2016), which showed prognostication of tumor response and survival for HCC treated with ^{90}Y-labeled glass microspheres; a study by Chiesa et al. (2015), showing dose thresholds for tumor response and liver toxicity for HCC treated with ^{90}Y-labeled glass microspheres; and finally a study by Flamen et al. (2008), showing the prediction of metabolic response evaluated by multimodality imaging for metastatic liver tumors and ^{90}Y-labeled resin microspheres. Interestingly, besides providing correlation between dose and effect, these investigations pointed out apparent and unexpected differences in tolerability and response associated with glass vs. resin ^{90}Y microspheres, which could not be resolved by invoking the mere concept of absorbed dose, at least if only the mean absorbed dose at macroscopic level was considered.

As a whole, the findings up to now (circa 2016) highlight the need for more refined models to improve dosimetry information, and especially, the importance of radiation biology to define the effect of radioembolization on tissue.

As a natural consequence, the first attempts at defining radioembolization dose–effect relationships have been extensively compared with the data from external beam radiotherapy (EBRT). Of course, in EBRT, the use of radiobiological models is well established and enables the possibility of comparing effects due to different irradiation modalities,

dose rate, dose distribution, organ structure, volume effect, radiosensitivity, combined therapies, and risk factors all at the same mean dose. In the clinical context of radioembolization, outcomes such as the volume effect, the influence of the functional reserve and/or concomitant therapies, and higher tolerability of retreatment have been empirically observed. This reflects a further similarity with EBRT, although differences with EBRT exist and must be taken in mind. Moreover, three-dimensional (3D) voxel dosimetry methods at both a macroscopic and microscopic levels recently applied to radioembolization have provided dose distribution maps and dose–volume histograms (DVH) that permit the development of more refined radiobiological models.

This chapter presents the basic aspects of radiation biology, developed for EBRT and subsequently adapted to nuclear medicine therapy. The most widely used radiobiological models will be discussed, including the linear quadratic model with the biological effective dose (BED) concept, tumor control probability (TCP), and the normal tissue complication probability (NTCP) models. These are the first of many acronyms common in radiation biology which will be introduced in this chapter. A summary of acronyms is provided in Table 8.1.

Table 8.1 Abbreviations and acronyms

BED = biological effective dose	N_0 = initial number of clonogenic cells
BED$_{50}$ = BED value for 50% complication probability	NTCP = normal tissue complication probability
BED$_{EBRT}$, BED$_{RE}$ = BED in EBRT and in RE treatments	QUANTEC = Quantitative Analyses of Normal Tissue Effects in the Clinic
CTCAE = Common Terminology Criteria for Adverse Events	RE = radioembolization
D = absorbed dose	RECIST = Response Evaluation Criteria in Solid Tumors
d = absorbed dose delivered in one EBRT fraction	SF = surviving fraction of irradiated cells
DVH = dose volume histograms	T = total duration of exposure
EASL = European Association for the Study of the Liver	T^* = effective time reached when BED = 0
EBRT = external beam radiotherapy	TD$_5$, TD$_{50}$ = tolerated absorbed dose for 5% or 50% complication probability (at 5 years if specified by TD$_{5,5}$, TD$_{50,5}$)
E_{rel} = relative effectiveness (E_{rel} = BED/D)	T_{av} = doubling time of proliferation
EUD = equivalent effective dose	TCP = tumor control probability
EUBED = equivalent uniform biological effective dose	T_{eff} = effective half-life
fr = fraction of EBRT	T_{phys} = physical half-life of ^{90}Y
FSU = functional subunits	T_{rep} = repair half-life
HCC = hepatocellular carcinoma	V = tumor volume
LKB model = Lyman–Kutcher–Burman model	V_{eff} = effective volume
msA = microsphere-specific activity	λ_{eff} = effective decay constant
	λ_{phys} = physical decay constant
	μ_{rep} = repair constant = $0.693/T_{rep}$

Finally, some explanatory examples on how radiation biology can guide radioembolization planning are also given; the most relevant studies on these issues are briefly reported as well.

8.2 BASIC ASPECTS OF RADIATION BIOLOGY

Radiation biology is the study of the effect of ionizing radiation on biological tissues. These complex effects involve physics, chemistry, and biology concepts. The most important biological factors that play a role during irradiation, affecting to some extent the outcome of the treatment, are summarized by the so-called "4Rs of radiation biology": repair of DNA damage, redistribution of cells in the cell cycle (spanning a few hours), repopulation (spanning 5–7 weeks), and reoxygenation of hypoxic tumor areas (spanning a few hours to few days) Pajonk et al. (2010). Certainly tumor response is modulated by many additional factors, and some authors add the intrinsic radiosensitivity of individual cancer stem cells as the fifth R, although this may vary during radiation therapy and needs deeper investigation.

8.2.1 REPAIR

Ionizing radiation cell killing is a consequence of unrepairable DNA double-strand breaks. Most radiation-induced DNA injury is, however, sublethal and may be repaired depending on factors including the type and energy of radiation, the dose rate, and the phase in the cell cycle. If radiation is delivered with a low dose rate, the repair possibility increases, while at increasing dose rates, sublethal lesions can rapidly accumulate without full repair, contributing to lethality. Based on the different ability of normal tissues and tumors to repair radiation damage, therapy fractionation is a recurrent strategy in EBRT to spare normal tissues.

This concept has also been extrapolated to some nuclear medicine therapies, including radioembolization (Cremonesi et al., 2008), dividing the treatment in cycles to reduce toxicity to late responding organs at risk. The repair probability is described as an exponentially decaying function over time, with half-time (T_{rep}) varying from minutes to few hours.

8.2.2 REPOPULATION

Both tumor and normal tissues have a characteristic proliferation rate that allows tumor growth and regeneration of some normal tissues (e.g., the bone marrow and the liver). During irradiation, repopulation counteracts the cell killing induced by radiation, so repopulation is desirable following irradiation of normal tissues to limit side effects, while it is unwanted in the case of tumors due to the potential impairment of treatment.

Keeping in mind that radiation-induced cell killing includes the loss of the reproductive capability of the cell, it follows that damage becomes visible when the cell reaches the phase of mitosis: at that stage, if lethal DNA damage has occurred, cell replication is prevented. Therefore, radiation response, namely tumor control or organ failure, arises after a latency time that is linked to the proliferation rate in that tissue.

In the case of tumors, fast growing tumors may show a decrease in their growth rate within a few days after irradiation, while more indolent tumors may need weeks or months (HCC) to reduce in size, but still have the possibility to respond completely to therapy (Withers et al., 1988; Withers, 1992). Similarly, quickly proliferating tissues such as bone marrow, skin, and intestinal mucosa may give a warning sign of damage very early after the beginning of the treatment. Conversely, slowly proliferating tissues such as kidney, liver, lung, and bone may manifest injury after months or years.

8.2.3 REOXYGENATION

The oxygen enhancement ratio (OER) refers to the enhancement of a therapeutic or detrimental effect of ionizing radiation due to the presence of oxygen, and is quantitatively defined as the ratio between the absorbed dose in hypoxic conditions and normal conditions for a same biological effect. Oxygen is a potent modifier of radiosensitivity and hypoxic cells are typically two to three times more resistant to radiation. Tumors typically contain regions of transient acute and/or chronic hypoxia, especially in the center due to vascularization. These tumors have often been shown to be associated with a poor prognosis. Moreover, there is evidence that the duration of hypoxic conditions and the extent of the hypoxia are influential factors. Although the underlying mechanisms are to be further clarified, cells irradiated shortly after reoxygenation or after long-term exposure to hypoxia are more radiosensitive compared with those irradiated after 4–24 hours of hypoxia.

Reoxygenation between dose fractions is generally believed to ease the sterilization of hypoxic cells by increasing tumor radiosensitivity. Reoxygenation mechanisms span a few hours to few days and typically occur a few days after the beginning of irradiation, when the depopulation of the more radiosensitive cells from the bulk tumor enables more hypoxic cells to reach the blood vessels, stimulating oxygenation.

8.2.4 REDISTRIBUTION

Cells exhibit a different radiosensitivity at different phases of the cell cycle, with cells in the early S- and late-G2/M phase being most sensitive to ionizing radiation, while cells in late S-phase are the most resistant. During fractionated radiation therapy, cells in the G2/M-phase are preferentially killed, leading to a block of cells in the G2 phase and to a resulting synchronization of cells in the radioresistant S-phase. The time interval between fractions allows resistant cells from the S-phase of the cell cycle to unsynchronize, redistributing into phases in which cells of both tumors and normal tissues are more radiosensitive, thereby increasing the radiation damage. Redistribution effects span several hours and play an important role in EBRT, particularly when fractions are spaced out by several hours. In fact, redistribution during fractionated irradiation allows the sparing of normal tissues that have few rapidly cycling cells compared with tumors containing many cells with rapid turnover.

The effects summarized by the 4Rs have a leading role for the success and optimization of radiation therapy, including radionuclide therapy, and have represented a landmark for the development of radiobiological models describing the survival of cells after irradiation.

8.2.5 FURTHER CONSIDERATIONS SPECIFIC TO THE 4Rs IN RADIOEMBOLIZATION

Radiation therapy using EBRT and radioembolization differ in various aspects. First, in EBRT the treatment is generally delivered by fractionating the total dose, e.g., at 2 Gy per fraction (fr), while in radioembolization treatment is delivered in a single session and the dose rate decreases over time due to the physical decay of ^{90}Y. Furthermore, resin microspheres produce an additional embolic effect that may contribute to tumor control.

In particular, the use of fractionation in EBRT permits the redistribution of cells in the hypoxic and oxic compartment, which generally occurs in the interval between fractions. This phenomenon arises during dose delivery in radionuclide therapy but in the case of HCC, cell loss could require treatments separated by months as reported in the clinical report assessing tumor response (Kong and Hong, 2015). The possibility to increase the oxic levels of tumoral cells present in the hypoxic area before therapy could be possible during radionuclide therapy.

From another point of view, large hypoxic areas contribute to the stability of tumor volume over time. Stability is considered as a form of tumor control for liver cancer. Moreover, some authors report that chronic hypoxic areas are more radiosensitive than oxic ones, while the contrary is true for acute hypoxic areas. In general, from the radiobiological point of view, a tumor can be considered to be constituted of both chronic and acute hypoxic cells as reported in Strigari et al. (2010). This means that chronic/acute hypoxic areas could be damaged by radiation and further that damage may be only partially repaired due to the local absence of the oxygen.

A positive oxygen effect could occur in radioembolization strategies that plan more than one cycle. The first cycle of radioembolization should provoke a partial tumor shrinkage, potentially increasing the vascular perfusion to the remaining tumor. This should facilitate oxygenation and thus increase radiosensitivity, enhancing the effects of a subsequent cycle. Such an approach, although clinically more complex, could improve the response of tumors, compared with a single radioembolization therapy. Increased efficacy is likely, especially in the case of tumors large in size and with areas poorly vascularized but not necrotic, or in cases where potential toxicity to normal tissues is of concern.

Furthermore, radiotherapy using both external beam and internal emitters is a therapeutic strategy based on the oxidative stress. Both treatment modalities are specifically designed to increase reactive oxygen levels in tumor cells to elicit their death through sudden and intense oxidative stress (Manda et al., 2015). Cancer cells present a high intrinsic oxidative activity, so less additional reactive oxygen species are required compared with normal cells for triggering cell death. Levels of reactive oxygen species that are cytotoxic for cancer cells induce less drastic effects in normal cells, which have a lower oxidative status and are endowed with efficient systems to repair injuries induced by reactive oxygen species—within certain limits. Nevertheless, precise targeting of dose to the diseased tissue is a priority, aiming to spare normal tissues against the deleterious action of "therapeutic" reactive oxygen species. The sparing of normal tissue is guaranteed in EBRT by the possibility of using advanced delivery techniques, while in radioembolization it is achieved by selective or superselective administration of radioactive sources within the liver.

In addition, the macroembolic effect of resin microspheres is accompanied by a greater lack of oxygen resulting in ischemia, and therefore, enhanced efficacy. On the other hand, a shortage of oxygen might also diminish the tumoricidal effect of ionizing radiation due to a lack of oxygen radicals. In other words, the embolic effect and the potential reduction of oxygen free radicals are opposite phenomena and the resulting final net effect of these processes is still unclear.

The process of tumor control or normal tissue damage is complex with respect to the currently proposed modeling (Strigari et al., 2011; Cremonesi et al., 2014). Increased cell damage after repair signaling can cause mitotic catastrophe and cell death with an associated inflammatory tissue response. Tissue changes due to significant cell death can alter oxygenation status (reoxygenation) and trigger accelerated repopulation and redistribution in the cell cycle. Finally, tumor heterogeneity derives also from the nonuniform spatial distribution of microenvironmental stresses, such as hypoxia,

acidosis, oxidative stress, and nutrient deprivation (Mitsuishi et al., 2012). Unfortunately, these factors are only partially included in the radiobiological models. Additional studies are needed to further address these issues.

8.3 RADIOBIOLOGICAL MODELS

8.3.1 THE LINEAR QUADRATIC MODEL

Among radiobiological models describing the survival of cells after irradiation, the linear quadratic model is the most well known.

A cell survival curve describes the relationship between the fraction of surviving cells (SF), i.e., the fraction of irradiated cells that maintain reproductive integrity, and the absorbed dose. Its shape depends on several factors, including the type of radiation, the type of cells, and the radiation dose rate.

The linear quadratic model describes the effect induced by radiation in a cell population as a function of the dose delivered, and from that, the SF. The main hypothesis assumed by the linear quadratic model is that the SF of cells receiving an instantaneous absorbed dose D (as occurs, e.g., in EBRT) follows the equation:

$$\ln(\text{SF}) = (-\alpha D - \beta D^2) \tag{8.1}$$

or, equivalently,

$$\text{SF} = \exp(-\alpha D - \beta D^2) \tag{8.2}$$

The radiation-induced damage is described by the sum of two terms, αD and βD^2, respectively, representing

- DNA irreparable events (double-strand breaks) in which both strands in the double helix are simultaneously severed. The number of such events is proportional to D by the factor α, which represents the intrinsic radiosensitivity. α is tissue specific and describes the initial, linear slope of the SF curve.
- Two independent DNA reparable events (single-strand breaks), occurring close enough in time and space on the DNA filament to generate cellular death. The number of such events

is proportional to D^2 (since it is a combination of two independent events, each having an occurrence proportional to D) by a factor β, smaller than α, which represents the potential sparing capacity. β is also tissue specific and describes the quadratic slope of the SF curve.

The ratio α/β represents the dose at which the linear and quadratic components of cell killing are equal and describe the shape of the SF curve. The features of the linear quadratic model are described visually in Figure 8.1.

Repair mechanisms play a role in the quadratic component of DNA damage. In fact, a single-strand break that is not repaired can potentially combine with another single-strand break occurring close to it in space and time, and result in damage. Conversely, if the first single-strand break is repaired before the occurrence of another single-strand break, no lethal effect will occur. This is more likely to happen if radiation is delivered with a low dose rate, which means that there is a longer time interval between any two consecutive single-strand breaks. Thus, the probability of cell death increases with the increasing dose rate. This is particularly true for tissue characterized by low

α/β values, corresponding to a marked impact of the quadratic term; the opposite occurs for tissues that exhibit high α/β values, with lower relevance of the quadratic term ($-\alpha D - \beta D^2 \cong -\alpha D$). The effect of dose rate on SF for dose rates typical in common radiation therapy modalities, including ^{90}Y radioembolization, is illustrated visually in Figure 8.2.

To account for this in the formalism, the quadratic term needs to include an explicit dependence on the dose, dose rate, repair half-time T_{rep}, and the total duration of exposure, T. The Lea–Catcheside factor is a positive, dimensionless function $g(T)$ ranging between 0 and 1 (Millar, 1991; Brenner et al., 1998; Baechler et al., 2008), which is coupled to the quadratic component to express reduced cell killing due to increased repair of sublethal damage during a continuous irradiation and/or between fractions:

$$SF(D) = e^{\left(-\alpha \cdot D - g(T) \cdot \beta \cdot D^2\right)} \tag{8.3}$$

In the case of irradiation with a decaying source having an effective half-life T_{eff}, if the dose delivery is protracted for a time T significantly longer than the repair half-time ($T \gg T_{rep}$), then $g(T)$ can be approximated by

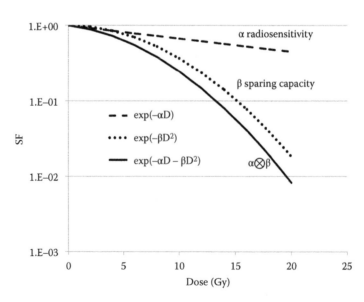

Figure 8.1 Curves describing the surviving fraction (SF) of cells vs. absorbed dose. The dashed curve represents the linear component of the SF related to DNA irreparable lesions resulting from a single hit, with the parameter α associated to the radiosensitivity of the specific tissue. The dotted curve describes the quadratic component of the SF resulting from two DNA sublethal lesions occurring close enough in time and space to create lethal damage. β is associated to the sparing capacity for the specific tissue. The continuous curve represents the SF curve with both α and β components.

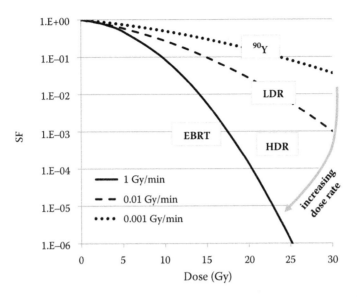

Figure 8.2 Typical trend for SF of cells following irradiation at different dose rates (dotted curve: 0.001 Gy/min; dashed curve: 0.01 Gy/min; continuous curve: 1 Gy/min). The areas characterized by dose rate values typically applied for EBRT, high-dose-rate brachytherapy (HDR), low-dose-rate brachytherapy (LDR), and ^{90}Y radioembolization (^{90}Y) are also indicated.

$$g(T \gg T_{rep}) \approx \frac{\lambda_{eff}}{\lambda_{eff} + \mu} = \frac{T_{rep}}{T_{rep} + T_{eff}} \quad (8.4)$$

where $\mu = 0.693/T_{rep}$ and $\lambda_{eff} = 0.693/T_{eff}$ (Strigari et al., 2011).

Typically, α/β is assumed to be 2.5 Gy for normal tissues and 10 Gy for tumors. For radioembolization, T_{eff} equals the physical half-life of ^{90}Y (T_{phys} = 64.24 hours, i.e., $\lambda_{eff} = \lambda_{phys} = 0.0108$ hour^{-1}), while T_{rep} is equal to 2.5 hours ($\mu_{rep} = 0.28$ hour^{-1}) for normal liver and 1.5 hours ($\mu_{rep} = 0.53$ hour^{-1}) for tumor (Cremonesi et al., 2008, 2014). A continuous low dose rate, such as is the case in radioembolization, confers the same benefits as fractionation, sparing the normal tissues—with lower α/β value—more than the tumor characterized by higher α/β value.

8.3.2 THE BIOLOGICAL EFFECTIVE DOSE

The biological effective dose (BED) is defined by the natural logarithm of the surviving fraction function versus absorbed dose as in the following expression:

$$\ln(SF) = -\alpha \cdot BED \quad (8.5)$$

$$SF = \exp(\alpha \cdot BED) \quad (8.6)$$

By comparison with Equation 8.3, BED can also be defined as the product of D and a modifying factor (named relative effectiveness, E_{rel}) that includes dose rate effects depending on radiosensitivity and repair of damage:

$$BED = D \cdot E_{rel} = D(1 + g(T) \cdot D \cdot \beta/\alpha) \quad (8.7)$$

In the case of EBRT, with a total dose D_{EBRT} delivered in N equal fractions ($d_{i,EBRT} = D_{EBRT}/N$), from the comparison of Equations 8.1 and 8.5 it follows that the BED for each fraction, $BED_{i,EBRT}$, is

$$\alpha \cdot BED_{i,EBRT} = \alpha \cdot d_{i,EBRT} + \beta \cdot d_{i,EBRT}^2 \quad (8.8)$$

For the whole treatment, the total SF will be the product of the single fractions SF_i, so

$$\alpha \cdot BED_{EBRT} = \Sigma_i (\alpha \cdot d_{i,EBRT} + \beta \cdot d_{i,EBRT}^2) = \alpha \cdot N \cdot d_{i,EBRT} + \beta \cdot N \cdot d_{i,EBRT}^2 = \alpha \cdot D_{EBRT} + \beta \cdot D_{EBRT} \cdot (D_{EBRT}/N)$$

and

$$(E_{rel})_{EBRT} = \left(1 + \frac{D_{EBRT}/N}{\alpha/\beta}\right) \quad (8.9)$$

In the case of radioembolization, with a total dose delivered, D_{RE}:

$$\alpha \cdot BED_{RE} = \alpha \cdot D_{RE} + g(T) \cdot \beta \cdot D_{RE}^2 \quad (8.10)$$

and including the expression of $g(T)$ from Equation 8.4:

$$(E_{rel})_{RE} = D_{RE} \cdot \left(1 + \frac{D_{RE} \cdot \lambda_{eff}}{(\mu_{rep} + \lambda_{eff}) \cdot \alpha / \beta}\right) \quad (8.11)$$

or

$$(E_{rel})_{RE} = D_{RE} \cdot \left(1 + \frac{D_{RE} \cdot T_{eff}}{(T_{rep} + T_{eff}) \cdot \alpha / \beta}\right) \quad (8.12)$$

For tumor tissue, an additional term can be added to the BED expression accounting for repopulation. This phenomenon competes with cell killing, so it appears as a term, the biological repopulation factor (BRF), which reduces the effect of irradiation:

$$BED = DE_{rel} - BRF \quad (8.13)$$

BRF is the biological equivalent of tumor repopulation that occurs during treatment and which effectively "wastes" some of the delivered dose (Dale, 1996; Dale et al., 2000). BRF increases with the duration of the treatment, T, during which the repopulation occurs. Considering exponential cell proliferation with doubling time T_{av}, Equation 8.10 becomes (Dale, 1996; Antipas et al., 2001)

$$BED = DE_{rel} - KT = DE_{rel} - \frac{\ln 2}{\alpha T_{av}} T \quad (8.14)$$

For EBRT the whole duration of treatment, T, is the overall treatment time.

In radionuclide therapy, with continuously decaying dose rate, the treatment is assumed to be concluded after a so-called "effective time," T^*, which is reached when the BED of Equation 8.10 equals zero. After the effective time, the repopulation exceeds the cell killing and the dose delivered is considered to be wasted.

It can be demonstrated (Antipas et al., 2001) that

$$T^* = -\frac{1}{\lambda_{eff}} \cdot \ln\left(\frac{\ln 2}{\alpha \cdot R_0 T_{av}}\right) \quad (8.15)$$

The BED is widely used to compare different radiation therapy modalities, for example, EBRT delivered with different fractionation schemes or EBRT versus radionuclide therapy. Two different radiation modalities delivering the same BED are expected (by definition) to produce the same effect. The equivalent dose (EQ) is the absorbed dose delivered with EBRT in N fractions of dose d each (EQ$d = Nd$), leading to a BED (BED$_{EBRT}$) that is the same as a radioembolization treatment (BED$_{RE}$).

The following equations relate EQ to BED$_{EBRT}$ and BED$_{RE}$:

$$BED_{EBRT} = EQ\left(1 + \frac{d}{\alpha / \beta}\right) = BED_{RE} \quad (8.16)$$

thus

$$EQ = \frac{BED_{RE} \cdot \alpha / \beta}{d + \alpha / \beta} \quad (8.17)$$

8.3.3 TUMOR CONTROL PROBABILITY (TCP)

Considering a tumor uniformly irradiated with a certain BED, the function that describes the tumor control probability is defined (Strigari et al., 2011) by

$$TCP(BED) = \exp[-N_0 \cdot SF(BED)]$$
$$= \exp[-N_0 \cdot \exp(-\alpha \cdot BED)] \quad (8.18)$$

where N_0 is the initial number of clonogenic cells and α is the radiosensitivity parameter previously defined. N_0 is the typically proportional to the tumor volume V, so $N_0 = \rho V$ is the strongly intra-tumor and intertumor dependent (ρ is the density number of clonogenic cells) (Wigg, 2001).

The mathematical formulation expressed by Equation 8.18 is valid under the assumption that radiation-induced cell killing is ruled by Poisson statistics. Poisson statistics describe the observations of random events, each having a small probability of success, for which the average number of successes (ς) after an "experiment" involving a large number of events has been defined. Also, the events must be independent, namely, the outcome of one event does not affect the outcome of the others. Under these assumptions, the probability of observing ϕ successful events after a large number of events is given by

$$P(\varphi) = \frac{e^{-\varsigma} \cdot \varsigma^\varphi}{\varphi!} \quad (8.19)$$

In our case, the "experiment" is represented by the irradiation of tumor cells, the (large) number of events is quantified by the absorbed dose D or BED. The "successful" event is represented by a single cell surviving after being hit by radiation, and the average number of "successful events" after the "experiment," ς, is given by the average number of cells surviving after irradiation:

$$\varsigma = N_0 \cdot SF(BED) = N_0 \cdot \exp(-\alpha \cdot BED)) \quad (8.20)$$

The hypothesis of event independence is represented by the assumption that the death of a cell due to radiation does not affect the successive probability of cell killing in any other cell.

TCP is represented by the probability of having no surviving cells, which, under the previous formalism, is the probability of having no "successful" events. According to Equation 8.19 TCP becomes

$$TCP = P(0) = e^{-\varsigma} \tag{8.21}$$

which is equivalent to Equation 8.18 after combining Equations 8.20 and 8.21.

In the case of nonuniform irradiation of the tumor, the same formalism in Equation 8.18 can be maintained after calculation involving the absorbed dose (or BED) distribution over the tumor volume, which must be known. If we assume that the BED distribution in the tumor volume is described by a probability function, $P(BED)$, then for each BED value, $P(BED)$ indicates the fraction of the total tumor volume receiving a biological effective dose equal to BED. The SF value over the entire tumor volume is obtained by calculating the surviving fraction associated with each BED value ($SF(BED) = \exp(-\alpha BED)$) and by opportunely weighting each value according to its probability of occurrence:

$$SF = \int_0^\infty P(BED) \cdot e^{-\alpha \cdot BED} \, dBED \tag{8.22}$$

A new parameter has been defined, the equivalent uniform biological effective dose (EUBED) (O'Donoghue, 1999; Jones and Hoban, 2000), representing the BED, which, if given uniformly over the whole tumor volume would produce the same fraction of surviving cells as the nonuniform BED distribution described by P(BED) in Equation 8.22. Following Equation 8.5

$$EUBED = -\frac{1}{\alpha} \ln(SF) = -\frac{1}{\alpha}$$
$$\ln(\int_0^\infty P(BED) \cdot e^{-\alpha \cdot BED} \, dBED) \tag{8.23}$$

In clinical practice, the BED distribution is not expressed by a continuous function P(BED), but a discrete formulation {N_j, BED$_j$}, which can be easily obtained if the 3D absorbed dose distribution is available and converted into a 3D BED map.

After dividing the possible BED range into a number M of intervals, N_i represents the number of voxels receiving a BED value included in the ith interval and BED$_i$ is the average BED value in that same interval. With N defined as the total number of voxels constituting the tumor, the EUBED becomes

$$EUBED = -\frac{1}{\alpha} \left[\ln \sum_i^M \frac{SF_i}{N} \right]$$
$$= -\frac{1}{\alpha} \left[\ln \sum_i^M \frac{\exp(-\alpha \cdot BED_i)}{N} \right] \tag{8.24}$$

In the extreme case of infinitely small intervals, each BED value represents an interval itself and Equation 8.24 becomes

$$EUBED = -\frac{1}{\alpha} \left[\ln \sum_i^N \frac{\exp(-\alpha \cdot BED_i)}{N} \right] \tag{8.25}$$

In this case, the TCP is calculated by

$$TCP(\{N_i, BED_i\}) = TCP(EUBED)$$
$$= \exp(-N_0 \cdot \exp(-\alpha \cdot EUBED)) \tag{8.26}$$

Examples of application of the TCP concept to radioembolization clinical data can be found in Strigari et al. (2010) and Chiesa et al. (2015), for the treatment with resin and glass microspheres, respectively.

Strigari et al. (2010) defined the tumor response according to the Response Evaluation Criteria in Solid Tumors (RECIST) and European Association for the Study of the Liver (EASL) criteria accounting for tumor necrosis when evidenced by non-enhanced areas. Equation 8.18 was adopted, including a further term able to take into account possible variation in the density of clonogenic cells in the patient's tumor:

$$TCP(D) = \exp \left(-\sum_i \eta_i \cdot N_0 \cdot SF(BED) \right) \tag{8.27}$$

where

$$\eta_i = \frac{1}{\sqrt{2\pi}} \cdot \frac{1}{\sigma_{\ln(N)}} \cdot \exp \left\{ -\frac{1}{2} \left(\frac{\ln(N_i) - \ln(N_0)}{\sigma_{\ln(N)}} \right)^2 \right\} \tag{8.28}$$

η_i is the proportion of population having a number of clonogenic cells equal to N_i, derived from a Gaussian distribution of $\ln(N)$ with a mean value of $\ln(N_0)$ and a standard deviation $\sigma_{\ln(N)}$. The fit of experimental data indicated the presence of two subpopulations, a more radioresistant one with $\alpha = 0.001$ Gy^{-1} and $\ln(N_0) = 23$, and a less radioresistant one with $\alpha = 0.005$ Gy^{-1} and $\ln(N_0) = 6.9$ (Strigari et al., 2010). According to both criteria (RECIST and EASL), an average absorbed dose to the tumor higher than 110–120 Gy guaranteed partial or complete response to 50% of the patients. However, following a more detailed analysis, the two criteria for tumor response produced different results in some cases, and the authors suggested a combination of both with the further inclusion of additional criteria based on 18-fluoro-2-deoxyglucose positron emission tomography/computed tomography (^{18}F-FDG-PET) imaging. The authors did not account for the heterogeneity of dose distribution in their analysis, but remarked the need to adopt a more advanced mathematical formalism to consider this issue as well.

Chiesa et al. (2015) also classified the tumor response according to the radiological EASL criteria. CT scans were performed every third month and the best tumor response over time was considered. Tumor control was defined considering both complete responding and partial responding lesions. TCP data were plotted as a function of the average absorbed dose to the lesion, and the parameters α, N_0, and $g(T) \cdot \beta$ were obtained from the fit of experimental data with two simplified expressions of Equation 8.26:

$$\text{TCP} = \exp(-N_0 \cdot \exp(-\alpha \cdot D - g(T) \cdot \beta \cdot D^2)) \quad (8.29)$$

or neglecting the quadratic term (i.e., in the assumption of negligible influence of the dose-rate effect):

$$\text{TCP} = \exp(-N_0 \cdot \exp(-\alpha \cdot D)) \quad (8.30)$$

The values obtained for both N_0 and α were much lower than the corresponding values reported for EBRT: $\alpha = 0.002$ Gy^{-1} versus the typical EBRT value of around 0.01 Gy^{-1}, and N_0 ranging from 2.7 to 3.4. The apparent lower radiosensitivity exhibited following radioembolization should be interpreted—according to the authors—as a consequence of the high heterogeneity of microspheres and hence

dose distribution at the microscopic level, which necessitates higher average doses to reach an effect comparable with EBRT. Chiesa et al. (2015) also investigated the possible variation of TCP parameters with tumor size, dividing the whole pool of data in two groups. Substantially different trends were observed: the absorbed dose yielding a 50% TCP was around 250 Gy for lesions smaller than 10 cm^3, while 50% TCP was achieved at 1300 Gy in large volumes.

8.3.4 NORMAL TISSUE COMPLICATION PROBABILITY

The literature offers several NTCP models, either developed to fit the phenomenological curves from EBRT and then extended/adapted to other treatment modalities or based directly on concepts derived from radiation biology, physics, and physiology.

The modeling of NTCP for the liver is of special interest in the context of radioembolization since irradiation of normal liver can become a limiting factor for therapy. Three main models have been used to interpret dose–effect correlations for the liver: the Lyman–Kutcher–Burman (LKB) model, derived from EBRT (Lyman, 1985; Kutcher and Burman, 1989); the parallel architecture model (Withers et al., 1988; Yorke et al., 1992; Jackson et al., 1993; Niemierko and Goitein, 1993); and the model by Walrand, based on dosimetric simulations at the microscopic level (Walrand et al., 2014a, 2014b) for two types of microspheres providing different microsphere-specific activity (msA) of 0.05 and 2.5 kBq per microsphere for resin and glass microspheres, respectively.

Here we present the LKB and the parallel architecture models. Please refer to Chapter 9, Section 9.12, for a deep description of the model by Walrand.

8.3.4.1 The Lyman (and LKB) model

The first formulation of the Lyman model (Lyman, 1985) was developed to predict the effect after uniform EBRT irradiation of whole organs. Clinical data were collected with reference to a specific endpoint (i.e., complication) and the following expression was proposed to fit the absorbed dose–response data:

$$\text{NTCP}(D) = \frac{1}{\sqrt{2\pi}} \int_{-\infty}^{t} \exp\left(-\frac{x^2}{2}\right) dx \quad (8.31)$$

with

$$t = \frac{D - TD_{50}}{m \cdot TD_{50}} \qquad (8.32)$$

where D is the uniform absorbed dose in the organ (delivered at a specified dose per fraction), TD_{50} is the absorbed dose value at the same dose per fraction for which 50% of the population exhibited complications at a fixed time point after the irradiation (e.g., 5 years), and m is a parameter representing the steepness of the dose–effect curve.

For each effect (complication) considered, the $TD_{50,5}$ and m parameters have been experimentally obtained from epidemiological data. Similar curves can be obtained for different fractionation schemes, different endpoints, or at different time points, with each scenario described by its own TD_{50} and m parameters. In case of partial irradiation of an organ, for example, when only a fraction v of the whole volume is irradiated, but still with uniform absorbed dose in the irradiated area, the same formalism can be adopted for NTCP estimation (Equation 8.31) provided use of properly adapted value of TD_{50}:

$$TD_{50}(v) = TD_{50}(1) \cdot v^{-n} \qquad (8.33)$$

where n is a parameter, in case of partial irradiation, that quantifies the behavior of the organ and to what extent NTCP is affected by the amount of the irradiated volume v. Usually, n is restricted to the 0–1 interval, where an n value close to zero indicates that NTCP presents a weak variation for different irradiated volumes, with the converse true for n approaching 1. In liver, the values of n found by various authors are quite close to 1 (0.95–1.1), indicating a notable volume effect (see Figures 8.4 and 8.6 and Example 8.2). The power law of Equation 8.33 is not based on a biological or physiological rationale but is a simple expression that adequately describes clinical observations (Yorke et al., 2001).

The Kutcher and Burman reduction scheme (Kutcher and Burman, 1989) extends the Lyman model to the case of nonuniform irradiation, with fractions of volumes v_i uniformly irradiated with an absorbed dose D_i. In this case, the NTCP formalism (Equation 8.31) can still be maintained as long as the effective volume method or the equivalent uniform dose (EUD) concept and DVH reduction are applied.

Using the effective volume method, NTCP can be calculated using Equation 8.31 assuming that a uniform dose equal to the maximum dose in the DVH, D_{max}, is delivered to a volume called effective volume V_{eff}, calculated by

$$V_{eff} = \sum v_i (D_i / D_{max})^{1/n} \qquad (8.34)$$

NTCP can be calculated by Equations 8.31 and 8.32 considering $D = D_{max}$ and TD_{50} (V_{eff}/V_{tot}) calculated with Equation 8.33.

Alternatively, the nonuniform irradiation $\{v_i, D_i\}$ can be reduced to a whole organ irradiation with uniform dose equal to EUD, calculated by

$$EUD = (\sum v_i (D_i)^{1/n})^n \qquad (8.35)$$

NTCP can be calculated by Equations 8.31 and 8.32 considering $D = EUD$ and $TD_{50}(1)$. To adapt this model to radioembolization, it would be necessary first to calculate the radioembolization BED volume histogram, then convert the absorbed dose values into the equivalent dose of EBRT as described by Equation 8.17. EUD or V_{eff} could then be computed using the newly obtained DVH and finally, the NTCP could be estimated considering EUD/V_{eff} and curve parameters ($TD_{50,m}$), which refer to the same fractionation scheme used for the DVH conversion.

The very first publication providing values for the parameters of the Lyman model was in 1991 by Burman et al. (1991) based on the partial liver tolerance data published by Emami et al. (1991). Subsequently, more updated sets of parameters were presented by Dawson et al. (2001, 2002) and Dawson and Ten Haken (2005). More recently, radiation liver injury and dose–volume data have been reviewed in the framework of the Quantitative Analyses of Normal Tissue Effects in the Clinic (QUANTEC) project (Pan et al., 2010).

It is important to remember that the model has been conceived and developed in the framework of EBRT, which originally aimed to deliver a nearly uniform irradiation to the target, with an accepted nonuniformity within a few percent. The adoption in radioembolization of the same parameter values (m, TD_{50}) for EBRT should be avoided, or at least made with caution, even after appropriate conversion of radioembolization absorbed dose into the equivalent EBRT value. This is because the dose delivery during radioembolization can be very heterogeneous, as described in Chapter 9. When considering the healthy parenchyma only (thus excluding the tumor), the LKB model exhibits high sensitivity to the presence of small high-dose regions (Yorke et al., 1999), possibly leading to unrealistic NTCP estimates due to the heterogeneity of dose distribution in radioembolization. This will be illustrated by some examples in Section 8.4.5.2 (Figures 8.14 and 8.15).

Moreover, the heterogeneity of dose delivery in the case of radioembolization may be present on both a microscopic and macroscopic scale (Walrand et al., 2014b) due to the pattern of microspheres distribution in the arteries; in light of this, even for an apparent homogeneous dose delivery, as shown by macroscale posttreatment imaging, microscopic inhomogeneities will still create discordance with EBRT. Nonetheless, it is possible to adopt the LKB formalism (Equations 8.31 and 8.32) to fit the radioembolization dose-toxicity data and derive a dedicated set of parameters suitable to make predictions for patients with comparable clinical characteristics, treatment modalities and dosimetry protocols.

As for TCP, examples on this can be found in Strigari et al. (2010) for RE with resin microspheres and Chiesa et al. (2015) for glass spheres.

Strigari et al. (2010) classified toxicity according to the Common Terminology Criteria for Adverse Events (CTCAE) and all events with toxicity grade equal or higher than 2 were recognized as complications. Average BED to the liver was calculated and plotted with the observed complication probability associated to each BED value; data fit performed according to Equation 8.31 resulted in a BED_{50} estimate equal to 93 Gy and m equal to 0.28 (assuming $n = 1$ in case of partial irradiation, Equation 8.34). This value for BED_{50} is higher than the values reported for late effect induction following EBRT, which is 72 Gy according to Emami et al. (1991) and 64 Gy according to Dawson et al. (2002). The distribution of microspheres in the vasculature of the vessels determines the absorption of high doses in small volumes, with a possible loss of biologic effect.

Chiesa et al. (2015) reported treatment-related liver disease according to international guidelines, avoiding those clearly attributable to tumor progression. Complication probability was plotted as a function of different dosimetric and radiobiological quantities, including average absorbed dose to the uninvolved liver, average BED, EUD, and EUBED; two different sets of radiobiological parameters (either gathered from EBRT experience or self-derived from analysis of radioembolization patients' data) and two different methods for volume delineation (SPECT-based or CT-based) were adopted. For each combination of these variables, a fit of the experimental data was performed according to Equation 8.31 and the TD_{50} and m parameters were derived. In all cases, the TD_{50} values obtained were significantly higher than those adopted for EBRT. For example, when considering the absorbed dose, $TD_{50}(1)$ ranged from 97 to 106 Gy, much higher than the EBRT values, which are typically lower than 50 Gy (Dawson and Ten Haken, 2005; Pan et al., 2010). For radioembolization using glass microspheres, Chiesa recommended an average absorbed dose to the normal liver equal to 75 Gy corresponding to an accepted complication probability equal to 15% (TD_{15}) to guide the treatment planning.

8.3.4.2 The parallel architecture model

The architecture model assumes that the organ is organized in multicellular entities, called functional subunits (FSU), that respond to irradiation according to the cell radiosensitivity and to the minimum number of cells that needs to be alive to keep the FSU functioning (Withers et al., 1988).

The occurrence of complication in the whole organ after irradiation depends on the radiation distribution among the different FSUs, the radiosensitivity of each FSU, and on how the different FSUs collaborate to guarantee organ function.

Two basic architectures are defined to represent the organization of cells, the serial and the parallel architecture. The serial architecture assumes that the different FSUs work in series and the impairment of even a single subunit is able to compromise the whole organ functionality. Conversely, the parallel architecture describes the organ as many FSUs working in parallel, each nearly independent from the others, and states that the whole organ functionality is preserved until the fraction of impaired subunits is lower than a certain fixed amount, referred to as a "functional reserve."

Focusing on the liver, its functionality can be described by a parallel architecture model (Jackson et al., 1995), which we will now illustrate. A dose–response function is defined for each subunit, with $p(d)$ the probability, p, of damaging a subunit after irradiation at a given dose d. A sigmoidal dose–response function has been proposed to describe the subunit response phenomenologically:

$$p(d) = \frac{1}{\left[1 + (d_{1/2} / d)^k \right]} \quad (8.36)$$

where $d_{1/2}$ represents the dose at which 50% of the subunits is damaged and k is the slope of the dose–response function.

The fraction of an organ damaged by the treatment, f, can be calculated from the healthy liver differential DVH assuming that the organ is composed of a large number of subunits which respond to radiation independently and that each FSU is small enough to consider their individual irradiation to be homogeneous. In addition, when the number of subunits is large, the total damage is demonstrated to be well approximated by the mean damage over the voxels, thus yielding:

$$f = \sum_i v_i p(d_i) \quad (8.37)$$

with v_i being the fraction of voxels receiving dose d_i.

To calculate the probability that the damage of a fraction f of the organ actually results in a complication, a comparison is made to the distribution of functional reserve in the patient population. Assuming that the cumulative distribution is described by a displaced error function H with a mean value of the functional reserve fr_{50} and width σ_{fr}, the complication probability (NTCP) associated to the damage of a fraction f is given by

$$NTCP = H(f)$$

$$= \frac{1}{\sqrt{2\pi\sigma_{fr}^2}} \int_0^f d(fr) \exp[-(fr - fr_{50})^2 / 2\sigma_{fr}^2] \quad (8.38)$$

In other words, when a patient suffers the damage of an organ fraction equal to f, the probability of observing a complication is estimated as the proportion of population having a functional reserve lower than f. The four parameters of the model (d_{50}, k, fr_{50} and σ_{fr}) have been obtained from the analysis of clinical complication data of patients undergoing EBRT (Jackson et al., 1995). The best-fit parameters, together with the 68% confidence interval, are $fr_{50} = 0.497 \pm 0.043$; $\sigma_{fr} = 0.047 \pm 0.027$; $d_{50} = (41.62 \pm 3.5)$ Gy delivered at 1.5 Gy fractions; $k = 1.95 \pm 0.77$.

Yorke et al. (1999) investigated the possibility of using this model to explain clinical observations after liver radioembolization using glass microspheres, where the absence of complication was observed after estimated absorbed doses to the normal liver up to 150 Gy. First, to apply the model to radioembolization, the $d_{1/2}$ parameter reported above must be converted into a BED_{50} value according to Equations 8.7 and 8.9, and the radioembolization DVH needs to be converted in a BED volume histogram according to Equations 8.7 and 8.11. Such BED values should replace the d values in Equations 8.36 and 8.37. Yorke et al. (1999) found that the value of T_{rep} strongly influenced the NTCP estimate and that only short repair times (≤ 1 h) allowed consistency between the model and clinical data. Second, it was shown that for a same average dose to the whole liver, different DVHs result in radically different NTCP estimations. For this reason, the question was raised about which DVH is adequate to describe the absorbed dose distribution during radioembolization, which exhibits steep absorbed dose gradients that may not be described by the 3D dose distribution obtained from posttreatment imaging. In the Yorke study (Yorke et al., 1999), when the parallel architecture model was applied using DVH derived from autoradiography, consistency with clinical data was found for model parameters varying in the range indicated by Jackson et al. (1995). When the parallel architecture model was applied using DVH derived from a simulation of the microsphere distribution superimposed to the FSU structure (the liver lobule), consistency with clinical data was found for a wider range of parameters. The authors concluded that the parallel architecture model could be adequate to describe clinical observations on hepatic toxicity, but that a deeper investigation on microsphere distribution and mechanisms of damage repair was needed.

The dose-toxicity model recently presented by Walrand (Walrand et al., 2014a, 2014b) and described in Chapter 9 has important commonalities with the parallel architecture model adapted to radioembolization by Yorke et al. (1999). These common features include the concept of the FSU, the use of an EBRT-derived sigmoid dose–response function for the irradiation of a single FSU, the development a microscopic model simulating the FSU structure, and the microsphere distribution to derive the dose distribution at the lobule level. Walrand's model parameters are derived from clinical data, and, notably, also incorporate the dependence on the mSA, allowing predictions for both resin and glass radioembolization therapy.

8.4 APPLICATION OF THE MODELS AND EXAMPLES

In radioembolization, treatment-planning techniques suggested by the manufacturers are based essentially on empiric or, at best, rough dosimetry algorithms (more details provided in Chapter 5). In the latter case, threshold dose limits were proposed to preserve the liver, namely 120 Gy to the whole liver or the treated lobule according to a monocompartmental model proposed for glass microspheres, and 80 Gy to normal liver, according to a multicompartmental model proposed for resin microspheres. Such threshold values were included in the clinical trials for both radioembolization products and are still routinely used.

Other researchers preferred to embrace a cautionary approach and have performed feasibility studies with escalating dose (Cremonesi et al., 2008; Sangro et al., 2008). The basic idea was to extrapolate constraints from EBRT and establish an absorbed dose limit to the healthy liver. A limit of 40 Gy in radioembolization for a whole-liver treatment was initially identified as a good compromise (see Example 8.1). The first clinical cases of liver toxicity manifested at doses higher than 40–50 Gy for whole-liver treatments with resin microspheres seemed to verify this limit, under the condition of whole-liver treatments with resin spheres (Sangro et al., 2008).

A subsequent step was the comparison of radioembolization with EBRT through the BED, EUD, and EUBED concepts used as a guide to define thresholds not only for whole-liver treatments but also for lobar treatments and multicycle approaches. Considering inherent differences between the two kind of therapies, EBRT models were derived for uniform dose delivery, while radioembolization dose distribution is typically nonhomogeneous not only at macroscopic level (i.e., at the voxel level shown by nuclear medicine imaging) but also microscopically. In addition, resin and glass microspheres exhibit very different patterns of dose distribution at the microscopic level due to the differences in the number of microspheres. Resin microspheres, with a lower activity load, 0.05 kBq/sphere, are much more numerous and can distribute more regularly in capillaries, providing an overall increase in dose uniformity compared with glass microspheres. The microdosimetric behavior of glass and resin microspheres has been illustrated by Walrand (Walrand et al., 2014a).

The use of EUD and EUBED concepts represents a first attempt to account for nonuniformity, but they are limited to the macroscopic level. In certain circumstances, only a microscopic modeling could better describe the results, as discussed further in Chapter 9.

Here, follow some examples that show various steps of the comparison between RE and EBRT, allowing use of EBRT evidence as a first guide for the interpretation of the outcomes observed in RE.

Unless otherwise specified, all of the examples provided here consider the following set of radiobiological parameters taken from the literature (Dawson and Ten Haken, 2005; Cremonesi et al., 2008):

1. $\alpha/\beta = 2$ Gy; $T_{1/2\,rep} = 2.5$ hours; $T_{1/2\,eff} = T_{1/2\,phys}$ $_{Y\text{-}90} = 64.2$ hours, for normal liver.
2. $\alpha/\beta = 10$ Gy; $T_{1/2\,rep} = 1.5$ hours; $T_{1/2\,eff} = T_{1/2\,phys}$ $_{Y\text{-}90} = 64.2$ hours, and $T_{av} = 30$ days for tumors.

Figure 8.3a illustrates the BED curve as a function of the absorbed dose for radioembolization and EBRT for the uninvolved liver considering 1.5 Gy/fraction, (Dawson et al., 2002; Pan et al., 2010). Figure 8.3b highlights the trends in Figure 8.3a at lower doses and shows the crossing of the two curves at a dose D^* (~40 Gy). Of note, for a same dose $D < D^*$, BED_{EBRT} is greater than BED_{RE}, meaning that for the same effect in normal liver tissue, a lower dose of EBRT would be necessary compared with radioembolization, while the opposite holds for $D > D^*$.

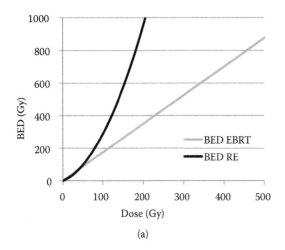
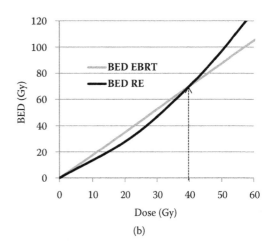

(a)

(b)

Figure 8.3 BED as a function of absorbed dose of radioembolization and EBRT. (a) The trend in a large interval of dose up to 500 Gy; (b) the dose interval up to 60 Gy. D^* represents the dose for which BED_{RE} and BED_{EBRT} have the same value. For a same dose lower than D^*, $BED_{RE} < BED_{EBRT}$, while the opposite occurs for doses higher than D^*.

Table 8.2 Absorbed dose tolerance from EBRT (TD) delivered at 2 Gy/fr (Emami et al., 1991), and correspondent BED ($BED_{EBRT} = DED_{RE}$) and absorbed dose values (D_{RE}) for radioembolization

Field	Doses or BED (Gy)	1/1 whole-liver irradiation	Two-thirds liver irradiation	One-third liver irradiation
EBRT	$TD_{5/5, EBRT}$	30	35	50
	$TD_{50/5, EBRT}$	40	45	55
EBRT = RE	$BED_{5/5,EBRT=RE}$	60	70	100
	$BED_{50/5,EBRT=RE}$	80	90	110
RE	$D_{5/5, RE}$	36	40	51
	$D_{50/5, RE}$	44	48	54

Example 8.1: Tolerance doses from EBRT extrapolated to RE

In the last 25 years, several authors have presented outcomes of liver tolerability after EBRT, including the historical data from Emami et al. (1991), the analysis of Dawson et al. (2002), and the study of Pan et al. (2010) reported in the QUANTEC papers. The tolerance doses derived for EBRT by Emami are reported in Table 8.2 in terms of doses delivered at 2 Gy/fr and associated to 5% ($TD_{5/5}$) and 50% ($TD_{50/5}$) probability to cause severe hepatitis/liver failure within 5 years. These values are also converted into the corresponding absorbed doses in radioembolization by means of the BED concept. First, EBRT absorbed doses are converted into BED using the following (from Equations 8.7 and 8.9):

$$BED_{EBRT} = D_{EBRT}\left(1 + \frac{2}{\alpha/\beta}\right) = D_{EBRT}\left(1 + \frac{2}{2}\right)$$

$$= 2 \cdot D_{EBRT} = 2 \cdot D_{EBRT}$$

By imposing $BED_{EBRT} = BED_{RE}$, the corresponding radioembolization absorbed dose is derived by solving the following equation including the radiobiology parameters previously specified (from Equations 8.7 and 8.11):

$$BED_{EBRT} = BED_{RE} = D_{RE}\left(1 + \frac{D_{RE} \cdot T_{rep}}{(T_{rep} + T_{eff}) \cdot \alpha/\beta}\right)$$

$$= D_{RE}\left(1 + \frac{D_{RE} \cdot 2.5}{(2.5 + 64.2) \cdot 2}\right) = D_{RE}\left(1 + \frac{D_{RE}}{53.4}\right)$$

The tolerance doses of EBRT for two-thirds and one-third of liver irradiation could be considered as possible values to guide radioembolization treatments of the right and left lobe, respectively.

Alternatively, if a 40 Gy absorbed dose threshold is set for whole-liver radioembolization treatment, the corresponding BED and absorbed dose in EBRT would be

$$BED_{RE} = D_{RE}\left(1 + \frac{D_{RE}}{53.4}\right) = 40 \cdot \left(1 + \frac{40}{53.4}\right) = 70\,Gy$$

$$BED_{RE} = BED_{EBRT} = 70\,Gy = 2\,D_{EBRT}\ thus,\ D_{EBRT} = 35\,Gy$$

Example 8.2: Lyman NTCP curves derived for EBRT and extrapolated for RE

Dawson et al. (2002) used the LKB NTCP model to interpolate clinical outcomes obtained for an EBRT scheme of 1.5 Gy, twice a day, for whole-liver irradiation and for partial volume irradiation as well (Lyman, 1985; Kutcher and Burman, 1989). In the case of whole-liver irradiation, the fit yielded the following parameter values: $TD_{50}(1) = 43.3$ Gy, $m = 0.18$, $n = 1.1$. In case of partial irradiation of two-thirds and one-third of the liver, Equation 8.33 leads to

$TD_{50}(2/3) = 43.3 \cdot (2/3)^{-1.1} = 67.6$ Gy, and $TD_{50}(1/3) = 43.3 \cdot (1/3)^{-1.1} = 145$ Gy.

It is important to note that such values, referred to 1.5 Gy/fr schema, cannot be directly compared with the Emami values without appropriate conversion through the BED concept because of the difference in dose per fraction delivered (Emami: 2 Gy/fr; Dawson: 1.5 Gy/fr). Similarly, the NTCP curves corresponding to radioembolization are derived applying the BED conversion from EBRT doses, e.g., for $D_{EBRT} = 30$ Gy,

$$BED = D_{EBRT}\left(\frac{dose / fr}{\alpha / \beta}\right) = 30\left(1 + \frac{1.5}{2}\right) = 52.5$$

$$= D_{RE}\left(1 + \frac{D_{RE} \cdot T_{rep}}{(T_{rep} + T_{eff}) \cdot \alpha / \beta}\right)$$

$$= D_{RE}\left(1 + \frac{D_{RE} \cdot 2.5}{(2.5 + 64.2) \cdot 2}\right)$$

$$= D_{RE}\left(1 + \frac{D_{RE}}{53.4}\right)$$

thus, $D_{RE} = \{[(26.7)^2 + 53.4 \times BED]^{1/2}\} - 26.7 = 33$ Gy, and so on.

Figure 8.4 illustrates the NTCP curves for the Lyman model as a function of both EBRT and radioembolization absorbed dose, while Figure 8.5 shows the NTCP curves as a function of BED. The curves for full, two-thirds, and one-third of irradiated liver volume can be used for irradiation of the whole liver, the right lobe, and left lobe, respectively.

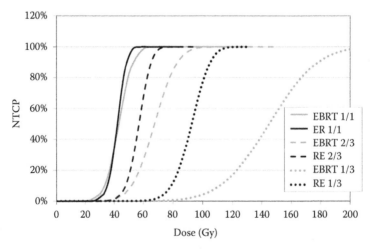

Figure 8.4 NTCP curves as a function of absorbed dose for whole and partial irradiation of the liver for EBRT and radioembolization. Uniform dose distribution within irradiated tissue is assumed. Continuous lines represent whole liver irradiation (1/1), dashed lines represent the irradiation of 2/3 of the liver volume, and dotted lines represent the irradiation of 1/3 of the liver volume. Black curves refer to radioembolization, gray curves refer to EBRT.

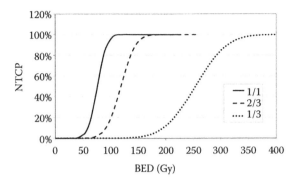

Figure 8.5 NTCP curves as a function of BED assuming uniform dose distribution. Continuous line represents whole liver irradiation (1/1), dashed line represents the irradiation of 2/3 of the liver volume, and dotted line represents the irradiation of 1/3 of the liver volume.

In the case of more selective treatments involving a fraction v of the whole liver, the NTCP can be estimated either after calculating the corresponding $TD_{50}(v)$ value (Equation 8.33) or after calculating the EUD (Equation 8.35) and using the $TD_{50}(1)$ value. Consider, for example, the irradiation of a segment of 100 g of normal liver tissue using radioembolization, with a mean absorbed dose of 150 Gy and with a 2100 g total liver mass:

$$BED = D_{RE} + D_{RE^2}/53.4 = 150(1+150/53.4)$$

$$= 150 \cdot (1+0.0187 \cdot 150) = 572 \text{ Gy}$$

which corresponds to a EBRT dose (at 1.5 Gy/fr) of

$$D_{EBRT} = EQ_{1.5} = BED/(1+1.5/2) = 327 \text{ Gy}$$

and this to an EUD of $(100/2100) \cdot D_{EBRT} = 0.048 \cdot 327 = 15.6$ Gy, which is associated to 0% NTCP.

For such a small liver involvement, the corresponding TD values are

$$TD_{50}(100/2100) = 43.3 \times (0.048)^{-1.1} = 1233 \text{Gy}$$

$$TD_5(100/2100) = 30.5 \cdot (0.048)^{-1.1} = 868 \text{ Gy}$$

that correspond to absorbed doses in RE of 314 and 259 Gy, respectively.

This illustrates that in the extreme, when very low liver fractions are involved, the deliverable ("tolerated") absorbed doses may reach very high values. This reflects the absence of toxicity encountered in selective/superselective approaches and a behavior similar to that of liver resection, provided that a minimum volume of normal liver is not excised. This is attributed to the extraordinary recovery capacity of the liver as described in Chapter 6.

Studies based on surgical resection have identified the minimum remnant volume of the liver necessary to maintain function to be approximately 40% in patients without chronic liver disease (Kubota et al., 1997). This figure would indicate the potential for the safe irradiation of one-third (33%) of the liver (e.g., left lobe). This represents a discrepancy with the possible risk at high doses represented in the curves of the Lyman model extrapolated to radioembolization (see Figures 8.4 and 8.5, curve "1/3"). The clinical experience of radioembolization seems to confirm the absence of injury when small portions of the liver, and even the left lobe, receive high doses. Therefore, the Lyman model may be too conservative in the case of segmental and left-lobar treatments.

A final important point to highlight regarding the outcomes provided by the Lyman model in EBRT is that two distinct curves are obtained if a distinction is made between primary and secondary tumors. Each curve is described by a different set of parameters: $m = 0.12$, $n = 0.97$, and $TD_{50,EBRT}$ is 40 Gy for HCC (primary tumor); $m = 0.18$, $n = 1.1$, and $TD_{50,EBRT}$ is 46 Gy for metastases (secondary tumor), in the case of whole-liver irradiation. These values indicate a shift and slight deformation of the NTCP curves versus higher doses in the case of metastases as compared with HCC. This is expected considering that the lower liver functional reserve of the normal liver often present in patients with HCC.

The NTCP curves generated by these parameters are illustrated in Figure 8.6, with the inclusion of the corresponding curves extrapolated for radioembolization, for which $TD_{50,RE}$ in the case of whole-liver irradiation is 40 Gy for HCC and 44 Gy for metastases.

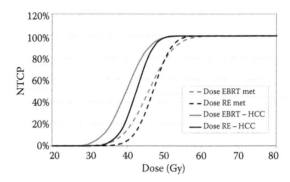

Figure 8.6 NTCP curves vs. absorbed dose of EBRT for HCC (black, continuous curve) and metastases (black, dashed curve) and curves derived for radioembolization for HCC (gray continuous curve) and metastases (gray dashed curve).

In summary, examples have been provided on how to convert the EBRT parameters/doses involved in the NTCP Lyman model into the corresponding values for radioembolization. It is important to emphasize that the method shown to derive radioembolization values from EBRT data can be nothing more than a guidance, which may be useful in the planning of radioembolization treatments aiming to remain on the side of safety. However, the model suffers from some limitations due to the differences between the two irradiation modalities. For example, in some cases the Lyman model predicts high toxicity if very high doses are delivered even to a small portion of the liver (see Figure 8.4, Example 8.2; Figures 8.11 and 8.13), which does not agree with clinical observations. Other models are able to account for the absence of toxicity provided a minimum portion of the liver is involved, as highlighted in Figures 8.8 and 8.9 and Example 8.4.

For this reason, now that clinical safety data are widely available, correlation studies need to be performed based on radioembolization dosimetry and clinical data, rather than extrapolation from EBRT, allowing the derivation of self-consistent parameter values for the NTCP models.

Example 8.3: Multiple cycle approach with the linear quadratic model
After the first clinical examples of hepatic toxicity were observed following whole-liver radioembolization, less aggressive treatment methodologies were investigated. Besides lobar or selective approaches, multiple-cycle strategies have been proposed, separated by a time interval sufficient to allow liver damage recovery. In most cases, clinicians have opted for large intervals (e.g., at least 1 month) to allow the evaluation of several clinical parameters and the regeneration of liver functionality before proceeding with additional treatment cycles.

In the case of normal liver, radiobiological models can describe the different effects obtained by dividing the treatment into different numbers of cycles at different doses per cycle; however, the model does not account for the time between two consecutive cycles or the total treatment time. For tumor tissue, the total treatment time is accounted for in the repopulation term (see Equation 8.14).

In general, when a treatment is composed of multiple cycles (N), each (i) giving an absorbed dose D_{Li} to normal liver tissue and D_{Ti} to the tumor, the overall BED (BED$_L$ for normal liver, BED$_T$ for tumor) is the sum of the BED values of each cycle:

$$\mathrm{BED_L} = \sum_i \mathrm{BED}_i = \sum_i D_{Li}\left(1 + \frac{D_{Li} \cdot T_{rep}}{(T_{rep} + T_{eff}) \cdot \alpha/\beta}\right)$$

$$= \sum_i D_{Li}\left(1 + \frac{D_{Li}}{53.4}\right)$$

$$\mathrm{BED_T} = \sum_i \mathrm{BED}_{Ti} = \sum_i \left\{ D_{Ti} \cdot \left(1 + \frac{D_{Ti} \cdot T_{rep}}{(T_{rep} + T_{eff}) \cdot \alpha/\beta}\right) \right.$$

$$\left. - \mathrm{K} \cdot \mathrm{T} \right\}$$

In a general case of metastases ($\alpha = 0.3$ Gy^{-1}) and a time interval of 30 days between cycles:

$$\text{BED}_T = \sum_i \left\{ D_{Ti} \cdot \left(1 + \frac{D_i \cdot 1.5}{(1.5 + 64.2) \cdot 10} \right) - \frac{0.693}{0.3 \cdot 30} T \right\}$$

$$= \sum_i \left(D_{Ti} \cdot (1 + D_{Ti} \cdot 0.0023) \right) - 2.3 \cdot (N - 1)$$

where KT = 2.3 Gy represents the effect of tumor repopulation (i.e., the "wasted dose"). So, at least with the parameters considered, the low value of the wasted dose seems not to greatly influence tumor response. Alternatively, if we consider HCC assuming $\alpha = 0.01$ Gy^{-1}, the repopulation factor is much higher: KT = 0.693/0.01·30/30 ~69 Gy.

Such modeling can be used when the treatment is intentionally divided into multiple cycles at the planning stage, as well as in the case of a retreatment, not previously foreseen but representing a further option after a first radioembolization therapy. However, contrary to other therapies with radiopharmaceuticals that may often apply three or more cycles, the general experience of radioembolization demonstrates the reasonable clinical feasibility of two cycles.

Below we describe three examples (Cases 1, 2, 3) showing a whole liver, a left lobar, and a right lobar treatment repeated in two cycles under the hypotheses of same administered activity and uniform dose distribution at each cycle. The questions to be answered include the possible advantages of whole and lobar approaches by means of liver sparing for a fixed dose or a possible increase of tumor irradiation for a fixed liver effect (BED). The Lyman (or LKB) model has been applied below to derive the NTCP prediction.

Case 1 (Table 8.3):
Radioembolization of the whole liver in two cycles delivering 22 Gy/cycle to normal liver tissue is compared with a single treatment delivering the same total dose of 44 Gy.

For the single treatment, $\text{BED}_L = 44 \cdot (1 + 44/53.4) = 80$ Gy, while for two cycles $\text{BED}_L = 2 \cdot 22 \cdot (1 + 22/53.4) = 62$ Gy, with a BED sparing of ~18 Gy (30%). At each cycle, $\text{EQ}_{1.5} = 31.1/1.75 = 18$ Gy, leading to a total EUD = $2 \cdot \text{EQ}_{1.5} = 36$ Gy, associated with an NTCP of 16%—compared with the 46 Gy EUD value of the single treatment approach, with NTCP = 62%. Therefore, the use of a two cycle therapy allows for a notable advantage in the sparing of normal liver tissue.

Case 2 (Table 8.4):
In this case, we consider radioembolization of the left liver lobe in two cycles delivering 45 Gy/cycle to normal liver tissue compared with a single treatment delivering the same total dose of 90 Gy.

The calculations follow the same process as the previous case, but with an additional calculation for EUD evaluation given by EUD = $v_1^n (\text{EQ}_{1.5,1}) + v_2^n (\text{EQ}_{1.5,2}) = 2 \cdot (0.33)^{1.1} \cdot 47.4 = 28$ Gy for two cycles, and EUD = $v^n (\text{EQ}_{1.5}) = 0.33^{1.1} \cdot 138.2 = 41$ Gy for the single treatment.

Also in this case, there is a notable advantage for normal liver tissue, with BED and EUD sparing of ~46% (23 Gy for BED and 13 Gy for EUD) and a consequent NTCP reduction from high (40%) to acceptable (3%) risk.

Case 3 (Table 8.5):
This example describes radioembolization of a right liver lobe in two cycles delivering a dose per cycle to normal liver tissue that corresponds to the

Table 8.3 Comparison of the radiobiological quantities for the treatment of the whole liver in two cycles as compared with one single cycle with a same total absorbed dose of 44 Gy (case 1).

Case 1: Whole liver	First cycle	Second cycle	Single treatment
Volume %	100%	100%	100%
Dose RE (Gy)	22	22	44
BEDi (Gy)	31	31	80.3
EQ1.5,i = EUDi (Gy)	18	18	46
EUD tot (Gy)	36 ($\Delta = -29\%$)		46
BED tot (Gy)	62 ($\Delta = -29\%$)		80
NTCP %	16%		63%

Table 8.4 Radiobiological quantities for the treatment of the left liver lobe in two cycles as compared with one single cycle with a same total absorbed dose of 90 Gy (case 2).

Case 2: Left lobe	1st cycle	2nd cycle	Single treatment
Volume (%)	33	33	33
Dose RE (Gy)	45	45	90
BEDi (Gy)	83	83	242
EQ1.5,i (Gy)	47	47	138
EUDi (Gy)	14	14	41
EUD tot (Gy)	28 ($\Delta = -46\%$)		41
BED tot (Gy)	50 ($\Delta = -46\%$)		72
NTCP (%)	3%		40%

Table 8.5 Comparison of the radiobiological quantities for the treatment of the right liver lobe in two cycles as compared with one single cycle with a same total absorbed dose of 48 Gy (case 3).

Case 3: Right lobe		1st cycle	2nd cycle	Single treatment
	Volume (%)	67	67	67
	Dose RE,i (Gy)	30	30	48
	BEDi (Gy)	46	46	91
Liver	D_{tot} RE (Gy)	59 (+23%)		48
	EQ1.5,i (Gy)	26	26	52
	EUDi (Gy)	17	17	34
	EUD$_{\text{tot}}$ (Gy)	33.5		33.5
	NTCP %	10%		10%
	Dose RE,i (Gy)	92	92	150
Tumor	BEDi (Gy)	112	112	201
	D_{tot} RE (Gy)	185 ($\Delta = 23\%$)		
	BED$_{\text{tot}}$ (Gy)	221 ($\Delta = 10\%$)		201

same BED obtained for a single treatment with an absorbed dose of 48 Gy.

A single dose to the liver lobe of 48 Gy corresponds to a BED = 48·(1+48/53.4) = 91 Gy, which when split in two cycles corresponds to BED = 46 Gy/cycle and a dose of 30 Gy/cycle. For the same effect (risk) to the liver, the cumulative dose to the liver lobe is 59 Gy as compared with 48 Gy of a single treatment, indicating a possible increase of the cumulative activity by a factor of 1.23, and consequently a same dose increase to the tumor.

To quantify the effect on tumor, consider a delivery of a dose D_T = 150 Gy to the tumor for a single treatment approach. If the dose increase per cycle previously derived is applied, a dose per cycle of $D_{T,1\text{cycle}}$ = 92 Gy/cycle is obtained. Setting an interval of 30 days between the two cycles, the repopulation term is TK = 2.3 Gy and the correspondent BED values are BED$_T$ = 150·(1 +

150·0.0023) = 201 Gy for the single treatment, and BED$_{T, \text{tot}}$ = 2·92.3·(1 + 92.3·0.0023) − 2.3 = 221 Gy in two cycles.

The BED gain is, therefore, 20 Gy corresponding to a gain of 10% in BED. This is lower than the increase in the total dose, which is 23%.

Figure 8.7 illustrates the possible cumulative dose increase as a function of the absorbed dose delivered to the liver by radioembolization when splitting the treatment in two or three cycles. Although three cycles may be clinically impractical, it is interesting to see the trend and to have a rough idea of the benefit realized by this technique.

Example 8.4: Various models compared
The previous examples described the extrapolation of useful information from the EBRT experience (Lyman model) obtaining a first estimate of the response to radioembolization. As previously

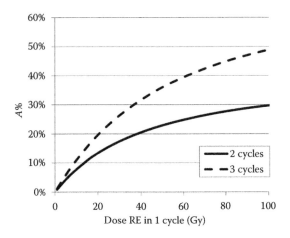

Figure 8.7 Possible cumulative dose increase as a function of the absorbed dose delivered to the liver by radioembolization when splitting the treatment in 2 or 3 cycles.

evidenced, this procedure may lead, in some cases, to improbable predictions. Other models may be more suited to describe radioembolization outcomes from clinical data.

In the ensuing discussion, the LKB, the parallel architecture, and the Walrand models are considered comparing their predictions in exemplificative cases and highlighting the volume effect that each incorporates. Of special interest is the irradiation of small portions of the liver and the different effects related to the use of resin as compared with glass spheres characterized by a higher tolerability and lower responsiveness.

Assuming for simplicity a uniform dose distribution D in a volume fraction v of the normal liver tissue, the parameters needed to apply for the different models consist of the following:

Lyman model:

$$EUD = v^n \times D$$

$$NTCP = \frac{1}{2} - \frac{1}{2}\mathrm{erf}\left(\frac{EUD - TD_{50}}{mTD_{50}\sqrt{2}}\right)$$

with $n = 1.1$, $m = 0.18$, $TD_{50} = 43$ Gy, erf = error

$$\text{function} = \frac{2}{\sqrt{\pi}} \int_0^x e^{t^2} dt$$

Parallel architecture model:

$$NTCP = \frac{1}{2} \cdot [\mathrm{erf}(a) + \mathrm{erf}(b)] \quad \text{if } f > 0.497 \text{ or}$$
$$NTCP = \frac{1}{2} \cdot [\mathrm{erf}(a) - \mathrm{erf}(b)] \quad \text{if } f < 0.497$$

where $f = v/(1 + d_{50}/D)^k$ is the fraction of organ damaged and

$$a = \frac{\mathrm{fr}_{50}}{\sigma_{\mathrm{fr}}\sqrt{2}} = 7.48$$

$$b = \frac{f - \mathrm{fr}_{50}}{\sigma_{\mathrm{fr}}\sqrt{2}}$$

with the fit parameters: $\mathrm{fr}_{50} = 0.497$, $\sigma_{\mathrm{fr}} = 0.047$, $d_{50} = 43$ Gy (absorbed dose in RE), $k = 3$. The last two parameters were obtained by imposing that, for same $BED = D_{EBRT} \cdot E_{rel,EBRT} = D_{RE} \cdot E_{rel,RE}$, $p(D_{EBRT}) = p(D_{RE})$.

Walrand model

$$NTCP = 1/(1 + TD_{50}/D)^\gamma$$

where

$$TD_{50} = (25.2 + 22.1 \cdot (1 - \exp(-2.74 \cdot \mathrm{msA})))/(v - 0.4)^{0.584}$$
$$\gamma = 13.7 \cdot (v)^2 + 30.6 \cdot v - 8.41$$

D is the mean dose (regardless of uniformity) and msA is the microsphere-specific activity (0.05 kBq for resin microspheres, 2.5 kBq for glass microspheres).

Figure 8.8 illustrates the NTCP curves of the parallel architecture model as a function of dose D for the case of uniform irradiation of different liver portions. Similarly, Figure 8.9a and 8.9b shows the same NTCP curves for the Walrand model, for msA of 0.05 kBq (resin spheres, Figure 8.9a) and 2.5 kBq (glass spheres, Figure 8.9b). Despite some differences, the volume effect is quite evident in both cases, and for irradiation of 40% (or less) of the total volume both models predict a null NTCP irrespective of the absorbed dose. Thus, cases with less than 40% of liver involvement will not be presented in this section because only the Lyman

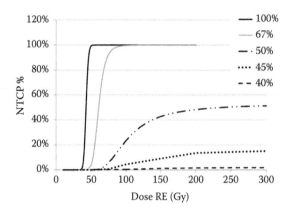

Figure 8.8 NTCP curves for the parallel model in the case of uniform dose to the entire volume of normal liver tissue (black curve), 67% of normal liver (gray curve), 50% of normal liver (dot-dashed curve), 45% of the normal liver (dotted curve), and 40% of the normal liver (dashed curve).

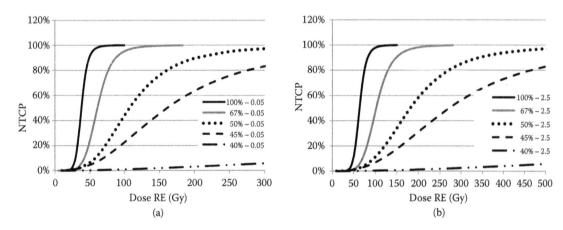

(a)   (b)

Figure 8.9 NTCP curves for the Walrand model as a function of mean dose, irrespective of uniformity, delivered to the whole liver (black continuous curve), 67% of normal liver (gray curve), half of the liver (dotted curve), 45% of the normal liver (dashed curve), and 40.1% of the liver (dot-dashed curve). (a) curves for msA= 0.05 kBq (resin spheres) and (b) those for 2.5 kBq (glass spheres).

model still shows curves predicting the possible toxicity if a high dose is delivered. In particular, an example already discussed is the radioembolization of the left lobe for which Lyman still indicates a possible complication risk (see Figures 8.4 and 8.5 for one-third of the liver volume).

8.4.1 NTCP PREDICTIONS UNDER THE HYPOTHESIS OF UNIFORM IRRADIATION

Figure 8.10 directly compares the NTCP curves of the different models for a uniform dose delivered to 67% (Figure 8.10a) and to 50% (Figure 8.10b) of normal liver tissue. Interestingly, the models of Lyman, Parallel, and Walrand for 0.05

kBq are quite similar in the case of 67% of liver involvement, while the model of Walrand for 2.5 kBq shows a higher tolerance as compared with resin spheres, in agreement with clinical observations. On the contrary, at lower volume fractions, such as for irradiation of 50% of normal liver, the predictions of the different models diverge.

Table 8.6 summarizes the parameters, the absorbed doses, and the evaluations of risk derived for several representative cases using the three NTCP models, under the hypothesis of uniform dose. In particular, Cases 1A–D represent whole-liver treatments delivering different doses to the parenchyma: 27 Gy, where all models predict no toxicity; 40 Gy, which is considered

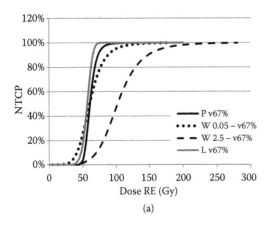
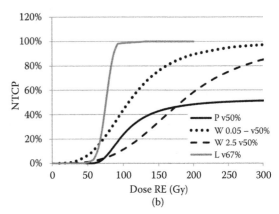

Figure 8.10 NTCP curves of the different models under the hypothesis of a uniform dose delivered to 67% (a) and to 50% (b) of normal liver tissue. The black curve represents the parallel model and the gray curve represents the Lyman model. The dotted and dashed curves are the model by Walrand for 0.05 kBq and 2.5 kBq microspheres, respectively.

Table 8.6 Parameters, absorbed doses (uniform), and evaluations of risk derived for several representative cases using the different NTCP models

Irradiated liver portion		Whole liver				Right lobe			Left lobe		Segment
Cases		1A	1B	1C	1D	2A	2B	2C	3A	3B	4B
Legends in the figure		WL27	WL40	WL58	WL80	RL 37	RL 50	RL105	LL37	LL80	S250
V% of NTL involved		100%	100%	100%	100%	67%	67%	67%	33%	33%	10%
NTL dose (Gy)		27	40	58	80	37	50	105	37	80	250
LKB model	EUD (RT)	23	40	69	114	23	35	153	11	40	61
	NTCP	**0%**	**34%**	**100%**	**100%**	**0%**	**16%**	**100%**	**0%**	**33%**	**99%**
Parallel	Damaged V%	20%	45%	71%	87%	26%	41%	64%	13%	30%	9%
model	**NTCP**	**0%**	**14%**	**100%**	**100%**	**0%**	**2.8%**	**100%**	**0%**	**0%**	**0%**
Walrand	TD_{50}	38	38	38	38	61	61	61	n.e.	n.e.	n.e.
model msA 0.05 kBq	**NTCP**	**5%**	**62%**	**97%**	**100%**	**5%**	**24%**	**96%**	**0%**	**0%**	**0%**
Walrand	TD_{50}	63.7	64	64	64	102	102	102	n.e.	n.e.	n.e.
model msA 2.5 kBq	**NTCP**	**0%**	**2%**	**31%**	**87%**	**0%**	**1%**	**54%**	**0%**	**0%**	**100%**

Abbreviations: WL27, WL40, WL58, WL80, whole liver receiving a mean dose of 27, 40, 58, 80 Gy, respectively; RL37, RL50, RL105, right liver lobe receiving a mean dose of 37, 50, 105 Gy, respectively; LL37, LL80, left liver lobe receiving a mean dose of 37, 80 Gy, respectively; S250, liver segment receiving a mean dose of 250Gy; V%, liver volume%; mSA (0.05 kBq for resin, 2.5 kBq for glass microspheres); n.e. is not evaluable; NTL, normal liver.

a threshold for possible toxicity by some authors (Cremonesi et al., 2008); 58 Gy, which is associated to different toxicity probabilities by the models; and 80 Gy, which has been suggested as a limit for safety by the empiric approach in the operators' manual of the resin spheres. Cases 2A–C consider the irradiation of the right lobe at doses of 37 Gy, indicating safety, 50 Gy, associated with acceptable risk, and 105 Gy, with high risk foreseen by all models. Cases 3A and

3B refer to treatment of the left lobe delivering a low dose of 37 Gy, with no risks foreseen by any model, and a dose of 80 Gy, where only the Lyman model shows nonnegligible risk. Finally, Case 4 reports the predicted toxicity for a highly irradiated segment (250 Gy).

Details are presented in Table 8.6, while Figure 8.11 illustrates a visual comparison of these results.

For whole-liver treatment, all the models predict a maximum risk at 80 Gy. Maximum risk

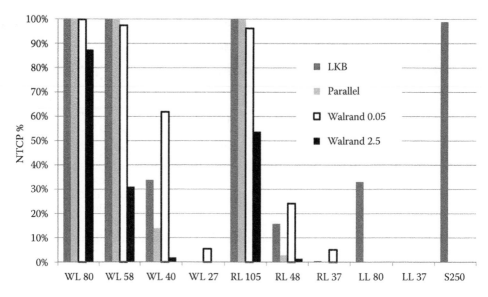

Figure 8.11 NTCP predictions by the Lyman model (dark gray bar), the Parallel model (light gray bar), the Walrand model for msA of 0.05 (white bar) and for msA of 2.5 (black bar), according to the cases included in Table 8.6 (WL80, WL58, WL40, WL27 refer to whole liver with uniform dose of 80, 58, 40, and 27 Gy, respectively; RL105, RL48, RL37 refer to right lobe irradiation with uniform dose of 105, 48, and 37 Gy, respectively; LL80, LL37 refer to left lobe irradiation with uniform dose of 80, and 37 Gy, respectively; S250 refers to a segment irradiation with uniform dose of 250 Gy).

is also seen at 58 Gy, except in the case of glass microspheres due to their decreased toxicity. Of note, at 40 Gy and even 27 Gy the Walrand model for resin microspheres indicates higher risk compared with the other models. The Walrand model for resin spheres is in fact more conservative than even the Lyman model for large liver volumes irradiated at low doses. This relationship can also be observed for right lobe radioembolization up to ~50 Gy, as illustrated in Figure 8.10a. Also for 50% liver irradiation and low doses, this model predicts nonnegligible NTCP values higher than the other models up to ~65 Gy. However, for higher doses more agreement is seen. For example, in the case of right lobe radioembolization and a dose of 105 Gy, all models indicate unacceptable risks, being lower in the case of glass microspheres, but higher than would normally be clinically acceptable (54%). Finally, as already pointed out, for left lobe radioembolization and segmental approaches, only the Lyman model indicates the presence of risk, with NTCP values as high as 100%. However, as already outlined, the Lyman model is not reliable for left lobe or segmental radioembolization.

8.4.2 NTCP PREDICTIONS FOR CLINICAL CASES WITH NONUNIFORM IRRADIATION

When considering real clinical situations, the dose distribution—as evidenced by posttreatment imaging—is always nonuniform. To include this nonuniformity in the models, the DVH $\{v_i, D_i\}$ needs to be derived and used to calculate EUD for the Lyman model:

$$EUD = (\Sigma_i v_i \cdot (D_i)^{1/n})^n s$$

and the fraction of damaged organ for the parallel architecture model:

$$f = \Sigma_i v_i / (1 + 43 / D_i)^3$$

Oddly, the Walrand model requires only the evaluation of the mean absorbed dose to the volume treated, without taking into account the macroscopic absorbed-dose uniformity.

Below, the DVH of four patients has been used to compare the predictions of the models. Patients 1 and 2 underwent right lobe radioembolization with similar mean doses to the treated lobe

(48 and 44 Gy, respectively), but different dose distributions and thus, different NTCP. Patient 3 underwent right lobe radioembolization with a mean dose of 105 Gy, and patient 4 underwent a left lobe treatment with a mean dose of 80 Gy. Figure 8.12a and 8.12b shows the DVHs of patients 1 and 4, respectively. Table 8.7 gathers the results of the evaluations using each model with the mean dose specified in place of uniform dose. Figure 8.13 gives an immediate comparison of the NTCP predictions.

Figure 8.14 compares the NTCP predictions in patients 1, 3, and 4 accounting for the nonuniformity observed by posttreatment imaging (patient 1: RL Dm48; patient 3: RL Dm105; patient 4: LL Dm80) to NTCP predictions neglecting nonuniformity, i.e., accepting the hypothesis of a uniform dose equal to the mean absorbed dose (patient 1: RL,D48; patient 3: RL,D105; patient 4: LL,D80).

As previously evidenced, the models by Walrand predict the same values with or without uniformity. Using the other models, the histogram shows the typical effect of nonuniformity. In the case of the right lobe with a mean dose of 105 Gy, the parallel architecture model changes from suggesting 100%

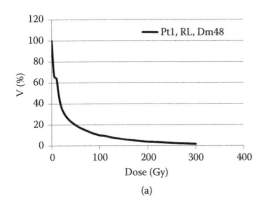

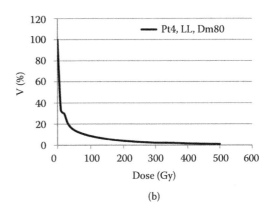

(a)

(b)

Figure 8.12 DVHs of patient 1 (a), who received a mean dose of 48 Gy to the right lobe (Pt1, RL, 48 Gy), and patient 4 (b), who received a mean dose of 80 Gy to the left lobe (Pt4, LL, 80 Gy).

Table 8.7 NTCP evaluations from different models from the DVHs of patients

Irradiated liver portion			Right lobe		Left lobe
Patients and legends (Figure 8.13)		Pt1	Pt2	Pt3	Pt4
		RL, Dm48	RL, Dm44	RL, Dm105	LL, Dm80
NTL V% involved		67	67	67	67
NTL mean dose RE (Gy)		48	44	105	80
LKB model	EQ1.5 (EBRT)	70	55	252	86
	EUD (EBRT)	60	48	216	68
	NTCP	98%	73%	100%	100%
Parallel model	Damaged V%	22%	22%	36%	16%
	NTCP	0%	0%	0%	0%
Walrand model	TD$_{50}$	61	61	61	n.e.
(0.05 kBq)	NTCP	20%	20%	96%	0%
Walrand model	TD$_{50}$	102	102	102	n.e.
(2.5 kBq)	NTCP	1%	1%	54%	0%

Abbreviations: Pt1, RL, Dm48 = patient 1, right liver lobe irradiation with a mean dose of 48 Gy; Pt2, RL, Dm44 = patient 2, right liver lobe irradiation with a mean dose of 44 Gy; Pt3, 2/3, Dm105 = patient 3, right liver lobe irradiation with a mean dose of 105 Gy; Pt4, 1/3, Dm48 = patient 4, left liver lobe irradiation with a mean dose of 80 Gy; EQ1.5 is the corresponding EBRT dose released at 1.5 Gy/fr; V%, the liver volume%; mSA (0.05 kBq for resin, 2.5 kBq for glass microspheres); NTL, the normal liver; n.e., not evaluable.

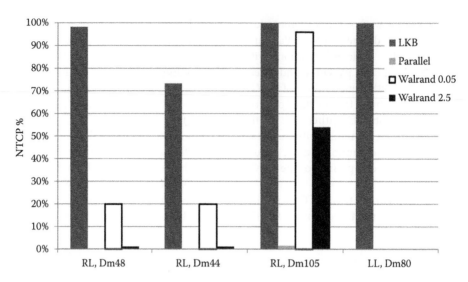

Figure 8.13 Comparison of the NTCP predictions for patient 1 (RL, Dm48: right lobe, mean dose 48 Gy), patient 2 (RL, Dm44: right lobe, mean dose 44 Gy), patient 3 (RL, Dm105: right lobe, mean dose 105 Gy), and patient 4 (LL, Dm80: left lobe, mean dose 80 Gy). The dark gray bars are associated to the Lyman model, the light gray bars to the Parallel model, the white bars to the Walrand model for msA = 0.05 kBq, and the black bars to the Walrand model for msA = 2.5 kBq.

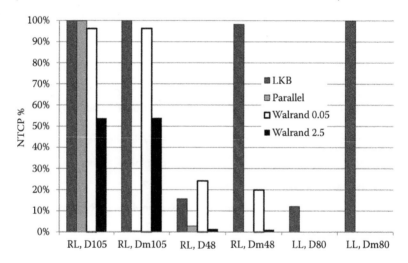

Figure 8.14 NTCP predictions by different models accounting for non-uniformity or accepting the hypothesis of a uniform dose equal to the mean absorbed dose. Data shown for patient 1 (RL, D48: right lobe, uniform dose of 48 Gy; RL,Dm48: right lobe, same mean dose of 48 Gy but accounting for non-uniformity), patient 3 (RL,D105: right lobe, uniform dose of 105 Gy; RL,Dm105: right lobe, same mean dose of 105 Gy but accounting for non-uniformity), and patient 4 (LL,D80: left lobe, mean dose 80 Gy; LL,Dm80: left lobe, same mean dose of 80 Gy but accounting for non-uniformity). The dark gray bars are associated to the Lyman (LKB) model, the light gray bars to the parallel model, the white bars to the Walrand model for msA = 0.05 kBq, and the black bars to the Walrand model for msA = 2.5 kBq.

NTCP to a feasible therapy when nonuniformity is taken into account. However, the Lyman model maintains a 100% NTCP because of its conservative characteristics, especially at high doses. The other two cases, however, show an unexpected effect for the Lyman model: for left lobe radioembolization with a mean dose of 80 Gy, the EUD for uniform dose is 34 Gy, while accounting for nonuniformity the EUD reaches ~68 Gy. This leads to the prediction of feasibility in the case of uniform dose distribution

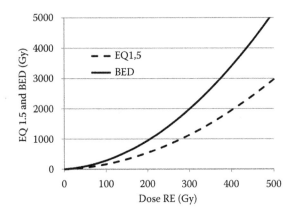

Figure 8.15 BED and EQ1.5 curves derived as a function of dose of radioembolization. The continuous line represents BED, and the dashed line represents EQ1.5, i.e., the dose of EBRT released at of 1.5 Gy/fraction.

and high risk with nonuniform distribution. This is the opposite of what was observed with the parallel architecture model for the right lobe in both the 105 and 48 Gy cases. The predictions of the Lyman model are also contrary to clinical observations, which suggest that while hot spots of nonuniformity raise the mean dose, they should lower the global injury and decrease NTCP.

Figure 8.15 should help to explain this point by showing the BED and EQ1.5 (the correspondent EBRT dose released at 1.5 Gy/fr) as a function of radioembolization dose. The calculation of the EUD includes the conversion of the radioembolization absorbed dose to each voxel into the corresponding EBRT dose (via the BED). However, the DVH of real patients can show high doses following radioembolization (e.g., up to ~300 Gy for patient 1, RL, Dm48; up to ~500 Gy for patient 4, LL, Dm80), that correspond to extremely high doses of EBRT. For example, 200 Gy in radioembolization should correspond to 1000 Gy in EBRT, and 500 Gy following radioembolization corresponds to 5000 Gy of EBRT. The validity of the Lyman model, which is a phenomenological derivation, can only be supported at much lower ranges of doses of EBRT and with nonuniformities that are not extreme.

8.5 SUMMARY FOR CLINICAL APPLICATIONS

The aim of this section is to assemble the main radiobiological issues described in the previous sections and provide direct insight into their effect on the clinical practice of radioembolization. While medical physicists and scientists will be able to apply aforementioned radiobiologic models to clinical therapy, the broader concepts presented below will be appreciated by other members of the radioembolization treatment team, including interventional radiologists and nuclear medicine physicians.

1. **Liver hypertrophy and resectability:** Lessons from surgery show that liver tissue is able to regenerate if a certain portion of the liver is embolized (i.e., excluded from blood circulation) or even excised. This property of the liver is used to increase the efficacy of liver resection by pretreatment portal-vein embolization, which artificially augments the future remnant liver volume. The final result is a lower occurrence of hepatic complications/insufficiency after partial hepatectomy.

   There are many studies from surgery regarding liver resectability, in particular the minimum portion of liver remnant necessary to avoid hepatic damage or failure. In summary, the literature reports that, in a major hepatectomy, a remnant: (1) ≤20–25% could be at high risk of complication in patients with normal liver function (Chun et al., 2008; Lin et al., 2014; Truant et al., 2015; Vauthey et al., 2000); (2) ≥40% should guarantee safety in patients without chronic liver disease (Kubota et al., 1997); (3) ≤ 40% and 55% would represent considerable risk for failure in patients heavily treated by chemotherapy, and patients with cirrhotic liver, respectively (Narita et al., 2012; Lin, 2014). These guidelines can be aptly applied for partial liver radioembolization (see point 4 below).

2. **Hypertrophy and radioembolization:** The presence of hypertrophy emerges also in the follow-up imaging of many patients who have underwent radioembolization as an intrinsic defense against the radiation damage to liver cells. Such a property increases the liver function, offering a major resource that can be fully exploited when partial liver irradiation or multiple-cycle strategies are applied.

The regeneration capability of the liver is not included in any of the mathematical models discussed in this chapter but should appear in phenomenological observations, with curves describing higher tolerability than expected. So, in the parallel and the Walrand theoretical models, it might be necessary to add a further element that takes into account this phenomenon. Alternatively, as more toxicity data becomes available, some parameters might be adjusted to account for a radiosensitivity lower than predicted. In other words, slightly less conservative models might better reflect observed results, with a shift toward higher doses for liver damage.

3. **Multiple-cycle approach and time interval between cycles:** A multicycle radioembolization strategy is certainly of help to reduce the risk of toxicity or to increase the dose delivered to the tumor. The gain in terms of absorbed dose or BED has been shown in Example 8.4 for the linear quadratic model.

In initial multicycle therapy trials, the time interval between cycles was an open question. Some authors proposed just a few days between cycles, which could be a retreatment of a same target volume or a separately treatment of right and left lobes. This short time could be acceptable from a radiobiological perspective, as the repair of radiation damage occurs in a time frame on the order of hours (see Section 8.2.1). However, the liver can do more than repair injured cells; it is capable of regeneration, and so short intervals have been replaced by at least 30- to 40-day intervals. Longer intervals are needed to induce a countervailing hypertrophy, with the intent to recover as much of the liver functionality as possible between treatment cycles.

4. **Safety of lobar and selective radioembolization:** According to current models, for segmental or selective radioembolization for the treatment of tumors with minimal involvement, NTCP is negligible at reasonable clinical absorbed doses. Such models adhere to clinical observations in both radioembolization and surgery (see point 1 of this section). In particular, the minimum recommended remnants for a safe liver resection in patients with different disease status/pathologies offer guidelines for partial liver radioembolization. In fact, the worst consequence induced by irradiation is cell killing, which can be associated with tissue removal, i.e., surgery/resection. Therefore, the data summarized in point 1 above can be related to safe partial-liver irradiation with radioembolization, which could involve 60% of the liver (i.e., sparing 40% of normal liver tissue) for patients with normal liver function. In patients heavily treated with chemotherapy and in cirrhotic patients, 60% (i.e., 40% spared) and 45% (i.e., 55% spared) thresholds can be appropriately applied from surgical resection data. Even adding a margin of prudence, the barrier of no risk imposed by the parallel and the Walrand models at volume fractions lower than 40%, irrespective of the dose, is nicely coherent with the experience of surgery. This refers to patients with normal liver function or minimal previous chemotherapy treatment, while for patients with liver disease (HCC, substantial previous chemotherapy) a further margin for safety may be taken into account.

5. **Sparing effect of nonuniform dose:** Nonuniformity weakens the effect of radiation. This comes from a common logic and from the EBRT data (Kassis and Adelstein, 2005). Since the worst insult resulting from radiation is cell death, as radiation dose increases beyond the threshold necessary for cell killing, there is no biological impact of further increasing the radiation dose. Therefore, in the case of nonuniformity, there may be tissue areas with wasted energy delivered (due to an absorbed dose well above the threshold for cell killing), and other areas where the dose is insufficient to provoke the same damage as the mean dose.

The formalism of the Lyman and the parallel models include the possibility of accounting for nonuniformity. In fact, the NTCP values associated with uniform and nonuniform doses for a same mean dose do differ by these models. However, it must be emphasized that the parallel model correctly provides lower NTCP values in the case of nonuniformity, while the Lyman model can fail in certain cases, predicting higher NTCP values for uniform irradiation when very high doses exist in the dose map (see the discussion of the results for pt1 in Figure 8.14 [RL,D48 vs. RL,Dm48] and Figure 8.15).

6. **Conservativeness and risks to be added:** While conservatism is often taken in treatment planning for radioembolization of patients with HCC, there are factors that suggest a precautionary attitude for all patients may be warranted. Most patients receiving radioembolization are not naïve to therapy but have already received at least two lines of hepatotoxic chemotherapy. With this in mind, the differences between NTCP curves related to HCC and to metastatic patients (Figure 8.6) could be less than predicted. This consideration relates also to what has been reported in point 1 above. Thus, a distinction between two NTCP curves lower than that provided by Dawson and Pan (Dawson et al., 2002; Pan et al., 2010) for EBRT should not be unexpected.

7. **Toxicity evaluation:** The collection of toxicity data represents a milestone for correlation analysis with doses and radiobiological quantities. To further refine radiobiologic models for radioembolization moving forward, reliable methods with appropriate timing, completeness, and uniformity among researchers is needed. In particular, the analysis of the cholinesterase is strongly recommended as a method to evaluate liver function damage, for patient screening and follow-up (Meng et al. 2013). This is a most reliable indicator for liver injury, commonly used in surgical and hepatological disciplines, although less common in nuclear medicine and interventional radiology. The evaluation of indocyanine or xenobionts is also good alternative methods.

8.6 CONCLUSIONS

In this chapter, the basic principles of radiation biology have been illustrated, together with the most important models that could be used to describe the outcomes of radioembolization. Some models are derived from the experience of EBRT and can be very useful so long as limitations in their applicability are considered. Other models have been developed in the context of radioembolization, and can be more suitable to describe clinically observed effects.

In general, many concepts have been illustrated to give the reader useful instruments and confidence with the different models and formalisms available in the literature. Once assimilated, all these concepts should be applied to clinical radioembolization data with a critical and conscientious attitude. With increasing dosimetric data and clinical evidence, it will be possible to build more robust models allowing better predictivity and personalization of radioembolization treatments, following the example of EBRT. In this sense, continued collection of toxicity and efficacy data should be encouraged, as well as the performance of personalized pre- and posttreatment 3D dosimetry with the highest possible level of accuracy.

All models are wrong, but some are useful (George E. P. Box)

REFERENCES

Antipas, V., Dale, R.G., Coles, I.P. (2001). A theoretical investigation into the role of tumour radiosensitivity, clonogen repopulation, tumour shrinkage and radionuclide RBE in permanent brachytherapy implants of 125I and 103Pd. *Phys Med Biol* 46:2557–2569.

Baechler, S. et al. (2008). Extension of the biological effective dose to the MIRD schema and possible implications in radionuclide therapy dosimetry. *Med Phys* 35(3):1123–1134.

Brenner, D.J. et al. (1998). The linear-quadratic model and most other common radiobiological models result in similar predictions of time-dose relationships. *Radiat Res* 150:83–91.

Burman, C. et al. (1991). Fitting of normal tissue tolerance data to an analytic function. *Int J Radiat Oncol Biol Phys* 21:123–135.

Chiesa, C. et al. (2015). Radioembolization of hepatocarcinoma with 90Y glass microspheres: development of an individualized treatment planning strategy based on dosimetry and radiobiology. *Eur J Nucl Med Mol Imaging* 42(11):1718–1738.

Chun, Y.S. et al. (2008). Comparison of two methods of future liver remnant volume measurement. *J Gastrointest Surg* 12(1):123–128.

Cremonesi, M. et al. (2008). Radioembolisation with 90Y-microspheres: dosimetric and radiobiological investigation for multi-cycle treatment. *Eur J Nucl Med Mol Imaging* 35(11):2088–2096.

Cremonesi, M. et al. (2014). Radioembolization of hepatic lesions from a radiobiology and dosimetric perspective. *Front Oncol* 4:210.

Dale, R.G. (1996). Dose-rate effects in targeted radiotherapy *Phys Med Biol* 41:1871–1884.

Dale, R.G., Jones, B., Sinclair, J.A. (2000). Dose-equivalents of tumour repopulation during radiotherapy: the potential for confusion *Br J Radiol* 73:892–894.

Dawson, L.A., Lawrence, T.S., Ten Haken, R.K. (2001). Partial liver irradiation. *Semin Radiat Oncol* 11:240–246.

Dawson, L.A. et al. (2002). Analysis of radiation-induced liver disease using the Lyman NTCP model. *Int J Radiat Oncol Biol Phys* 53(4):810–821.

Dawson, L.A., Ten Haken, R.K. (2005). Partial volume tolerance of the liver to radiation. *Semin Radiat Oncol* 15(4):279–283.

Emami, B. et al. (1991). Tolerance of normal tissue to therapeutic irradiation *Int J Radiat Oncol Biol Phys* 21:109–122.

Flamen, P. et al. (2008). Multimodality imaging can predict the metabolic response of unresectable colorectal liver metastases to radioembolization therapy with Yttrium-90 labeled resin microspheres. *Phys Med Biol* 53(22):6591–6603. Erratum in: *Phys Med Biol* 2014;59(10):2549–2551.

Garin, E. et al. (2012) Dosimetry based on 99mTc-macroaggregated albumin SPECT/CT accurately predicts tumor response and survival in hepatocellular carcinoma patients treated with 90Y-loaded glass microspheres: preliminary results. *J Nucl Med* 53(2):255–263.

Garin, E. et al. (2015). Personalized dosimetry with intensification using 90Y-loaded glass microsphere radioembolization induces prolonged overall survival in hepatocellular carcinoma patients with portal vein thrombosis. *J Nucl Med* 56(3):339–346.

Garin, E. et al. (2016). Clinical impact of (99m) Tc-MAA SPECT/CT-based dosimetry in the radioembolization of liver malignancies with (90Y)-loaded microspheres. *Eur J Nucl Med Mol Imaging* 43(3):559–575.

Jackson, A. et al. (1995). Analysis of clinical complication data for radiation hepatitis using a parallel architecture model. *Int J Radiat Oncol Biol Phys* 31:883–891.

Jackson, A., Kutcher, G.J., Yorke, E.D. (1993). Probability of radiation-induced complications for normal tissues with parallel architecture subject to non-uniform irradiation. *Med Phys* 20:613–625.

Jones, L.C., Hoban, P.W. (2000). Treatment plan comparison using equivalent uniform biologically effective dose (EUBED). *Phys Med Biol* 45(1):159–170.

Kassis, A.I., Adelstein, S.J. (2005). Radiobiologic principles in radionuclide therapy. *J Nucl Med* 46(Suppl 1):4S–12S.

Kong, M., Hong, S.E. (2015). Optimal follow-up duration for evaluating objective response to radiotherapy in patients with hepatocellular carcinoma: a retrospective study. *Chin J Cancer* 34(2):79–85.

Kubota, K. et al. (1997). Measurement of liver volume and hepatic functional reserve as a guide to decision-making in resectional surgery for hepatic tumors. *Hepatology* 26(5):1176–1181.

Kutcher G., Burman, C. (1989). Calculation of complication probability factors for non-uniform normal tissue irradiation: the effective volume method *Int J Radiat Oncol Biol Phys* 16(6):1623–1630.

Lin, X.J., Yang, J., Chen, X.B., Zhang, M., Xu, M.Q. (2014). The critical value of remnant liver volume-to-body weight ratio to estimate posthepatectomy liver failure in cirrhotic patients. *J Surg Res* 188(2):489–495.

Lyman, J.T. (1985). Complication probability as assessed from dose-volume histograms. *Radiat Res* 104:S13–S19.

Manda, G. et al. (2015). The redox biology network in cancer pathophysiology and therapeutics. *Redox Biol* 5:347–357.

Meng, F. et al. (2013). Assessment of the value of serum cholinesterase as a liver function test for cirrhotic patients. *Biomed Rep* 1(2):265–268.

Millar, W.T. (1991). Application of the linear-quadratic model with incomplete repair to radionuclide directed therapy. *Br J Radiol* 64:242–251.

Mitsuishi, Y., Motohashi, H., Yamamoto, M. (2012). The Keap1–Nrf2 system in cancers: stress response and anabolic metabolism. *Front Oncol* 2. Available from: http://dx. doi. org/10.3389/fonc.2012.

Narita, M. et al. (2012). What is a safe future liver remnant size in patients undergoing major hepatectomy for colorectal liver metastases and treated by intensive preoperative chemotherapy? *Ann Surg Oncol.* 19(8):2526–2538.

Niemierko, A., Goitein, M. (1993). Modeling of normal tissue response to radiation: the critical volume model. *Int J Radiat Oncol Biol Phys.* 25:135–145.

O'Donoghue. (1999). Implications of nonuniform tumor doses for radioimmunotherapy: equivalent uniform dose. *J Nucl Med* 40:1337–1341.

Pajonk, F., Vlashi, E., McBride, W.H. (2010). Radiation resistance of cancer stem cells: the 4 Rs of radiobiology revisited. *Stem Cells* 28(4):639–648.

Pan, C.C. et al. (2010). Radiation-associated liver injury. *Int J Radiat Oncol Biol Phys* 76(3 Suppl):S94–S100.

Sangro, B. et al. (2008). Liver disease induced by radioembolization of liver tumors: description and possible risk factors. *Cancer* 112(7):1538–1546.

Strigari, L. et al. (2010). Efficacy and toxicity related to treatment of hepatocellular carcinoma with 90Y-SIR spheres: radiobiologic considerations. *J Nucl Med* 51(9):1377–1385.

Strigari, L. et al. (2011). Dosimetry in nuclear medicine therapy: radiobiology application and results. *Q J Nucl Med Mol Imaging* 55(2):205–221.

Truant, S. et al. (2015). Liver function following extended hepatectomy can be accurately predicted using remnant liver volume to body weight ratio. *World J Surg* 39(5):1193–1201.

Vauthey, J.N. et al. (2000) Standardized measurement of the future liver remnant prior to extended liver resection: methodology and clinical associations. *Surgery* 127(5):512–519.

Walrand, S., Hesse, M., Jamar, F., Lhommel, R. (2014a). A hepatic dose-toxicity model opening the way toward individualized radioembolization planning. *J Nucl Med* 55(8):1317–1322.

Walrand, S. et al. (2014b). The low hepatic toxicity per Gray of 90Y glass microspheres is linked to their transport in the arterial tree favoring a nonuniform trapping as observed in post-therapy PET imaging. *J Nucl Med* 255:135–140.

Wigg, D.A. (2001). *Applied Radiobiology and Bioeffect Planning*. Madison, WI: Medical Physics Publishing.

Withers, H.R., Taylor, J.M.G., Maciejewski, B. (1988). Treatment volume and tissue tolerance. *Int J Radiat Oncol Biol Phys.* 15:751-759.

Withers, H.R. (1992). Biological basis of radiation therapy for cancer. *Lancet* 339(8786):156–159.

Yorke, E.D. et al. (1992). Probability of radiation-induced complications in normal tissues with parallel architecture under conditions of uniform whole or partial organ irradiation. *Radiother Oncol* 26:226–237.

Yorke, E.D. et al. (1999). Can current models explain the lack of liver complications in Y-90 microsphere therapy? *Clin Cancer Res* 5:3024s–3030s.

Yorke, E.D. et al. (2001). Modeling the effects of inhomogeneous dose distributions in normal tissues. *Semin Radiat Oncol* 11(3):197–209.

Microsphere deposition, dosimetry, radiobiology at the cell-scale, and predicted hepatic toxicity

STEPHAN WALRAND

9.1 INTRODUCTION

Over the past decade, it has become well established that hepatic toxicity per Gy is significantly different between 50 Bq/sphere resin and 2500 Bq/sphere glass ^{90}Y microspheres (Kennedy et al., 2007). An overview of the similarities and differences between resin and glass microspheres is presented in Chapter 1. The hepatic toxicity per unit absorbed dose of glass microspheres is about one-third of that observed in external beam radiotherapy (EBRT) (Dawson et al., 2001).

Gulec et al. (2010) performed the first simulation of cell-scale dosimetry applied to compare the effects of hepatic radioembolization using resin and glass microspheres. Gulec et al. (2010) used electron Monte Carlo (MC) tracking. Because MC electron transport is computationally burdensome, Gulec et al. (2010) assumed that all the hepatic lobules shared the same microsphere trapping pattern enabling the use of a fast reflective boundary technique. In this translation invariant setup, the simulation did not clearly establish a difference in hepatic toxicity per Gy between the two microsphere devices.

However, experimental microscopy studies of microsphere distribution have revealed strongly nonuniform microsphere trapping (Pillai et al., 1991; Roberson et al., 1992; Campbell et al., 2000; Kennedy et al., 2004). Chiesa et al. (2011) suggested that the lower hepatic toxicity per Gy observed with glass microspheres could be due to a more nonuniform microsphere distribution owing to their lower number, resulting in sparing of more regions of normal hepatic parenchyma.

This chapter reviews the recent experimental and theoretical developments that have provided a better understanding of this difference. Some predictions of the hepatic toxicity in liver radioembolization as a function of microsphere number and target liver volume fraction are also provided.

The author would like to emphasize that if these developments prove the hepatic toxicity per unit absorbed dose decreases with decreasing microspheres number, then the tumor response per unit dose is also theoretically expected to decrease. This fact is confirmed with clinical observation. As a result, the theoretical study of the impact of microsphere number on therapy efficacy requires modeling of the microsphere distribution in tumor, which is a challenging problem due to the anarchic nature of tumor vasculature.

9.2 SCALES IN LIVER RADIOEMBOLIZATION

The human adult liver is a lattice of $\approx 10^6$ independent functional subunits called lobules (Gulec et al., 2010). Each lobule is a hexagonal prism of ≈ 1.5 mm

length and ≈ 1.2 mm diameter (Figure 9.1). A portal triad, consisting of a bile duct, a portal venule, and a few arteries (2.4 on average; Crawford et al., 1998), is located at each corner of the prism. The six portal triads are each shared by three lobules, resulting in a total number of triads $\approx 2 \times 10^6$. The hepatic arterial tree consists of approximately 21 vessel bifurcations or about $2^{21} \approx 2 \times 10^6$ terminal arterioles. Compared with all other tissues, the hepatic lobules have the unique feature to be fed by both arterial and venous sources. After injection via a hepatic artery branch, the microspheres that are larger than the intralobule arteriole diameter are predominantly trapped in linear clusters in the triad arteries. Thus, the most uniform activity distribution already exhibits a heterogeneity periodic pattern of ≈ 1.5 mm scale corresponding to the length of each lobule.

Blood outflow in normal hepatic structure is ensured by a single vein located at the center of the lobule—the central vein. As the primary venous drain, integrity of the central vein is essential to preserve the blood flow and thus lobule viability. On the other hand, as the nutrient and oxygen diffusion range in soft tissue is approximately 500 µm, one or two preserved portal triads are likely enough to keep the lobule alive, although this has not yet

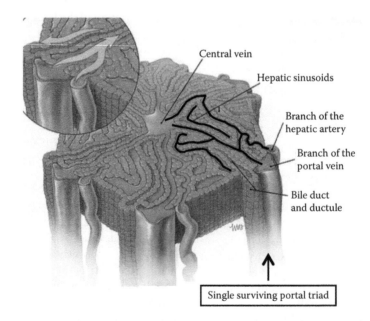

Central vein

Hepatic sinusoids

Branch of the hepatic artery

Branch of the portal vein

Bile duct and ductule

Single surviving portal triad

Figure 9.1 Schematic representation of lobule. (Courtesy of Will McAbee, Educational Resource Center, College of Veterinary Medicine at University of Georgia, Athens, GA.) Because hexagon side-to-side distance is about 1200 µm, almost all lobule structures are closer than 500 µm from small intralobule arterioles or venules (highlighted) still transporting blood from only one surviving triad.

been validated in an experimental *in vivo* model. The absorbed dose in the vicinity of a microsphere reaches several hundred Gy; therefore, a portal triad trapping one or more microspheres will suffer from local microscale radiation necrosis.

The dose delivered by a ^{90}Y loaded microsphere quickly decreases with the distance, by a factor ≈ 1000 from the microsphere boundary up to 0.5 mm. This might suggest that the central vein or portal triad empty of microspheres is quite preserved from lethal irradiation due to distance alone. However, the maximal range of the ^{90}Y β-particle is about 11 mm, and consequently, all the lobule structures are also irradiated by the microspheres trapped in the 600 closest surrounding lobules. The following sections will show that the mean absorbed dose in a lobule free of a microsphere, but surrounded by lobules containing a constant number of microspheres, is only 20% lower than if that lobule contained the same number of microspheres. As a result, microsphere trapping heterogeneities on the centimeter scale, rather than the millimeter scale, are thus required to preserve lobule structures from lethal radiation.

Using typical radiation dosages and number of infused microspheres for both resin and glass microspheres (Chapter 1), the average number of spheres per triad is 16 resin and 1 glass microsphere, delivering about 40 and 120 Gy to the liver parenchyma, respectively. As a result of the random transport of the microspheres through the arterial tree by the flow of blood, microsphere cluster sizes are distributed around the mean number of microspheres per triad. In comparison, a typical 250 MBq 18F-fludeoxyglucose positron emission tomography (^{18}FDG-PET) scan corresponds to 10^5 ^{18}FDG molecules per lobule (assuming 4.5% of uptake in the liver; Mettler and Guiberteau, 2012). In contrast to radioembolization, this high number of FDG molecules per lobule dramatically smooth transport fluctuations. This explains why the FDG distribution in liver appears, and is, much more uniform than that of microspheres.

In contrast to other tissues, liver regeneration is not dependent on a small group of stem cells, but is carried out by proliferation of its intact mature cells (Michalopoulos and DeFrances, 1997). Hepatocytes can proliferate almost without limit, but more remarkably they have the capacity to proliferate while simultaneously performing all essential functions needed for homeostasis. This

explains why living donor liver transplantation (LDLT) can safely survive when only 33% of the liver remains while 90% of the cells in the residual liver undergo proliferation or mitosis (Haga et al., 2008). This also explains why liver is one of the most radioresistant tissues.

Is it safe to kill two-thirds of the liver volume by irradiation? Obviously not! Partial liver irradiation in EBRT teaches us that killing 60% and 40% of the liver volume by irradiation gives a normal tissue complication probability (NTCP) of 99% and 50%, respectively (Dawson et al., 2001). The major difference with surgical resection is that immediately after irradiation the surviving liver volume has no free space to regenerate, has to handle toxins released by dying cells, and has to recycle necrotic tissue while maintaining homeostasis.

9.3 INTRODUCTION TO MONTE CARLO METHODS

MC methods often appear quite obscure to the non-physicist. MC methods are based upon repeated, numerous random drawings (or sampling) according to a specific probability distribution in order to numerically solve a mathematical or a physical problem. Let us illustrate this concept with a simple example.

Typically, to assess the value of π, one would start from the relation $\pi / 4 = \tan^{-1}(1)$ to develop the inverse tangent function in the Taylor series and to numerically compute the terms of the series. More sophisticated series expansions of π have been developed allowing fast computation of trillions of decimal digits. Besides this computational method, there are two simple experimental methods to estimate π.

The first one, often performed in elementary school, is to surround a disc by a rope and to compute the ratio between the length of the rope and the diameter of the disk. A drawback of this method is that it requires an accurate length measurement.

A second method which was one of the first applications of MC is (1) draw equidistant and parallel lines on a floor by moving a pen along the side of a rectangular rule, the opposite side being successively shifted to the last drawn line, (2) set a wood stick along the small side of the rule and cut it off at the same width. Using this setup, GL Leclerc Comte de Buffon mathematically proved that if the stick were randomly dropped, the probability that the stick will lie across a line is $2/\pi$ (Schroeder, 1974). By dropping the stick numerous times and counting how many times it crossed a line, it is thus possible to get an estimation of π (Figure 9.2). The elegance of this experimental MC estimation of π is that it involves no measurement of length.

The modern success of MC methods originated from the development of uniform pseudorandom number generators, i.e., algorithms for generating a sequence of numbers that cannot be distinguished from a true random sequence, such as that obtained using a roulette wheel. Historic random generators were based on the recurrence equation (Press et al., 2007):

$$n_{i+1} = (a\,n_i + c)\%m \qquad (9.1)$$

where a, c, m are positive integers and $\%$ represents the modulus operation, i.e., the remainder of the integer division by m. This generator has a period p, which is lower than m, i.e., after p generations the sequence of p numbers is repeated. Although the random number stream produced using Equation 9.1 is deterministic, a set of generated numbers less than its period cannot be distinguished from a true random set using simple statistical tests.

Nowadays, fast uniform pseudorandom number generators are based on Mersenne twister theory (Saito et al., 2008, http://www.agner.org/random/). They are fast, have an extremely long period (up to $2^{11213}-1$), succeed in all known statistical tests, and are thus well adapted to MC methods. Specific probability distribution can be obtained from the uniform pseudorandom generator using the transformation of variable or rejection methods (Press et al., 2007).

MC methods have the capacity to increase the speed of some numerical computations and to allow simulation of physical process governed by probabilistic laws. Various codes are used in nuclear medicine (Ljungberg et al., 2013). In this

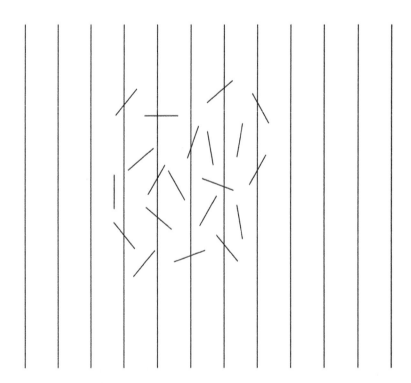

Figure 9.2 Schematic representation of Buffon's needle experiment showing the lying positions of a stick randomly dropped 20 times on the ground. When the length of the stick matches the distance in between the lines, the fraction of positions crossing a line tends to 2/π when the dropping number increases.

chapter, MC methods are used for three purposes: (1) computation of the absorbed dose delivered around a β point source, (2) growing of a synthetic arterial tree, and (3) microsphere transport through the arterial tree.

9.4 DOSE DEPOSITION AROUND A β SOURCE

Since electron trajectories are governed by the probabilistic laws of quantum electrodynamics (QED), the only way to compute the dose delivered from a β source is to simulate numerous electron trajectories, electron by electron, and to sum the energy spatially deposited by each trajectory. Nearly all MC codes simulate electron trajectories in discrete small steps (Figure 9.3).

Figure 9.4a shows the dose $D(r)$ delivered in a medium as a function of the distance r to a small ^{90}Y source computed using different MC codes, except the dashed line that represents an analytical

approximation using the simple equation (Russell et al., 1988):

$$D_R(r) = D_0 \left(1 - \frac{r}{R}\right) \frac{1}{r^2} \qquad (9.2)$$

The dose $D(r)$ quickly decreases with the distance r. However, when the activity is spread in a large region of tissue, it is better to consider the function $4\pi r^2 D(r)$. This function describes the dose received in a point from all the activity located at a distance r when the uptake is uniform. In this case, one can clearly see that the mean absorbed dose coming from the activity localized within the 0.7 mm radius surrounding region (dark gray area in Figure 9.4b) is much smaller than that coming from the farther activity (light gray area in Figure 9.4b).

The small differences between the different MC computations mainly arise from the different setups: Cross et al. (1992) modeled a ^{90}Y point source in water, Gulec et al. (2010) modeled a 32-μm-diameter spherical activity distribution inside a soft tissue equivalent medium, and Paxton et al. (2012) also modeled the actual microsphere density.

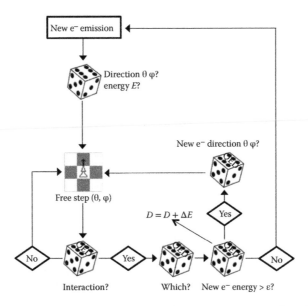

Figure 9.3 Symbolic representation of MC simulation of electron histories. After the random drawing of an emission direction, the e⁻ freely travels a small step in this direction, then the occurrence of an interaction with the medium and eventually the new properties of the e⁻ are randomly drawn according to the cross sections computed from quantum electrodynamics (QED) theory. The energy deposited by the interaction is summed to the absorbed dose, and when the residual e⁻ energy becomes negligible (≤ε) the tracking of the electron is stopped and a new e⁻ emission is initiated. For the sake of clarity, emissions of secondary particles (such as photons produced by bremsstrahlung, recoil electrons, electron–positron pair creations, etc.) were not represented. Dices are loaded according to the QED rules.

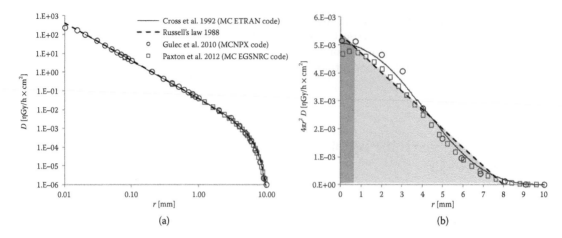

Figure 9.4 **(a)** Dose deposition around a small 50 Bq ^{90}Y source in soft medium. **(b)** Dose delivered in $r = 0$ coming from spherical shell of radius r in a uniform ^{90}Y distribution. Cross et al. (1992): ETRAN simulation for point source in water. Gulec et al. (2010): MCNPX simulation for a 30 µm spherical source in soft tissue equivalent medium. Paxton et al. (2012): EGSNRC simulation for glass 25 µm microsphere surrounded by liver tissue.

9.5 VOXEL-BASED ABSORBED DOSE

Assuming a liver of uniform density, a fast and easy method to compute the voxel-based absorbed dose from a known voxel based activity is to convolve it by a voxel dose kernel K. The mean absorbed dose D_{ijk} in a voxel (i,j,k) is simply

$$D_{ijk} = \sum_{mns} K_{|i-m|,|j-n|,|k-s|} \; N_{mns} \tag{9.3}$$

where N_{mns} is the number of decays occurring in the voxel (m,n,s). The kernel element $K_{|u|,|v|,|w|}$ represents the mean absorbed dose delivered into the voxel (u,v,w) when one decay occurred in a position averaged in the whole voxel $(0,0,0)$.

The voxel dose kernel K can be computed by MC or more easily by integrating Russell's law (Equation 9.2):

$$K_{u,v,w} = \frac{1}{V} \int\!\!\int\!\!\int_{\vec{y}\in V_{uvw}} \mathrm{d}\vec{y} \; \frac{1}{V} \int\!\!\int\!\!\int_{\vec{x}\in V_{ooo}} \tag{9.4}$$

$$\mathrm{d}\vec{x} \; D_{R}\left(\left|\vec{y}-\vec{x}\right|\right)$$

where V_{uvw} is the voxel (u,v,w) domain while V is the volume of the voxels.

Equation 9.4 is not analytically computable and one could be tempted to numerically compute it as

$$K_{u,v,w} = \frac{1}{N^6} \sum_{m=0}^{N-1}\sum_{n=0}^{N-1}\sum_{s=0}^{N-1} \; \sum_{i=0}^{N-1}\sum_{j=0}^{N-1}\sum_{k=0}^{N-1}$$
$$D_{R}\left(\sqrt{\left(uH+(m-i)h\right)^2+\left(vH+(n-j)h\right)^2+\left(wH+(s-k)h\right)^2}\right) \tag{9.5}$$

where H is the voxel size and $h = H/N$.

However, much more efficiently, Equation 9.5 can be reduced into a triple summation:

$$K_{u,v,w} = \frac{1}{N^6} \sum_{i=-N}^{N}\sum_{j=-N}^{N}\sum_{k=-N}^{N}\left(N-|i|\right)\left(N-|j|\right)\left(N-|k|\right)$$
$$D_{R}(\sqrt{\left(uH+ih\right)^2+\left(vH+jh\right)^2+\left(wH+kh\right)^2}) \tag{9.6}$$

Other methods of determining voxel dose kernels are presented in Chapter 12.

9.6 INTRALOBULE DOSIMETRY FROM MONTE CARLO SIMULATIONS IN TRANSLATION INVARIANT TRAPPING

Gulec et al. (2010) computed the intralobule dose distribution by MC for a translation invariant ^{90}Y loaded microsphere distribution in which all the lobules of the liver trap microspheres with the same pattern. Two microsphere distributions were modeled: 24 × 50 Bq microspheres in each portal triad and 1 × 2500 Bq microsphere in every other portal triad, both distributions corresponding to a mean liver absorbed dose of ≈64 Gy.

To boost the computation speed, the reflective boundary method was used: only one lobule is modeled in the MC simulation and when a tracked electron reached the lobule boundary, it underwent a reflection back into the lobule in order to account for the cross-fire effect of the surrounding lobules. Similarly to Figure 9.4b, the MC simulations showed that four-fifths of the absorbed dose to the central vein and about one-half of the absorbed dose to the triad vein came from the microspheres trapped in the surrounding lobules.

Table 9.1 shows the mean absorbed dose to the lobule structures derived from the Gulec et al. (2010) MC simulations. In order to better correspond to the clinical practice, the number of

microspheres was roughly rescaled to get a mean liver absorbed dose of 40 and 120 Gy for the 50 Bq and 2500 Bq microsphere models, respectively (Kennedy et al., 2004; Lau et al., 2012). For the triad structures, the absorbed doses are given for a triad containing, or not, a 2500 Bq microsphere.

The results of these simulations show that in both models, all the portal triads, including those not containing any microspheres, receive a lethal mean absorbed dose which is in discrepancy with the low toxicity observed in clinical therapy.

9.7 INTRALOBULE DOSIMETRY FROM RUSSELL'S LAW IN TRANSLATION INVARIANT TRAPPING

Even to compute the absorbed dose distribution by MC electron tracking in a single lobule, Gulec et al. (2010) boosted the computation speed by using the reflective boundary method that required identical microsphere trapping in all lobules. Computing the absorbed dose distribution by tracking the electrons inside all the 10^6 lobules, each surrounded by lobules having different trapping patterns, is far beyond the capacity of the state-of-the-art computing technologies.

A much faster alternative is to use the Russell dose distribution. Table 9.2 shows the comparison of the intralobule dosimetry obtained by Gulec et al. (2010) by MC electron tracking and that obtained using the Russell dose distribution

Table 9.1 Mean absorbed dose in the different hepatic structures

Tissue	D for 50 Bq/ms, 15 ms per triad (Gy)	D for 2500 Bq/ms, 2 ms every other triad (Gy)
Liver	40	120
Hepatocytes	39	120
Central vein	37	109
Triad bile duct	70	109–320
Triad vein	68	109–501
Triad artery	118	109–636

Table 9.2 Comparison of mean absorbed doses computed with MC and from the Russell law

Tissue	24 × 50 Bq ms in each triad artery		1 × 2500 Bq ms in every other triad artery	
	MC (Gulec et al., 2010) (Gy)	Russell (Walrand et al., 2014a) (Gy)	MC (Gulec et al., 2010) (Gy)	Russell (Walrand et al., 2014a) (Gy)
Liver	64	63	64	65
Hepatocytes	63	63	64	65
Central vein	59	58	58	60
Triad bile duct	112	118	58–171	60–187
Triad vein	109	113	58–167	60–182
Triad arteriole	188	206	58–339	60–377

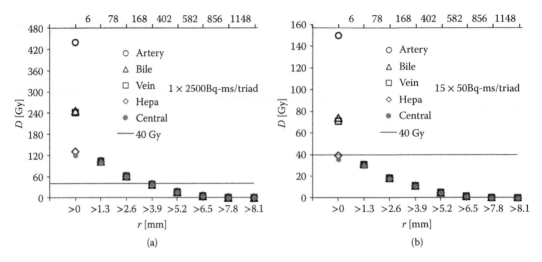

Figure 9.5 Absorbed dose to the different lobule structures coming from the microspheres trapped in triad arteries located farther than $i \times 1.3$ mm to the lobule center. **(a)** 1×2500 Bq microsphere in each triad artery. **(b)** 15×50 Bq microspheres in each triad artery. The number of portal triads included between the two concentric shell of radius $i \times 1.3$ mm and $(i + 1) \times 1.3$ mm is indicated on the upper axis.

(Equation 9.2). The relative deviation is about 4% for hepatocytes and central vein and about 10% for structures containing or in contact with microspheres.

Figure 9.5 shows the computed absorbed doses to the different lobule structures coming from the microspheres trapped in the triad arteries located at a distance farther than $i \times 1.3$ mm. Two scenarios were simulated: 1×2500 Bq microsphere per triad artery corresponding to a mean liver absorbed dose of 130 Gy, typical of a clinical therapy with glass microspheres (Figure 9.5a), and 15×50 Bq microspheres per triad artery corresponding to a mean liver absorbed dose of 40 Gy, typical of a clinical therapy using resin microsphere (Figure 9.5b).

This simulation shows that, for a surrounding translation invariant microspheres trapping, a 8-mm- and a 3-mm-diameter sphere free of microspheres are required to keep the absorbed dose inside the central lobule below 40 Gy for the 120 and 40 Gy scenarios, respectively.

The fact that the needed heterogeneity scale is larger in the setup in which fewer microspheres are injected is compatible with statistical fluctuations resulting in the transport dynamics of the microspheres from the catheter tip up to the terminal triad arteries. Prediction of these fluctuations requires a model of the hepatic arterial tree.

9.8 MICROSPHERES BIODISTRIBUTION STUDIES IN LIVER

Early studies (Pillai et al., 1991; Roberson et al., 1992) on microspheres biodistribution were performed in explanted rabbit livers after radio-embolization with 27-μm-diameter polystyrene microspheres. At a mean number of four micro-spheres per triad, i.e., in between clinical resin and glass microspheres radioembolization, Pillai et al. (1991) found some clusters larger than 25 micro-spheres. Some reports followed on cluster gather-ings studied in two-dimensional (2D) sections of explanted human liver tumors (Campbell et al., 2000; Kennedy et al., 2004).

Recently, Högberg et al. (2014, 2015a, 2015b) conducted a first real three-dimensional (3D) scanning of the microsphere clusters in 16 biopsies of a normal human liver tissue explanted 9 days after radioembolization using resin microspheres. The autoradiography of the explanted tissue (Figure 9.6a) showed an extremely nonuniform microsphere distribution with 1-cm-scale subregions of very low or of high microsphere densities. The 16 biopsies displayed 125 single microspheres and 277 clusters containing a total of 3736 microspheres. Two different types of clusters

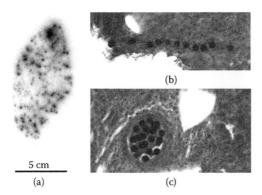

Figure 9.6 **(a)** Autoradiography of normal liver tissue explanted 9 days after resin microspheres radio-embolization. (b and c) Slice of 2 of the 275 clusters found in the 16 biopsies performed in the liver tissue. **(b)** Linear cluster of microsphere sequentially trapped in terminal triad artery. **(c)** Central slice of a globular cluster with microspheres gathered inside a larger artery. (Courtesy of Dr. Högberg and of Dr. Bernhardt.)

were identified: linear clusters (Figure 9.6b), corresponding to microspheres sequentially trapped in a terminal triad artery, and globular clusters (Figure 9.6c), corresponding to microspheres gathered in larger arteries, likely at branching nodes where the artery splits into two smaller arteries. The mean cluster size (or equivalently the mean number of microspheres per triad) in the biopsies was 9.2. The largest cluster had a globular shape and contained 453 microspheres. Large globular clusters were found in artery generations 13–19, with a maximal frequency in the 17th and 18th generations.

9.9 HEPATIC ARTERIAL TREE MODELING

The spatial resolution of current human *in vivo* computed tomography (CT) is a little less than 0.5 mm. Although this spatial resolution continuously improves, *in vivo* imaging of the whole hepatic arterial tree down to 40 μm (diameter of triad arteries) will likely remain inaccessible, especially considering the inability to completely eliminate motion effects. *Ex vivo* corrosion casting is a powerful technique that, in theory, should be able to achieve this goal, especially when combined with a high-dose industrial CT capable of submicrometer spatial resolution (Cnuddea and Boone, 2013).

However, detailed simulation of the microsphere transport dynamics requires an accurate assessment of the vessels' diameter and curvature,

and thus a small imaging voxel. Using a 5 μm voxel size, the reconstruction of the whole liver will have to be performed using a $\left(6\times10^4\right)^3$ matrix, which is still far beyond the limits of conventional computers. In addition, automatic analysis programs still exhibit segmentation issues that have to be manually addressed when two vessel branches touch. This alone could represent a monumental task when one considers the 4×10^6 vessel branches to be segmented. However, Debbaut et al. (2012, 2014) obtained a very impressive vascular tree segmentation up to the sinusoid level but on a limited 2 mm × 2 mm × 2 mm sample size.

Significant improvements were obtained this last decade in the mathematical modeling of the hepatic vascular tree. Three different approaches are competing: constrained constructive optimization (CCO), deterministic geometric construction, and angiogenesis-based construction (the reader can find an exhaustive literature survey and a discussion of these three approaches in Schwen and Preusser, 2012).

Currently, CCO, introduced by Schreiner and Buxbaum (1993), is a very promising approach (Schwen and Preusser, 2012). Briefly, CCO is an MC process where the arterial tree is updated by randomly drawing in the liver a free node that is afterward connected to the closest branch of the arterial tree (see the video demo at http://www.mevis-research.de/~oschwen/research/talks/20120823-BerlinISMP-iCCO.pdf). The initial tree consists of a major hepatic vessel network obtained from CT arteriography. The optimization step consists of designing new branch bifurcations to minimize the total vascular volume taking into account that the vessel radii are, at each iteration step, constrained to ensure an equal blood flow to all the lobules.

Recently, Schwen and Preuser (2012) built a realistic arterial tree, but only supplied 10,000 nodes. Assuming the viscosity was independent to the vessel radius, which is only valid for radius larger than 150 μm, the workload for generating N nodes is of the order $O(N^2 \ln(N))$. Thus, the generation of the whole-liver arterial tree will require 60,000-fold more computation time. Taking into account the radius dependence of the viscosity will still significantly increase the workload.

9.10 MICROSPHERE TRANSPORT MODELING

Kennedy et al. (2010) and Basciano (2010) modeled fluid dynamics and microsphere transport in the four major branches of a hepatic arterial tree derived from the population-representative morphological data. The computations were performed under the hypothesis of dilute microsphere suspension, i.e., the presence of microspheres does not impact the fluid dynamics and the interaction between microspheres can be neglected. Simulations were performed not only in steady flow, but also in transient dynamics by introducing in the equations a hepatic pressure waveform also derived from population-representative data.

The simulations showed that the microsphere partition at an arterial node does not follow that of the blood flow. In addition, it depends on the microsphere position in the vessel lumen prior to the node, the flow acceleration phase, and the bifurcation angles of the daughter vessels. These simulations were confirmed in an experimental model (Richards et al., 2012, 2013).

After having crossed several bifurcations, one can expect that the particles are more or less evenly distributed in the vessel lumen. Microsphere injection is often performed slowly during several cardiac cycles, the impact of which is therefore averaged. In a steady state, Kennedy et al. (2010) showed that for a uniform inflow, the local partitions between the four daughter vessels (1, 2, 3, 4 in Figure 9.7) were (0.26, 0.20, 0.29, 0.25) and (0.14, 0.32, 0.36, 0.18) for the blood flow and resin microspheres, respectively. Thus, local microsphere partitioning in the nodes of daughter vessels (1,2) and (3,4) was (0.30, 0.70) and (0.67, 0.33), respectively.

These microsphere partitions must be corrected for small blood flow differences. At the first order, i.e., for the assumption that the microspheres follow the blood flow, the correction is

$$P_i^{ms} = \frac{p_i^{ms}}{p_i^{bf}} \left/ \left(\frac{p_i^{ms}}{p_i^{bf}} + \frac{\left(1 - p_i^{ms}\right)}{\left(1 - p_i^{bf}\right)} \right) \right. \tag{9.7}$$

where p_i^{ms} and p_i^{bf} are the simulated local partition of daughter i for the microspheres and for the blood flow, respectively. P_i^{ms} is the corrected partition, i.e., rescaled to equal daughter blood flow.

After correction using Equation 9.7, microsphere partitions become (0.25, 0.75) and (0.63, 0.37). Note

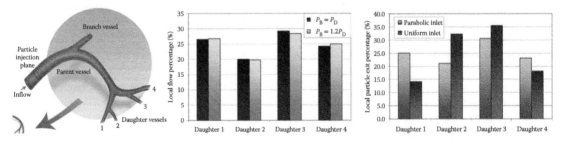

Figure 9.7 (Left) Arterial branches modeled. Middle: percentage of incoming blood flow exiting individual daughter vessels. (Right) Percentage of incoming particles exiting individual daughter vessels. (Reprinted from *Int J Radiat Oncol Biol Phys*, 76, Kennedy et al., Computer modeling of yttrium-90-microsphere transport in the hepatic arterial tree to improve clinical outcomes, 631–637, Copyright (2010), with permission from Elsevier.) Note that for the uniform inlet, even when the blood flow is lower, the particles preferably go into the two most curved bifurcation, i.e., daughters 2 and 3.

that for the two nodes (1,2) and (3,4), the microsphere partition is always greater in the bifurcating vessel.

Basciano (2010) reported a computing time of about 60 hours per microsphere tracked through the three nodes of the model using a quad core CPU. Simulating millions of microspheres through the 20 successive nodes of a liver will remain challenging for many years.

9.11 MICROSPHERE DISTRIBUTION SIMULATION

In order to achieve a reasonable computation time, Walrand et al. (2014a) built a full 3D hepatic arterial tree using a simplified CCO scheme, i.e., the total vessel length was optimized rather than the total vessel volume. Microsphere dynamics and transport were modeled by a simple random selection at each node of the daughter vessel crossed by the microsphere.

The main trunk, composed of the eight artery branches feeding the eight liver segments, was manually drawn according to the standard liver morphology. The 2×10^6 triad arteries were successively randomly selected in the liver volume and the closest existing vessel was identified (Figure 9.8). The position of the connection node in this vessel was constrained to be closer to the trunk than to the selected triad. This constraint avoids retrograde artery vessels that are not physiologically present.

Under this constraint, the node position and the folding of the existing vessel that minimizes the total length of the vessels were selected. Minimizing the total vessel length rather than the total vessel volume avoided the recomputation of all the vessel radii that is needed after each new lobule connection in order to ensure an equal blood flow to all the lobules, saving considerable computational time.

When the arterial tree is built, the blood flow of all vessel branches was computed to ensure an equal blood flow to all the terminal triad arteries. The probability of each terminal triad artery trapping a microsphere was computed by following, in reverse, the artery path from the triad to the catheter tip. At each node, the probability was multiplied by the local microsphere partition of the considered bifurcation, rescaled by its local blood flow partition using Equation 9.7.

Triad arteries were randomly populated under the dilute microsphere suspension assumption, i.e., microsphere by microsphere according to the probability associated with a given triad. After each microsphere delivery, the trapping probability of the triad was reduced on order to account for the reduction of blood flow by partial embolization. As lobule triads have on average 2.4 arteries, each 1300 μm in length, the reduction was designed such that the trapping probability linearly vanishes after 300 microspheres.

Figure 9.9 shows a slice comparison of simulated 2500 Bq microsphere distributions (Figure 9.9b and c) delivering 120 Gy to the liver versus a

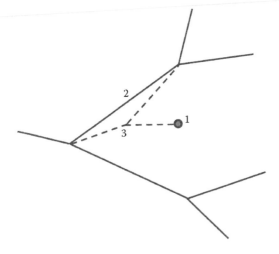

Figure 9.8 One iteration step of the simplified CCO arterial tree generation. 1. Random selection of a free lobule. 2. Identification of the closest existing vessel branch. 3. Determination of the new connection node position and of the existing branch folding which minimizes the total vessel length.

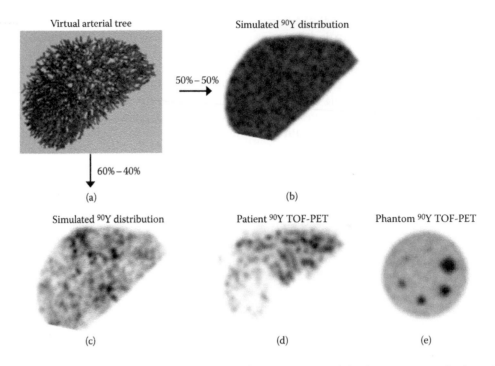

Virtual arterial tree

Simulated ^{90}Y distribution

50%–50%

60%–40%

(a)

(b)

Simulated ^{90}Y distribution

Patient ^{90}Y TOF-PET

Phantom ^{90}Y TOF-PET

(c)

(d)

(e)

Figure 9.9 **(a)** 3D rendering of virtual arterial tree after generation of the first 1500 vessels. (b and c) Glass microsphere distribution with a 120 Gy average liver dose from a virtual arterial tree using 50%–50% **(b)** and 60–40% **(c)** microsphere relative-partition probability between two daughter vessels. Both slices were convolved with a blurring kernel to match PET spatial resolution. **(d)** Typical ^{90}Y TOF PET slice in normal liver of a patient treated with glass microspheres at a 120 Gy average left liver dose. Note the similar granularity of glass microsphere distribution shown in (c) and (d). **(e)** TOF PET imaging of hot sphere phantom with the same acquisition time and same ^{90}Y-specific activity as shown in patient image in (d). (Reprinted in black and white from Walrand et al., *J Nucl Med*, 5, 135–140, 2014a.)

typical time of flight (TOF) ^{90}Y PET acquisition of a patient (Figure 9.9d) and of a hot spheres phantom (Figure 9.9e). The patient was treated with glass microspheres with a 120 Gy average dose to the left liver lobe, while the phantom was filled with an identical specific background activity. More information on ^{90}Y PET imaging can be found in Chapter 11.

Figure 9.10 shows the cumulated cluster size distribution observed by Högberg (2015b) from biopsies of normal liver tissue explanted 9 days after radioembolization with resin microspheres (see Figure 9.6a). The best agreement with the model of Walrand et al. (2014a) was obtained for an asymmetric microsphere partition probability of 64%–36% at the bifurcation nodes in line with the dynamic transport simulations (Basciano, 2010; Kennedy et al., 2010). Although the cluster size distribution is well predicted in this model, all of the microsphere clusters are located in the

terminal triad arteries, and globular clusters, shown in Figure 9.6c, are not present in the simulation. The largest cluster contained 158 microspheres, which is threefold less than that observed by Högberg (2015b).

In order to also simulate globular clusters, Högberg (2015b) developed an arterial tree including an exponentially decreasing diameter of arterial branches from the main trunk up to the terminal triad arteries as observed by Debbaut et al. (2012, 2014). Three variable parameters were optimized to obtain concordance between simulated and *in vivo* microsphere distributions: (1) a combined artery coefficient of variation (ACV) parameter for the inner diameter of all arterial generations throughout the virtual tree structure that controls the microsphere flow distribution at the nodes, (2) the hepatic tree distribution volume (HDV) parameter, and (3) the embolization (EMB) parameter that reduces the arterial diameter.

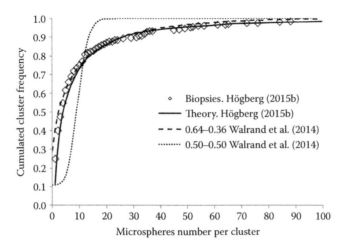

Figure 9.10 Cumulated cluster size distribution in biopsies of normal liver tissue explanted 9 days after radioembolization using resin microspheres: observed (diamonds) and predicted (straight line). (Derived from Figure 12A in Högberg, J., *Small-Scale Absorbed Dose Modelling in Selective Internal Radiation Therapy: Microsphere Distribution in Normal Liver Tissue*, MA: University of Gothenburg, 2015b). The mean microsphere number per triad is 9.2. Dotted and dashed lines: Predictions using the model of Walrand et al. (2014a) with a microsphere partition probability of 50%–50% and 64%–36% at the bifurcation nodes, respectively.

A good agreement was obtained for the cumulated cluster size distribution (straight line in Figure 9.10) and for the cluster frequency in the different artery generations as well (Figure 12A in Högberg, 2015b). Currently, this arterial tree model is in the form of a schematic two-dimensional (2D) arborescence. Additional assumptions on the spatial distribution of the clusters are needed in order to compute the absorbed dose distribution.

9.12 MICROSPHERE DISTRIBUTION AND HEPATIC TOXICITY

The first interesting quantitative result obtained from the microsphere transport simulations was to show that the typical therapy doses of 40 and 120 Gy delivered to the liver by using resin and glass ^{90}Y loaded microspheres, respectively, provide similar dose distribution to the portal triad—the critical radiosensitive structure in liver radioembolization (Walrand et al., 2014a).

After tissue irradiation, two processes occur: a fraction of cells are killed, followed by either a complete recovery or loss of the tissue. Due to the huge number of cells and of electron tracks involved, the first process is completely deterministic and monotonically dependent on the absorbed dose according to the well-known relation (Barendsen, 1962):

$$SF = e^{-\alpha D - \beta D^2} \tag{9.8}$$

where SF is the survival fraction, α and β are the linear and quadratic radiosensitivities, and D is the absorbed dose (assumed to be instantaneously delivered).

Organ recovery is characterized by a dose threshold that is tissue dependent. However, this threshold is also variable among individuals of the same species due to genetic differences and also variations in metabolism between individuals. Therefore, organ recovery frequency as a function of the absorbed does not exhibit a step shape, but rather a sharp sigmoid shape. This defines a region around the dose threshold where the complete recovery displays some random nature. The hepatic lobule is a functional tissue subunit acting as an independent organ on its own; therefore, the recovery of a population of lobules can thus be described by a sigmoid function.

In science, we strive to describe the behavior of a large set of observations by a single formalism or theory. For example, in the present case, we aim to develop a formalism to describe the hepatic toxicity observed in EBRT and that observed in liver radioembolization. The two modalities are characterized by a different dose rate: instantaneous 1.5–2 Gy doses spaced in time (>8 hours) in EBRT and exponentially decreasing dose rate in liver radioembolization (half-life = 2.7 days for ^{90}Y). The biological effective dose (BED) concept based on a linear-quadratic model (LQM) has been introduced to account for the dose rate (Fowler, 1989). This concept succeeded in unifying the renal toxicity observed in EBRT and in peptide receptor radionuclide therapy (PRRT) (Barone et al., 2005; Wessels et al., 2008) and is now also considered in liver radioembolization (Cremonesi et al., 2008; Strigari et al., 2010). Regarding the irradiation itself, both the photons used in EBRT and the β-particles used in liver radioembolization are low linear energy transfer (LET) particles and thus share the same radiobiology effectiveness (RBE) (ICRP, 2007).

Partial liver irradiation in EBRT (Dawson et al. 2001) showed that NTCP = 0.5 is obtained by killing ≈40% of the liver volume or by irradiating 100%, 80%, or 66% of the liver volume with a BED of 77, 95, or 115 Gy, respectively. These results can be described by the following the sigmoid curve for the lobule nonrecovery probability (Walrand et al. 2014b):

$$NR(bed) = \frac{1}{1 + \left(\dfrac{93.8}{bed}\right)^{2.12}} \qquad (9.9)$$

The liver NTCP as a function of the killed lobule fraction (KF) is (Dawson et al. 2001)

$$NTCP(KF) = \frac{1}{1 + \left(\dfrac{0.4}{KF}\right)^{8.29}} \qquad (9.10)$$

where KF can be computed using

$$KF = \sum_i v(bed_i)\, NR(bed_i) \qquad (9.11)$$

where $v(BED_i)$ is the fraction of lobules receiving a dose BED_i. Note that in EBRT, Equation 9.11 reduces to

$$KF = Vf\, NR(BED) \qquad (9.12)$$

where Vf is the irradiated liver fraction.

Walrand et al. (2014b) computed the $v(BED_i)$ fractions by convolving the Russell dose kernel

Equation 9.2 with the glass microsphere distribution obtained using the microsphere transport MC simulations corresponding to different BED delivered to the liver parenchyma. The best agreement with studies in glass and resin microsphere radioembolization (Chiesa et al., 2015; Strigari et al., 2010, respectively) was obtained using the asymmetric probability of 69%–31% for the microsphere bifurcation partition (Figure 9.11). This value is not far from the 64% to 36% value giving the best agreement with the cluster size distribution observed (Högberg, 2015). Note that a misunderstanding occurred in Walrand et al. (2014b) where for the MC simulations the median toxic dose (TD50) was given in the targeted liver region as done in EBRT, while clinical TD50 data observed in radioembolization were reported averaged on the whole liver (Chiesa et al., 2015; Strigari et al., 2010). This explains why 60%–40% was the previous optimal microsphere partition probability.

An important benefit of the MC simulation is the prediction of the hepatic toxicity as a function of various parameters. Figure 9.12 shows the WLTD50, i.e., the dose averaged over the whole liver giving NTCP = 0.5, as a function of the targeted liver fraction and of the microsphere-specific activity (msA). For resin microspheres, the WLTD50 is almost constant for a targeted liver fraction larger than 65%, reducing the drawback of mixing whole and right liver radioembolizations in the same NTCP reporting (Strigari et al., 2010). Note that Equations 9.9 and 9.10 derived from EBRT involve that the WLTD50 become infinite when the targeted liver fraction is lower than 40%.

9.13 APPLICATION OF THE HEPATIC TOXICITY MODEL

For convenient use, the WLTD(p,Vf,msA), i.e., the whole-liver dose providing a NTCP = p when targeting a liver volume fraction Vf using ^{90}Y loaded microspheres of specific activity msA, is fitted by

$$WLTD(p, Vf, msA) = 47.1\ Gy$$
$$\times\ \frac{(1 + 0.457\,p)\ F(msA)}{(Vf - Kf(p))^{0.869\ F(msA)}}\ Vf \qquad (9.13)$$

where the dimensionless scale factor F(msA) is

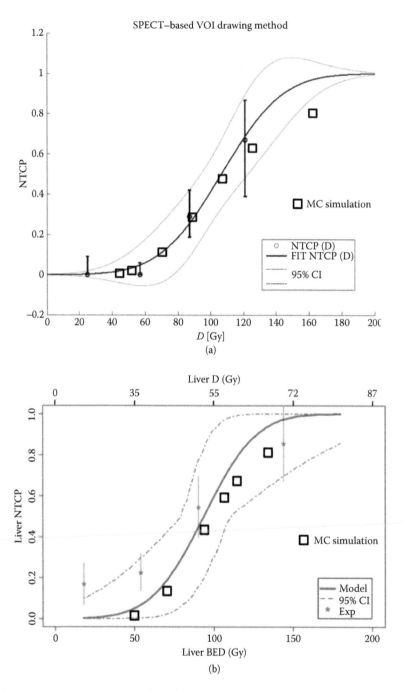

Figure 9.11 NCTP comparison between clinical observations and prediction (squares) from Equations 9.9 and 9.10 using 69%–31% as the microsphere partition probability in the MC calculation of $\nu(BED_i)$. **(a)** Glass microspheres in right liver radioembolizations ($Vf \approx 0.66$) (With kind permission from Springer Science + Business Media: *Eur J Nucl Med Mol Imaging*, Radioembolization of hepatocarcinoma with (90)Y glass microspheres: Development of an individualized treatment planning strategy based on dosimetry and radiobiology, 42, 2015, 1718–1738, Chiesa et al.). **(b)** Resin microspheres in mixed whole and right liver radioembolizations ($Vf \approx 0.80$). (Reprinted from Strigari et al., *J Nucl Med*, 51, 1377–1385, 2010. With permission of the Nuclear Medicine Society). Squares were added by the author. Doses are averaged over the whole liver parenchyma.

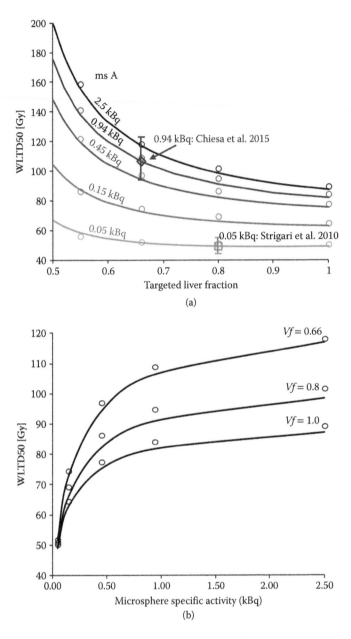

Figure 9.12 Circles: $WLTD_{50}$ (i.e., dose averaged on the whole liver parenchyma giving NTCP = 0.5) derived from Equations 9.9 and 9.10 using a 69%–31% microsphere partition probability in the MC simulation of $\nu(BED_i)$. Straight lines: fit with Equations 9.13 and 9.14. **(a)** $WLTD_{50}$ as a function of the targeted liver fraction for different microsphere-specific activities (msA). **(b)** $WLTD_{50}$ as a function of the msA for different targeted liver volume fractions (Vf).

$$F(msA) = \left(1 - e^{-\sqrt[3]{msA/0.0471\,kBq}}\right) \qquad (9.14)$$

Note that, for a fixed total radioembolization activity, the cube root in $F(msA)$ is a quantity strongly proportional to the mean intermicrosphere distance in the liver.

And $Kf(p)$ is the inverse of the relation (Equation 9.10), i.e.,

$$Kf(p) = 0.4 \; {}_{8.29}\!\sqrt{\frac{p}{1-p}} \qquad (9.15)$$

$Kf(0.5) = 0.4$ and $Kf(0.05) = 0.28$, which means that ablating by irradiation 40% and 28% of the liver induces a NTCP of 50% and 5%, respectively, as observed in EBRT (Dawson et al., 2001).

In the fitted domain {msA $\in$ (0.05,0.15, 0.45,0.94,2.5) kBq, p $\in$ (0.15,0.30,0.50), Vf $\in$ (0.55,0.66,0.80,1.00)} the mean absolute relative deviation of the fit versus MC simulations (i.e., $\frac{|fit - MC|}{MC}$) was 2.9% with extreme relative deviations (i.e., $\frac{fit - MC}{MC}$) of –4.97% and 4.60%. It is quite remarkable that such a simple expression of F accurately takes into account the impact of mSA.

As the microspheres are localized in the portal triads, the WLTD(p,Vf,msA) value when msA vanishes does not approach that of EBRT. Indeed, TD50 = 43 Gy is the dose of ^{90}Y corresponding to BED50 = 77 Gy observed in whole-liver EBRT. The simulation in Table 9.1 shows that, in the case of translation invariant microsphere distribution (which is expected when the microsphere number increases), 43 Gy in the triad arteries corresponds to a liver parenchyma dose of $40 \times 43/118 = 15$ Gy, i.e., BED = 19 Gy, fourfold smaller than the 77 Gy observed in EBRT.

As there is no clinical study available in order to validate the MC simulation for microsphere activity below 0.05 kBq, a conservative expression of F(msA) was chosen such as

$$\lim_{msA \to 0} WLTD(p,Vf,msA) = 0 \qquad (9.16)$$

Equations 9.13 and 9.14 could be used to compute the needed absorbed dose and also the activity in order to achieve a desired NTCP = p as a function of the targeted liver fraction volume and mSA. Note that this set of equations was obtained by tuning one parameter of the model (the microsphere partition probability at the arterial tree nodes) in order to fit the clinical hepatic toxicity observed in glass (Chiesa et al., 2015) and in resin (Strigari et al., 2010) radioembolization. Thus, the prediction accuracy of Equations 9.13 and 9.14 does not only depend on the goodness of the model but also on the goodness of the observed clinical hepatic toxicity.

9.14 CONCLUSIONS

MC simulations, using a very simple model of the hepatic arterial tree and microsphere transport,

reconcile the liver toxicities observed in EBRT and in radioembolization using glass and resin microspheres. More interestingly, these MC simulations can be fitted with an analytical formula (Equations 9.13 and 9.14) that allows performing of individualized radioembolization planning as a function of the liver fraction targeted and of the mSA used.

This model contains an adjustable parameter, i.e., the microsphere partition probability at the arterial nodes, which was fitted in order to reproduce the liver toxicities observed in clinical radioembolization studies (Strigari et al., 2010; Chiesa et al., 2015). However, the obtained partition probability 69%–31%: (1) is within the range of probabilities predicted by microsphere transport dynamic simulations (Kennedy et al., 2010), (2) is close to the value 64%–36% fitting the observed microsphere cluster distribution (Högberg et al., 2015), (3) fits both the hepatic toxicity reported in glass and resin microsphere radioembolization, and (4) unifies hepatic toxicity observed in radioembolization and in EBRT. These four facts support the coherence of the whole model and also of the reported clinical hepatic toxicity as well.

REFERENCES

Barendsen, G.W. (1962). Dose-survival curves of human cells in tissue culture irradiated with alpha-, beta-, 20-kV x- and 200-kV x-radiation. *Nature* 193:1153–1155.

Barone, R. et al. (2005). Patient-specific dosimetry in predicting renal toxicity with (90) Y-DOTATOC: Relevance of kidney volume and dose rate in finding a dose-effect relationship. *J Nucl Med* 46:99S–106S.

Basciano, C.A. (2010). *Computational Particle-Hemodynamics Analysis Applied to an Abdominal Aortic Aneurysm with Thrombus and Microsphere-Targeting of Liver Tumors.* Raleigh, NC: North Carolina State University.

Campbell, A.M., Bailey, I.H., Burton, M.A. (2000). Analysis of the distribution of intra-arterial microspheres in human liver following hepatic yttrium-90 microsphere therapy. *Phys Med Biol* 45:1023–1033.

Chiesa, C. et al. (2011). Need, feasibility and convenience of dosimetric treatment planning in liver selective internal radiation therapy with

(90)Y microspheres: The experience of the National Tumor Institute of Milan. *Q J Nucl Med Mol Imaging* 55:168–197.

Chiesa, C. et al. (2015). Radioembolization of hepatocarcinoma with (90)Y glass microspheres: Development of an individualized treatment planning strategy based on dosimetry and radiobiology. *Eur J Nucl Med Mol Imaging* 42:1718–1738.

Cnuddea, V., Boone, M. (2013). High-resolution X-ray computed tomography in geosciences: A review of the current technology and applications. *Earth Sci Rev* 123:1–17.

Crawford, A.R., Lin, X.Z., Crawford, J.M. (1998). The normal adult human liver biopsy: A quantitative reference standard. *Hepatology* 28:323–331.

Cremonesi, M. et al. (2008). Radioembolisation with 90Y-microspheres: Dosimetric and radiobiological investigation for multi-cycle treatment. *Eur J Nucl Med Mol Imaging* 35:2088–2096.

Cross, W.G., Freedman, N.O., Wong, P.Y. (1992). *Tables of Beta-Ray Dose Distributions in Water.* Report no. AECL-10521. Ontario, Canada: Atomic Energy of Canada, Ltd.

Dawson, L.A., Ten Haken, R.K., Lawrence, T.S. (2001). Partial irradiation of the liver. *Semin Radiat Oncol* 11:240–246.

Debbaut, C. et al. (2012). Perfusion characteristics of the human hepatic microcirculation based on three-dimensional reconstructions and computational fluid dynamic analysis. *J Biomech Eng* 134:011003.

Debbaut, C. et al. (2014). Analyzing the human liver vascular architecture by combining vascular corrosion casting and micro-CT scanning: A feasibility study. *J Anat* 224:509–517.

Fowler, J.F. (1989). A review: The linear quadratic formula and progress in fractionated radiotherapy. *Br J Radiol* 62:679–675.

Gulec, S.A. et al. (2010). Hepatic structural dosimetry in 90Y microsphere treatment: A Monte Carlo modeling approach based on lobular microanatomy. *J Nucl Med* 51:301–310.

Haga, J. et al. (2008). Liver regeneration in donors and adult recipients after living donor liver transplantation. *Liver Transpl* 14:1718–1724.

Högberg, J. et al. (2014). Heterogeneity of microsphere distribution in resected liver and tumour tissue following selective intrahepatic radiotherapy. *EJNMMI Res* 4:48.

Högberg, J. et al. (2015a). Increased absorbed liver dose in Selective Internal Radiation Therapy (SIRT) correlates with increased sphere-cluster frequency and absorbed dose inhomogeneity. *EJNMMI Phys* 2:10.

Högberg, J. (2015b). *Small-Scale Absorbed Dose Modelling in Selective Internal Radiation Therapy: Microsphere Distribution in Normal Liver Tissue.* Gothenburg: University of Gothenburg.

ICRP. (2007). The 2007 recommendations of the international commission on radiological protection. ICRP publication 103. *Ann ICRP* 37:1–332.

Kennedy, A.S. et al. (2004). Pathologic response and microdosimetry of (90)Y microspheres in man: Review of four explanted whole livers. *Int J Radiat Oncol Biol Phys* 60:1552–1563.

Kennedy, A. et al. (2007). Recommendations for radioembolization of hepatic malignancies using yttrium-90 microsphere brachytherapy: A consensus panel report from the radioembolization brachytherapy oncology consortium. *Int J Radiat Oncol Biol Phys* 68:13–23.

Kennedy, A.S., Kleinstreuer, C., Basciano, C.A., Dezarn, W.A. (2010). Computer modeling of yttrium-90-microsphere transport in the hepatic arterial tree to improve clinical outcomes. *Int J Radiat Oncol Biol Phys* 76:631–637.

Lau, W.Y. et al. (2012). Patient selection and activity planning guide for selective internal radiotherapy with yttrium-90 resin microspheres. *Int J Radiat Oncol Biol Phys* 82:401–407.

Ljungberg, M., Strand, S.-E., King, M.A. (2013). *Monte Carlo Calculations in Nuclear Medicine. Applications in Diagnostic Imaging.* Boca Raton, FL: CRC Press.

Mettler, F.A., Guiberteau, M.J. (2012). *Essentials of Nuclear Medicine Imaging.* Philadelphia, PA: Saunders.

Michalopoulos, G.K., DeFrances, M.C. (1997). Liver regeneration. *Science* 276:60–66.

Paxton, A.B., Davis, S.D., Dewerd, L.A. (2012). Determining the effects of microsphere and surrounding material composition on 90Y dose kernels using egsnrc and mcnp5. *Med Phys* 39:1424–1434.

Pillai, K.M. et al. (1991). Microscopic analysis of arterial microsphere distribution in rabbit liver and hepatic VX2 tumor. *Selective Cancer Ther* 7:39–48.

Press, W.H., Teukolsky, S.A., Vetterling, W.T., Flannery, B.P. (2007). *Numerical Recipes in C: The Art of Scientific Computing*. Cambridge, England: Cambridge Press.

Richards, A.L. et al. (2012). Experimental microsphere targeting in a representative hepatic artery system. *IEEE Trans Biomed Eng* 59:198–204.

Richards, A.L. (2013). *Experimental Investigations Into the Targeted Delivery of Microspheres in Radioembolization Therapy*. PhD dissertation. Raleigh, NC: North Carolina State University, 191 pp.

Roberson, P.L. et al. (1992). Three-dimensional tumor dosimetry for hepatic yttrium-90-microsphere therapy. *J Nucl Med Official Publ* 33:735–738.

Russell, J.L., Carden, J.L., Herron, H.L. (1988). Dosimetry calculation for yttrium-90 used in the treatment of liver cancer. *Endocriether Hyperther Oncol* 4:171–186.

Saito, M., Matsumoto, M. (2008). SIMD oriented fast Mersenne twister: A 128-bit pseudorandom number generator. In

Keller A, Heinrich S, Niederreiter H (Eds.) *Monte Carlo and Quasi-Monte Carlo Methods 2006*. Berlin: Springer-Verlag.

Schreiner, W., Buxbaum, P.F. (1993). Computer-optimization of vascular trees. *IEEE Trans Biomed Eng* 40:482–491.

Schroeder, L. (1974). Buffon's needle problem: An exciting application of many mathematical concepts. *Mathematics Teacher* 67:183–186.

Schwen, L.O., Preusser, T. (2012). Analysis and algorithmic generation of hepatic vascular systems. *Int J Hepatol*. doi:10.1155/2012/357687

Strigari, L. et al. (2010). Efficacy and toxicity related to treatment of hepatocellular carcinoma with 90Y-SIR spheres: Radiobiologic considerations. *J Nucl Med* 51:1377–1385.

Walrand, S. et al. (2014a). The low hepatic toxicity per Gray of 90Y glass microspheres is linked to their transport in the arterial tree favoring a nonuniform trapping as observed in posttherapy PET imaging. *J Nucl Med* 5:135–140.

Walrand, S., Hesse, M., Jamar, F., Lhommel, R. (2014b). A hepatic dose-toxicity model opening the way toward individualized radioembolization planning. *J Nucl Med* 55:1317–1322.

Wessels, B.W. et al. (2008). MIRD pamphlet No. 20: The effect of model assumptions on kidney dosimetry and response—Implications for radionuclide therapy. *Nucl Med* 49:1884–1899.

PART 4

Following Patients Treated with Radioembolization

Postradioembolization imaging using bremsstrahlung ^{90}Y SPECT/CT

S. CHEENU KAPPADATH

10.1 BACKGROUND AND RATIONALE

^{90}Y radioembolization is used in the management of patients with unresectable primary and metastatic liver cancers. ^{99m}Technetium-labeled macroaggregated albumin (^{99m}Tc MAA) has been successfully used, at least in terms of patient safety, as a surrogate radiopharmaceutical for treatment planning since the inception of the therapy, as described in Chapter 4. Planar scintigraphy of ^{99m}Tc MAA has been used primarily to evaluate the lung shunt fraction and estimate the mean absorbed dose to lung after the ^{90}Y radioembolization treatment (Ho et al., 1997). MAA uptake in the lung consequently affects the prescription of administered ^{90}Y microsphere activity to prevent radiation pneumonitis. MAA distribution *in vivo* is also used to assess extrahepatic distribution and judge the adequateness of tumor perfusion from the catheter placement. It has been demonstrated that the assessment of MAA distribution with single-photon emission computed tomography/CT (SPECT/CT) is superior compared with SPECT, which in turn is superior to planar imaging (Ahmadzadehfar et al.,

2010). Overall accuracies of 72%, 79%, and 96% for planar, SPECT, and SPECT/CT, respectively, have been reported (Hamami et al., 2009).

There are a number of qualitative studies that have suggested that MAA distributions observed during planning often match the ^{90}Y microsphere distributions after therapy. Concordance between MAA and ^{90}Y has been reported (Chiesa et al., 2015) that increases in confidence when selective segmental or lobar therapies are planned (Kao et al., 2013).

However, as reviewed in Chapter 4, several studies have also shown that the distribution of MAA during treatment planning may not be a consistent and reliable indicator of the distribution of the ^{90}Y microspheres after the administration of treatment (Ilhan et al., 2015). Differences of greater than 20% uptake between ^{99m}Tc MAA and ^{90}Y have been reported in 43% (97/225) of cases (Wondergem et al., 2013).

Furthermore, especially in the past decade, *in vivo* ^{99m}Tc MAA distributions have begun to be used in conjunction with dosimetry models such as the medical internal radiation dose (MIRD) (Gulec et al., 2006) or partition (Ho et al., 1996) models to calculate the tumor and normal liver doses (see, e.g., Chiesa et al., 2015; Garin et al., 2013).

However, the potential discrepancies in distribution between planning ^{99m}Tc MAA and treatment ^{90}Y radioembolization argue for the need for posttreatment ^{90}Y imaging to assess the delivered distribution of ^{90}Y treatment. Investigations have demonstrated the positive role of ^{90}Y planar and ^{90}Y bremsstrahlung SPECT/CT in the management of patients after radioembolization (Ahmadzadehfar et al., 2012). ^{90}Y SPECT/CT with an overall accuracy of 99% was shown to be superior to SPECT and planar imaging in the prediction of gastrointestinal (GI) ulcers.

In addition, ^{90}Y SPECT/CT is also being investigated for the calculation of tumor and normal liver doses using the MIRD or partition dosimetry models, as described in Chapter 5. Furthermore, as described in Chapter 12, quantitative ^{90}Y SPECT/CT also facilitates dosimetry at the voxel level that allows for investigations into the volume distribution of absorbed doses in tumors and normal liver (Kappadath et al., 2014). The role of postradioembolization imaging, whether it is performed with ^{90}Y SPECT/CT or ^{90}Y PET/CT, is absolutely critical when it comes to assessing tumor dose–response relationships and radioembolization-induced liver disease. The role of ^{90}Y PET/CT is reviewed in Chapter 11. However, this chapter will focus on the technical aspects related to ^{90}Y bremsstrahlung SPECT/CT for posttreatment imaging.

10.2 CHALLENGES ASSOCIATED WITH ^{90}Y SPECT IMAGING

As discussed in Chapter 1, ^{90}Y decays via beta emission to ^{90}Zr with a half-life of 64.1 hours and with a maximum beta energy of 2.28 MeV. In contrast with the majority of radionuclides used in nuclear medicine imaging, ^{90}Y is effectively a pure beta emitter, i.e., it lacks discrete energy photon emissions, such as gamma and/or characteristic fluorescence X-rays; therefore, imaging of ^{90}Y with gamma cameras is challenging.

^{90}Y activity distribution *in vivo* is traditionally assessed by imaging the bremsstrahlung photon—produced from interactions of the energetic beta particles with soft tissue—using a gamma camera or by SPECT/CT. Bremsstrahlung photons are not considered to be well suited for imaging because

of the continuous nature of the energy spectrum and the lack of readily identifiable spectral characteristics, which result in high levels of scatter. The lack of photo-peak emissions has been a major barrier for the standardization of a ^{90}Y bremsstrahlung imaging procedure; consequently, there is a wide variability in image quality among different facilities.

Historically, only planar imaging of the ^{90}Y bremsstrahlung emission, if at all, was performed after ^{90}Y therapies to confirm the ^{90}Y microsphere distribution within the liver compartment; however, the nonuniform uptake of microspheres in the liver cannot be visualized because of poor image contrast. ^{90}Y bremsstrahlung SPECT, if performed, used a wide energy window and reconstructed using filtered back-projection (FBP) with no compensations for scatter or attenuation. Therefore, historically, ^{90}Y-SPECT images have been nonquantitative and have poor contrast and resolution.

10.3 APPROACHES TO ^{90}Y BREMSSTRAHLUNG IMAGING

The technical approaches for improving ^{90}Y bremsstrahlung imaging have included evaluation using phantoms, simulations, and patient data. The major parameters investigated with respect to data acquisition have been energy window selection for imaging, and choice of collimation. The major parameters investigated with respect to data processing for SPECT imaging have been correction techniques for scatter and attenuation, iterative reconstructions, partial volume correction, and calibration for quantitative SPECT. There has also been some early work done on image filtration for noise suppression in planar images.

10.3.1 PLANAR IMAGING

The ^{90}Y bremsstrahlung planar acquisition parameters of collimator, sensitivity, and energy window have been empirically evaluated using phantoms. The spectral components of ^{90}Y bremsstrahlung images have been investigated and improved spatial resolution and image contrast reported for ^{90}Y bremsstrahlung imaging with a medium-energy (ME) collimator when compared with a low-energy collimator with an energy window of 55–285 keV (Shen et al., 1994a). The marginal improvement in resolution observed with the high-energy collimator was offset by the large decrease in its sensitivity. Early attempts at quantitative ^{90}Y planar imaging have focused on image restoration using filtration; both Wiener (King et al., 1983; Shen et al., 1994b) and wavelet-based neural network (Qian et al., 1994b) filters have been studied. These approaches use image filtration to deconvolve scatter and septal penetration and have shown improvement in energy resolution and image contrast. Filtered bremsstrahlung geometric-mean images (that partially compensate for attenuation) yielded individual activities within 17% (Shen et al., 1994a). However, filter-based deconvolutions are image dependent and require phantom calibration for quantification (Shen et al., 1994b). In addition, they can result in image artifacts and overcompensation of the system response functions (King et al., 1991). The superpositioning of overlapping ^{90}Y sources (tumor and normal liver) inherent in planar images outweigh their quantitative capabilities; therefore, quantitative planar techniques have not been applied to clinical ^{90}Y imaging studies.

10.3.2 SPECT AND SPECT/CT IMAGING

Historically, only ad hoc arguments about the various kinds of photons have been proposed to optimize the acquisition energy window for ^{90}Y bremsstrahlung imaging. Recent improvements in computational speed and processing power have enabled Monte Carlo (MC) simulation-based investigations of ^{90}Y bremsstrahlung spectral decomposition to elucidate the spectral composition of ^{90}Y bremsstrahlung (see e.g., Heard et al., 2004; Minarik et al., 2008; Rault et al., 2010; Elschot et al., 2013). Heard et al. (2004) used the EGSnrc MC code to simulate the Philips/ADAC Forte gamma camera, and Rault et al. (2010) used the Geant4/GATE MC to simulate the Philips AXIS/IRIS gamma camera. These studies corroborate that ME collimators and a 100–150 keV energy window provided optimal image contrast, as previously suggested by Shen

et al. (1994a). These studies also showed that a wide energy range of 50–200 keV could increase the sensitivity without substantial loss of spatial resolution or contrast.

Simulation studies have shown that ^{90}Y bremsstrahlung emission spectrum from the liver can be considered to be a summation of the following spectral components: primary bremsstrahlung, object scatter, camera backscatter, collimator scatter and penetration, and lead X-rays from the collimator. The photons within the energy spectrum between 70 and 100 keV have large contributions of characteristic X-rays originating primarily from the lead collimator. The energy spectrum between 200 and 300 keV was shown to predominantly arise from backscatter, while photons higher than 300 keV were from collimator scatter and septal penetration. At any given energy window, the ratio of primary bremsstrahlung to the total photons detected is typically less than 15%, with the highest primary fraction around 80–180 keV (Heard et al., 2004). Some previous studies have suggested using acquisition energy windows with a lower threshold than 100 keV because of the large number of detected photons (Shen et al., 1994; Heard et al., 2004). However, this appears to be not a good choice for accurate quantitative imaging unless the high-order scatter can be modeled accurately in the projector (Rong et al., 2012a).

Other studies have also reported on ^{90}Y bremsstrahlung SPECT imaging using phantoms. Minarik et al. (2008) evaluated ^{90}Y bremsstrahlung SPECT for ^{90}Y ibritumomab tiuxetan with a General Electric VH/Hawkeye system using CT-based attenuation correction, scatter-kernel-based scatter correction, model-based collimator response, and iterative reconstruction. SPECT data of an anthropomorphic torso phantom with liver insert was acquired over a relatively wide 105–195 keV energy range and reconstruction accuracies of 10%–16% were reported. They substantiate expectations that phantom-based sensitivity calibration can yield accurate quantification of ^{90}Y SPECT images. However, they reported results for only one phantom experiment and the scatter kernel was computed only for one configuration. Their use of a single linear-attenuation coefficient for CT-based attenuation correction over a wide 105–195 keV energy window could, however, potentially introduce errors from the large variation of linear attenuation across the wide energy window.

Rong and Frey published a series of papers that critically investigated energy windows for imaging ^{90}Y bremsstrahlung SPECT/CT (Rong et al., 2012a, 2012b). Rong et al. (2012a) developed an optimization method that takes into account both the bias and the variance of the activity estimates for optimizing acquisition energy window for quantitative ^{90}Y bremsstrahlung SPECT imaging. They used the weighted root mean squared error of volume of interest (VOI) activity estimates, which took into account both the bias due to model mismatch and the variance of the VOI activity estimates, as the figure of merit for optimization. They concluded that the optimal acquisition energy window for ^{90}Y SPECT imaging was 100–160 keV. To obtain the optimal acquisition energy window for general situations of interest in clinical ^{90}Y microsphere imaging, they generated phantoms with multiple tumors of various sizes and various tumor-to-normal activity concentration ratios using a digital phantom that realistically simulates human anatomy. ^{90}Y microsphere imaging was then simulated with a clinical SPECT system and typical imaging parameters using a previously validated MC code and a previously proposed method for modeling the image degrading effects in quantitative SPECT reconstruction.

In another study, Rong et al. (2012b) used simulations of imaging nuclear detectors (SIMIND) to accurately model image degrading factors such as object attenuation, scatter, and the collimator–detector response, all of which are energy dependent but essential to obtain quantitatively accurate images. Their approach used a single, wide acquisition window (100–500 keV), with separate treatment of photons in various energy ranges and in various logical categories during the modeling process. In particular, they separated photons into eight categories based on energy and logical category. Primary and scattered photons (i.e., photons not scattered in the body) were separated into four categories according to their emission energies: 0–250, 250–500, 500–1000, and 1000–2000 keV. They demonstrated a good agreement between the experimental measurement and Monte Carlo simulation. In the extended cardiac-torso (XCAT) phantom simulation, the proposed

method achieved excellent accuracy in modeling photon counts (errors ~1%) and the quantitative accuracy of activity estimates for all organs (errors were below ~12%). The net percent errors in activity estimates from physical geometrical phantom experiments were shown to be 7%–10%, and in the simulated patient data corresponded to mean organ dose errors of ~10% for the lung and 4%–5% for the liver.

In a different technique, the Utrecht Monte Carlo Simulator was adapted for ^{90}Y and incorporated into a statistical reconstruction algorithm by Elschot et al. (2013). Photon scatter and attenuation for all photons sampled from the full ^{90}Y energy spectrum were modeled during reconstruction by Monte Carlo simulations. The energy- and distance-dependent collimator–detector response was modeled with precalculated convolution kernels. For the purpose of computationally efficient modeling of the distance- and energy-dependent collimator-detector response (CDR), the updated photon intensities were binned voxel-wise in eight energy-dependent, three-dimensional (3D) scatter maps according to their energy after the last scatter event; the energy of photons ranged from 50 to 2000 keV. The energy used for the generation of the final image was 50–250 keV. The quantitative accuracy of ^{90}Y bremsstrahlung SPECT was substantially improved by Monte Carlo-based modeling of the image degrading factors. The International Electrotechnical Commission (IEC) image quality phantom was used to quantitatively evaluate the performance of their approach in comparison with those of clinical SPECT reconstruction. Their approach demonstrated substantially improved image contrast in patient scans from 25% to 88% for the 37 mm sphere and decreased the mean residual count error in the lung insert from 73% to 15% at the cost of higher image noise.

While there have been a number of studies to improve ^{90}Y bremsstrahlung imaging both qualitatively and quantitatively, most of the published approaches require some sort of Monte Carlo simulation, which cannot be easily implemented in routine clinical practice. This has led some groups to develop a simple ^{90}Y bremsstrahlung SPECT/CT imaging protocol based on energy-window-based scatter compensation and CT-based attenuation correction that is readily implemented in commercial SPECT/CT systems, yet improves ^{90}Y bremsstrahlung SPECT/CT image quality and quantification (Siman et al., 2016). Based on phantom experiments, subsequently verified on patient images, they determined that energy windows 90–125 keV and 310–410 keV were suitable imaging and scatter-estimate energy windows, respectively. For clinical images, the scatter correction factor was determined to be 0.5–0.6 with a coefficient of variation of ~10%; hence using a single-scatter correction factor was considered adequate. They acknowledge that their proposed energy-window-based scatter-correction method only partially corrects for backscatter, septal penetration, and septal scatter components, a claim supported by the simulation studies for components that dominate the scatter-estimate energy window.

SPECT calibration factor is defined as the ratio of the total activity in the field of view (FOV) to the total counts in the FOV. For ideal image reconstruction (with accurate corrections for scatter, attenuation, and collimator–detector response), calibration with a point source in air will suffice. In practice, however, the image reconstruction is not perfect; therefore, the calibration factor is usually derived from phantom images with all the necessary corrections applied. Posttherapy ^{90}Y SPECT/CT scan presents a unique condition where the total ^{90}Y activity inside the liver (and hence inside the SPECT FOV) can be determined with uncertainty <10%. Clinical images acquired under such conditions can be used to calibrate the SPECT/CT imaging system. In clinical studies, the total SPECT counts observed were found to be proportional to the ^{90}Y activity in the FOV. SPECT images with CT attenuation correction and scatter correction can be used to accurately quantify the activity present in the FOV with an average absolute deviation ≤5% (Siman et al., 2016).

There have also been some very innovative approaches into specifically imaging ^{90}Y using nonclinical gamma cameras. Walrand et al. (2014) used MC simulations of energy spectra and showed that a camera based on a 30-mm-thick bismuth germanium oxide (BGO) crystal and equipped with a high-energy pinhole collimator was well adapted to bremsstrahlung imaging. The total scatter contamination is reduced by a factor of 10 versus a conventional NaI camera equipped with a high-energy parallel hole collimator, enabling acquisition using an extended energy window ranging from 50 to 350 keV. By using the recorded event energy in the reconstruction method, shorter acquisition time and reduced orbit range will be feasible allowing the design of a simplified mobile gantry.

10.4 SUMMARY AND CONCLUSIONS

There have been a lot of advances toward improving imaging of ^{90}Y bremsstrahlung emissions using SPECT/CT (Figure 10.1). The general consensus appears to be that an imaging window around 90–160 keV is suitable for qualitative SPECT/CT imaging using ME collimators (Table 10.1). Quantitative SPECT/CT is also feasible when additional corrections such as attenuation, scatter, and collimator-response modeling are also incorporated during reconstruction. In fact, when energy-dependent corrections are incorporated the energy window width can be expanded to around 250 keV.

The potential for discrepancies in distribution between planning ^{99m}Tc MAA and treatment ^{90}Y radioembolization strongly argues for the need for posttreatment ^{90}Y imaging to assess the delivered distribution of ^{90}Y treatment. The investigations summarized in this chapter readily facilitate the qualitative assessment of posttherapy ^{90}Y microsphere distributions. In fact, quantitative ^{90}Y SPECT/CT can also be achieved without too much effort using commercial SPECT/CT scanners. Quantitative ^{90}Y SPECT/CT facilitates dosimetry at the voxel level that allows for investigations into the volume distribution of absorbed doses in tumors and normal liver (see Chapter 12).

The advancement of ^{90}Y radioembolization from what is usually considered to be palliative into a more frontline therapy will rest heavily on the ability to determine the absorbed doses to tumor necessary to elicit a response. Once the tumor dose response following radioembolization is known, ^{90}Y therapy planning can be based on delivering tumoricidal doses. This advancement in treatment planning will be essential in the promotion of radioembolization. The ability to accurately determine tumor doses stems from the ability to quantify ^{90}Y SPECT/CT, or by using

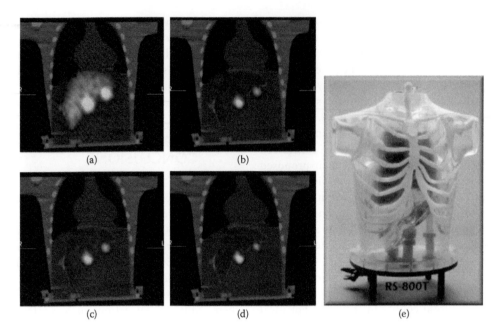

(a)   (b)

(c)   (d)   (e)

Figure 10.1 An illustration of different ⁹⁰Y SPECT/CT image quality based on the data acquisition and analysis schemas used. The fused SPECT/CT images (energy window 90–150) of the anthropomorphic torso phantom through the same coronal slices showing the liver insert and two hot spheres but with different SPECT reconstruction schemes: **(a)** filter backprojection (FBP), the technique of choice when SPECT-only systems (i.e., no CT) are used; **(b)** ordered subsets maximum likelihood expectation maximization (3D-OSEM) with neither attenuation nor scatter corrections; **(c)** 3D-OSEM with only attenuation but no scatter correction; **(d)** 3D-OSEM with both attenuation and scatter corrections; and **(e)** a photograph of the torso phantom. The series of images shows the dramatic improvements in image quality achieved using 3D-OSEM iterative reconstruction with both attenuation and scatter corrections; the background signal outside the liver is largely accounted for while the spheres are visualized with better contrast. The SPECT display scale is 0%–60% of maximum SPECT counts.

other quantitative posttreatment imaging modalities such as ⁹⁰Y PET/CT, discussed in Chapter 11.

Table 10.1 Summary of energy window ranges for imaging ⁹⁰Y bremsstrahlung emissions using SPECT/CT together with the associated publications

Energy range (keV)	References
105–195	Minarik et al. (2008)
100–160	Rong et al. (2012a)
90–125	Siman et al. (2016)
100–500[a]	Rong et al. (2012b)
50–250[a]	Elschot et al. (2013)

[a] The energy ranges indicate that energy-dependent corrections are necessary when using those energy windows.

REFERENCES

Ahmadzadehfar, H. et al. (2010). The significance of 99mTc-MAA SPECT/CT liver perfusion imaging in treatment planning for 90Y-microsphere selective internal radiation treatment. *J Nucl Med* 51:1206–1212.

Ahmadzadehfar, H. et al. (2012). The significance of bremsstrahlung SPECT/CT after yttrium-90 radioembolization treatment in the prediction of extrahepatic side effects. *Eur J Nucl Med Mol Imaging* 39:309–315.

Chiesa, C. et al. (2015). Radioembolization of hepatocarcinoma with 90Y glass microspheres: Development of an individualized treatment planning strategy based on dosimetry and radiobiology. *Eur J Nucl Med Mol Imaging* 42:1718–1738.

Elschot, M. et al. (2013). Quantitative Monte Carlo–based 90Y SPECT reconstruction. *J Nucl Med* 54:1557–1563.

Garin, E. et al. (2013). Boosted selective internal radiation therapy with 90Y-loaded glass microspheres (B-SIRT) for hepatocellular carcinoma patients: A new personalized promising concept. *Eur J Nucl Med Mol Imaging* 40:1057–1068.

Gulec, S.A., Mesoloras, G., Stabin, M. (2006). Dosimetric techniques in 90Y-microsphere therapy of liver cancer: The MIRD equations for dose calculations. *J Nucl Med* 47:1209–1211.

Hamami, M.E. et al. (2009). SPECT/CT with 99mTc-MAA in radioembolization with 90Y microspheres in patients with hepatocellular cancer. *J Nucl Med* 50:688–692.

Heard, S., Flux, G.D., Guy, M.J., Ott, R.J. (2004). Monte Carlo simulation of 90Y bremsstrahlung imaging. *IEEE Nucl Sci Symp Conf Record* 6:3579–3583.

Ho, S. et al. (1996). Partition model for estimating radiation doses from yttrium-90 microspheres in treating hepatic tumours. *Eur J Nucl Med* 23:947–952.

Ho, S. et al. (1997). Clinical evaluation of the partition model for estimating radiation doses from yttrium-90 microspheres in the treatment of hepatic cancer. *Eur J Nucl Med* 24:293–298.

Ilhan, H. et al. (2015). Predictive value of 99mTc-MAA SPECT for 90Y-labeled resin microsphere distribution in radioembolization of primary and secondary hepatic tumors. *J Nucl Med* 56:1654–1660.

Kappadath, S.C., Mikell, J., Mourtada, F., Mahvash, A. (2014). Voxel-based dosimetry and radiobiological modeling of HCC tumor response after 90Y microsphere therapy. *J Nucl Med* 55:151P.

Kao, Y.H. et al. (2013). Post-radioembolization yttrium-90 PET/CT—Part 2: Dose–response and tumor predictive dosimetry for resin microspheres. *EJNMMI Res* 3:57.

King, M.A., Coleman, M., Penney, B.C., Glick, S.J. (1991). Activity quantitation in SPECT: A study of prereconstruction Metz filtering and use of the scatter degradation factor. *Med Phys* 18(2):184–189.

King, M.A., Doherty, P.W., Schwinger, R.B., Penney, B.C. (1983). A Wiener filter for nuclear medicine images. *Med Phys* 10(6):876–880.

Minarik, D., Gleisner, K.S., Ljungberg, M. (2008). Evaluation of quantitative 90Y SPECT based on experimental phantom studies. *Phys Med Biol* 53:5689–5703.

Qian, W., Clarke, L.P. (1996). A restoration algorithm for P-32 and Y-90 bremsstrahlung emission nuclear imaging: A wavelet-neural network approach. *Med Phys* 23(8):1309–1323.

Rault, E. et al. (2010). Fast simulation of yttrium-90 bremsstrahlung photons with GATE. *Med Phys* 37:2943–2950.

Rong, X., Du, Y., Frey, E.C. (2012a). A method for energy window optimization for quantitative tasks that includes the effects of model-mismatch on bias: Application to Y-90 bremsstrahlung SPECT imaging. *PhysMed Biol* 57(12): 3711–3725.

Rong, X. et al. (2012b). Development and evaluation of an improved quantitative 90Y bremsstrahlung SPECT method. *Med Phys* 39:2346–2358.

Shen, S. et al. (1994a). Planar gamma camera imaging and quantitation of yttrium-90 bremsstrahlung. *J Nucl Med* 35:1381–1389.

Shen, S., DeNardo, G.L., DeNardo, S.J. (1994b). Quantitative bremsstrahlung imaging of yttrium-90 using a Wiener filter. *Med Phys* 21(9):1409–1417.

Siman, W., Mikell, J., Kappadath, S.C. (2016). Energy window-based scatter compensation method to improve image quality and quantification of 90Y bremsstrahlung imaging. *Med Phys* 43:.

Walrand, S. et al. (2014). Optimal design of anger camera for bremsstrahlung imaging: Monte Carlo evaluation. *Front Oncol* 4:149.

Wondergem, M. et al. (2013). 99mTc-macroaggregated albumin poorly predicts the intrahepatic distribution of 90Y resin microspheres in hepatic radioembolization. *J Nucl Med* 54:1294–1301.

11

Quantitative postradioembolization imaging using PET/CT

MARCO D'ARIENZO, LUCA FILIPPI, AND ORESTE BAGNI

11.1 INTRODUCTION

Over the last decade, the transarterial radioembolization with ^{90}Y microspheres has emerged as a mainstream treatment modality for inoperable hepatic malignancies. Radioembolization with ^{90}Y is a liver-directed therapy with the potential of delivering a high radiation dose directly to liver tumors in the form of localized β radiation, capable of sparing healthy tissue. Due to its long β particle range, ^{90}Y allows for reasonably uniform irradiation of large tumors commonly expressing heterogeneous perfusion and hypoxia.

^{90}Y is a pure β-emitter that decays to the ground state of ^{90}Zr with a maximum beta energy of 2279.8 (17) keV and a half-life of 2.6684 (13) days (Bé et al.,

2006), where the uncertainties associated with the beta energies are at $k = 1$. The average energy of β⁻ emissions from ^{90}Y is 926.7 (8) keV (Bé et al., 2006), with a mean tissue penetration of 2.5 mm and a maximum of 11 mm.

The reader is directed to refer Chapter 1 for a comprehensive overview of the clinical aspects of radioembolization with ^{90}Y microspheres. However, it is worth pointing out again that there are presently two clinically available ^{90}Y microsphere devices: glass microspheres, sold under the commercial name of TheraSpheres (TheraSphere®, Nordion Inc. for BTG International, Ottawa, ON, Canada), and resin microspheres, sold under the commercial name of SIR-Spheres (SIR-Spheres®, Sirtex Medical Limited, North Sydney, Australia).

Although ^{90}Y has been traditionally considered as a pure β–emitter, the decay of this radionuclide has a minor branch to the 0$^+$ first excited state of stable ^{90}Zr at 1.76 MeV, which is followed by a β$^+$/β$^-$ emission with an extremely small branching ratio. In recent years, a number of authors showed that ^{90}Y internal pair production can be imaged by positron emission tomography (PET), with results superior to that of bremsstrahlung single-photon emission computed tomography (SPECT) for evaluating the ^{90}Y microsphere biodistribution after therapeutic radioembolization. The major issue associated with ^{90}Y PET imaging is the extremely small emission probability of the β$^+$ particles. Therefore, in order to obtain acceptable image quality, a high concentration of the ^{90}Y is required. In liver radioembolization, typical injected activity ranges from one to several GBq and the total amount of radioactivity is concentrated within the liver or even in hepatic segments. Hence, high ^{90}Y concentrations may be obtained with this technique and the PET imaging of ^{90}Y has proven to be a viable imaging option, possibly leading to the accurate evaluation of *in vivo* dosimetry.

Quantitative imaging with ^{90}Y PET/CT is a relatively new imaging strategy and imaging capabilities are strongly related to PET scanner performance. Despite the number of phantom and patient studies that have proven feasibility and superior quality of quantitative imaging using PET/CT, to date no standardized imaging protocol has been proposed. This chapter gives insight into the major issues related to quantitative postradioembolization imaging techniques using PET/CT.

11.2 EMISSION OF β$^+$ PARTICLES VIA INTERNAL PAIR PRODUCTION IN THE 0$^+$–0$^+$ TRANSITION OF ^{90}Zr

In the past, there has been a great interest in electric monopole transitions (E0) in certain nuclei (e.g., ^{16}O, ^{40}Ca, ^{72}Ge, ^{90}Zr), occurring when there is no angular momentum change between initial and final nuclear states and no parity change. For spin-zero to spin-zero transitions, single gamma emission is strictly forbidden and three alternative processes may occur: (1) a transition may give

rise to the transfer of radiation energy to an atomic electron in the orbital cloud by internal conversion; (2) a transition may occur via electron–positron internal pair production (if the energy of the process is greater than $2m_e c^2$, i.e., 1.022 MeV, where m_e is the mass of the electron); (3) two-photon emission, which generally is associated with a negligibly small yield. In 1955, Ford first predicted an excited state (0$^+$ state) of ^{90}Zr (Ford, 1955) that was experimentally proven by Johnson et al. (1955) in the same period. Using a ^{90}Y source beta-decaying to ^{90}Zr, the authors discovered a transition at 1.76 MeV followed by a β$^+$/β$^-$ emission with an extremely small branching ratio (Figure 11.1). These authors also reported the probability of pair creation per beta decay as $w_p / w_\beta = (2 \pm 1) \times 10^{-4}$.

One year later, Greenberg and Deutsch (1956) evaluated the entity of internal pair creation by assessing the number of positron emissions relative to the main beta spectrum. From their experiment, the positron branching ratio was determined to be $w_p / w_\beta = (3.6 \pm 0.9) \times 10^{-5}$. Later, Langhoff and Hennies (1961) determined, with a scintillation coincidence spectrometer, the positron branching ratio for ^{90}Y to be $w_p / w_\beta = (3.4 \pm 0.4) \times 10^{-5}$. In recent years, Selwyn et al. (2007) used a high-purity germanium detector to determine the internal pair

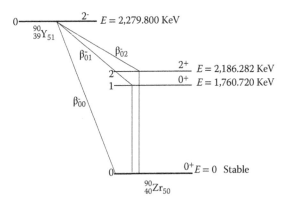

Figure 11.1 Decay scheme of ^{90}Y. ^{90}Y undergoes β$^-$ decay to ^{90}Zr with a half-life of 2.6684 (13) days and maximum beta energy of 2279.8 (17) keV. (From Bé et al., *Monographie BIPM-5.*, 2006.) In addition, ^{90}Y has a minor branch to the 0$^+$ first excited state of stable ^{90}Zr at 1.76 MeV, which is followed by a β$^+$/β$^-$ emission with an extremely small branching ratio, namely $(3.186 \pm 0.047) \cdot 10^{-5}$. (From Selwyn et al., *Appl Radiat Isot.*, 65, 2007.)

Table 11.1 Probability of pair creation per beta decay measured in previous and recent literature studies

References	w_p / w_β	Detector
Johnson et al. (1955)	$(2\pm1)\times10^{-4}$	NaI
Greenberg and Deutsch (1956)	$(3.6\pm0.9)\times10^{-5}$	NaI
Langhoff and Hennies (1961)	$(3.4\pm0.4)\times10^{-5}$	NaI
Selwyn et al. (2007)	$(3.186\pm0.047)\times10^{-5}$	HPGe

production branching ratio of the 0^+–0^+ transition of ^{90}Zr. The basic measurement technique required counting the gross number of gammas detected within a 511 keV (annihilation) peak and subtracting the bremsstrahlung continuum, environmental continuum, and environmental peak at 511 keV. The authors identified the branching ratio to be $w_p / w_\beta = (3.186\pm0.047)\times10^{-5}$ with a near 10-fold increase in precision compared with previous measurements. A detailed description of the experiments performed in the past to study the emission of β^+ particles via internal pair production in the 0^+–0^+ transition of ^{90}Zr can be found in D'Arienzo (2013). For decades, this transition of ^{90}Y was not exploited in nuclear medicine. Only recently, a number of authors have used the small positronic emission of ^{90}Y to obtain high-resolution PET images of ^{90}Y-labeled microspheres. Table 11.1 reports the experimental values for the internal pair production branching ratio of the $0^+/0^+$ transition of ^{90}Zr.

The accuracy of quantitative ^{90}Y PET/CT imaging is strongly related to the knowledge of the internal pair production branching ratio. It is essential in the near future that additional measurements of the β^+ emission probability will be performed in order to validate the intensity of the branching ratio and possibly reduce the related standard uncertainty (currently about 1.5%; Selwyn et al., 2007).

11.3 CURRENT PRACTICE IN POSTRADIOEMBOLIZATION QUANTITATIVE IMAGING WITH ^{90}Y

The first phantom study on ^{90}Y PET/CT imaging was performed in 2004 by Nickles et al. (2004). The authors assessed the ^{90}Y distribution on a Derenzo phantom using a micro-PET scanner equipped with bismuth germanate (BGO) crystals. The study showed the remarkable resolution and quantitative accuracy of positron tomography with ^{90}Y. The first clinical study based on the PET detection of ^{90}Y internal pair production was carried out by Lhommel et al. (2009). Using a Philips Gemini TF LYSO crystal ToF PET/CT scanner, the authors demonstrated the feasibility of *in vivo* imaging of ^{90}Y-labeled resin microspheres in a patient treated with radioembolization for colorectal liver metastases. With a 30-minute acquisition, the authors obtained high-resolution images of the radiopharmaceutical, clearly surpassing the image quality of traditional bremsstrahlung SPECT. Of note, the authors used a 2.5-mm copper ring to avoid detector saturation due to bremsstrahlung radiation. In a similar study in 2010, the same authors demonstrated the possibility of performing ^{90}Y ToF PET-based dosimetry *in vivo* in a patient receiving radioembolization with resin microspheres for metastatic liver disease (Lhommel et al., 2010). Their pioneering studies led to an explosion in scientific research in this field. Werner et al. (2010) performed similar phantom studies using a non-ToF PET/CT scanner (Biograph Hi-Rez 16, Siemens Healthcare, Knoxville, TN, USA) showing for the first time that ^{90}Y PET/CT scan was possible even without a ToF. Furthermore, they did not encounter issues with detector saturation. A similar study by Wissmeyer et al. (2011) demonstrated the feasibility of ^{90}Y imaging following liver radioembolization using a Philips Gemini ToF PET/MR hybrid scanner. Encouraging results with non-ToF scanners were obtained by subsequent studies performed by Gates et al. (2011), with a Siemens Biograph 40 with cerium-doped lutetium orthosilicate (LSO crystal) detectors, by D'Arienzo et al.

(2012) and Bagni et al. (2012) using a GE Discovery ST PET/CT scanner provided with BGO crystals and by Fourkal et al. (2013). Other recent studies confirmed the possibilities of perfoming ^{90}Y acquisitions with non-ToF PET scanners (Ng et al., 2013; Tapp et al., 2014). Nonetheless, extensive research to date has provided ample evidence that quantitative ^{90}Y PET imaging is likely to improve if ToF PET/CT scanners are used (van Elmbt et al., 2011; Willowson et al., 2012, 2015; Carlier et al., 2013, 2015; Attarwala et al., 2014; Martí-Climent et al., 2014). This is explained by the higher sensitivity and spatial resolution associated with ToF PET/CT systems, resulting in the potential to increase the image signal-to-noise ratio (Lewellen, 1998; Conti, 2009; Surti, 2014; Surti and Karp, 2016). Broadly speaking, image generation in non-ToF 3D PET imaging is based on the detection of coincident lines-of-response (LORs) at many angles. Tomographic images are then generated through using iterative reconstruction methods. The concept of ToF is based on the capability of fast electronics and scintillators to measure the difference in arrival times between two coincident photons, thus providing additional information related to the location of the decay of a coincidence event. While the reconstruction techniques associated with ToF are outside the scope of this chapter, the end result is improved image quality and quantification with shorter scan times in the case of traditional oncologic imaging with fludeoxyglucose (^{18}F-FDG). For ^{90}Y imaging, due to the low branching fraction for positron emission, ToF imaging allows for acceptable image quality and quantification.

The outcome of a recent multicenter comparison of quantitative ^{90}Y PET/CT for dosimetric evaluation after radioembolization with resin microspheres (quantitative uptake evaluation in SIR-Spheres therapy [QUEST] phantom study, Willowson et al., 2015) provided support for the hypothesis that ToF PET/CT scanners are capable of achieving higher accuracy in quantitative ^{90}Y imaging. The QUEST phantom study investigated and compared the quantitative accuracy of ^{90}Y imaging across different generation ToF and non-ToF PET/CT scanners. Quantitative accuracy was assessed by 47 international sites (for a total of 69 scanners, 37 with ToF mode) following a strict experimental and imaging protocol based on acquisitions of the NEMA 2007/IEC 2008 PET body phantom. Each center was asked to perform ^{90}Y PET acquisitions over a 7-day period (activity range: 0.5–3.0 GBq). Imaging consisted of two overlapping bed positions, each of 15- to 20-minute duration. Data were analyzed at a core laboratory for consistent processing. Based on these data, the authors concluded that GE Healthcare and Siemens ToF systems are suitable for quantitative postradioembolization imaging using PET/CT. Quantitative accuracy non-ToF scanners from GE Healthcare and Siemens was inferior to ToF systems. Greater deviations were obtained on the Philips systems at low count rates (Willowson et al., 2015).

It is worth noting that most of the current generation ToF PET scanners are equipped with lutitium-based crystals, e.g., cerium-doped LYSO or cerium-doped LSO. Lutitium-based compounds have desirable properties for ToF PET imaging due to the high detection efficiency (density ranging from 6.7 to 8.3 g/cm^3) and excellent temporal resolution (Conti et al., 2009). The major drawback with these scintillators is that the presence of the naturally occurring isotope ^{176}Lu, which gives rise to background count rates within the crystal. While this is not likely to be an issue with traditional ^{18}FDG-PET imaging due to the high positron abundance, it may hinder accurate quantification under conditions of low counts and high random fraction, as in ^{90}Y PET/CT imaging. This issue will be further discussed in Section 11.5.5.

Recent research has shown that quantitative postradioembolization imaging using PET/CT may pave the way for accurate *in vivo* dosimetry studies. In particular, a number of authors suggested that dose–volume histograms (DVH) obtained from ^{90}Y PET/CT images may be an important predictor of dose response (D'Arienzo et al., 2013; Kao et al., 2013, Fowler et al., 2016). In another study, Ng et al. (2013) extended the analysis of DVHs obtained from quantitative data to derive the biologically effective dose in patients treated with ^{90}Y radioembolization.

Despite encouraging results reported by the aforementioned studies, it is worth noting that quantitative postradioembolization imaging using ^{90}Y PET/CT is subject to inherent limitations, mostly attributable to acquisition at low count rates with a high random fraction. In addition to the presence of background radiation due to ^{176}Lu in lutitium-based crystals, several authors pointed out a possible reconstruction bias due to the production of negative pixels resulting from random-coincidence correction. In fact, commercial software packages truncate the negative pixel values before iterative reconstruction, thus leading to a significant positive bias in ^{90}Y PET imaging (Tapp et al., 2014; Walrand et al., 2015).

Of additional concern is the lower contrast recovery obtained with ^{90}Y PET imaging compared with ^{18}FDG PET, which remains at least partly unexplained. A deeper insight into this problem is provided in Section 11.5.10 (Willowson et al., 2012; Carlier et al., 2013, 2015).

Another aspect to consider is that many PET workstations do not offer ^{90}Y as a viable radionuclide choice for PET scans. In such a case, acquisitions can be performed selecting one of the positron emitters available in the software package and accounting for the correct branching ratio and half-life (Fourkal et al., 2013; Pasciak et al., 2014a; Carlier et al., 2015).

However, in order to set up a routine acquisition protocol for quantitative postradioembolization imaging using ^{90}Y PET/CT, preliminary phantom studies are required, aimed to characterize and optimize the performance of the scanner being used. Sections 11.4–11.6 offer insight into the procedures needed to setup a clinical protocol for quantitative postradioembolization imaging using PET/CT.

11.4 ACTIVITY MEASUREMENTS FOR QUANTITATIVE IMAGING

Accurate and precise activity measurements are an essential prerequisite of therapy with ^{90}Y microspheres. The International Atomic Energy Agency (IAEA) Basic Safety Standards states that (IAEA, 1996): "the calibration of sources used for medical exposure shall be traceable to a Standard dosimetry laboratory" and "unsealed sources for nuclear medicine procedures shall be calibrated in terms of activity of the radiopharmaceutical to be administrated, the activity being determined and recorded at the time of administration." Furthermore, performance characteristics of a PET scanner are generally assessed through dedicated phantom studies

where a calibrated amount of radiopharmaceutical is inserted into the phantom. Therefore, accurate activity measurements at a clinical level are the backbone of quantitative imaging as any uncertainties in the initially measured activity concentration will propagate to an uncertainty in final clinical quantification.

One of the major drawbacks of quantitative imaging with ^{90}Y microspheres is related to the quick microsphere sedimentation over time. Therefore, in order to have a homogeneous solution, phantom studies dedicated to ^{90}Y PET/CT imaging are generally performed using ^{90}Y chloride (^{90}YCl$_3$) instead of ^{90}Y microspheres. In this section, the issues related to activity measurements of ^{90}Y at a clinical level (both in form of ^{90}Y chloride and ^{90}Y microspheres) are described.

11.4.1 MEASUREMENTS OF ^{90}Y CHLORIDE

The instrument typically used to measure the administered activity to patients in nuclear medicine procedures is the radionuclide dose calibrator. Recent (Fenwick et al., 2014; Ferreira et al., 2016; Kossert et al., 2016) and previous (Woods et al., 1996) findings reported the difficulties of measuring ^{90}Y chloride and other beta emitters using clinically available ionization chambers. This is because dose calibrators available in medicine perform activity measurements of beta emitting radionuclides indirectly by detecting bremsstrahlung emissions. Bremsstrahlung production is highly dependent on the source material, its container, and the calibrator chamber wall. The ionization current also depends on the probability of electron detection within the chamber, which varies with electron energy and individual dose calibrator construction. Moreover, slight variations in the container wall thickness, solution volume, or location within the well can lead to an increase in the overall assay uncertainty when using the manufacturer supplied calibration factor, which is typically traceable to national standards.

A proper quality control program should be in place for any clinically used radionuclide dose calibrator, including a track record of calibrations and consistent daily quality assurance measurements. For activity measurements of ^{90}YCl$_3$ at a clinical level, it is expected that radionuclide dose calibrators provide accuracy within

±5% (at $k = 2$ level) (Gadd et al., 2006; AAPM, 2012). However, if the activity is determined by a National Metrology Institute, uncertainty on the activity concentration can be reduced significantly. Primary activity standards for ^{90}Y are widely available and expanded uncertainties ($k = 2$ or two standard deviations) of less than 1% can be achieved (Zimmerman and Ratel, 2005; Dezarn et al., 2011).

11.4.2 MEASUREMENTS OF ^{90}Y MICROSPHERE

Clinical measurement of beta particles emitted by ^{90}Y microspheres poses additional problems related to measurement geometry and homogeneity of the sample (microspheres in solution settle over time, with measurements affected as spheres settle). Currently, there is no traceability to national and international standards for ^{90}Y microspheres. As a consequence, there is an urgent need to establish a capability for accurately measuring the activity of ^{90}Y microspheres to traceable measurement standards.

A number of recent studies have been dedicated to the standardization of ^{90}Y and determination of calibration factors for ^{90}Y microspheres. The reader is referred to Lourenço et al. (2015), Ferreira et al. (2016), and Thiam et al. (2015). At present, BTG participates in the NIST Radioactivity Measurement Assurance Program (NRMAP) and NIST maintains a secondary measurement standard for the routine calibration of TheraSphere. SIR-Spheres do not have a NIST traceable calibration, however, activity measurements of ^{90}Y SIR-Spheres have been performed at the Australian Nuclear Science and Technology Organization (ANSTO) and the Australian Radiopharmaceuticals and Industrials (ARI) (Dezarn et al., 2011).

At present, each vial of resin microspheres is calibrated individually within a ±10% range. When new clinical centers start using ^{90}Y resin microspheres, the manufacturer provides an activity from the batch report for the first three microsphere vials shipped. This is needed to allow the clinical user to normalize the ionization chamber available on-site to the same calibration as the manufacturer. However, activity measurements of ^{90}Y microspheres at a clinical level remain quite critical. Recent literature findings showed that a total dose

delivery uncertainty on the order of 20% can be expected. One of the largest components of the total uncertainty is related to the initial activity measurement of the treatment dosage. (Dezarn et al., 2011).

At a clinical level, activity measurements of ^{90}Y are made using radionuclide calibrators traceable to a national standards laboratory for the geometry being measured. However, dose calibrator response to ^{90}Y radioembolization is far from being ideal due to the variation of microsphere distribution in the vial, sample geometry, and possible variations in the container wall thickness. Furthermore, a number of studies showed that in the absence of a well-defined local calibration procedure, variations in activity measurements on the order of 10% can occur (Dezarn and Kennedy, 2007a, 2007b). In particular, microsphere sedimentation is a major issue during the activity measurement. Activity measurements performed on a sample of settled microspheres will likely differ from activity measurements performed with the same activity of $^{90}YCl_3$ uniformly dispersed in the aqueous solution. The effect of microsphere sedimentation on the activity measurement is shown in Figure 11.2. The dashed curve (indicated with triangles) was obtained using a clinical dose calibrator available at IFO-Regina Elena Hospital, Rome, Italy. A vial containing a 5 mL solution of sterile water uniformly mixed with 3 GBq of ^{90}Y resin microspheres was measured after shaking the vial (microspheres resuspended in the sample). The same measurement (dotted curve, indicated with boxes) was repeated using the NPL-CRC ionization chamber radionuclide calibrator available at the Italian National Institute of Ionizing Radiation Metrology (INMRI). Finally, the solid line (indicated with circles) was obtained after the removal of the liquid buffer from the sample (only microspheres present in the vial, no sedimentation). This provides support for the hypothesis that a reliable and reproducible measurement should be performed after at least a 200-second waiting time to allow the complete sedimentation of microspheres. If the activity measurement is performed at the very beginning of the measurement, it is likely that the total activity would be underestimated, resulting in possible radiopharmaceutical overdose to the patient. This finding was recently confirmed by Ferreira et al. (2016).

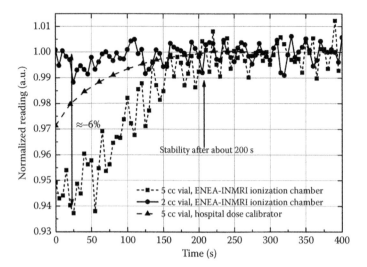

Figure 11.2 Measurements of ^{90}Y microspheres. The dashed line was obtained using the clinical dose calibrator (IFO-Regina Elena Hospital, Rome). A vial containing 5 mL solution of sterile water uniformly mixed with 3 GBq of 90Y-microspheres was measured after shaking the vial (microspheres resuspended in the sample). The same measurement (dotted line) was repeated using the NPL-CRC ionization chamber radionuclide calibrator available at ENEA-INMRI. The solid line was obtained after the removal of liquid buffer from the sample (only microspheres present in the vial, no sedimentation). ENEA-INMRI, Italian National agency for new technologies, Energy and sustainable economic development-National Institute of ionizing Radiation Metrology; IFO, Istituti Fisioterapici Ospitalieri; NPL-CRC, National Physics Laboratory-Capintec Radionuclide Calibrators.

11.5 PERFORMANCE CHARACTERISTICS OF PET SCANNERS

A major objective of PET studies with ^{90}Y PET/CT is to obtain good quality images and a significant level of detail despite the very low β^+ emission probability of ^{90}Y. Reaching this goal depends strongly on how well the scanner performs in the image formation. Several parameters associated with the scanner may have a significant impact on the image formation. A description of the major parameters possibly influencing quantitative post-radioembolization imaging using PET/CT is given below.

11.5.1 PET SENSITIVITY

The sensitivity of a PET device is defined as the rate of true coincidence events per unit time detected by the scanner per unit of radioactivity concentration present in a given source phantom. Sensitivity is generally expressed in counts per second, per

bequerel. Sensitivity depends on a number of physical and geometrical factors, among which are intrinsic efficiency, geometric efficiency, window settings, and the dead time of the system. It is worth noting that detection efficiency depends on the scintillation decay time, density, atomic number, and thickness of the detector material. The sensitivity of a scanner is highest at the center of the axial FOV and gradually decreases toward the periphery (Cherry et al., 2012). Sensitivity measurements are generally performed according to the NEMA NU 2-2007 procedure (NEMA, 2007). Since the emitted positrons annihilate with electrons to create a pair of γ-rays, a sufficient amount of material must surround the source to ensure annihilation. However, the surrounding material also attenuates the emitted γ-rays. Therefore, in order to arrive at an attenuation-free value of the sensitivity, successive measurements must be repeated with a uniform line source surrounded by a number of known absorbers. The sensitivity with no absorber is then extrapolated from these measurements. To this purpose, a National Electrical Manufacturers Association (NEMA)

PET Sensitivity Phantom™ can be used, consisting of a set of six concentric aluminum tubes.

PET acquisitions can be performed either in two-dimensional (2D) or in three-dimensional (3D) acquisition mode. In 2D acquisition mode, axial collimation is obtained by positioning tungsten septa between the detector rings. Septa are 1–2 mm thick and extend approximately 8–12 cm radially and are generally used to improve the image quality by reducing the detection of scatter and random coincidence events (Peller et al., 2012). However, most true coincidence events are also reduced. As a consequence, the PET scanner sensitivity will decrease when scans are performed in the 2D acquisition mode. To maximize the sensitivity of a PET scanner, septa can be removed with the aim to increase the number of detected events. This is known as 3D PET mode. In the 3D mode, the system sensitivity is significantly higher than the 2D acquisition mode, with a sensitivity greatest at the axial center of the system.

Sensitivity measurements for ^{90}Y PET studies can be performed following the NEMA NU 2-2007 procedure, as reported by multiple authors. In a recent work by Martí-Climent et al. (2014), the authors used a polyethylene tube (1 mm internal diameter) filled with a ^{90}Y solution of 879 MBq and inserted into the NEMA PET Sensitivity Phantom. Each acquisition lasted 300 s. The system sensitivity was determined by the ratio of true rate events without absorbing material, obtained by extrapolation, with respect to the activity of the source. The absolute sensitivity for ^{90}Y was 0.403 and 0.388 counts-per-second (cps)/MBq with the line source centered on the FOV and 10-cm off-center, respectively. Acquisitions were performed using a Biograph mCT-TrueV scanner with ToF. In another work, Bagni et al. (2012) measured the PET scanner sensitivity to ^{90}Y following the same procedure. PET images were acquired with a GE Discovery ST PET/CT scanner. The NEMA sensitivity phantom made of six fillable tubes was used, with the inner tube homogenously filled with water mixed with 300 MBq calibrated activity of ^{90}Y. The system sensitivity was measured at two radial positions, at a 0- and 10-cm radial offset from the center of the transaxial FOV. The measured absolute sensitivity of the scanner for detecting annihilation photons is 0.409 and 0.577 cps/MBq at 0 and 10 cm offset, respectively. In the same study, the authors assessed the sensitivity of the PET scanner during the 2D acquisition mode. The authors found that for 2D acquisitions the scanner sensitivity is about one order of magnitude lower than 3D mode (0.076 and 0.077 cps/MBq at 0 and 10 cm radial offset from the center of the transaxial FOV, respectively). The different impact of 2D and 3D acquisition modes in ^{90}Y PET imaging is shown in Figure 11.3. The 3D sensitivity values obtained by Bagni et al. (2012) can be compared with those found by Ng et al. (2013), 0.32 cps/MBq, and Werner et al. (2010), 0.72 cps/MBq using a GE Discovery STE PET/CT scanner

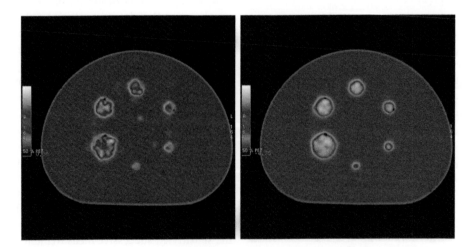

Figure 11.3 International Electrotechnical Commission (IEC) body phantom imaged with 2D (left) and 3D (right) acquisition modes. The image is highly contrasted as it is obtained with a lesion-to-background ratio of 30:1. With such concentrations, the difference in the signal-to-noise ratio between the 2D and the 3D mode can be clearly seen, with 2D acquisitions showing dishomogeneity areas and blurred margins. (From Bagni et al., *Nucl Med Commun.*, 33, 2012. With permission.)

and a Siemens Biograph 16 HiRez PET scanner, respectively.

Ultimately, it is worth stressing that because of the very low emission probability of positrons from ^{90}Y, sensitivity is greatly reduced compared with ^{18}F-FDG PET, which is several orders of magnitude greater both in 2D mode ($\approx$1–2 cps/kBq) and in 3D mode ($\approx$5–15 cps/kBq) (Peller et al., 2012). In particular, Werner et al. (2010) found that the sensitivity for ^{90}Y PET is reduced by a factor of 3.4e-5 in comparison with ^{18}F PET. Similar results were confirmed by D'Arienzo et al. (2012) who found a system sensitivity to ^{90}Y ($\approx$0.5 cps/MBq) about four orders of magnitude lower than that of ^{18}F ($\approx$9 cps/kBq). Because of the decreased sensitivity of ^{90}Y PET compared with ^{18}F, 2D imaging mode is generally not compatible with ^{90}Y PET imaging.

11.5.2 PET ABSOLUTE ACTIVITY CALIBRATION

PET absolute activity calibration is also often referred to as "well counter calibration." Absolute activity calibration factors are required to convert pixel values into a measure of absolute activity per voxel. Following the absolute activity calibration, the voxel intensity in any ^{90}Y PET image is divided by the calibration factor to obtain calibrated images in terms of kBq/cm^3. A standard source configuration is generally recommended consisting of a phantom containing a known and homogeneous activity concentration. The latter can be measured with the on-site dose calibrator. Traceability to national standards laboratory for the geometry being measured is essential for activity determination and for uncertainty reduction. However, if activity is determined by a national laboratory, the final uncertainty can be reduced significantly. Calibration to absolute radioactivity concentration is generally accomplished by scanning a cylinder or a phantom with large volume. The calibration factor, f, is defined as (Cherry et al., 2012)

$$ f = \frac{\text{counts per pixel}}{\text{activity concentration (kBq} / \text{cm}^3)} $$

This procedure is well validated for PET imaging with ^{18}F. However, such a straightforward calibration method is not applicable to ^{90}Y microspheres as most scanners do not support ^{90}Y as a viable radionuclide option. Therefore, other radionuclides are generally used for PET absolute activity calibration. In order to obtain ^{90}Y activity concentration in terms of kBq/cm^3, counts need to be ultimately rescaled by the ratio of the β^+ emission probability of the used radionuclide ($w_{\beta^+}^X$) and that of ^{90}Y ($w_{\beta^+}^{^{90}Y}$). This procedure allows a new calibration factor, $f = w_{\beta^+}^{^{90}Y} / w_{\beta^+}^X$, to be obtained for correct ^{90}Y quantification. A number of surrogate radionuclides have been used in published literature (^{22}Na, ^{86}Y, ^{68}Ge, ^{18}F), with ^{22}Na being the most straightforward choice. Of course, an adjusted decay constant must be introduced in order to account for the different half-life of the selected radionuclide, and that of ^{90}Y. An extensive description of this calibration procedure is provided in Pasciak et al. (2014a) along with a list of adjusted decay constants for selected radionuclides. Of note, a recent study by Fourkal et al. (2013) identified the measurement of the calibration factor as being one of the major sources of uncertainties in the dose measurements (together with the uncertainty related to the positive bias due to the intrinsic radioactivity of scanner's crystals). In the study, the relative standard deviation of the calibration factor was found to be $\approx$12%. Therefore, it is expected, in the near future, that more accurate measurements of the ^{90}Y β^+ branching ratio will be published.

11.5.3 SPATIAL RESOLUTION

The spatial resolution of a system represents its ability to distinguish between two points after image reconstruction. The NEMA guideline NU 2-2007 (NEMA, 2007) describes a standard procedure for the measurement of the spatial resolution. According to the NEMA procedure, spatial resolution measurements are performed by imaging point sources in air, and then reconstructing images with no smoothing filters (i.e., using a ramp filter). Spatial resolution has to be measured in the axial slice and in the transverse slice, the latter both radially and tangentially. Measurements are performed with a point source consisting of a small quantity of concentrated activity inside a glass capillary with an inside diameter of 1 mm (or less) and an outside diameter of less than 2 mm. Reconstruction should be performed using filtered backprojection with no smoothing and pixel size should be set below one-third of the expected full width at half maximum (FWHM) in all three dimensions. The spatial resolution in each direction is then determined in terms of FWHM of one-dimensional response functions of the point source. Although spatial resolution measurements should be performed in nonrealistic clinical condition (e.g., absence of scatter, attenuation and smoothing filters), it provides a best-case comparison among scanners, indicating the highest achievable performance.

The spatial resolution in ^{90}Y-PET/CT imaging has been assessed by a number of authors with different PET scanners. However, it is worth noting that despite the existence of the NEMA standard procedure for the assessment of spatial resolution, it was not used in all literature studies dedicated to ^{90}Y quantitative imaging.

Werner et al. (2010) found a resolution of 5.2 ± 0.6 mm (336 × 336 matrix, 8 iterations, 16 subsets) and 7.8 ± 0.5 mm (128 × 128 matrix, 4 iterations, 8 subsets) using a non-ToF Siemens Biograph PET scanner with LSO detector elements. In a recent study by Martí-Climent et al. (2014), the spatial resolution under a number of conditions using a Siemens Biograph mCT-TrueV ToF scanner with LSO crystals was measured. The authors obtained spatial resolution values in the range 2.2–12.1 mm. In another study, D'Arienzo et al. (2012) performed spatial resolution measurements using a non-ToF BGO PET. With an acquisition matrix of 256 × 256, at a 1 cm radius, ^{90}Y PET transverse and axial spatial resolutions were found to be 5.8 ± 0.9 and 5.0 ± 0.6 mm, respectively. When the source was placed at a 10 cm radius, transverse radial, transverse tangential, and axial resolutions were found to be 5.5 ± 0.9, 5.7 ± 0.9, and 7.3 ± 1.0 mm, respectively. Similar values were obtained by Kao et al. (2013) using a LYSO GE Discovery 690 (10 mm) and by van Elmbt et al. (2011) using different PET scanners, i.e., Philips Gemini with GSO crystals (10 mm), Philips Gemini TF with LYSO crystals (9.3 mm), and Siemens Ecat Exact HR with BGO detector (10.6 mm). A summary of the spatial resolution values obtained in published literature is given in Table 11.2.

11.5.4 RECOVERY COEFFICIENTS

The use of recovery coefficients (RCs) is a simple and widely used tool for the correction of partial volume effects (PVE). RCs are defined as the ratio of measured activity to true activity in the object, from simple objects of known geometry (e.g., spheres). Under clinical conditions, the real activity value can be then obtained by dividing the measured activity in the region of interest by the RC. RCs are therefore a function of the object geometry, size, object-to-background activity concentration ratio, and position in field of view. RCs are typically assessed using a NEMA IEC image quality body phantom consisting of a water-filled cavity with six spherical inserts suspended by plastic rods of volumes: 0.5, 1.2, 2.6, 5.6, 11.5, and 26.5 mL (inner diameters of 10, 13, 17, 22, 28, and 37 mm). Measurements of RCs in ^{90}Y PET/CT imaging have been described by several authors. As a general rule, partial volume effects were evident in all but the largest NEMA sphere (the lower the object size, the lower the RC). All ^{90}Y PET studies show that RCs obtained with ^{90}Y PET/CT imaging are poorer than those obtained with ^{18}F PET/CT imaging (Werner et al., 2010; D'Arienzo et al., 2012; Willowson et al., 2012). Furthermore, ToF-PET scanners are likely to improve contrast of hot spheres and increase RCs (Willowson et al., 2012). The recent QUEST study (Willowson et al., 2015) aimed to investigate and compare the quantitative accuracy of ^{90}Y imaging across different PET/CT scanners. The results of the study clearly confirmed that partial volume effects dominate spheres of diameter <20 mm when ^{90}Y PET quantitative imaging is performed with current generation ToF scanners from GE, Philips, and Siemens (Figure 11.4). For spheres >20 mm in diameter, activity concentrations were consistently underestimated by about 20%. Of note, non-ToF scanners from GE Healthcare and Siemens were capable of producing accurate measures, but with inferior quantitative recovery compared with ToF systems. Recovery of activity concentration measured in the hot spheres on day 0 of imaging is shown in Figure 11.5. In particular for a 37-mm-diameter object average underestimates of −34% and −27% were found for GE Healthcare and Siemens scanners, respectively (Willowson et al., 2015).

In the presence of hot background, image quality can be assessed using the hot contrast recovery coefficient (CRC$_{hot}$), as reported by the NEMA guidelines (Daube-Witherspoon et al., 2002):

$$CRC_{hot} = \frac{(C_{hot} - C_{bkgd}) - 1}{(a_{hot} - a_{bkgd}) - 1}$$

Table 11.2 Image reconstruction parameters and spatial resolution values in quantitative postradioembolization imaging studies using ^{90}Y PET/CT

Reference	^{90}Y microspheres	Scanner manufacturer	Detector crystal	Acquisition mode	Reconstruction	Resolution
Lhommel et al. (2010)	Resin microspheres	Gemini Philips	LYSO	ToF	2 iterations, 33 substeps	–
Werner et al. (2010)	Resin microspheres	Biograph Hi-Rez 16 Siemens	LSO	Non ToF	8 iterations, 16 subsets and 4 iterations, 8 subsets	6.4 mm
Gates et al. (2011)	Glass microspheres	Biograph 40 Siemens	LSO	Non-ToF	3 iteration, 21 subsets	2.5–4 mm
Wissmeyer et al. (2011)	Glass microspheres	Philips Gemini PET/MR	LYSO	ToF	3 iterations, 33 subsets	–
Bagni et al. (2011)	Resin microspheres	Discovery ST GE	BGO	Non-ToF	2 iterations, 15 subsets	6.3 mm
Carlier et al. (2013)	Resin and glass microspheres	Biograph mCT 40 Siemens	LSO	ToF and Non-ToF	1 or 3 iterations, 21 or 24 subsets	–
Elschot et al. (2013)	Resin microspheres	Biograph mCT Siemens	LSO	ToF	3 iterations, 21 or 24 subsets	–
Kao et al. (2012)	Resin microspheres	Biograph WO Siemens	LSO	Non-ToF	2 iterations, 8 subsets	–
Kao et al. (2013)	Resin microspheres	Discovery 690 GE	LYSO	ToF	3 iterations, 18 subsets	10–12 mm
van Elmbt et al. (2011)	Resin microspheres	Philips Gemini TF	LYSO	ToF	3 iterations, 8 substeps	9.3 mm
van Elmbt et al. (2011)	Resin microspheres	Philips Gemini Power16	GSO	Non-ToF	3 iterations, 8 substeps	10 mm
van Elmbt et al. (2011)	Resin microspheres	Siemens Ecat Exact HRb	BGO	Non-ToF	3 iterations, 8 substeps	10.6 mm
Martí-Climent et al. (2014)	Resin microspheres	Biograph mCT-TrueV	LSO	ToF	1–3 iterations, 21–24 substeps	2.2–12.1 mm

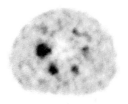

Figure 11.4 Acquisitions of the NEMA phantom used in the QUEST phantom study (Willowson et al., 2015) by Oxford University Hospitals NHS Foundation Trust. Images were acquired (from left to right) at days 0, 3, 5, 7 using a GE Discovery 710 ToF system. Total phantom activity at D0 was about 4.5 GBq. Reconstruction was performed with Q. Clear reconstruction algorithm (beta 4000), matrix size 246 × 256. (Courtesy of Lisa Rowley, Oxford University Hospitals NHS Foundation Trust.)

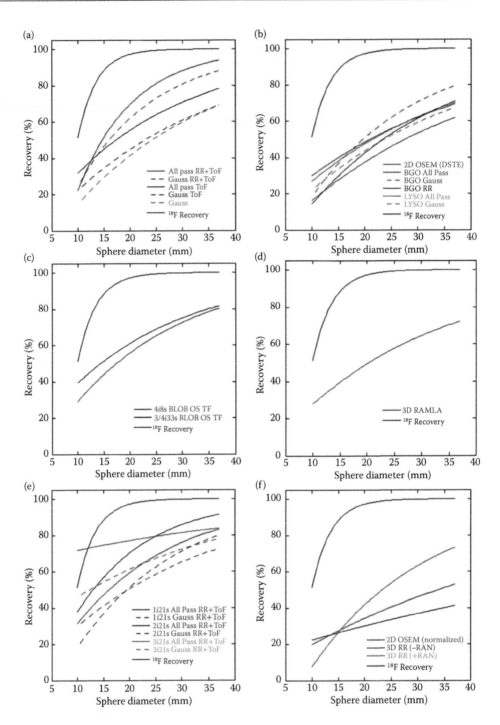

Figure 11.5 Recovery of activity concentration measured in the hot spheres from the QUEST phantom study. Lines of best fit ($y = a + bx$) for recovered concentrations in the largest hot sphere at different concentrations for (a) GE Healthcare ToF systems, (b) GE Healthcare non-ToF systems, (c) Philips ToF systems, (d) Philips non-ToF systems, (e) Siemens ToF systems, (f) Siemens non-ToF systems (where +RAN and -RAN correspond to data acquired in "PROMTS + RANDOMS" and "NETTRUES" mode, respectively, and where -RAN was normalized for analysis). Figure reproduced from Willowson et al. (2015) under the terms of the Creative Common Attribution 4.0 International License (https://creative-commons.org/licenses/by/4.0/).

where C_{hot} and C_{bkgd} are the average of the counts measured in the hot sphere region of interest (ROI) and the average of the counts in all background ROIs, respectively, a_{hot} is the true activity concentration in the hot sphere, and a_{bkgd} is the true activity concentration in the background. ROIs with diameters equal to the physical inner diameters of the spheres must be drawn on the spheres and throughout the background. A number of authors have assessed CRC_{hot} to quantify the image quality in ^{90}Y PET acquisitions. Willowson et al. (2012) found that ToF acquisition mode improves the recovered contrast of hot spheres, an effect which increases with decreasing sphere diameter. Van Elmbt et al. (2011) corroborated these findings pointing out that ToF led to a greater improvement in hot contrast for smaller diameters. Furthermore, if ToF reconstructions are used, the hot contrast recovery coefficient remains almost constant even for shorter acquisitions.

Finally, it is worth noting that contrast recovery obtained with ^{90}Y imaging is lower than those obtained with ^{18}F. Van Elmbt et al. (2011) analyzed PET/CT acquisitions performed with phantoms filled both with ^{90}Y and ^{18}F. He found that despite scatter fractions being approximately the same, contrast recovery with ^{18}F is superior to ^{90}Y. This is partly still unexplained. However, some explanations were proposed among which was (1) the effect of high image noise on ordered-subset expectation maximization (OSEM) image reconstruction algorithm. Negative pixel values are truncated before iterative reconstruction leading to a significant positive bias in ^{90}Y PET imaging. Another explanation was (2) possible coincidences arising from pair production in the scanner crystals by the bremsstrahlung x-rays above 1.022 MeV. Due to the large atomic numbers of the crystals, the probability of pair production in the PET crystals is high (Van Elmbt et al., 2011). These pair-production events may have an impact on random corrections. However, this issue has not yet been fully addressed and needs to be further investigated in the future.

11.5.5 DEGRADING FACTORS IN ^{90}Y PET QUANTITATIVE IMAGING

It is well known that PET imaging of ^{90}Y provides poor quality images compared with ^{18}F imaging.

The major issues in quantitative postradioembolization imaging using PET/CT are related to low counting statistics and a high random fraction. These effects are due to a number of factors attributable both to the physics of the decay and to the detection crystal of a PET scanner.

Regarding the limitations related to the physics of the decay, ^{90}Y PET images are inherently noisy due to the extremely small positron emission branching fraction of ^{90}Y, resulting in low true coincidence count rate. This is especially true in nontarget anatomical regions, where the ^{90}Y activity concentration will be much lower than in treated liver tissue. As a consequence, longer scan times than a traditional positron-emitting radioisotope are generally required to obtain satisfactory quality images (by comparison, ^{18}F has a branching fraction of 967 per 1000 decays). As an example, acquisition times in ^{18}FDG PET/CT imaging are 2–5 minutes per bed position depending on the amount of injected radioactivity, body mass index, and scanner sensitivity. While longer scans improve image quality, acquisition times for ^{90}Y PET patient studies can't be increased arbitrarily and a trade-off between patient comfort and image quality is required. Current literature studies report acquisition times in the range from 10 minutes per bed position using a ToF scanner (Tapp et al., 2014) to 40 minutes per bed position with non-ToF scanner (Werner et al., 2010). The imaging protocol proposed in the recent QUEST study consisted of two overlapping bed positions (to mitigate the triangular axial sensitivity profile of the scanner) each of 15–20 minutes duration, in 3D mode.

The high-energy primary β^- particle emission of ^{90}Y decay generates a continuous bremsstrahlung radiation spectrum (Stabin et al., 1994) that has the potential to degrade both the image quality and the quantitative accuracy of ^{90}Y PET imaging (maximum energy of β^- emissions and, therefore, bremsstrahlung photons from ^{90}Y is 2.28 MeV). In particular, the large flux of bremsstrahlung photons results in a singles count rate largely exceeding the true coincidence count rate. The highest bremsstrahlung yield is at energies below 20 keV, which is significantly attenuated by the patient. However, a significant portion of higher energy bremsstrahlung photons are emitted within the acceptance window of PET scanners and has the potential to saturate the PET detectors. In a previous study, Lhommel et al. (2009) used a homemade 2.5-mm-thick copper ring to reduce bremsstrahlung radiation emerging from a patient administered with 1.3 GBq of ^{90}Y. Of note, typical administered activities may be in the order of several GBq, therefore, possible detector saturation may not be ignored *a priori*.

Other degrading factors that significantly contribute randomly are (1) the detection of random coincidences from two bremsstrahlung photons emitted simultaneously; (2) detection of coincidences from an annihilation 511 keV photon and a bremsstrahlung photon emitted simultaneously; and (3) the presence of the naturally occurring isotope ^{176}Lu within the detection crystals (i.e., cerium-doped LYSO or cerium-doped LSO) of PET imaging systems that may generate background count rates. There is ample evidence in the literature that bremsstrahlung radiation together with the LSO background radiation greatly increases the random fraction in quantitative ^{90}Y PET imaging.

Lutetium (Lu)-based scintillators such as LSO and LYSO are widely used in current generation PET detectors (especially in ToF scanners) due to their relatively high stopping power for 511 keV gamma rays, high light yield, and short decay time. However, 2.6% of naturally occurring Lu is the isotope ^{176}Lu ($T_{1/2} \sim 3.6 \times 10^{10}$ year), a long-lived radioactive element undergoing beta decay (maximum energy 596 keV) and three major simultaneous gamma decays at energies 88 keV (15%), 202 keV (78%), and 307 keV (94%) (Browne and Junde, 1998). While the presence of ^{176}Lu is generally not an issue with traditional PET radionuclides due to the high true coincidence count rate (e.g., ^{18}F), this phenomenon is likely to introduce nonnegligible random events during ^{90}Y PET acquisition, thereby affecting system performance.

In a PET detector, the β particles emitted from Lu-based crystals, given the short range, deliver most of their energy in the same crystal. On the other hand, γ-rays can be detected not only in the same crystal where they are generated but also in other detector elements. Therefore, the background radiation generated by the radionuclide ^{176}Lu can contribute to the amount of random and true coincidences. The most likely event is the detection of coincidences originating from the detection of the β$^-$ in the crystal in which the ^{176}Lu decay occurred and one of the prompt γ-rays in another detector crystal (Goertzen et al., 2009). As a general rule, in

order to assess the impact of natural ^{176}Lu radioactivity on the image quality, long acquisition with no radioactivity present in the field of view is performed to determine the ^{176}Lu background count rate.

One last confounding factor in ^{90}Y PET quantitative imaging is attributable to scatter correction. At very low counts, PET images are very noisy and the resulting scatter correction might lead to heavy under- or overestimation of the scatter contribution.

To conclude, bremsstrahlung photons and prompt gammas are likely to result in a very high random fraction in imaging ^{90}Y on the order of 80% (Willowson et al., 2015) or even higher (Carlier et al., 2015) compared with a typical FDG scan of 30%–40%. The combination of high random fraction, extremely low true coincidences, and problematic scatter modeling for low counting statistics results in very noisy true coincidence sinograms. In addition, a well-known problem in PET imaging is the introduction of a positive bias after correction for random events (Ahn and Fessler, 2004; Rahmim et al., 2005; Li and Leahy, 2006). The most common method of correcting for random coincidences is the real-time or offline subtraction of a delayed coincidence time window from the prompt signal. In a scenario with low true coincidences and high random fraction (as in ^{90}Y PET imaging), negative sinogram ray-sum values can be produced. These negative sinogram values are often truncated in commercial software packages before iterative reconstruction (i.e., become zeroed) thus introducing a positive bias. This bias does not have a significant impact for clinical imaging with ^{18}F, but may become important in ^{90}Y PET imaging. This bias was observed by several authors (Tapp et al., 2014; Carlier et al., 2015) and it is possibly responsible for hot contrast recovery obtained with ^{90}Y being inferior to that obtained with ^{18}F.

It is worth mentioning that presently a number of literature studies provided ample evidence that neither detector saturation (D'Arienzo et al., 2012; Bagni et al., 2012; Carlier et al., 2013) nor intrinsic natural ^{176}Lu (Carlier et al., 2013) radioactivity represents a major issue in ^{90}Y PET quantification at activity concentrations commonly encountered in liver radioembolization. In particular, the presence of natural ^{176}Lu radioactivity produces a measurable but not limiting contribution. Carlier et al. (2013) suggested that emissions from the radionuclide

[176]Lu may significantly contribute to random coincidences when ^{90}Y radioactivity concentration is below 1 MBq mL^{-1} in the presence of high tumor to background activity concentration ratios.

An extensive study on the limitations and the accuracy achievable under conditions of low counts and high random fraction can be found in Carlier et al. (2015).

11.5.6 IMAGE RECONSTRUCTION

Iterative reconstruction has become the standard for routine clinical PET imaging. However, iterative algorithms are resource intensive, especially for ToF data, and OSEM algorithms are, therefore, commonly used to accelerate reconstruction. As a general rule, the image noise in the reconstructed images increases as the number of iterations proceeds (Figure 11.6). On the other hand, image quality is also degraded when OSEM is used with a large number of subsets. As a consequence, there is a tradeoff between the number of iterations/subsets and reconstructed image quality.

The impact of image reconstruction on ^{90}Y PET quantification has been widely investigated by a number of literature studies with varying reported success. The general consensus is that the best reconstruction technique will depend on the scanner and the acquisition modality (ToF or non-ToF). Both Willowson et al. (2012) and Carlier et al. (2013) found that one iteration provided the most accurate quantification on a ToF Siemens BioGraph mCT. On the other hand Bagni et al. (2012) and D'Arienzo et al. (2012) performed acquisitions on a BGO GE discovery ST scanner using two and three iterations, respectively. In another study van Elmbt et al. (2011) used three iterations (eight substeps + Gaussian filter) both on a Philips Gemini Power 16 and a Siemens Ecat Exact HR, while Lhommel et al. (2009) and Werner et al. (2010) used two iterations (33 substeps, ToF, RR) and eight iterations (16 substeps) on a Philips Gemini TF and a Siemens BioGraph Hi-Rez 16, respectively. A summary of the image reconstruction parameters obtained in published literature is given in Table 11.2.

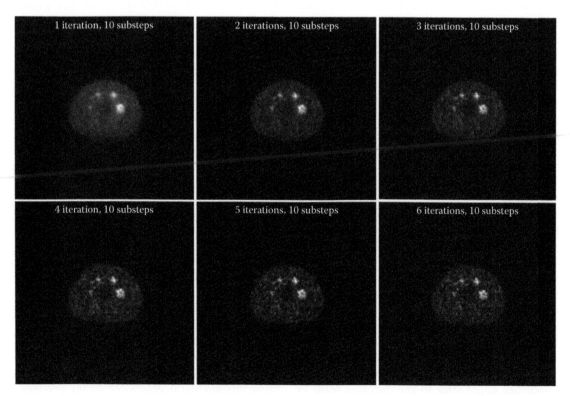

Figure 11.6 Effect of number of iterations on the NEMA phantom with six fillable spherical inserts. Image noise in the reconstructed images increases as the number of iterations proceeds. Images were obtained using a BGO GE Discovery ST PET scanner.

The major outcome of the QUEST phantom study is that ^{90}Y imaging capability appears to be optimal for Siemens systems using two iterations and 21 subsets with ToF and resolution recovery (RR) and an all-pass filter. For GE Healthcare systems, the use of an all-pass filter in conjunction with RR and ToF (2 iterations and 24 substeps) provided the best quantification results. Regarding non-ToF generation scanners, measures of background provided average deviations to within 1%, 5%, and 2% for GE Healthcare (all-pass filter, RR + ToF), Philips (4i8s ToF), and Siemens (2i21s all-pass filter, RR + ToF) ToF systems, respectively.

11.6 PREPARATION OF A CALIBRATED PHANTOM FOR ^{90}Y-PET STUDIES

As previously mentioned, most quantitative imaging studies make use of phantoms and require very accurate knowledge of both the activity and the volume of liquid solution inserted into the phantoms. The general purpose of quantitative imaging studies with calibrated phantoms is (1) to assess uncertainties in the imaging process; (2) to quantify accuracy of correction factors and reconstruction algorithms; and (3) to evaluate scanner performance over time. A recent study (Sunderland et al., 2015) demonstrated that technical error in phantom filling is one of the primary reasons for exclusion of PET/CT scanners from clinical trials. Section 11.4 was dedicated to the importance of accurate ^{90}Y activity measurements. Here, we describe further issues that need to be addressed for an optimal phantom preparation dedicated to quantitative analysis with ^{90}Y PET.

Adsorption of radionuclides on the inner walls of plastic phantoms may lead to inhomogeneous radionuclide distribution that can negatively affect quantitative imaging studies (Park et al., 2008). Therefore, the preparation of a carrier solution is recommended. The use of tap water should be avoided as minerals and other chemical impurities might stick to the phantom walls or combine with the radiopharmaceutical changing the radionuclide distribution. Carrier solutions should always be prepared using laboratory grade chemicals and purified water. Therefore, it is important to ensure that a favorable chemistry is used throughout the calibration procedure in order to have a uniform and stable solution. For ^{90}Y PET studies, ^{90}YCl$_3$ in an aqueous solution of 0.1 mol dm^{-3} hydrochloric acid also containing inactive yttrium at a concentration of about 50 µg g^{-1} can be used as a carrier solution. Alternatively, diethylenetriaminepentaacetic acid (DTPA) or ethylenediaminetetraacetic acid (EDTA) at a concentration of about 50 µg·g^{-1} can be used to prevent radioactive ^{90}Y from sticking to the phantom walls and to guarantee a homogeneous radionuclide solution.

It is recommended that all containers be prefilled with carrier 12 hours prior to addition of radioactive ^{90}YCl$_3$. This will help to "seal" the surface and reduce sticking or plating of activity. All containers should be emptied, dried, and the used carrier discarded before activity is added.

Activity concentration should be determined using a radionuclide dose calibrator traceable to a national standards laboratory for the geometry being measured, as previously discussed. Uncertainty for a typical well-calibrated field instrument can be expected to be in the region of ~5% for low-energy gamma emitters (<100 keV) and pure beta emitters, such as ^{90}Y. It is worth noting that if activity is determined by a national laboratory then this uncertainty can be reduced significantly.

As a general rule, preparation of a stock solution is recommended. Radioactive ^{90}Y provided by a supplier should be diluted using the carrier solution to desired volume and concentration. Activity concentration should be determined by measuring an aliquot of the stock solution in terms of activity per unit mass (or volume). This can then be used to determine the activity of all subsequent sources produced from this stock solution.

Filling of the phantoms should be performed using a calibrated (preferably four decimal places) analytic scientific scale and with routine double or triple weighing of the sources. The overall uncertainty in the activity concentration determined using this method is dependent on the precision of the scale being used as well as the accuracy of the method used to determine the activity concentration of the solution. Radioactivity should be dispensed using calibrated pipette devices or syringes and ensuring that no air bubbles remain in the phantom. Either way, it is suggested to flush the needle to remove all activity from the syringe or the pipette. Again the uncertainty is dependent on the precision of the volume measurement as well as the activity determined.

If large background volumes are used for calibration purposes, the phantom can be filled with nonradioactive water to measure the fillable volume (and to confirm the phantom is watertight with no leaks). When filling large phantom volumes, a funnel should be used. When the phantom is nearly full, the funnel can be removed and a syringe used to complete the filling process thereby preventing the spillage of radioactive water from the background compartment.

11.7 DISCUSSION AND SUMMARY

PET imaging of ^{90}Y microspheres is a rapidly evolving field whose features and attributes have not yet been fully addressed. At present, the general consensus is that accurate quantification is possible on a variety of PET scanner models, with or without ToF, although ToF is likely to improve the accuracy at lower activity concentrations. Quantitative accuracy has been investigated by a number of authors using different phantoms with varying reported success. Regardless of the scanner used, partial volume effects play a major role in activity quantification showing a steady decline for spheres with diameter below 37 mm. It is expected that quantification in small hot regions may be underestimated with all current generation scanners to a consistent degree of 15%–20% for a 37-mm-diameter object (Willowson et al., 2015). Higher quantification accuracy can be achieved for large regions uniformly filled with ^{90}Y. Non-ToF generation GE Healthcare and Siemens scanners are capable of recovering activity concentrations on the order of 300 kBq/mL within 2% and 9% of true values, while the deviation for ToF GE Healthcare and Siemens scanners can be expected to be in the range 1%–5% (Willowson et al., 2015). Furthermore, the recovery of activity concentration in hot spheres is inferior to that obtained with ^{18}F, probably due to effects of image noise on the OSEM reconstruction algorithms, which present an inherent nonnegativity constraint. While the intrinsic radioactivity present in lutetium-based crystals is a potential limitation, it has been shown to have a negligible impact at the high activity concentrations typically found in ^{90}Y radioembolization.

Regarding the acquisition time, a 40-minute acquisition is recommended in a clinical scenario, acquired as two bed positions, 20 minutes each. Figure 11.7 shows different acquisitions performed with a GE Discovery ST BGO scanner 2 hours after the microsphere administration (single bed acquisition, 20 minutes). An excellent match between microsphere accumulation and tumor areas was observed.

Last, the accuracy of quantitative ^{90}Y PET/CT imaging is strongly related to the measurement precision of the internal pair production branching ratio. At present, the β^+ emission probability of ^{90}Y is known with an uncertainty around 1.5% ($(3.186\pm0.047)\times10^{-5}$; Selwyn et al., 2007). A reduction of such uncertainty will lead to a reduction of the overall uncertainty in the activity quantification. This is especially important in light of the current uncertainty in the clinical measurement of ^{90}Y activity using a dose calibrator. In the near future, new experimental measurements of the internal pair production branching ratio are desirable in order to achieve more accurate quantification in postradioembolization imaging using PET/CT.

11.8 CONCLUSIONS

Currently, a number of clinics have developed their own quantitative imaging procedures using ^{90}Y PET/CT. Very often there is a lack of harmonization of these procedures, most likely because quantitative imaging with ^{90}Y PET/CT is a relatively new imaging strategy and further because imaging capabilities are strongly related to PET scanner performance. Ultimately, it is worth stressing that scanner performance is optimized for ^{18}F imaging and dedicated imaging protocols are required for accurate ^{90}Y PET/CT studies. Careful choice of reconstruction

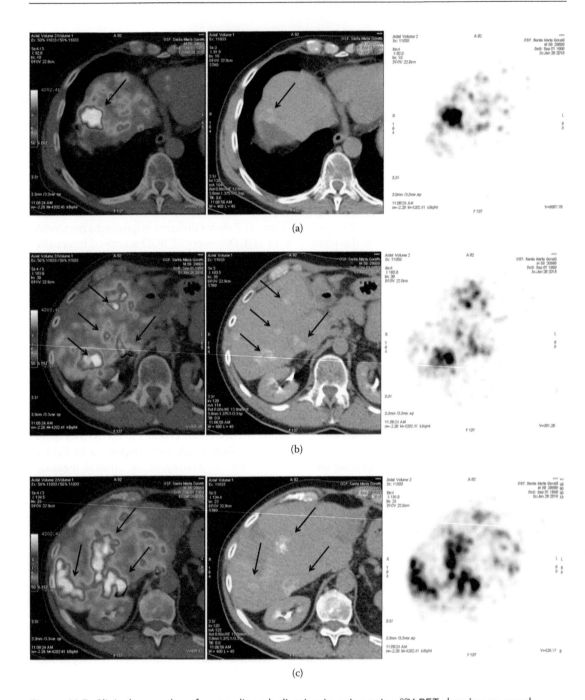

(a)

(b)

(c)

Figure 11.7 Clinical examples of postradioembolization imaging using ^{90}Y-PET showing an excellent match between microsphere accumulation on hepatic lesions. (Left) ^{90}Y-PET fused with CT data. (Central) CT of the liver parenchyma clearly showing tumor regions. (Right) ^{90}Y-PET. Figure parts (a) through (c) show different axial levels of the same patient.

parameters has the potential to increase the quantitative accuracy. However, the final outcome will depend on the scanner and reconstruction software.

Comprehensive guidance has yet to be presented in this field and there is no doubt that an internationally endorsed protocol on the ^{90}Y PET/

CT quantitative imaging would lead to further advances in this area.

REFERENCES

AAPM Task Group 181. (2012). The selection, use, calibration, and quality assurance of radio-nuclide calibrators used in nuclear medicine report of AAPM Task Group 181. College Park, MD: American Association of Physicists in Medicine.

Ahn, S., Fessler, J.A. (2004). Emission image reconstruction for randoms-precorrected PET allowing negative sinogram values. *IEEE Trans Med Imaging* 23:591–601.

Attarwala, A. et al. (2014). Quantitative and qualitative assessment of yttrium-90 PET/CT imaging. *PLoS One* 9:e110401.

Bagni, O. et al. (2012). 90Y-PET for the assessment of microsphere biodistribution after selective internal radiotherapy. *Nucl Med Commun* 33:198–204.

Bé, M. et al. (2006). Table of radionuclides. *Monographie BIPM-5*. Sèvres, Paris: Bureau International des Poids et Mesures, ISBN 92-822-2218-7.

Browne, E., Junde H. (1998). Nuclear data sheets for A = 176. *Nucl Data Sheets* 1998;84:337–486.

Carlier, T. et al. (2013). Assessment of acquisition protocols for routine imaging of Y-90 using PET/CT. *EJNMMI Res* 3:11.

Carlier, T., Willowson, K., Fourkal, E., Bailey, D., Doss, M., Conti, M. (2015). 90Y-PET imaging: Exploring limitations and accuracy under conditions of low counts and high random fraction. *Med Phys* 42:4295–4309.

Cherry, S., Sorenson, J., Phelps, M. (2012). *Physics in Nuclear Medicine*. Philadelphia, PA: Elsevier/Saunders.

Conti, M. (2009). State of the art and challenges of time-of-flight PET. *Phys Med* 25:1–11.

Conti, M., Eriksson, L., Rothfuss, H., Melcher, C. (2009). Comparison of fast scintillators with TOF PET potential. *IEEE Trans Nucl Sci* 56:926–933.

D'Arienzo, M. (2013). Emission of β+ particles via internal pair production in the 0+–0+ transition of 90Zr: Historical background and current applications. *Nucl Med Imaging Atoms* 1:2–12.

D'Arienzo, M. et al. (2012). 90Y PET-based dosimetry after selective internal radiotherapy treatments. *Nucl Med Commun* 33:633–640.

D'Arienzo, M. et al. (2013). Absorbed dose to lesion and clinical outcome after liver radio-embolization with 90Y microspheres: A case report of PET-based dosimetry. *Ann Nucl Med* 27:676–680.

Daube-Witherspoon, M.E. et al. (2002). PET performance measures using the NEMA NU 2-2001 standard. *J Nucl Med* 43:1398–1409.

Dezarn, W. et al. (2011). Recommendations of the American Association of Physicists in Medicine on dosimetry, imaging, and quality assurance procedures for 90Y microsphere brachytherapy in the treatment of hepatic malignancies. *Med Phys* 38:4824–4845.

Dezarn, W., Kennedy, A. (2007a). SU-FF-T-380: Significant differences exist across institutions in 90Y activities compared to reference standard. *Med Phys* 34:2489.

Dezarn, W., Kennedy, A. (2007b). Resin 90Y microsphere activity measurements for liver brachytherapy. *Med Phys* 34:1896–1900.

Elschot, M. et al. (2013). Quantitative comparison of PET and bremsstrahlung SPECT for imaging the in vivo yttrium-90 microsphere distribution after liver radioembolization. *PLoS One* 8:e55742.

Fenwick, A., Baker, M., Ferreira, K., Keightley, J. (2014). Comparison of Y-90 measurements in UK hospitals. NPL Report IR 20. United Kingdom: National Physics Laboratory.

Ferreira, K., Fenwick, A., Arinc, A., Johansson, L. (2016). Standardisation of 90Y and determination of calibration factors for 90Y microspheres (resin) for the NPL secondary ionisation chamber and a Capintec CRC-25R. *Appl Radiat Isotopes* 109:226–230.

Ford, K. (1955). Predicted 0+ level of Zr90. *Phys Rev* 98:1516–1517.

Fourkal, E. et al. (2013). 3D inpatient dose reconstruction from the PET-CT imaging of 90Y microspheres for metastatic cancer to the liver: Feasibility study. *Med Phys* 40:081702.

Fowler, K. et al. (2016). PET/MRI of hepatic 90Y microsphere deposition determines individual tumor response. *Cardiovasc Intervent Radiol* 39:855–864.

Gadd, R. et al. (2006). Protocol for establishing and maintaining the calibration of medical radionuclide calibrators and their quality control. *Measurement Good Practice Guide No. 93*. Teddington, Middlesex, UK: National Physical Laboratory.

Gates, V. et al. (2010). Internal pair production of ^{90}Y permits hepatic localization of microspheres using routine PET: Proof of concept. *J Nucl Med* 52:72–76.

Goertzen, A., Stout, D., Thompson, C. (2009). A method for measuring the energy spectrum of coincidence events in positron emission tomography. *Phys Med Biol* 55:535–549.

Greenberg, J.S., Deutsch, M. (1956). Positrons from the decay of P^{32} and Y^{90}. *Phys Rev* 102:415–421.

IAEA. (1996). International basic safety standards for protection against ionizing radiation and for the safety of radiation sources. International Atomic Energy Agency, Vienna, Austria.

Johnson, O., Johnson, R., Langer, L. (1955). Evidence for a 0$^+$ first excited state in Zr^{90}. *Phys Rev* 98:1517–1518.

Kao, Y. et al. (2013). Post-radioembolization yttrium-90 PET/CT—Part 2: Dose-response and tumor predictive dosimetry for resin microspheres. *EJNMMI Res* 3:57.

Kao, Y. et al. (2012). Yttrium-90 internal pair production imaging using first generation PET/CT provides high-resolution images for qualitative diagnostic purposes. *Br J Radiol* 85:1018–1019.

Kossert, K. et al. (2016). Comparison of 90Y activity measurements in nuclear medicine in Germany. *Appl Radiat Isot* 109:247–249.

Langhoff, H., Hennies, H. (1961). Zum experimentellen Nachweis von Zweiquantenzerfall beim 0$^+$–0$^+$ Ubergang des Zr^{90}. *Z Phys* 164:166–173.

Lewellen, T. (1998). Time-of-flight PET. *Semin Nucl Med* 28:268–275.

Lhommel, R., Goffette, P., Van den Eynde, M., Jamar, F., Pauwels, S., Bilbao, J., Walrand, S. (2009). Yttrium-90 TOF PET scan demonstrates high-resolution biodistribution after liver SIRT. *Eur J Nucl Med Mol Imaging* 36:1696–1696.

Lhommel, R. et al. (2010). Feasibility of ^{90}YTOF PET-based dosimetry in liver metastasis therapy using SIR-spheres. *Eur J Nucl Med Mol Imaging* 37:1654–1662.

Li, Q., Leahy, R.M. (2006). Statistical modeling and reconstruction of randoms precorrected PET data. *IEEE Trans Med Imaging* 25:1565–1572.

Lourenço, V. et al. (2015). Primary standardization of SIR-Spheres based on the dissolution of the 90Y-labeled resin microspheres. *Appl Radiat Isot* 97:170–176.

Martí-Climent, J. et al. (2014). PET optimization for improved assessment and accurate quantification of ^{90}Y-microsphere biodistribution after radioembolization. *Med Phys* 41:092503.

Mo, L. et al. (2005). Development of activity standard for 90Y microspheres. *Appl Radiat Isot* 63:193–199.

National Electrical Manufacturers Association. (2007). *NEMA NU 2-2007 Performance Measurements of Positron Emission Tomographs* Arlington, VA: NEMA.

Ng, S.C. et al. (2013). Patient dosimetry for ^{90}Y selective internal radiation treatment based on 90Y PET imaging. *J Appl Clin Med Phys* 14:212–221.

Nickles, R.J. et al. (2004). Assaying and PET imaging of yttrium-90: 1>>34 ppm>0. *IEEE Nucl Sci Symp Rec* 6:3412–3414.

Okuda, K. et al. (1985). Natural history of hepatocellular carcinoma and prognosis in relation to treatment study of 850 patients. *Cancer* 56:918–928.

Park, M., Mahmood, A., Zimmerman, R., Limpa-Amara, N., Makrigiorgos, G., Moore, S. (2008). Adsorption of metallic radionuclides on plastic phantom walls. *Med Phys* 35:1606.

Pasciak, A., Bourgeois, A., Bradley, Y. (2014a). A comparison of techniques for 90Y PET/CT image-based dosimetry following radioembolization with resin microspheres. *Front Oncol* 4:121.

Pasciak, A. et al. (2014b). Radioembolization and the dynamic role of ^{90}Y PET/CT. *Front Oncol* 4:38.

Peller, P., Subramaniam, R., Guermazi, A. (2012). *PET-CT and PET-MRI in Oncology*. Berlin: Springer.

Rahmim, A. et al. (2005). Statistical dynamic image reconstruction in state-of-the-art high resolution PET. *Phys Med Biol* 50:4887–4912.

Selwyn, R. et al. (2007). A new internal pair production branching ratio of ⁹⁰Y: The development of a non-destructive assay for ⁹⁰Y and ⁹⁰Sr. *Appl Radiat Isot* 65:318–327.

Stabin, M.G., Eckerman, K.F., Ryman, J.C., Williams, L.E. (1994). Bremsstrahlung radiation dose in yttrium-90 therapy applications. *J Nucl Med* 35:1377–1380.

Sunderland, J., Christian, P., Kiss, T. (2015). PET/CT scanner validation for clinical trials-reasons for failure, recipes for success: The Clinical Trials Network (CTN) experience. *J Nucl Med* 56(3):1737.

Surti, S. (2014). Update on time-of-flight PET imaging. *J Nucl Med* 56:98–105.

Surti, S., Karp, J. (2016). Advances in time-of-flight PET. *Phys Med* 32:12–22. doi: 10.1016/j.ejmp.2015.12.007.

Tapp, K. et al. (2014). The impact of image reconstruction bias on PET/CT 90Y dosimetry after radioembolization. *J Nucl Med* 55:1452–1458.

Thiam, C., Bobin, C., Lourenço, V., Chisté, V., Amiot, M., Mougeot, X., Lacour, D., Rigoulay, F., Ferreux, L. (2016). Investigation of the response variability of ionization chambers for the standard transfer of SIR-Spheres®. *Appl Radiat Isotopes* 109:231–235.

van Elmbt, L. et al. (2011). Comparison of yttrium-90 quantitative imaging by TOF and non-TOF PET in a phantom of liver selective internal radiotherapy. *Phys Med Biol* 56:6759–6777.

Walrand, S. et al. (2015). The impact of image reconstruction bias on PET/CT ⁹⁰Y dosimetry after radioembolization. *J Nucl Med* 56:494–495.

Wissmeyer, M. et al. (2011). ⁹⁰Y time-of-flight PET/MR on a hybrid scanner following liver radioembolisation (SIRT). *Eur J Nucl Med Mol Imaging* 38:1744–1745.

Werner, M. et al. (2009). PET/CT for the assessment and quantification of 90Y biodistribution after selective internal radiotherapy (SIRT) of liver metastases. *Eur J Nucl Med Mol Imaging* 37:407–408.

Werner, M. et al. (2010). PET/CT for the detection and quantification of the β-emitting therapeutic radionuclide yttrium-90 after liver SIRT [abstract]. *J Nucl Med* 51(Suppl 2):341.

Willowson, K. et al. (2012). Quantitative ⁹⁰Y image reconstruction in PET. *Med Phys* 39:7153.

Willowson, K., Tapner, M., Bailey, D. (2015). A multicentre comparison of quantitative 90Y PET/CT for dosimetric purposes after radioembolization with resin microspheres. *Eur J Nucl Med Mol Imaging*, 42:1202–1222.

Woods, M. et al. (1996). Calibration of the NPL secondary standard radionuclide calibrator for 32P, 89Sr and ⁹⁰Y. *Nucl Instrum Methods Phys Res A* 369:698–702.

Zimmerman, B., Ratel, G. (2005). Report of the CIPM key comparison CCRI(II)-K2Y-90. *Metrologia* 42:06001.

12

Image-based three-dimensional dosimetry following radioembolization

ALEXANDER S. PASCIAK AND S. CHEENU KAPPADATH

12.1 INTRODUCTION

The general dosimetric principles of ^{90}Y radioembolization have been discussed in detail in several chapters in this book. For example, Chapter 5 discusses a commonly used formula that allows for the calculation of the absorbed dose to a volume of tissue given a uniform distribution of ^{90}Y activity. While this is useful for determining the average absorbed dose to the liver, lobe, or segment treated, it is of limited use in the estimation of biological effect. More accurate methods for determining absorbed dose have been suggested such as the partition model, discussed in Chapters 4 and 5. However, the partition model cannot account for inhomogeneity of microsphere distribution in tumor or normal liver. Determining the absorbed dose following ^{90}Y radioembolization in all areas of the tumor, uninvolved liver, and extrahepatic tissues is a critical future component of managing patient

follow-up. For example, if radiation dose and, therefore, toxicity to normal or extrahepatic tissues can be determined immediately following radioembolization, prompt administration of prophylaxis can be considered which may decrease the severity of side effects. Furthermore, undertreated areas of tumor could be identified and alternative or adjuvant therapies could be prescribed, increasing the potential efficacy of ^{90}Y radioembolization in some patients.

The ability to utilize posttreatment dosimetry in the aforementioned manner depends on the availability of several important pieces of data. First, one must understand the radiation biology and dose–response properties of the normal liver and the tumor, which may vary substantially with tumor size, type, and other factors. The radiation biology of radioembolization at a macroscopic and a microscopic level has been discussed in Chapters 8 and 9, respectively. However, there are still many unknowns yet to be addressed. Once the radiation biology is understood for a patient with a given history, the second missing piece of data is a complete description of the absorbed dose following treatment. Postradioembolization three-dimensional (3D) image-based dosimetry is already capable of providing this information in a routine clinical scenario.

Chapters 10 and 11 have laid out the principles behind ^{90}Y posttreatment imaging using bremsstrahlung single-photon emission computed tomography (SPECT)/computed tomography (CT) or ^{90}Y positron emission tomography (PET)/CT with an emphasis on quantification. However, dosimetric calculation for internal emitters based on SPECT or PET imaging is not a new idea—it was suggested long before "quantitative imaging" was a routine component of clinical vocabulary (Loevinger et al., 1989; Bolch et al., 1999). It is now standard practice to perform compartment-based dosimetry for radiolabeled agents *in vivo* using SPECT or PET. Given a single injected bolus of a radioactive agent, pharmacokinetic parameters can describe a complex relationship of uptake followed by clearance into the tissue of interest. Increasing this complexity, the time–activity curve can vary from patient to patient, especially in cancer therapy necessitating a patient-specific evaluation. As a result, dosimetry typically involves imaging at multiple time points in order to determine the activity concentration, as a function of time, in both the tissue of interest and the surrounding tissues.

Fortunately, serial imaging is not required for image-based dosimetry following ^{90}Y radioembolization. All currently used radioembolization products are unique from pharmacologic radioactive agents in that they are brachytherapy devices, not drugs. As previously mentioned, both glass and resin ^{90}Y microspheres, as well as ^{166}Ho microspheres, form a nonbiodegradable permanent implant upon infusion, where they release their radiation burden locally (Chapter 1). As discussed in Chapter 7, little or no systemic release of ^{90}Y from the microspheres is seen *in vivo*. A single-session posttreatment quantitative scan via either bremsstrahlung SPECT/CT or ^{90}Y PET/CT can be used to compute the committed absorbed dose of the ^{90}Y as it decays. Of course, the limitations of the imaging system (resolution, quantification accuracy, and noise) will be directly translated into errors in the 3D dose map.

In this chapter, we will discuss the four primary methodologies for ^{90}Y image-based dosimetry following radioembolization.

12.2 FULLY 3D MONTE CARLO TRANSPORT FOR IMAGE-BASED HEPATIC DOSIMETRY

A brief introduction to the mathematics of the Monte Carlo method has been included in Chapter 9. Monte Carlo methods are the *de facto* standard for radiation transport in medicine and are used extensively in radiation oncology and radiology, both clinically and in research. Even with modern advancements in computer technology, the computational burden associated with fully 3D Monte Carlo simulation is significant. Its advantage, however, is that accurate dosimetric calculations can be performed for radiation traversing any heterogeneous material structure. If this material structure is a patient, voxelized patient-specific phantoms from CT image sets can be incorporated into the Monte Carlo simulation. However, the utility of this for radioembolization is limited.

When radiation penetrates a patient, whether it is x-ray, γ-ray, or β-radiation, it does not discern among tissue types as one might initially think. Radiation will interact differently in different tissue types if there is (1) a substantial difference in electron density (electrons/cm^3) or (2) a substantial difference in the atomic number of the atoms in the tissue. Unlike almost everything else in medicine, the behavior of radiation is unaffected by biochemical structure. Radiation will travel through healthy tissue, diseased tissue, and venous/arterial tissue in exactly the same way. Even adipose tissue has little effect on radiation from a dosimetric standpoint, less a small change in density. In fact, accurate transmission of radiation through a patient can be performed by lumping the patient's tissues into three broad categories: soft tissue, bone tissue, and lung tissue. Since the liver is one of the most physically homogeneous organs in the body, the advantage to using Monte Carlo radiation transport methods is, therefore, lost in most cases. For this reason, other methods for image-based dosimetry following ^{90}Y radioembolization are often employed.

12.3 DOSE-POINT KERNEL CONVOLUTION FOR IMAGE-BASED HEPATIC DOSIMETRY

Multiple authors have discussed application of dose-point kernel (DPK) convolution as an approach to absorbed-dose determination in voxelized phantoms (Bolch et al., 1999; Strigari et al., 2006; Pasciak and Erwin, 2009; Kennedy et al., 2011; D'Arienzo et al., 2013; Elschot et al., 2013). DPK convolution was desirable in past decades since Monte Carlo-based transport, particularly for high-energy electrons, carried a substantial computation burden. Of course, because of the inherent 3D homogeneity of liver tissue combined with the limited range of ^{90}Y β emissions, there is a little difference between DPK convolution and voxel-based fully 3D Monte Carlo transport (Pasciak and Erwin, 2009). Dome lesions at the apex of the liver may be a potential exception to this general rule (Mikell et al., 2015). Image-based dosimetry following ^{90}Y posttreatment imaging using a ^{90}Y DPK is defined as follows:

$$D(x,y,z) = \frac{1}{\lambda}(A \otimes \mathrm{DPK})(x,y,z)$$

$$= \frac{1}{\lambda}\sum_{x'}\sum_{y'}\sum_{z'} A(x',y',z') \cdot \quad (12.1)$$
$$\mathrm{DPK}(x-x',y-y',z-z')$$

where λ is the decay constant of the radionuclide and $A(x,y,z)$ is the 3D activity concentration matrix determined from quantitative imaging. While more computationally efficient than 3D Monte Carlo simulation, convolution of the activity concentration matrix, A, by the DPK can still be computationally demanding in certain circumstances. Mathematical transformations can ease this burden, such as the fast Hartley transform, which is an integral transformation similar to Fourier methods, except that it eliminates complex solutions. Following Fourier or Hartley transformation, multiplication of A and DPK in the frequency domain carries equivalence to convolution with a substantially decreased computational burden. However, for DPK convolution applied to ^{90}Y image-based dosimetry, speed is less of an issue for two reasons. First, because of the limited resolution of SPECT and PET, $A(x,y,z)$ matrix sizes are usually small. Further, unlike gamma emitters, ^{90}Y β emission has a limited range and the DPK does not need to cover more than 11 mm for any single octant. Therefore, the small matrix sizes defining A and DPK permit relatively fast 3D convolution.

By convention, a ^{90}Y DPK is computed using a validated Monte Carlo code for a predefined voxel size. ^{90}Y source specification is often uniformly distributed through the origin voxel, sometimes with the inclusion of microsphere composition and size (Paxton et al., 2012). ^{90}Y β spectra have been reported by a number of sources and are available from Eckerman et al. (1994) with energy bins beginning at 0.1 keV up to the maximum energy of the emission. DPKs are computed for a single transformation of the radionuclide of interest, necessitating additional scaling factors to convert the activity concentration matrix, A, into the total number of transformations in each voxel, integrated from time 0 to infinity. This is accomplished with the 1/λ constant in Equation 12.1.

In general, any Monte Carlo code capable of 3D transport of electrons through voxel geometries can be used to compute a DPK. The majority of modern Monte Carlo codes use a condensed history approximation to decrease the sizeable computational burden associated with track-structure electron transport simulation. Using the condensed history method, the combined energy losses along the track from multiple interactions (including excitation and electron impact ionization) are summed into a larger pseudointeraction (Berger, 1963). While there are some instances where the condensed history technique cannot provide sufficient accuracy, these instances are not found in the computation of ^{90}Y DPKs for clinical hepatic dosimetry. The following codes have been used for ^{90}Y DPK determination and point dosimetry: GEANT4 (Pacilio et al., 2009; Guimarães et al., 2010), MCNP5 (Paxton et al., 2012), MCNP4C (Pacilio et al., 2009), MCNPX (Dieudonne et al., 2010), EGSNRC (Strigari et al., 2006; Pacilio et al., 2009; Paxton et al., 2012), and EGS4 (Strigari et al., 2006); however, this is not an exhaustive list. Although not specific to ^{90}Y, some detailed analyses of the differences in electron transport dose profiles with codes utilizing different electron transport algorithms have been published (Uusijarvi et al., 2009). While the aforementioned manuscript has found minor variations in the results of different Monte Carlo codes for ^{90}Y dosimetry, the clinical importance of these differences should be put into perspective. After reading the discussion on the local deposition method (LDM) later in this chapter, it will become apparent that minor differences in DPK are likely to carry a negligible clinical impact. Instead, a Monte Carlo code which the user is comfortable with and finds easy to use should be selected. For example, the authors of this chapter find EGSNRC to be substantially easier to use than EGS4.

The majority of published ^{90}Y DPKs are for liver tissue only. Although not generally a part of routine clinical ^{90}Y radioembolization, there may be some scenarios where 3D image-based lung dosimetry might be performed using DPK convolution. For example, such calculations may be useful for root cause analysis following the occurrence of clinical side effects such as radiation pneumonitis after treatment. In these cases, the Monte Carlo calculation of lung DPKs can be computed. Alternatively, one can perform a first-order correction to existing DPKs for liver tissue using density scaling to account for differences between lung and soft tissue. Mikell et al. (2015) discuss this further.

In the previous chapters of this book, it has been emphasized that ^{90}Y is predominantly a pure β emitter. However, high-energy β particles generate bremsstrahlung x-rays as they slow down in an absorber, creating a coupled photon–electron spectrum within the patient. Bremsstrahlung x-rays have the propensity to penetrate through substantially greater tissue thickness than ^{90}Y β particles, which may seem to confound 3D dosimetry. Certainly, coupled photon–electron transport using Monte Carlo methods can be performed with high accuracy. However, the dose contribution of bremsstrahlung photons will not be accounted for when using ^{90}Y DPKs, which cover only the maximum β range. However, as demonstrated by Stabin et al. (1994), the hepatic absorbed dose contribution from bremsstrahlung x-rays is three orders of magnitude below that of the ^{90}Y β emission. Therefore, bremsstrahlung x-rays will have a negligible impact on the accuracy of DPK and other non-Monte Carlo methods for 3D dosimetry discussed in the remainder of this chapter.

12.4 THE VOXEL S-VALUE MIRD APPROACH FOR IMAGE-BASED HEPATIC DOSIMETRY

The voxel S-value (VSV) Medical Internal Radiation Dose (MIRD) approach to image-based dosimetry is an alternative to DPK convolution with a formalism analogous to MIRD S-values, traditionally used for organ → organ dosimetry of internal emitters. The VSV method was first described by Bolch et al. (1999). The relationship for determining the absorbed dose in target voxel ($D_{\text{Voxel T}}$) using VSVs is given in

$$D_{\text{Voxel T}} = \frac{1}{\lambda} \sum_N A_{\text{Voxel S}} \cdot S\left(\text{Voxel}_T \leftarrow \text{Voxel}_S\right) \quad (12.2)$$

where S is the average energy deposition in the target voxel, Voxel$_T$, from a single transformation in the source voxel, Voxel$_S$. Corresponding to the maximum range of a ^{90}Y β-emission, VSVs for all source voxels (N) within 11 mm of the target voxel should be summed. As with DPK convolution, the $1/\lambda$ constant converts activity in a voxel to the total number of transformations. While DPKs are stored and applied in 3D matrix form, S-values are tabulated for each target voxel as a function of source voxel location.

The VSV MIRD approach is defined differently than DPK convolution; however, many properties are identical. Like DPK convolution, inhomogeneous tissues cannot be accommodated that, again, is generally not an issue for hepatic ^{90}Y radioembolization. Further, VSVs and DPKs are interconvertible and when applied to the same activity concentration matrix will yield the same dosimetric solution. An advantage in the computational simplicity of the VSV method compared with DPK convolution can be realized when calculating the absorbed dose to a subset of voxels in the volume, for example, to calculate isodose curves. However, when computing the absorbed dose at every voxel (i.e., in the entire 3D image set), the computational burden of the VSV MIRD approach is equivalent to that of DPK convolution.

One problem with both VSVs and DPK convolution is that in a modern clinical environment, voxel size often varies with both scanner and image reconstruction parameters. Under normal circumstances, this makes the routine clinical use of these techniques challenging since tabulated DPKs and VSVs are available for only a limited array of ^{90}Y voxel sizes. However, additional methods have been proposed for quickly generating VSVs for an arbitrary sized voxel from predetermined data. Fernandez et al. (2013) used Monte Carlo methods to perform electron transport for several radionuclides, including ^{90}Y at a voxel pitch of 0.5 mm. These data have been made publically available; however, what is unique is their proposed method to analytically rescale these data for use in problems with arbitrary voxel size, without rerunning the Monte Carlo simulation. Fernandez et al. has validated their rescaling method by comparing it to native Monte Carlo simulations performed for the following voxel sizes: 0.5, 0.7, 1.0, 1.23, 1.5, 1.8, 2.0, 2.4, 3.0, 4.8, 6.0, 8.0, and 10.0 mm. When used for image-based dosimetry, the Fernandez rescaling method introduces less than 1.5% error compared with calculating VSVs directly with the Monte Carlo method (Fernández et al., 2013). To this end, several other authors have suggested methods for on-the-fly rescaling of VSVs for use in image-based dosimetry that are worthy of consideration (Dieudonne et al., 2011; Amato et al., 2012).

As previously mentioned, the dosimetric solutions computed from both the VSV MIRD approach and DPK convolution will be equivalent. The ensuing discussions will compare the accuracy of alternative computation techniques to DPK convolution, and it should be understood that these comparisons apply to the VSV MIRD method as well.

12.5 LDM FOR IMAGE-BASED EPATIC DOSIMETRY

The LDM is the final method that we will discuss for performing ^{90}Y postradioembolization image-based dosimetry. As with prior techniques, quantitative postradioembolization imaging from either bremsstrahlung SPECT/CT or ^{90}Y PET/CT is required as a starting point for dosimetry. Image sets must be quantified to activity concentration (Bq/mL) of ^{90}Y at the time of infusion in order to perform dosimetry using the LDM.

The LDM is based on the premise that the entirety of the energy released from each decay of ^{90}Y within a voxel is deposited locally, within that same voxel. This assumption reduces all mathematics and computation associated with 3D postradioembolization dosimetry to a simple scaling factor. The following derivation describes the scaling factor used in the LDM.

Under the aforementioned assumptions, the absorbed dose, $D_{^{90}Y}$ (Gy), within a voxel can be determined according to

$$D_{^{90}Y}(\text{Gy}) = \frac{A_0(\text{Bq}/\text{mL}) \cdot 4.998 \times 10^{-8} (\text{J} \cdot \text{s})}{\rho_{\text{liver}}(\text{kg}/\text{mL})} \quad (12.3)$$

where A_0 is the activity concentration of ^{90}Y (Bq/mL) within the voxel and ρ is the density of liver tissue in kg/mL. The constant factor 4.998×10^{-8} J·s is the energy released per unit activity of ^{90}Y, previously derived in both Chapters 5 and 7. If the density of liver tissue as defined by the International Commission on Radiation Units and Measures, ICRU (1992) is selected, the constants in Equation 12.3 can be grouped into a single conversion factor K_{90Y}, valid only for ^{90}Y. Note that Equation 12.3 is independent of voxel volume. This powerful feature allows LDM to be utilized regardless of reconstruction matrix size or isotropicity following bremsstrahlung SPECT/CT or ^{90}Y PET/CT posttreatment imaging. However, the accuracy of LDM still rests on the assumption that the entirety of the energy from each decay is absorbed by the voxel where the decay occurred. This assumption may be violated as voxel volume decreases or as charged-particle equilibrium is lost between voxels. This will be discussed in detail in Section 12.5.1.

$$D_{^{90}Y}(\text{Gy}) = A_0\left(\frac{\text{Bq}}{\text{mL}}\right) \cdot K_{^{90}Y}\left(\text{Gy} \cdot \frac{\text{mL}}{\text{Bq}}\right) \quad (12.4)$$

Equation 12.4 is identical to Equation 12.3, except that constants have been replaced by the K_{90Y} factor previously mentioned. For any quantitative ^{90}Y image set, with units of ^{90}Y activity concentration at the time of microsphere infusion, a K_{90Y} factor of 4.782×10^{-5} (Gy-mL/Bq) can be used to determine the absorbed dose in every voxel using the LDM.

As described in Chapter 10, some work is required to quantify ^{90}Y SPECT/CT images using bremsstrahlung SPECT/CT; therefore, it is expected that as part of this effort appropriate activity concentration units (Bq/mL of ^{90}Y) will be determined. However, in many cases ^{90}Y PET/CT will produce quantitative posttreatment images without any additional effort by the end user. As described in Chapter 11, this is often true even if the PET/CT manufacturer never intended direct imaging of ^{90}Y. As a result, when scanning a patient postradioembolization using PET/CT ^{90}Y may not be an available option on the scanner console, although several PET/CT manufacturers have made an effort to include it on recent software updates.

At a basic level, quantification for different positron-emitting radionuclides by PET/CT systems is performed with a simple rescaling of pixel values based on the half-life and branching ratio for positron emission of the injected nuclide. Therefore, this process can easily be integrated into the K scaling factor in Equation 12.4 for systems that do not directly support ^{90}Y. This is shown in Equation 12.5:

$$D_{90_Y}(\text{Gy}) = A_0\left(\frac{\text{Bq}}{\text{mL}}\right) \cdot K_x\left(\text{Gy} \cdot \frac{\text{mL}}{\text{Bq}}\right) \cdot e^{\lambda_c} \quad (12.5)$$

If ^{90}Y is not directly supported by the PET/CT system in question, but phantom experiments with ^{90}Y suggest accurate quantification (Chapter 10), the LDM can still be used. Scanning a postradioembolization patient in these instances as either ^{22}Na or ^{68}Ge will also allow for posttreatment dosimetry of ^{90}Y under the above conditions with the LDM. K_x in Equation 12.5 is a modified scaling factor that includes the branching ratio for positron emission of the radionuclide selected on the scanner (i.e., ^{22}Na or ^{68}Ge) relative to the branching ratio of ^{90}Y. The exponential term and the adjusted decay constant (λ_c) accounts for the difference in half-life between ^{90}Y and the radionuclide selected. However, for posttreatment imaging obtained on the day of treatment, the exponential can be omitted with minimal introduced error so long as ^{22}Na or ^{68}Ge are used:

$$K_x\left(\text{Gy} \cdot \frac{\text{mL}}{\text{Bq}}\right) = K_{90_Y}\left(\text{Gy} \cdot \frac{\text{mL}}{\text{Bq}}\right) \cdot \frac{B_X}{B_{90_Y}} \quad (12.6)$$

$$\lambda_c = \lambda_{90_Y} - \lambda_X \quad (12.7)$$

Equations 12.6 and 12.7 illustrate the derivation of K_x and λ_c for radionuclide X based on the relative branching ratio, β, and decay constant, λ. Values for K_X and λ_c have been computed for ^{22}Na and ^{68}Ge in Table 12.1. Note that K_X is equivalent to $K_{90_Y} K_{90Y}$ in Table 12.1.

12.5.1 VALIDATION OF THE LDM

As previously mentioned, the LDM rests upon the assumption that all energy released in the decay of radioactivity within a voxel contributes to the absorbed dose in the same voxel. Several authors have utilized LDM for image-based ^{90}Y dosimetry. An indirect approach was explored by Chiesa et al. (2012) and Mazzaferro et al. (2013), who both used Tc-99m macroaggregated albumin (^{99m}Tc-MAA) to perform dosimetry based on LDM. Use of LDM with ^{99m}Tc-MAA SPECT/CT was first suggested by Pasciak and Erwin (2009). While LDM is fully capable of producing 3D dosimetry based on ^{99m}Tc-MAA, its accuracy rests on the additional assumption that the relative activity distribution of MAA mirrors that of ^{90}Y radioembolization. This assumption may or may not be true as discussed in Chapters 10 and 11.

LDM has been used clinically for dosimetry by several authors following ^{90}Y PET/CT (Kao et al., 2013; Bourgeois et al., 2014; Srinivas et al., 2014) and its use has been validated specifically for postradioembolization imaging based on PET (Pasciak et al., 2014) and SPECT (Pacilio et al., 2015). One important consideration that is often ignored in

Table 12.1 ^{90}Y dosimetry using PET/CT and the local deposition method

Isotope used for imaging	Branching ratio for positron emission	Decay constant λ (hours^{-1})	Adjusted decay constant λ_c (hours^{-1})	Conversion factor K_X (Gy-mL/Bq)
^{90}Y	0.000032	1.083×10^{-2}	0	4.782×10^{-5} (K_{90Y})
^{22}Na	0.905	3.038×10^{-5}	1.080×10^{-2}	1.353
^{68}Ge	0.890[a]	1.066×10^{-4}	1.072×10^{-2}	1.330

[a] Adapted from Pasciak, A.S. et al. (2014). *Front Oncol* 4:121. doi: 10.3389/fonc.2014.00121. Branching ratio of daughter, ^{68}Ga, which is in secular equilibrium with the parent radionuclide, ^{68}Ge.

postradioembolization image-based dosimetry is that the bremsstrahlung SPECT or ^{90}Y PET/CT images used as a starting point are not perfect representations of the true activity concentration. If a ^{90}Y imaging technique is said to be quantitative, then it can be used to determine the activity concentration at the center of a large phantom homogeneously filled with ^{90}Y. However, the activity concentration of ^{90}Y microspheres in tumor is not homogeneous and because of the imperfect resolution of the imaging method, the true activity concentration will be convolved (blurred) by the point-spread function (PSF) of the imaging modality:

$$A(x,y,z) = T(x,y,z) \otimes PSF \qquad (12.8)$$

where A is the 3D activity concentration reported by quantitative imaging and T is the true 3D activity concentration. If we revisit Equation 12.1 where we defined the relationship used for DPK convolution in image-based 3D dosimetry, we can now see that it contains an inaccuracy. The inaccuracy in Equation 12.1 is that $A(x,y,z)$ is convolved by the DPK. In actuality, the true 3D absorbed dose (D_{true}) is described in:

$$D_{true}(x,y,z) = \frac{1}{\lambda}\left(T(x,y,z) \otimes DPK\right) \quad (12.9)$$

Unfortunately, because $T(x,y,z)$ cannot be determined in a patient, the aim is to find the best method for approximating D_{true} given the reported activity concentration, $A(x,y,z)$.

The PSF measured at full-width and half-maximum (FWHM) for ^{90}Y PET/CT ranges between 3.1 and 10.5 mm, depending on the scanner and reconstruction parameters, as discussed in Chapter 11. For bremsstrahlung SPECT, expected PSFs will be substantially higher than ^{90}Y PET/CT as discussed in Chapter 10. However, because of the relatively poor resolution of both ^{90}Y bremsstrahlung SPECT and ^{90}Y PET, the blur in the data introduced by the scanner PSF is often greater than the spread in energy deposition computed by convolution with the DPK. Therefore, in some circumstances use of the LDM, which introduces no additional blur, may result in an absorbed dose map that is closer to D_{true} than the absorbed dose map calculated using DPK convolution. This is qualitatively illustrated in Figure 12.1.

Indeed, the qualitative example in Figure 12.1 suggests that LDM might more accurately estimate D_{true} than DPK convolution. Under ideal circumstances and in the absence of image noise, the accuracy of LDM and DPK can be directly compared as a function of scanner resolution, defined by PSF at FWHM. Pasciak et al. performed this ideal mathematical comparison, as well as one based on phantom data with a particular focus on ^{90}Y PET/CT (Pasciak et al., 2014). The ideal approach involves a precise mathematical phantom convolved with a range of 3D Gaussian blur kernels to simulate the shift-invariant PSF representative of either ^{90}Y PET/CT or SPECT. For the results in Figure 12.2, the mathematical phantom simulated was the NEMA IEC body phantom, with hot spheres ranging from 10 to 37 mm in size. Gaussian kernels with FWHM ranging from 2.0 to 15.0 mm in increments of 0.25 mm were applied to create simulated activity concentrations, $A(x,y,z)$. Either DPK convolution or the LDM was applied to $A(x,y,z)$ for subsequent comparison with the $D_{true}(x,y,z)$. In this case, since $T(x,y,z)$ is known, D_{true} can be computed exactly as shown in Equation 12.9. Figure 12.2 shows the results of a voxel-by-voxel comparison of absorbed dose using the LDM and DPK convolution in three of the phantom hot spheres as a function of the scanner PSF. It can be seen in Figure 12.2 that LDM has the best accuracy with a scanner FWHM of just over 4 mm, where the PSF closely approximates the shape of the ^{90}Y DPK. At a FWHM of less than 4 mm, the accuracy of LDM worsens while the accuracy of DPK convolution continues to improve until perfect scanner resolution is obtained (FWHM = 0 mm). However, for PSF FWHM greater than 4 mm, LDM always carries a slight advantage. It should be noted that the shape of the PSF and reconstructed voxel size will affect the relationships in Figure 12.2.

Any accuracy gained with the use of LDM depends on image reconstruction parameters, scanner PSF, voxel size, image noise, and also, patient motion. Several authors have investigated some or all of these factors and their effect on the accuracy of LDM relative to DPK convolution for both ^{99m}Tc-MAA SPECT and ^{90}Y bremsstrahlung SPECT-based dosimetry (Pasciak and Erwin, 2009; Ljungberg and Sjögreen-Gleisner, 2011; Pacilio et al., 2015) as well as ^{90}Y PET/CT (Pasciak et al., 2014). A common consensus between all of these reports is that with the current ^{90}Y imaging technology, either bremsstrahlung SPECT or ^{90}Y PET/CT, neither DPK convolution nor the VSV MIRD formalism provides

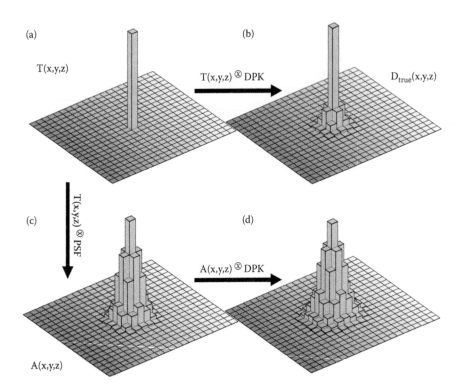

Figure 12.1 A qualitative example illustrating the effect of scanner PSF and DPK convolution on the shape of the absorbed dose profile for a point source of ^{90}Y occupying a single voxel. The voxel size in this example is 2.0 mm^3. The scanner PSF is approximated by a 6.5-mm FWHM Gaussian filter, which has been shown to be a reasonable approximation for ^{90}Y PET/CT (Pasciak et al., 2014). **(a)** 2D representation of a point source of ^{90}Y activity, $T(x,y,z)$. This represents the true activity distribution, which is unknown for *in vivo* patient imaging; **(b)** convolution of the true activity concentration with the ^{90}Y DPK gives the true absorbed-dose distribution, D_{true}. This is also unknown for *in vivo* patient imaging. **(c)** The true activity concentration convolved by the PSF of the imaging system produces $A(x,y,z)$. Since LDM applies no additional burring to $A(x,y,z)$, the shape in (c) is the shape of the absorbed-dose profile if the LDM is applied. **(d)** $A(x,y,z)$ is convolved by the ^{90}Y DPK. Notice the additional spread in the absorbed-dose profile in (d) compared with (c). Both overestimate the spread of the true absorbed-dose profile (D_{true}), owing to the 6.0-mm FWHM PSF of the imaging system in this fabricated example.

an accuracy advantage over LDM. While it is possible the LDM may produce slightly more accurate results, this increased accuracy may not be clinically detectible. In light of the fact that the LDM can be applied with a simple scaling factor, it is highly recommended for routine ^{90}Y image-based dosimetry.

12.5.2 LDM AND ALTERNATIVE RADIONUCLIDES

While this book has been primarily focused on ^{90}Y radioembolization, ^{166}Ho radioembolization may become more widespread in the future owing to

its ability to be directly imaged by SPECT/CT and MRI. These characteristics have been reviewed in Chapters 1 and 15. ^{166}Ho decays with the emission of an 80.5 KeV γ-ray with a branching ratio of 6% in addition to its therapeutic β emission. While one might initially assume this γ emission would preclude the use of LDM since it will result in dose deposition beyond the voxel of interest, the γ emission is unlikely to be of significance. Both Traino et al. (2013) and Pacilio et al. (2015) previously showed the validity of the use of the LDM for both ^{90}Y and ^{131}I. Owing to the much larger branching ratios of γ-emissions in ^{131}I (81.5% for the 364.5 keV γ-ray) compared with ^{166}Ho, it is likely that the

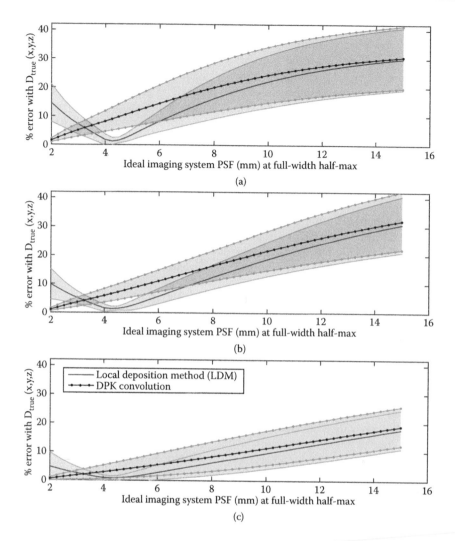

Figure 12.2 Error analysis of DPK convolution and LDM in comparison to D_{true} for mathematically simulated NEMA IEC body phantom, in the absence of image noise. Error is presented as a function of scanner PSF FWHM, with blur applied using a Gaussian kernel with a 2-mm isotropic voxel size. Every voxel in each hot sphere was compared with corresponding voxels in D_{true}, with average error, maximum, and minimum indicated with the central, upper, and lower bounds of the shaded regions, respectively. **(a)** 10 mm hot sphere; **(b)** 17 mm hot sphere; and **(c)** 37 mm hot sphere.

validity of LDM would also be extended to [166]Ho; however, this has not been explicitly explored.

12.6 INSTANCES WHERE DPK CONVOLUTION AND LDM MAY NOT BE APPROPRIATE

As discussed in Section 12.1, fully 3D Monte Carlo radiation transport calculations are rarely used for ^{90}Y hepatic radioembolization owing to the homogeneity of liver tissue. However, the inhomogeneous material density at the dome of the liver represents one case where the uniformity requirement of DPK may not be valid. Mikell et al. (2015) simulated the accuracy of the aforementioned image-based dosimetry techniques, including 3D Monte Carlo at the liver–lung interface as a function of scanner resolution. Mikell demonstrated that DPK convolution, LDM, and 3D Monte Carlo are within 10% of the true liver absorbed dose when deeper than 12 mm from the liver–lung interface; however, this distance is expected to increase as the scanner resolution becomes poorer. Similarly, DPK, LDM, and 3D Monte Carlo achieved similar accuracy in the lung when deeper than 39 mm from the interface. Monte Carlo-based dosimetry is required for the precise determination of liver or lung absorbed dose at the interface.

12.7 CONCLUSION

Quantitative postradioembolization imaging is feasible in routine clinical scenarios as shown in Chapters 10 and 11. This has largely been made possible not only by ^{90}Y PET/CT, but also by advancements in techniques to quantify ^{90}Y bremsstrahlung SPECT. The major barrier related to patient-specific image-based dosimetry for ^{90}Y has always been the lack of broadly available quantitative imaging. With these data, and combined with the fact that radioembolization is a permanent implant obviating the need for serial imaging, transformation into 3D absorbed dose becomes trivial. Methods used to generate absorbed-dose maps can range in complexity; however, for many clinical scenarios the application of a simple scaling factor (LDM) is all that is necessary for excellent results. With these techniques now becoming firmly established, the much more challenging problem of developing a standard-of-care protocol for applying these data toward the management of patient care can begin to be addressed.

REFERENCES

Amato, E. et al. (2012). An analytical method for computing voxel S values for electrons and photons. *Med Phys* 39:6808–6817.

Berger, M.J. (1963). Monte Carlo calculation of the penetration and diffusion of fast charged particles. *In Methods in Computational Physics*, Vol. I, Adler, B., Fernbach, S., Rotenberg, M. (Eds). New York, NY: Academic Press, 135–215.

Bolch, W.E. et al. (1999). MIRD pamphlet No. 17: The dosimetry of nonuniform activity distributions—Radionuclide S values at the voxel level. *J Nucl Med* 40:11S–36S.

Bourgeois, A.C. et al. (2014). Intra-procedural 90Y PET/CT for treatment optimization of 90Y radioembolization. *J Vasc Interv Radiol* 25:271–275.

Chiesa, C. et al. (2012). A dosimetric treatment planning strategy in radioembolization of hepatocarcinoma with 90Y glass microspheres. *Q J Nucl Med Mol Imaging* 56:503–508.

D'Arienzo, M. et al. (2013). Absorbed dose to lesion and clinical outcome after liver radioembolization with (90)Y microspheres: A case report of PET-based dosimetry. *Ann Nucl Med* 27:676.

Dieudonne, A. et al. (2010). Fine-resolution voxel S values for constructing absorbed dose distributions at variable voxel size. *J Nucl Med* 51:1600–1607.

Dieudonne, A. et al. (2011). Clinical feasibility of fast 3-dimensional dosimetry of the liver for treatment planning of hepatocellular carcinoma with 90Y-microspheres. *J Nucl Med* 52:1930–1937.

Eckerman, K.F., Westfall, R.J., Ryman, J.C., Cristy, M. (1994). Availability of nuclear decay data in electronic form, including beta spectra not previously published. *Health Phys* 67:338–345.

Elschot, M. et al. (2013). Quantitative comparison of PET and bremsstrahlung SPECT for imaging the in vivo yttrium-90 microsphere distribution after liver radioembolization. Villa E, ed. *PLoS One* 8:e55742.

Fernández, M. et al. (2013). A fast method for rescaling voxel S values for arbitrary voxel sizes in targeted radionuclide therapy from a single Monte Carlo calculation. *Med Phys* 40:082502.

Guimarães, C.C., Moralles, M., Sene, F.F., Martinelli, J.R. (2010). Dose-rate distribution of 32P-glass microspheres for intra-arterial brachytherapy. *Med Phys* 37:532–539.

ICRU. (1992). *Photon, Electron, Proton and Neutron Interaction Data for Body Tissues*. ICRU Report 46. Bethesda: International Commission on Radiation Units and Measurements.

Kao, Y.H. et al. (2013). Post-radioembolization yttrium-90 PET/CT—Part 2: Dose–response and tumor predictive dosimetry for resin microspheres. *EJNMMI Res* 3:1–1.

Kennedy, A., Dezarn, W., Weiss, A. (2011). Patient specific 3D image-based radiation dose estimates for 90Y microsphere hepatic radioembolization in metastatic tumors. *J Nucl Med Radiat Ther* 2:1–8.

Ljungberg, M., Sjögreen-Gleisner, K. (2011). The accuracy of absorbed dose estimates in tumours determined by quantitative SPECT: A Monte Carlo study. *Acta Oncol* 50:981–989.

Loevinger, R., Kassis, A.I., Burt, R.W. (Eds.). (1989). *The MIRD Perspective*. Washington, DC: The American College of Nuclear Physicians.

Mazzaferro, V., Sposito, C., Bhoori, S., Romito, R. (2013). Yttrium-90 radioembolization for intermediate-advanced hepatocellular carcinoma: A phase 2 study. *Hepatology* 57:1826–1837. doi: 10.1002/hep.26014. Epub 2013 Mar 22.

Mikell, J.K. et al. (2015). Comparing voxel-based absorbed dosimetry methods in tumors, liver, lung, and at the liver-lung interface for 90Y microsphere selective internal radiation therapy. *EJNMMI Phys* 2:947.

Pacilio, M. et al. (2015). Differences in 3D dose distributions due to calculation method of voxel S-values and the influence of image blurring in SPECT. *Phys Med Biol* 60:1945–1964.

Pacilio, M. et al. (2009). Differences among Monte Carlo codes in the calculations of voxel S values for radionuclide targeted therapy and analysis of their impact on absorbed dose evaluations. *Med Phys* 36:1543–1552.

Pasciak, A.S., Bourgeois, A.C., Bradley, Y.C. (2014). A comparison of techniques for 90Y PET/CT image-based dosimetry following radioembolization with resin microspheres. *Front Oncol* 4:121. doi: 10.3389/fonc.2014.00121.

Pasciak, A.S., Erwin, W.D. (2009). Effect of voxel size and computation method on Tc-99m MAA SPECT/CT-based dose estimation for Y-90 microsphere therapy. *IEEE Trans Med Imaging* 28:1754–1758.

Paxton, A.B., Davis, S.D., DeWerd, L.A. (2012). Determining the effects of microsphere and surrounding material composition on 90Y dose kernels using EGSnrc and mcnp5. *Med Phys* 39:1424–1434.

Srinivas, S.M. et al. (2014). Determination of radiation absorbed dose to primary liver tumors and normal liver tissue using post-radioembolization (90)Y PET. *Front Oncol* 4:255.

Stabin, M.G., Eckerman, K.F., Ryman, J.C., Williams, L.E. (1994). Bremsstrahlung radiation dose in yttrium-90 therapy applications. *J Nucl Med* 35:1377–1380.

Strigari, L., Menghi, E., D'Andrea, M., Benassi, M. (2006). Monte Carlo dose voxel kernel calculations of beta-emitting and Auger-emitting radionuclides for internal dosimetry: A comparison between EGSnrcMP and EGS4. *Med Phys* 33:3383.

Traino, A.C. et al. (2013). Dosimetry for nonuniform activity distributions: A method for the calculation of 3D absorbed-dose distribution without the use of voxel S-values, point kernels, or Monte Carlo simulations. *Med Phys* 40:042505.

Uusijarvi, H. et al. (2009). Comparison of electron dose-point kernels in water generated by the Monte Carlo codes, PENELOPE, GEANT4, MCNPX, and ETRAN. *Cancer Biother Radiopharm* 24:461–467.

Diagnostic reporting using postradioembolization imaging

YUNG HSIANG KAO

13.1 INTRODUCTION

As described in Chapter 1, yttrium-90 (^{90}Y) radioembolization is a point-source brachytherapy delivered by millions of radioactive microspheres permanently embedded within the target vascular territory and dispersed by intra-arterial injection. The aim of postradioembolization imaging is to verify the intended radiation plan. This consists of two components: first to confirm technical success and second to assess the likelihood of clinical success or serious toxicity based on tissue absorbed doses. Postradioembolization imaging may be achieved either indirectly using bremsstrahlung single-photon emission computed tomography with integrated computed tomography (CT) (SPECT/CT) or directly by positron emission tomography with integrated CT (PET/CT) of minuscule positron emission.

The term "technical success" refers to whether the treatment was carried out according to pretherapy planning expectations based on the qualitative assessment of angiography and subsequent microsphere biodistribution (Salem et al., 2011; Kao et al., 2013a). Technical success cannot be determined by angiographic findings alone due to irreconcilable biophysical and injection technique differences between soluble angiographic contrast molecules versus particulate microspheres (Kao et al., 2011; Jiang et al., 2012). Technical success should not be confused with "clinical success," which takes into consideration the tissue effects of microsphere radiobiology as revealed on follow-up imaging or biochemistry (Salem et al., 2011; Kao et al., 2013a).

While the focus of this chapter is limited to direct or indirect imaging of ^{90}Y postinfusion, Chapter 14 will focus on imaging techniques used in follow-up posttherapy.

13.2 BREMSSTRAHLUNG SPECT/CT

Today's minimum standard for postradioembolization imaging is bremsstrahlung SPECT/CT. The advantage of SPECT/CT over planar scintigraphy is the tomographic assessment of activity biodistribution with the added specificity of CT correlative anatomy and CT attenuation correction. Planar bremsstrahlung scintigraphy of the abdomen yields little clinically useful information and should be superseded by SPECT/CT. Even the visual assessment of hepatopulmonary shunting by planar bremsstrahlung scintigraphy is subjective and of limited clinical value beyond what has already been estimated by technetium-99m (^{99m}Tc) macroaggregated albumin (MAA).

As a form of scatter radiation of a continuous energy spectrum without a clear photopeak, bremsstrahlung scintigraphy is limited by inherently poor spatial resolution and quantitative inaccuracy. These factors limit its clinical utility as a diagnostic modality for qualitative and quantitative analyses of target and nontarget activity. Both bremsstrahlung planar and SPECT/CT images appear blurry and are generally of low quality, precluding reliable assessment of subcentimeter lesions. This means that if the majority of targeted tumors are small, technical success will be difficult to confirm. As a result, bremsstrahlung SPECT/CT is frequently indeterminate for subcentimeter lesions or nontarget activity (Kao et al., 2014a), for example, a portal vein tumor thrombosis (Figure 13.1). Furthermore, the thin visceral walls of the stomach, duodenum, and gallbladder are often inseparable from the adjacent liver tissue on bremsstrahlung SPECT/CT. Liver movement due to tidal breathing may also lead to SPECT/CT misregistration. Compounded together, all of these technical issues adversely affect the clinical utility of bremsstrahlung SPECT/CT for the assessment of target and nontarget activity.

Quantification of bremsstrahlung radiation is technically challenging and largely inaccurate. Special techniques are required to quantify these images, as detailed in Chapter 10. However, regardless of whether quantification methods have been employed, if a nontarget activity has unequivocally been detected qualitatively by bremsstrahlung SPECT/CT, it may imply a clinically significant amount of nontarget ^{90}Y activity. Depending on the nontarget organ involved and likelihood of complications, such cases may require urgent clinical attention to mitigate potential toxicity. The general technique of nontarget activity assessment by bremsstrahlung SPECT/CT is similar to that developed for ^{90}Y PET/CT (Table 13.1 and Figure 13.2). The topic of nontarget toxicity risk assessment is further discussed in Section 13.4.5 and in Chapter 14.

Overall, bremsstrahlung SPECT/CT can only provide a gross overview of the general microsphere biodistribution. It typically cannot confidently confirm the presence or absence of activity within subcentimeter lesions or confidently exclude nontarget activity, although there are exceptions. Preliminary data using pinhole collimators for bremsstrahlung scintigraphy may partially ameliorate some of these problems and may benefit from further research (Walrand et al., 2011). Currently, the qualitative and quantitative capabilities of bremsstrahlung SPECT/CT are suboptimal for diagnostic reporting and therefore an alternative modality for postradioembolization imaging should be sought.

13.3 ^{90}Y PET/CT

13.3.1 ADVANTAGES OVER BREMSSTRAHLUNG SPECT/CT

^{90}Y PET/CT is the most promising modality to replace bremsstrahlung SPECT/CT due to its superior qualitative and quantitative capability. Coincidence imaging of ^{90}Y PET is possible due to positron emission resulting from its natural decay (Nickles et al., 2004). Conventional PET/CT or PET/magnetic resonance (MR) scanners with time-of-flight (TOF) may be used for coincidence imaging of positrons from ^{90}Y decay without hardware modification (Lhommel et al., 2009; Wissmeyer et al., 2011). ^{90}Y PET/CT can obtain high-resolution images of microsphere biodistribution with spatial resolutions between 5 and 10 mm (Pasciak et al., 2014a). This means that ^{90}Y PET has the capability to assess for the presence or absence of activity within subcentimeter lesions (Figure 13.3).

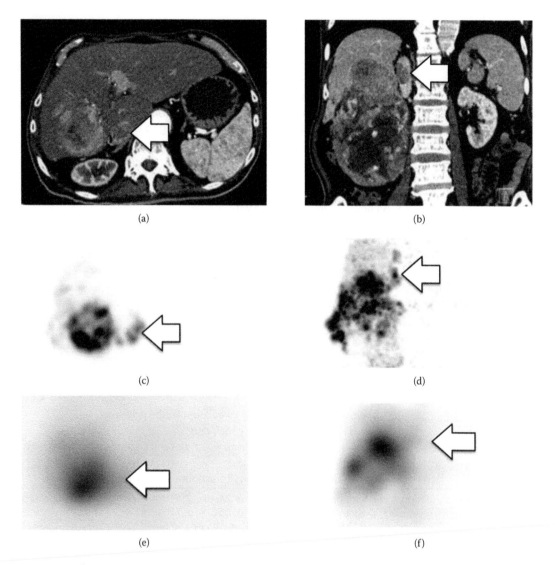

Figure 13.1 Inferior vena cava tumor thrombosis in multifocal hepatocellular carcinoma (arrow). **(a, b)** Triphasic computed tomography (CT) liver in the arterial phase in trans-axial and coronal planes; **(c, d)** yttrium-90 positron emission tomography with integrated CT (⁹⁰Y PET/CT) in trans-axial and coronal planes depict focal activity within the inferior vena cava tumor thrombus in high resolution; **(e, f)** bremsstrahlung single-photon emission computed tomography with integrated CT (SPECT/CT) in trans-axial and coronal planes show faint, indistinct activity in the same region in low resolution.

Quantification of ⁹⁰Y activity by PET obtains reasonably accurate tissue absorbed doses, which may be used to guide postradioembolization management (Willowson et al., 2015). ⁹⁰Y PET voxel dosimetry can also generate dose–volume histograms to graphically describe the heterogeneous nature of microsphere biodistribution, which cannot be accounted for using mean absorbed doses alone (Kao et al., 2013b).

13.3.2 TECHNICAL CHALLENGES OF ⁹⁰Y PET

Clinical ⁹⁰Y PET is currently in the early phases of development with many technical challenges to address. Most TOF PET scanners contain lutetium-176 within its crystal array that is naturally radioactive. Coupled with a scant number of positrons from ⁹⁰Y decay, the reconstructed images are

Table 13.1 Recommendations for diagnostic reporting of postradioembolization ^{90}Y PET/CT[a]

Continuity of care	Postradioembolization ^{90}Y PET/CT is best reported by the same attending nuclear medicine physician who has followed through the entire planning-therapy continuum from exploratory angiography and predictive dosimetry, to ^{90}Y radioembolization.
PET display threshold setting for nontarget ^{90}Y activity detection	For detection of nontarget ^{90}Y activity, the operator should actively adjust the upper PET visual display threshold setting to deliberately *increase* the background noise to moderate levels.
Criteria for technical success	1. ^{90}Y activity present in the majority of targeted tumors, or good overall activity coverage of large targeted tumors; and 2. The absence of clinically significant nontarget ^{90}Y activity; and 3. All findings are in keeping with pretherapy radiation planning expectations.
Criteria for a technically unsuccessful ^{90}Y radioembolization	1. The complete or near-complete absence of ^{90}Y activity in the majority of targeted tumors, or poor overall activity coverage of large targeted tumors; or 2. The presence of any nontarget ^{90}Y activity where ^{90}Y PET dose quantification predicts a high likelihood of clinically significant radiation toxicity; or 3. Any other situation where the ^{90}Y activity biodistribution is adversely inconsistent with pretherapy radiation planning expectations.
Criteria for nontarget ^{90}Y activity	1. Nonrandom pattern of activity distribution; and 2. Conforms morphologically to an untargeted anatomical structure on CT; with or without 3. A plausible vascular etiology to account for its presence.
Criteria for noise spikes	1. Small, discrete, ovoid activity foci; and 2. Random pattern of distribution, which does not conform to underlying anatomy on CT; and 3. No plausible vascular etiology to account for its presence.

Source: With kind permission from **Springer Science+Business Media:** *EJNMMI Res,* Post-radioembolization yttrium-90 PET/CT—Part 1: Diagnostic reporting, 3, (2013a), 56, Kao, Y.H. et al.

Note: ^{90}Y PET CT, yttrium-90 positron emission tomography with integrated computed tomography.

[a] Reproduced under the terms of the Creative Commons Attribution License for an Open Access article (Kao et al., 2013a).

(a) 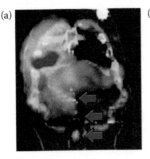 (b) 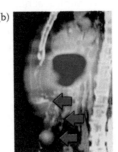 (c)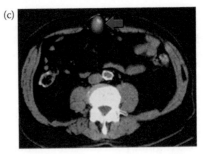

Figure 13.2 Nontarget activity along the falciform ligament to the umbilicus detected on bremsstrahlung SPECT/CT (arrows) depicted in **(a)** coronal, **(b)** sagittal, and **(c)** trans-axial planes. The patient underwent resin microsphere radioembolization of colorectal liver metastases, with bevacizumab 6 weeks prior. Widespread mild nontarget activity was detected throughout the celiac axis, presumed to be due to stasis and reflux. The patient experienced significant abdominal pain during and after radioembolization that was managed well with analgesia. ^{90}Y PET was not available. There were no clinical signs of radiomicrosphere dermatitis on follow-up over 2 months.

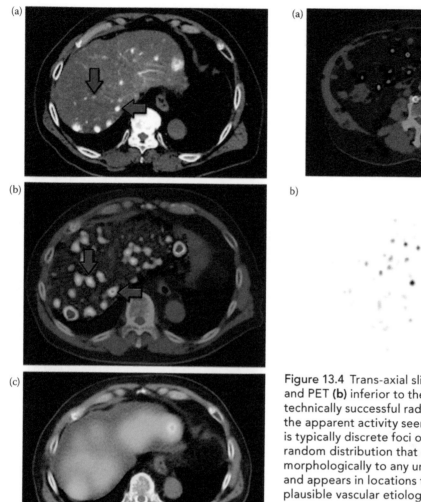

(a)

(b)

(c)

Figure 13.3 Multifocal hepatocellular carcinoma with multiple small tumors. **(a)** Catheter-directed CT of the proper hepatic artery demonstrates the hypervascularity of the multiple small tumors. Small tumors in the right lobe are apparent, several of which are subcentimeter in diameter. **(b)** ^{90}Y PET/CT depicts, in high resolution, discrete focal activity within the small tumors. **(c)** Bremsstrahlung SPECT/CT shows focal activity only in the two largest tumors; activities in the other small tumors were indistinguishable from nontumorous liver activity.

Figure 13.4 Trans-axial slice of ^{90}Y PET/CT **(a)** and PET **(b)** inferior to the level of the liver of a technically successful radioembolization. All of the apparent activity seen here is noise. Noise is typically discrete foci of variable intensity in a random distribution that does not correspond morphologically to any underlying anatomy and appears in locations that do not have any plausible vascular etiology.

vulnerable to high background noise (Figure 13.4). As can be expected, the severity of background noise is correlated to the ^{90}Y radioconcentration within the field-of-view, that is, the higher the ^{90}Y radioconcentration, the better the image quality and vice versa. Noise is worst if the entire liver was treated with a low total activity of ^{90}Y microspheres resulting in very low ^{90}Y radioconcentration within the field-of-view. The resultant noisy images may adversely affect the diagnostic accuracy of target and nontarget activity assessment.

Similarly, the quantitative accuracy of ^{90}Y PET in areas of low radioconcentration such as the nontumorous liver and lung are also vulnerable to noise. Another problem with lung activity quantification by ^{90}Y PET is the "spill-in" of activity from the liver dome into lung bases due to tidal breathing (Figure 13.5). This will result in overestimation of absorbed doses at the lung bases (Figure 13.6) and underestimation of the liver dose. Respiratory-gated ^{90}Y PET/CT may be a promising solution and further research is warranted (Mamawan et al., 2013).

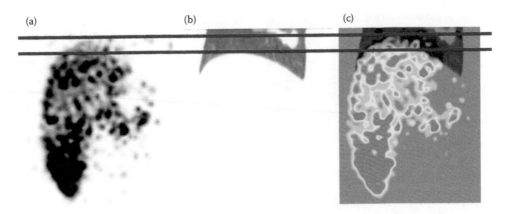

Figure 13.5 Nonrespiratory-gated ^{90}Y PET/CT demonstrating misregistration at the right diaphragm. Horizontal red lines depict the extent of misregistration between PET and CT components. Parts a, b, and c are coronal ^{90}Y PET, coronal CT, and coronal fused ^{90}Y PET/CT images, respectively.

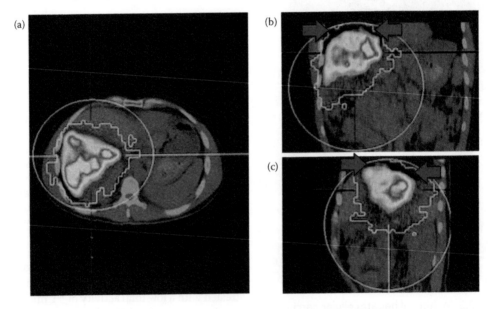

Figure 13.6 Activity "spill over" from the liver dome into the right lung base (arrows) depicted by volumetric isocontour thresholding of ^{90}Y PET/CT. This may affect the accuracy of activity quantification at the right lung base. Parts a, b, and c are axial, coronal, and sagittal reconstructions, respectively.

The benefits and challenges associated with quantitative ^{90}Y PET/CT imaging are described in additional detail in Chapter 11.

13.4 PRINCIPLES OF DIAGNOSTIC REPORTING

^{90}Y PET is vulnerable to noise due to the low–true coincidence rate and natural radioactivity from lutetium-based PET crystals. At the outset, the visual quality of the reconstructed ^{90}Y PET images is considerably noisier than conventional PET tracers and may seem uninterpretable. Fortunately, the problem of noise may be ameliorated using some simple qualitative techniques to facilitate diagnostic reporting.

Recommendations on the general technique for diagnostic reporting of postradioembolization ^{90}Y PET are summarized in Table 13.1 (Kao et al., 2013a). Many of its recommendations are also applicable for bremsstrahlung SPECT/CT, but within its inherent limitations of poorer spatial resolution and lack of quantitative accuracy.

13.4.1 CONTINUITY OF CARE

Radioembolization is a multistage continuum requiring close interdisciplinary coordination and communication. Case-specific technical complexities often influence the final angiographic approach and radiation plan. Each patient's planning-therapy continuum is unique and continuity of care is paramount for clinically meaningful diagnostic reporting of postradioembolization imaging. Therefore, it is preferred that the same members of the multidisciplinary team follow through the entire workflow from exploratory angiography to radioembolization. If any team member is different between the planning and treatment stages, handover may risk inadvertent omission of crucial technical details or their significance may not be fully appreciated by the new member.

For diagnostic reporting of postradioembolization imaging, the two key questions are whether the radioembolization was technically successful and whether any clinically significant nontarget activity was detected. To answer these questions meaningfully, the reporting doctor should have case-specific knowledge of the target arterial territories (e.g., whole-liver, lobar, segmental, subsegmental), treatment intent (e.g., Palliative, lobectomy, segmentectomy, sequential, intentional sparing), arteries-at-risk, and any last minute deviations from the intended angiographic plan due to unforeseen on-table events. In general, ^{90}Y activity biodistribution that is within pretherapy planning expectations may be considered as a technical success. It should be reiterated that "technical success" is a qualitative term that has little or no bearing on "clinical success" unless tissue absorbed doses are known by predictive dosimetry or ^{90}Y PET quantification.

Even prior to looking at the postradioembolization images, the reporting doctor should already have an expectation of what the ^{90}Y activity biodistribution *should* be, as was simulated by Tc-99m MAA SPECT/CT. If the Tc-99m MAA biodistribution appears unexpectedly discordant to the ^{90}Y activity biodistribution, the reporting doctor should attempt to explain this based on the case-specific knowledge to further enhance patient management.

13.4.2 TARGET ACTIVITY

The aim of target activity assessment is to determine whether the biodistribution of target ^{90}Y activity

is within pretherapy planning expectations. The first step is to manually adjust the PET upper display threshold to qualitatively suppress the visual appearance of noise (Figure 13.7). This is because target activity is usually more intense than nontarget activity and background noise. It is important to note that normal microsphere biodistribution is always heterogeneous at both the microscopic and macroscopic levels. Therefore, "clumps" of activity in tumor, nontumorous liver, or lungs should not be routinely disregarded as noise.

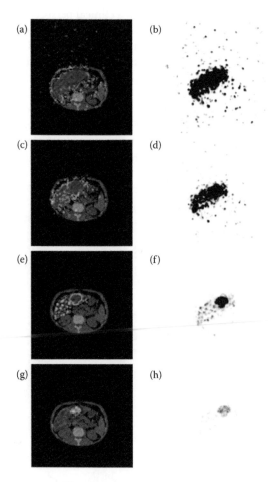

Figure 13.7 The importance of qualitative adjustment of the PET upper display threshold to minimize the visual appearance of noise for target activity assessment. This series of images depicts the same trans-axial slice of a ^{90}Y PET/CT over four different PET upper display thresholds: **(a, b)** 0% (40 kBq/mL); **(c, d)** 2% (190 kBq/mL); 20% (2,030 kBq/mL); and **(g, h)** 100% (10,430 kBq/mL). The left lobe tumor with heterogeneous ^{90}Y activity is best seen in (g, h).

For hypervascular tumors, ^{90}Y activity is normally more intense at its hypervascular periphery becoming less intense toward its center. For large or massive tumors, the heterogeneous microsphere biodistribution is often visually detectable as heterogeneous clumps of activity. In these cases, technical success requires good circumferential coverage of ^{90}Y activity throughout the tumor with minimal gaps of absent activity. It is a normal finding in massive tumors for the intensity of ^{90}Y activity to gradually decrease toward its core due to reduced arterial penetration and central necrosis (Figure 13.8). From a clinical perspective, this means that a complete response is difficult to achieve for massive tumors using radioembolization alone because the total prescribed ^{90}Y activity would have been dosimetrically constrained for safety to the nontumorous liver or lungs.

For multiple small- to medium-sized tumors, technical success requires ^{90}Y activity to be detected in the majority of targeted tumors, within the limitations of ^{90}Y PET spatial resolution. Hypovascular

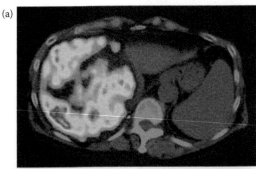

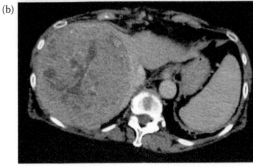

Figure 13.8 Massive hepatocellular carcinoma in the right lobe with central necrosis. **(a)** ^{90}Y PET/CT depicts in high resolution the inherently heterogeneous tumor activity biodistribution. Photopenic regions correspond to central necrosis. **(b)** CT liver in the portovenous phase for correlation.

tumors such as those partially treated by other modalities usually appear relatively photopenic as compared with its surrounding nontumorous liver parenchyma (Figure 13.9).

It may be possible to comment on ^{90}Y activity within subcentimeter tumors if the focally implanted radioconcentration is high. Within the limitations of ^{90}Y PET spatial resolution, it is sometimes necessary to comment on the presence or absence of activity within small but critical target lesions such as portal vein tumor thrombosis. For example, the absence of significant ^{90}Y activity within targeted portal vein tumor thrombosis is ominous for early progressive disease and warrants vigilant follow-up or adjuvant treatment modalities (Figure 13.10).

13.4.3 NONTARGET ACTIVITY VERSUS NOISE

Assessment of nontarget activity by ^{90}Y PET is challenging due to background noise, which may confuse the reporting doctor unless an appropriate diagnostic technique is applied. Until the arrival of better imaging and reconstruction protocols for ^{90}Y PET to minimize noise, the following simple qualitative techniques may improve the accuracy of nontarget activity detection.

First, the reporting doctor should be aware that noise spikes might be of greater visual intensity than nontarget activity. Noise typically appears as ovoid foci that are randomly distributed throughout the field-of-view and do not correlate with any angiographically plausible anatomical structure (Figure 13.4). The visual intensity of nontarget activity may be subtle or mild because it reflects the amount of implanted ^{90}Y microspheres. This means that the visual appearance of nontarget activity may vary widely; hence any suspicious activity should not be disregarded as noise based on visual intensity alone.

Nontarget activity should also be interpreted in conjunction with the reporting doctor's impression of the likelihood of nontarget tissue toxicity. The absence of clinical symptoms at the time of postradioembolization imaging does not exclude a qualitative diagnosis of nontarget activity. This is because clinical symptoms may not manifest until days, weeks, or months later depending on the organ involved and the nontarget absorbed dose.

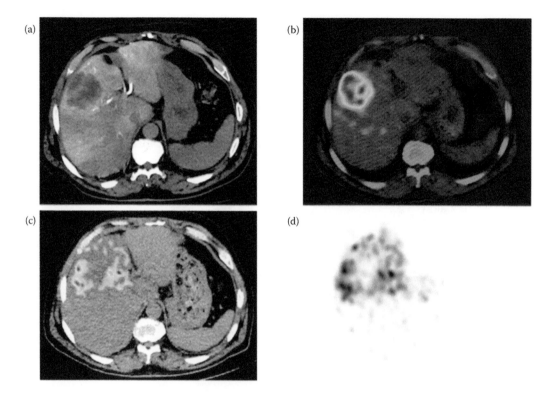

Figure 13.9 Hypovascular intrahepatic cholangiocarcinoma in the right lobe. **(a)** Catheter-directed CT of the proper hepatic artery shows that the targeted tumor is predominantly hypovascular. **(b)** 18-Fluorodeoxyglucose (FDG) PET/CT shows that the tumor is mostly viable with only a small amount of central necrosis. **(c, d)** ^{90}Y PET/CT after radiomicrosphere segmentectomy depicts in high resolution the heterogeneous activity biodistribution at the tumor periphery with low activity within the tumor mass, consistent with a hypovascular tumor.

As nontarget activity may be of lower visual intensity than background noise, the reporting doctor should first qualitatively adjust the PET display threshold to deliberately *increase* background noise to moderate levels. This counterintuitive technique facilitates the visual detection of any nonrandom pattern among a random noise background. The rotating maximum intensity projection (MIP) is useful to detect any nonrandom activity pattern protruding against target tissue activity. Any nonrandom activity pattern detected on the rotating MIP should be pursued on the PET/CT (or PET/MR) images in axial, coronal, and sagittal planes. The CT or MR images should be carefully examined for any corresponding anatomical structures conforming morphologically to the visual distribution of the suspected nontarget activity. This should also be supported by an angiographically plausible theory to explain the presence of nontarget activity. The converse is true: any activity pattern that does not correspond to an angiographically plausible

anatomical structure is unlikely to represent nontarget activity and is probably noise. The absence of a proven artery-at-risk by retrospective review of angiography does not exclude a diagnosis of nontarget activity because the culprit artery may not always be identified on angiography.

Nontarget activity in the stomach, duodenum, and gallbladder should appear in a linear pattern conforming to its walls (Figure 13.11). In untargeted liver, nontarget activity should conform morphologically to the parenchyma of untargeted lobe or segments (Figure 13.12). Suspicious but visually subtle activity is diagnostically challenging and often indeterminate for nontarget activity versus noise. Fortunately, such indeterminate cases are usually clinically insignificant because the nontarget absorbed dose is likely to be low.

Both ^{90}Y PET and bremsstrahlung SPECT/CT are usually not respiratory-gated and therefore vulnerable to misregistration. This problem is worst when assessing for nontarget activity in viscera that lie

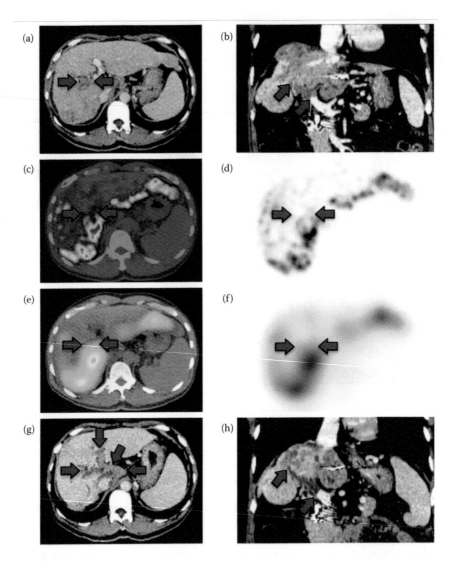

Figure 13.10 Multifocal hepatocellular carcinoma with portal vein tumor thrombosis. (a, b) Portal vein tumor thrombosis (arrows) depicted on portal venous phase of triphasic CT liver in trans-axial and coronal planes. (c, d) ^{90}Y PET/CT shows, in high resolution, subtle ^{90}Y activity within the portal vein tumor thrombus, which is unlikely to be effective. (e, f) Bremsstrahlung SPECT/CT was indeterminate for the presence or absence of focal activity within the portal vein tumor thrombus due to visual interference from adjacent liver activity. (g, h) Follow-up triphasic CT liver 3 months after radioembolization shows progression of the portal vein tumor thrombosis, depicted here in the portal venous phase in transaxial and coronal planes. This clinically validates the ^{90}Y PET/CT finding of subtle, ineffective activity within the portal vein tumor thrombosis.

closely adjacent to the liver serosa such as the gastric pylorus, proximal duodenum, or gallbladder fundus. Due to the poor spatial resolution of bremsstrahlung SPECT/CT, these areas are often indeterminate for the nontarget activity. The improved spatial resolution of ^{90}Y PET partially ameliorates this problem with additional quantitative capability.

13.4.4 TECHNICAL FAILURE

The fundamental premise of radioembolization is to balance safety and efficacy within a multidisciplinary framework. Technical failure has occurred when the ^{90}Y activity biodistribution is adversely inconsistent with pretherapy planning expectations.

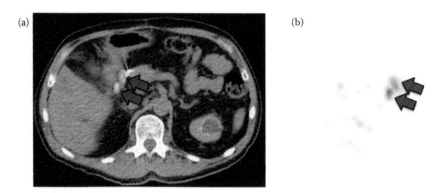

Figure 13.11 Nontarget ⁹⁰Y activity in the proximal duodenum. ⁹⁰Y PET/CT shows mild nontarget activity in a linear morphology corresponding to the proximal duodenal wall. Nontarget activity can be appreciated both on the fused ⁹⁰Y PET/CT **(a)** and ⁹⁰Y PET image **(b)**. ⁹⁰Y PET quantification was not performed. The patient was clinically asymptomatic on follow-up.

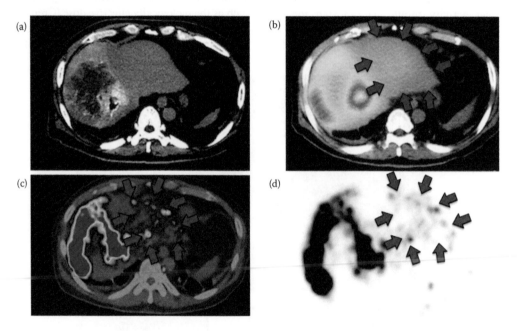

Figure 13.12 Nontarget ⁹⁰Y activity in the untargeted left lobe. **(a)** Large hypovascular hepatocellular carcinoma of the right lobe depicted by catheter-directed CT of the right hepatic artery. **(b)** Bremsstrahlung SPECT/CT shows, in low resolution, subtle diffuse bremsstrahlung activity in the untargeted left lobe (arrows). **(c, d)** ⁹⁰Y PET/CT shows, in high resolution, nontarget activity in a nonrandom distribution conforming to the anatomy of the untargeted left lobe. The nontarget activity was probably due to mild microsphere reflux, arterioportal shunting, or both.

In other words, significant irreversible technical complications had occurred during radioembolization with the potential to cause severe radiomicrosphere toxicity. Technical failure is, therefore, a serious diagnosis that should not be made without due consideration because it implies a high likelihood of adverse clinical outcomes and also negatively impacts team morale. A diagnosis of technical

failure should prompt an urgent clinical review for early management in terms of adjuvant treatment or mitigative action to minimize potential toxicity.

The specific definition of technical failure differs depending on the size and distribution of the targeted tumors and the overall treatment intent. For tumors, technical failure generally means very poor tumor ⁹⁰Y activity coverage and early disease progression is expected. For massive tumors, technical failure means that large regions of the tumor are devoid of ^{90}Y activity, excluding central necrosis.

For nontarget activity, technical failure means that a significant amount of ^{90}Y activity has been detected within nontarget tissue and severe toxicity is likely. However, visual assessment alone is subjective and unreliable for toxicity prediction. Therefore, any nontarget activity of clinical concern should be guided by ^{90}Y PET absorbed dose quantification.

13.4.5 QUANTIFICATION OF NONTARGET ABSORBED DOSE

For clinically meaningful diagnostic reporting, nontarget activity should be reported together with its likelihood of toxicity. To achieve this, ^{90}Y PET quantification of the nontarget absorbed dose

should be performed because visual assessment alone is subjective. The nontarget tissue absorbed dose provides an objective and radiobiologically rational basis to predict toxicity, which in turn impacts mitigative action. Nontarget activity may be assumed to be clinically insignificant only if visually subtle; all other cases should be objectively supported by absorbed dose quantification.

The likelihood and severity of nontarget toxicity depends on the organ involved and absorbed dose biodistribution and should always be assessed on a case-specific basis. Radiomicrosphere dose–response data for nontarget tissue toxicity are currently scarce. To fill this knowledge gap, dose–response experiences of external beam radiotherapy may serve as an interim guide using mathematical extrapolations such as the biologically effective dose (Cremonesi et al., 2014). However, such extrapolations must be cautiously used because the radiobiology of radioactive

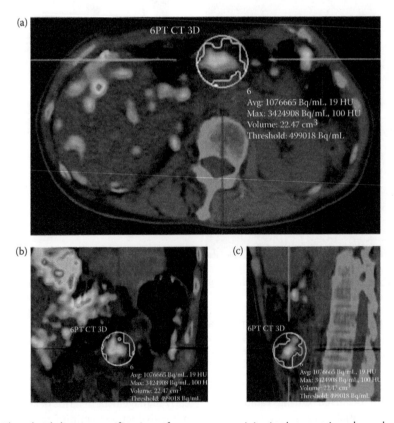

Figure 13.13 Absorbed dose quantification of nontarget activity in the gastric pylorus by ^{90}Y PET. A volume-of-interest was defined in the pylorus by volumetric isocontour thresholding and its mean radioconcentration obtained. After decay correction to the time of radioembolization, a mean absorbed dose of approximately 65 Gy was obtained. This patient developed chronic abdominal pain and pyloric ulceration seen on endoscopy 3 months later. Parts a, b, and c are axial, coronal, and sagittal reconstructions, respectively.

microspheres is different to that of external beam radiotherapy and also between different types of radioactive microspheres.

^{90}Y PET absorbed dose quantification in hollow viscus is challenging. A simple solution may be to define a volume-of-interest to obtain its mean radioconcentration. This is then decay corrected and the ^{90}Y absorbed dose coefficient (approximately 50 Gy per GBq/kg) applied to obtain its mean absorbed dose (Figure 13.13). Using this method, preliminary data for ^{90}Y resin microspheres found that approximately 49 Gy to a localized area of the gastric wall may cause gastritis, 65 Gy may result in ulceration, whereas less than 18 Gy may be asymptomatic; 53 Gy to the duodenum may cause duodenitis (Kao et al., 2013b). Due to the current paucity of data, further research into nontarget dose–response is warranted to guide toxicity prognostication.

13.4.6 VERIFICATION OF ABSORBED DOSES

Modern personalized radioembolization utilizes patient-specific tomographic parameters to optimize the brachytherapy radiation plan. If a scientifically sound and meticulous method of pretherapy radiation planning had been used for ^{90}Y activity prescription (e.g., "artery-specific SPECT/CT partition model"; Kao et al., 2012), then technical success means that the intended radiation plan may be assumed to be valid within the general limits of

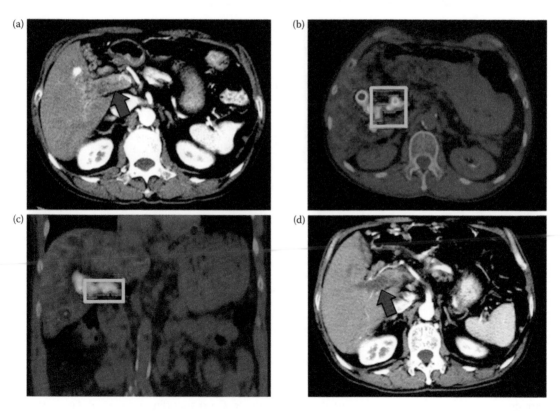

Figure 13.14 Absorbed dose quantification of a portal vein tumor thrombus by ^{90}Y PET. **(a)** Triphasic CT liver in the arterial phase demonstrates a large contrast-enhancing portal vein tumor thrombus (arrow). **(b, c)** A volume-of-interest approximating the activity boundaries of the portal vein tumor thrombus was defined by volumetric isocontour thresholding and its mean radioconcentration obtained. After decay correction to the time of radioembolization, a mean absorbed dose of approximately 248 Gy was obtained. **(d)** Follow-up triphasic CT liver in the arterial phase at 4 months postradioembolization shows a slight decrease in lesion size and a complete lack of contrast enhancement within the portal vein tumor thrombus (arrow), suggesting a complete response; this clinically validates the mean radiation absorbed dose quantified by ^{90}Y PET.

dosimetric uncertainty (Kao et al., 2013c; Song et al., 2015). This means that the treatment response is expected to be in accordance to the prescribed tissue absorbed doses. In such cases, postradio-embolization verification of absorbed doses by ^{90}Y PET quantification is unnecessary unless for quality control, research, or dose–volume histograms, or if the absorbed dose of a specific lesion is desired, for example, portal vein tumor thrombus (Figure 13.14).

One method of ^{90}Y PET quantification is to apply the "local deposition method" (Chapter 12) to obtain the mean absorbed dose and dose–volume histogram within a volume-of-interest (Figure 13.15) (Kao et al., 2013b; Pasciak et al., 2014). Using this analysis, predictive dosimetry

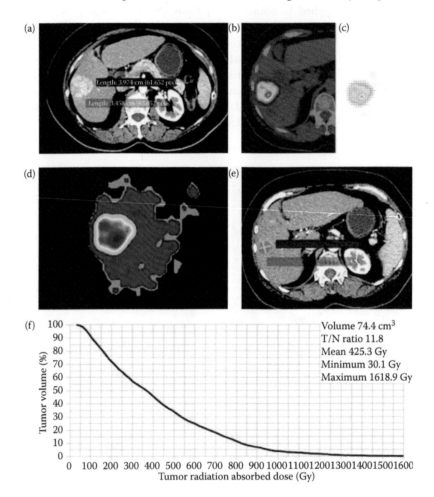

Figure 13.15 ^{90}Y PET tumor voxel dosimetry and dose–volume histogram using the local deposition method. **(a)** Triphasic CT liver in the arterial phase shows a right lobe hepatocellular carcinoma measuring 4.0 × 3.5 cm. **(b)** ^{90}Y PET/CT depicts activity biodistribution in high resolution, with intense tumor activity and low-grade activity in nontumorous liver. **(c)** Corresponding trans-axial slice of the ^{90}Y PET display used for manual contouring of tumor volume-of-interest for voxel dosimetry, indicated by small black dots. **(d)** Isodose map of the corresponding trans-axial slice of the right liver lobe provides a visual representation of dose heterogeneity within the target arterial territory and displays the full range of delivered dose from 0 Gy to >1600 Gy. **(e)** Follow-up triphasic CT liver in arterial phase 5.5 months later shows a noncontrast-enhancing hypodensity with significant size reduction to 2.1 × 1.8 cm, representing a complete response. **(f)** Dose–volume histogram generated by ^{90}Y PET voxel dosimetry from the tumor volume-of-interest shown in (c). Mean, minimum, and maximum tumor absorbed doses were 425 Gy, 30 Gy, and 1619 Gy, respectively; D_{70} was >210 Gy, where D_{70} is the minimum absorbed dose to 70% tumor volume.

based on Tc-99m MAA SPECT/CT was shown to be accurate for tumor absorbed doses with a low mean bias of +6.0% (95% confidence interval −1.2% to +13.2%) in a subset of highly select tumors (Kao et al., 2013b).

Tissue mean absorbed doses may also be calculated using simple count ratios to obtain the "true" tumor-to-normal liver (T/N) ratio, analogous to that estimated by Tc-99m MAA during pretherapy planning. Input of the true T/N ratio back into tissue masses and lung shunt fraction as per Medical Internal Radiation Dose (MIRD) macrodosimetry (i.e., "Partition Model") (Ho et al., 1996) will obtain more accurate tissue mean absorbed doses. Using this analysis, good correlations were found for mean absorbed doses by Tc-99m MAA compared with ^{90}Y PET for both tumor ($r = 0.64$; $p < .01$) and nontumorous liver ($r = 0.71$; $p < .001$) (Song et al., 2015).

Semiempirical ^{90}Y activity prescription such as the "body surface area method" for resin microspheres has no radiobiologically rational basis to establish any dose–response relationships, unless ^{90}Y PET quantification is retrospectively performed. As an inherent conceptual limitation of the semiempirical paradigm, patient-specific tissue absorbed doses are unknown at the time of ^{90}Y activity prescription. At the discretion of the treating team, ^{90}Y PET quantification may be retrospectively performed to discover what the tissue absorbed doses actually were, albeit too late for any absorbed dose modification. This situation is similar for glass microspheres, where ^{90}Y activity prescription is generally based on a mean absorbed dose averaged across the entire target arterial territory.

For the lung, absorbed dose verification by ^{90}Y PET is more technically challenging. First, tidal breathing may overestimate the lung absorbed dose at the lung bases due to "spill in" of activity from the liver dome (Figures 13.5 and 13.6). Respiratory gating may be a possible solution (Mamawan et al., 2013). Second, the lung radioconcentration within the PET field-of-view may be too low for accurate quantification. However, it may be theoretically possible to *indirectly* calculate the total lung activity as the difference between the total injected activity and whole-liver activity quantified by ^{90}Y PET. The lung mean absorbed dose may then be calculated using the patient-specific lung mass estimated by CT densitovolumetry (Kao et al., 2014b).

13.5 ECONOMICS OF ^{90}Y PET

A practical challenge of ^{90}Y PET is its relatively long acquisition time as compared with conventional PET tracers such as 18-fluorodeoxyglucose (FDG). Due to a very low ^{90}Y positron fraction, a longer acquisition per bed is preferred. Today's TOF scanners typically acquire ^{90}Y PET at 15–20 minutes per bed position. If the liver is markedly enlarged, two bed positions may be required, increasing the total PET acquisition time to 30–40 minutes. In comparison, a whole-body FDG PET/CT by a TOF scanner takes 20–30 minutes to complete. Hence, the economic impact of performing such long ^{90}Y PET scans for a single patient cannot be ignored, especially in high-throughput PET centers.

A total scan time of 40 minutes for two-bed positions is probably at the limit of tolerance for most patients. Any further increase in acquisition time may risk patient discomfort, movement, and misregistration (Figure 13.16). Unless future research can significantly shorten the PET acquisition time without compromising image quality, extending the PET field-of-view from the abdomen into the lungs will result in an impractically long total scan time of 45–80 minutes over three to four bed positions. Any respiratory gating will compound the total scan time even longer.

If ^{90}Y PET of both the lung and liver is clinically indicated, for example, patients with high

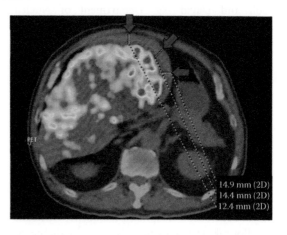

14.9 mm (2D)
14.4 mm (2D)
12.4 mm (2D)

Figure 13.16 Trans-axial PET/CT misregistration of up to 1.5 cm in the liver due to patient movement during a 15 minutes per bed ^{90}Y PET acquisition.

lung shunting where knowledge of the true lung absorbed dose is clinically relevant, a possible solution may be to break up the lung and liver into two separate acquisitions to allow the patient to rest between the two scans. All ^{90}Y PET quantification must be decay corrected to the time of radioembolization. The overall economic feasibility of ^{90}Y PET will vary depending on the healthcare financial model and research grant availability of each country and institution.

13.6 CONCLUSIONS

^{90}Y PET is technically superior to bremsstrahlung SPECT/CT and should be preferred where available. Continuity of care is central to clinically meaningful diagnostic reporting of postradioembolization imaging. Qualitative and quantitative ^{90}Y PET are interrelated and inseparable components that should be interpreted in the context of each other. Further research on ^{90}Y PET is required to improve the quality of reconstructed images and accuracy of absorbed dose quantification and to investigate dose–response relationships in microsphere radiobiology.

ACKNOWLEDGMENTS

The following members and their institutions are acknowledged for their contributions: Anthony Goh and David Ng, Department of Nuclear Medicine and PET, Singapore General Hospital, Singapore; Jeffrey Steinberg, Jianhua Yan, and David Townsend, Agency for Science Technology and Research—National University of Singapore Clinical Imaging Research Centre, Singapore; Mark Goodwin, Sze Ting Lee, and Andrew Scott, Department of Radiology, Department of Molecular Imaging and Therapy, Austin Health, Melbourne, Australia; Meir Lichtenstein, Department of Nuclear Medicine, The Royal Melbourne Hospital, Melbourne, Australia; and Richard Dowling, Department of Radiology, The Royal Melbourne Hospital, Melbourne, Australia; Jan Boucek, Department of Nuclear Medicine, Sir Charles Gairdner Hospital, Perth, Australia.

REFERENCES

Cremonesi, M. et al. (2014). Radioembolization of hepatic lesions from a radiobiology and dosimetric perspective. *Front Oncol* 2014;4:210.

Ho, S. et al. (1996). Partition model for estimating radiation doses from yttrium-90 microspheres in treating hepatic tumours. *Eur J Nucl Med* 23:947–952.

Jiang, M. et al. (2012). Segmental perfusion differences on paired Tc-99m macroaggregated albumin (MAA) hepatic perfusion imaging and yttrium-90 (^{90}Y) bremsstrahlung imaging studies in SIR-Sphere radioembolization: Associations with angiography. *J Nucl Med Radiat Ther* 3:122.

Kao, Y.H. et al. (2011). Imaging discordance between hepatic angiography versus Tc-99m-MAA SPECT/CT: A case series, technical discussion and clinical implications. *Ann Nucl Med* 25:669–676.

Kao, Y.H. et al. (2012). Image-guided personalized predictive dosimetry by artery-specific SPECT/CT partition modeling for safe and effective 90Y radioembolization. *J Nucl Med* 53:559–566.

Kao, Y.H. et al. (2013a). Post-radioembolization yttrium-90 PET/CT—Part 1: Diagnostic reporting. *EJNMMI Res* 3:56.

Kao, Y.H. et al. (2013b). Post-radioembolization yttrium-90 PET/CT—Part 2: Dose–response and tumor predictive dosimetry for resin microspheres. *EJNMMI Res* 3:57.

Kao, Y.H. (2013c). A clinical dosimetric perspective uncovers new evidence and offers new insight in favor of 99mTc-macroaggregated albumin for predictive dosimetry in 90Y resin microsphere radioembolization. *J Nucl Med* 54:2191–2192.

Kao, Y.H. et al. (2014a). Non-target activity detection by post-radioembolization yttrium-90 PET/CT: Image assessment technique and case examples. *Front Oncol* 4:11.

Kao, Y.H. et al. (2014b). Personalized predictive lung dosimetry by technetium-99m macroaggregated albumin SPECT/CT for yttrium-90 radioembolization. *EJNMMI Res* 4:33.

Lhommel, R. et al. (2009). Yttrium-90 TOF PET scan demonstrates high-resolution biodistribution after liver SIRT. *Eur J Nucl Med Mol Imaging* 36:1696.

Mamawan, M.D., Ong, S.C., Senupe, J.M. (2013). Post-90Y radioembolization PET/CT scan with respiratory gating using time-of-flight reconstruction. *J Nucl Med Technol* 41:42.

Nickles, R.J. et al. (2004). Assaying and PET imaging of yttrium-90: 1 >>34 ppm > 0. IEEE Nuclear Science Symposium Conference Record, October 16–22, 2004; Casaccia, Italy. New York, NY: IEEE; 2004:3412–3414.

Pasciak, A.S., Bourgeois, A.C., Bradley, Y.C. (2014). A comparison of techniques for (90) Y PET/CT image-based dosimetry following radioembolization with resin microspheres. *Front Oncol* 4:121.

Salem, R. et al. (2011). Research reporting standards for radioembolization of hepatic malignancies. *J Vasc Interv Radiol* 22:265–278.

Song, Y.S. et al. (2015). PET/CT-based dosimetry in 90Y-microsphere selective internal radiation therapy: Single cohort comparison with pretreatment planning on 99mTc-MAA imaging and correlation with treatment efficacy. *Medicine (Baltimore)* 94:e945.

Walrand S et al. (2011). Yttrium-90-labeled microsphere tracking during liver selective internal radiotherapy by bremsstrahlung pinhole SPECT: Feasibility study and evaluation in an abdominal phantom. *EJNMMI Res* 1:32.

Willowson, K.P., Tapner, M.; QUEST Investigator Team, Bailey, D.L. (2015). A multicentre comparison of quantitative (90)Y PET/CT for dosimetric purposes after radioembolization with resin microspheres: The QUEST Phantom Study. *Eur J Nucl Med Mol Imaging* 42:1202–1222.

Wissmeyer, M. et al. (2011). 90Y Time-of-flight PET/MR on a hybrid scanner following liver radioembolisation (SIRT). *Eur J Nucl Med Mol Imaging* 38:1744–1745.

14

The use of postprocedural imaging in the medical management of patients

AUSTIN C. BOURGEOIS, MARCELO S. GUIMARAES, YONG C. BRADLEY,
CHRISTOPHER HANNEGAN, AND ALEXANDER S. PASCIAK

14.1 INTRODUCTION

14.1.1 CHAPTER OVERVIEW

Hepatic radioembolization with yttrium-90 (^{90}Y) microspheres is unique in many ways compared with other methods of treating hepatic malignancy, including the use of radiation, size of the infused microsphere, the number of particles administered, and variable technical considerations involved in the infusion. While radioembolization imparts some of the same procedural risks to surgical interventional and other forms of liver-directed therapy (i.e., fulminant hepatic failure), it can result in a variable constellation of both normal treatment effects and possible complications. Thus, image interpretation following ^{90}Y can be confusing, particularly when performed in a vacuum of clinical information and without a robust understanding of radioembolization. This chapter provides a detailed discussion of the imaging and correlative clinical findings following ^{90}Y therapy with an emphasis on clinical relevance and underlying mechanisms.

It is important to consider interventional radiologists, radiation oncologists, nuclear medicine radiologists, and medical physicists may be involved in a ^{90}Y treatment planning and delivery. Each possesses some scope of formal training regarding imaging physics and interpretation. Because those involved in the care of ^{90}Y patients understand imaging, the integration of advanced imaging modalities into routine practice has evolved at a rapid pace. For example, a number of methods of directly imaging ^{90}Y have been implemented into clinical practice, allowing for early evaluation of efficacy and detection of potentially deleterious nontarget embolization (NTE). These have been discussed to some degree in Chapter 13. This chapter provides a summary of the imaging protocols that are routinely implemented in postprocedural surveillance and discusses the role of direct ^{90}Y imaging in its widely accepted and investigational settings.

14.1.2 BACKGROUND

As discussed in detail in Chapter 2, surgical resection and transplantation remain the mainstay of curative therapy for patients with hepatic malignancy (Poon et al., 2002). Unfortunately, the vast majority of patients with both hepatocellular carcinoma (HCC) and liver metastasis are not candidates for surgical cure upon presentation as a result of factors such as poor hepatic reserve, advanced tumors, tumor location, and/or presence of extrahepatic disease (Poon et al., 2002). For patients who are not surgical candidates, locoregional therapies (LRT) such as ablation, transarterial chemoembolization (TACE), and radioembolization with ^{90}Y microspheres play an important role in palliation and may provide a significant survival benefit (Salem et al., 2002; Higgins and Soulen, 2013; Salem et al., 2013b; Bargellini, 2014; Bester et al., 2014). LRTs may also be coupled with surgery in order to allow patients to maintain transplant candidacy while awaiting a donor liver, to debulk disease, or to downstage marginal surgical candidates to potentially curative surgical resection (Braat, 2014; Khan, 2014).

Transarterial LRTs including TACE and radioembolization play a particularly important role in the treatment of patients with extensive or multifocal hepatic tumor burden. These treatments capitalize on the dual blood supply to the liver, and the increased arterial perfusion to tumors relative to that of normal liver, that ranges from approximately 3:1 in colorectal metastasis to often greater than 5:1 for HCC, as described in Chapter 5. Whereas TACE derives a significant component of its efficacy from small vessel occlusion leading to tumor ischemia, ^{90}Y provides a localized radiation therapy with a less significant embolic effect (Kennedy et al., 2007). This fundamental difference provides the framework for interpreting the effects and complications of radioembolization, an emphasis of this chapter. Before proceeding, it is important to briefly review the basic principles of glass (Therasphere® BTG, Ontario, Canada) or resin (SIR-Spheres®, SIRTex Technology Pty, Lane Cove, Australia) ^{90}Y microspheres, as described in Chapter 1. In particular, the radiation biology of radioembolization (Chapter 8) will be important in understanding the discussion in this chapter, which will focus in part on adverse effects.

While the principle that radioembolization is a form of transarterial brachytherapy contributes to its efficacy and tolerability, it also directly contributes to its complications and adverse effects. In particular, NTE to normal liver tissue and extrahepatic soft tissues can be clinically significant, even when using the most rigorous preventative measures (Riaz, 2014). While Chapter 13 partially focused on identifying NTE using the posttreatment ^{90}Y positron emission tomography/computed tomography (PET/CT), this chapter describes the role of routine surveillance imaging in the continuing care of patients who have received radioembolization.

14.2 POSTPROCEDURAL IMAGING: FINDINGS AND SIGNIFICANCE

14.2.1 IMAGING SURVEILLANCE PROTOCOLS

It cannot be understated that the primary outcome in evaluating a cancer therapy is its impact on survival. Tumor response to a therapy is nonetheless an important surrogate marker of treatment success (Singh and Anil, 2013). In the context of ^{90}Y, follow-up imaging generally consists of contrast-enhanced CT, magnetic resonance imaging (MRI), or PET/CT. Ideally, follow-up examinations are performed identically to preprocedural imaging to allow for direct comparison. In clinical practice, the preferred imaging modality varies widely based on resource availability, as well as patient and cost considerations (Attenberger et al., 2015).

Contrast-enhanced CT (CECT) plays a primary role in surveillance in many institutions, given its widespread availability and relatively low costs (Boas et al., 2015). When compared with CT, contrast-enhanced MRI provides improved sensitivity for detecting hepatic tumors and may detect additional hepatic lesions in as many as 30% of patients (Kim et al., 2015). However, the use of MRI may be limited in patients with implantable electronic devices, poor pulmonary function who are unable to comply with required breathing instructions, and those unable to tolerate the relatively lengthy MRI acquisition time.

2-Deoxy-2-^{18}fluoroglucose PET/CT (^{18}FDG-PET/CT) may also be especially useful for characterizing classically hypermetabolic tumors such uveal melanoma metastasis (Eldredge-Hindy et al., 2014). PET imaging provides a quantitative reflection of cellular metabolism by imaging photons emitted from the glucose analog ^{18}FDG (Soydal et al., 2013). Using PET, important prognostic information regarding treatment response to radioembolization can be extrapolated in the preprocedural and early postprocedural periods (Piduru et al., 2012; Soydal et al., 2013; Annunziata et al., 2014; Cho et al., 2015). This will be discussed in detail in the ensuing section. PET/CT also shows improved detection of certain hepatic metastasis compared with CT alone and may aid in ^{90}Y treatment planning and patient selection (Annunziata et al., 2014). The relative efficacy of PET/CT compared with MRI is less clear. A recent prospective study showed that diffusion-weighted MRI (DW-MRI) provided better evaluation of early response of hepatic metastasis to ^{90}Y than PET/CT (Barabasch et al., 2015). However, these results should be taken in the context of the small patient series ($n = 35$) and its exclusion of HCC from the study cohort (Barabasch et al., 2015). In practice, the superiority of MRI over PET/CT is likely situational, depending on variable tumor histology and ^{18}FDG avidity, as well as the ability to acquire high-quality MRI images.

Posttreatment imaging after radioembolization conventionally begins 1 month after therapy, with serial examinations performed at 3-month intervals thereafter (Salem and Thurston, 2006). This methodology was adopted from experience with chemoembolization early in the investigation of radioembolization and remains in widespread use. While there is little foundational evidence for this or many other imaging follow-up regimens after ^{90}Y, posttherapy surveillance has been the subject of at least one large study. Boas et al. (2015) examined 1766 patients who underwent locoregional therapy for treatment of HCC. The authors found that disease recurrence occurs from 0 to 9 months after treatment in the vast majority of patients and peaks at 3 months. Using these data, the authors suggested that a "front loaded" surveillance protocol would provide the appropriate balance of cost optimization and reduction of diagnostic delay. The recommended protocol includes surveillance CT or MRI at 2, 4, 6, 8, 11, 14, 18, 24 months after ^{90}Y therapy (Boas et al., 2015). It should be emphasized that any imaging protocols should include a certain degree of flexibility and be adjusted according to patient-specific preprocedural risk factors, change in tumor markers, or clinical evidence of progressive disease or complication.

14.2.2 PREPROCEDURAL IMAGING WITH PROGNOSTIC SIGNIFICANCE

Before examining the imaging sequelae of radioembolization, it should be noted that several imaging findings on preprocedural imaging have been found predictive of treatment response.

Scintigraphic imaging of ^{99m}Tc macroaggregated albumin (MAA) administered during the ^{90}Y mapping is performed primarily to evaluate the lung shunt fraction (LSF) and reduce the incidence of radiation pneumonitis. However, emerging evidence suggests that the LSF may yield important prognostic information not directly related to extrahepatic radiation dose deposition. A study of 62 patients with colorectal carcinoma (CRC) metastasis found that LSF was an independent predictor of survival after radioembolization. Patients with LSF above the median value of 7.3% had significantly worse survival than those with LSF below 7.3% (Deipolyi et al., 2014). This could reflect a phenomenon of increased systemic shunting of circulating tumor cells in patients with increased LSF. Interestingly, hepatic tumor burden was not associated with poor outcomes in this cohort (Deipolyi et al., 2014).

Several imaging features on pretreatment CECT also correlate with survival. In patients with HCC, increased central hypervascularity and well-defined tumor margins are associated with improved survival (Salem et al., 2013a). In other words, infiltrative and necrotic tumors infer relatively worse prognosis in HCC. Similarly, centrally necrotic neuroendocrine tumor (NET) metastases have also been associated with to poor response to ^{90}Y therapy and poor prognosis (Neperud, 2013).

Changes in SUV_{max} as demonstrated on ^{18}FDG-PET/CT have been associated with early prediction of treatment response after radioembolization. However, the volume of hypermetabolic tumor seen on preprocedural ^{18}FDG-PET/CT has also been shown to be a solitary predictor of outcomes in patients with unresectable hepatic metastases from melanoma (Piduru et al., 2012). In one small series, patients with metabolic tumor burden (hypermetabolic tumor volume/total liver volume) greater than 7% had markedly reduced prognosis. However, it is unclear whether this finding is reciprocated in other tumor histologies.

14.2.3 NORMAL RESPONSE TO THERAPY

Based on prior research, it can be reasonably assumed that a technically successful radioembolization therapy will result in some degree of tumor response using standard dosimetry methods (Sangro et al., 2006; Sato et al., 2006; Salem et al., 2013b). However, the degree to which tumor response occurs, the associated imaging findings, and the time frame in which posttherapy changes evolve are often variable. This is probably a result of the unique mechanism of radioembolization, which imparts both radiation and some degree of microvascular embolization in the treatment zone (Sangro et al., 2006; Sato et al., 2006; Salem et al., 2013b). It should be emphasized that some standardized and widely accepted methods of reporting tumor response are optimized for reporting response to cytotoxic agents rather than radiation. Consequently, World Health Organization (WHO) criteria and Response Evaluation Criteria in Solid Tumors (RECIST) may misrepresent the effect of

^{90}Y therapy (Schlaak, 2013). Familiarity with the constellation of possible imaging findings in the post-^{90}Y patient is therefore important, so as to tailor an appropriate follow-up regimen and triage to adjuvant therapies when appropriate.

14.2.3.1 0–3 months: rim enhancement, necrosis, and pseudoprogression

One of the earliest findings after radioembolization is rim enhancement along the margins of the target tumor (Singh and Anil, 2013). Rim enhancement less than 5 mm in thickness is a common finding, occurring in approximately one-third of patients after ^{90}Y (Keppke et al., 2007). This is in contradistinction to normal imaging findings after TACE, where elimination of tumor vascularity is the desired treatment endpoint. For ^{90}Y, it is important to emphasize that the presence of thin rim enhancement is not necessarily indicative of recurrent or residual disease. Instead, this finding typically reflects granulation tissue along the margins of the treatment site. In fact, thin rim enhancement has been reported in a high percentage of patients found to have complete pathologic response to ^{90}Y following transplant (Kulik et al., 2006). In once study of 46 patients with HCC, 80% of patients with thin rim enhancement had response to therapy, whereas 13% had stable disease (Keppke et al., 2007). Thin rim enhancement is usually transient, occurring between 1 and 2 months and resolving between 4 and 5 months after therapy (Singh and Anil, 2013).

Thick rim enhancement surrounding a lesion after radioembolization is less specific. When observed in the early posttreatment time frame, this finding could reflect regional hyperemia related to therapy (Figure 14.1). This is a particularly prevalent normal finding following radiation segmentectomy. However, persistence of thick rim enhancement beyond 3 months or associated nodularity is concerning for residual disease (Kulik et al., 2006). In another study, residual nodular arterial phase enhancement seen on early surveillance imaging (mean = 55 days) was associated with progressive disease in a high proportion of patients (Keppke et al., 2007; Singh and Anil, 2013). In contrast, progressive low attenuation within the treatment site usually indicated

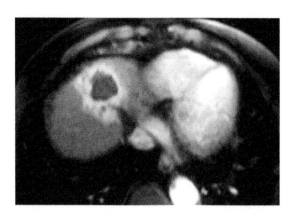

Figure 14.1 Thick rim enhancement is a nonspecific finding early after radioembolization, and may be a normal finding. This finding was indicative of regional hyperemia in this patient who underwent radiation segmentectomy. Nodular rim enhancement is a more worrisome finding. (Reproduced from Minocha, J. and Lewandowski, R., *Semin. Intervent. Radiol.*, 28, 226–229, 2011. With permission.)

necrosis, another common early imaging finding after ^{90}Y.

Approximately 95% of the radiation dose from ^{90}Y is delivered within four half-lives, or approximately 11 days. Consequently, early changes of radiation-induced coagulative necrosis and associated peritumoral edema can be frequently seen in the postprocedural period from 1 to 3 months(Singh and Anil, 2013). Necrosis manifests as an area of low attenuation on CECT with hypoenhancement on both CECT and MRI (Keppke et al., 2007; Miller et al., 2007). It is important to recognize necrosis as an indicator of early treatment response, as reduction in tumor size can take months to occur. In fact, median time to response has been demonstrated to be approximately 5–6 months using only size criteria, compared with 1 month when using both size and necrosis as criteria (Keppke et al., 2007; Miller et al., 2007; Singh and Anil, 2013; Salem et al., 2013b). This is in keeping with the 3 to 6-month delay that is generally expected to see reduction in tumor volume after external beam radiation therapy (EBRT) (Salem et al., 2013b). Patchy areas of regional hypoenhancement not meeting criteria for necrosis may also be transiently seen in the ^{90}Y treatment site. This finding has been noted on portal venous phase CT in approximately 40% of patients. These areas are without corresponding

mass effect and are usually distributed along lesional margins or in a vascular distribution (Miller et al., 2007). Like thin rim enhancement, this finding may mimic residual tumor. However, this finding is usually transient in nature, disappearing around 3 months following therapy (Miller et al., 2007; Singh and Anil, 2013) and is postulated to reflect the same radiation-induced inflammatory reaction commonly observed after EBRT (Atassi et al., 2008a; Salem et al., 2013b; Wang et al., 2013). However, as in TACE, a component of microvascular occlusion may contribute to these altered enhancement characteristics (Chung et al., 2010).

Each of the aforementioned findings reflects normal hepatic changes of radioembolization within the treatment size and may be seen on an unpredictable basis. When tumoral/peritumoral edema and peripheral enhancement are collectively present in the early (<3 months) posttreatment period, it may be particularly difficult to distinguish from disease progression. In this setting, lesional margins become indistinct and the tumor may appear to increase in size, a phenomenon commonly referred to as "pseudoprogression" (Salem et al., 2013b; Singh and Anil, 2013; Dhingra et al., 2014). However, it is important to note that tumors with this appearance decrease in average attenuation, indicating necrosis and peritumoral edema (Salem et al., 2013b; Dhingra et al., 2014). This finding helps to distinguish from early progression of disease, usually denoted by increased solid enhancing tumor (Salem et al., 2013b; Singh and Anil, 2013; Salem et al., 2013b; Dhingra et al., 2014). At times, progressive disease and pseudoprogression can be hard to distinguish and imaging findings must be taken in context of serum tumor markers, clinical status, and histologic risk factors. A follow-up exam at the 3- to 6-month interval after ^{90}Y can be helpful to confirm pseudoprogression if reduction in tumor volume is noted.

14.2.3.2 Beyond 3 months: reduction in tumor size and changes in liver volume

Reduction in tumor volume is an important endpoint in radioembolization (Singh and Anil, 2013). In those patients who will experience a therapeutic response, it is generally assumed that reduction in tumor volume occurs in a delayed fashion after treatment, as seen in EBRT (Salem et al., 2013b; Singh and Anil, 2013). While tumoral necrosis is often visible 0–3 months after ^{90}Y, reduction in tumor volume occurs more commonly from 4 to 6 months (Miller et al., 2007) (Figure 14.2). It should also be noted that radiation changes within the unaffected liver lobe or segment that underwent treatment also become visible at this time.

Although the unaffected liver tissue in the ^{90}Y treatment zone generally receives significantly reduced dose compared with neoplastic tissue, NTE to normal liver may present variable imaging sequelae on delayed (>3 months) follow-up imaging. Namely, changes in liver volume frequently occur, including ipsilateral lobar atrophy and compensatory contralateral hypertrophy (Jakobs et al., 2008). This phenomenon was examined in a cohort of patients who underwent glass radioembolization by variable infusion techniques (i.e., lobar, bilobar infusion). The authors noted an 11.8% decrease in total liver volume following bilobar infusion. In patients undergoing unilobar therapy, an ipsilateral volume loss of 8.9% and 21.2% increase in volume of the contralateral lobe was noted (Jakobs et al., 2008). Hepatic atrophy from radioembolization is characterized histologically by fibrosis, which may change enhancement characteristics on follow-up imaging. Although the degree of hepatic atrophy can be significant, this effect is commonly asymptomatic (Singh and Anil, 2013; Brown, 2014).

14.2.3.3 Radiation lobectomy and segmentectomy

Radiation lobectomy refers to the lobar infusion of relatively high ^{90}Y activity in patients with uni-lobar disease to provide the dual effect of treating the tumor and inducing contralateral lobe hypertrophy (Gaba et al., 2009). This method has a demonstrated ability to induce ipsilateral lobar atrophy and contralateral lobar hypertrophy averaging 52% and 40%, respectively, comparable with portal vein embolization (Gaba et al., 2009). Further, patients treated with radiation lobectomy have shown improved tumor response and survival, with a comparable 5 years to surgery (36.6 months) (Jakobs et al., 2008). The conceptual basis of radiation lobectomy has also been applied in a superselective fashion to a hepatic segment, in patients with localized disease but contraindications to ablation or surgery. This technique has been referred to as "radiation segmentectomy" and involves administration of high activity of ^{90}Y to a localized area of the liver, allowing absorbed doses greater than 1000 times than those delivered in EBRT (Salem et al., 2013b). Subsequently, the treated hepatic segment often experiences significant atrophy and may disappear on follow-up imaging. Additional information specific to radiation segmentectomy and lobectomy is available in Chapter 6.

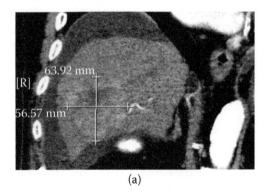
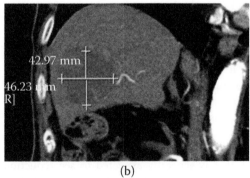

(a) (b)

Figure 14.2 **(a)** Coronal computed tomography (CT) showing large enhancing hepatocellular carcinoma that was subsequently treated with radioembolization; **(b)** complete response to therapy following radioembolization by modified Response Evaluation Criteria in Solid Tumors (mRECIST). Note that in spite of complete tumor necrosis, minimal decrease in tumor size was noted. (Reproduced from Singh, P. and Anil, G., *Cancer Imag.*, 13, 2013 under the terms of the Creative Commons Attribution 4.0 International License; http://creativecommons.org/licenses/by/4.0/.)

14.2.3.4 Sequela of portal hypertension after therapy

Although many patients may be asymptomatic, the delivery of significant radiation to the normal hepatic parenchyma can have clinical consequence, as a result of fibrotic remodeling. Just as lobar ^{90}Y infusion may incite contralateral lobar hypertrophy via induction of ipsilateral fibrosis and atrophy, this method may exacerbate portal hypertension. This is particularly true in patients with poor hepatic functional reserve. Consequences of worsened portal hypertension can often be seen in surveillance imaging (Lam et al., 2013a) (Figure 14.3). main portal vein (MPV) diameter has been shown to increase after both bilobar and unilobar infusion of glass microspheres (Jakobs et al., 2008). In patients who received bilobar treatment, delayed follow-up imaging showed 28% increase in splenic size, as well as increased diameter of the MPV, splenic vein, and superior mesenteric veins (Jakobs et al., 2008). Not surprisingly, these imaging findings correlate with worsening thrombocytopenia in some patients as a result of hepatic sequestration of platelets (Lam et al., 2013a). Increased number and size of porto-systemic collaterals vessels have also been noted following lobar infusion. These

imaging sequelae of worsening portal hypertension can be seen using both resin and glass microspheres and should be interpreted in context of their clinical relevance and underlying functional hepatic reserve.

Low-volume perihepatic ascites is a common early finding following radioembolization and does not necessarily represent evidence of portal hypertension or worsening hepatic function. Instead, this is thought to result from irritation of the Glisson capsule (Hilgard et al., 2010). A similar phenomenon may also occur in the adjacent lung base as a result of pleural irritation (Singh and Anil, 2013). In either case, findings are self-limited and infrequently warrant intervention.

14.2.3.5 Progression of disease

Disease progression manifests as new or enlarging hepatic tumor burden. It is often the result of new tumor formation or growth of microscopic rests of tumor cells unlikely to be effected by radioembolization, or from suboptimal dose delivery to the tumor (Sangro et al., 2006). It is important to note that median time to progression (TTP) occurs relatively late in patients without portal vein thrombus, appearing around 12–16 months after therapy (Hilgard et al., 2010; Salem et al., 2010). In all patients, TTP ranges from approximately 10–12 months (Hilgard et al., 2010; Salem et al., 2010). The determination of disease progression is usually augmented clinical and serologic information, as well as baseline tumor histologic risk factors. Identifying early disease progression is important, as triage to additional/alternative liver-directed therapy or systemic agents may help to prolong survival.

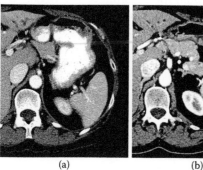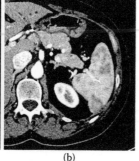

(a)   (b)

Figure 14.3 This image illustrates sequelae of worsening portal hypertension following radio-embolization. **(a)** Prior to radioembolization and **(b)** 3 months postradioembolization. This 57-year-old patient underwent bilobar therapy for colorectal metastases. Spleen volume increased 84.8% after therapy and platelet volume decreased 45.2%. (Reproduced from Lam, M.G.E.H. et al., *Cardiovasc. Intervent. Radiol.*, 37, 1009–1017, 2013a. With permission.)

14.2.4 FOLLOW-UP IMAGING ASSESSMENT CRITERIA IN PREDICTING TREATMENT RESPONSE

The WHO criteria and RECIST offered the first standardized methods of assessing the effect of an oncologic therapy. Both of these guidelines were initially optimized for reporting response to systemic cytotoxic therapy, accounting only for changes in tumor size (Lencioni and Llovet, 2010). As locoregional therapies aimed at devascularizing liver tumors gained use, it became clear that WHO and RECIST underestimate response rates in HCC (Miller et al., 1981; Therasse et al., 2000; Lencioni and Llovet, 2010) (Figure 14.4). This is due in large part to the early appearance of tumor necrosis in response to locoregional therapies such as radioembolization, a finding that occurs months before reduction in tumor size (Figure 14.2). Assessment systems incorporating tumor enhancement characteristics were therefore thought to more accurately represent TTP of liver tumors. Consequently, the WHO and RECIST criteria were modified in 2000 and 2008, respectively, to account for tumor enhancement characteristics. These new criteria, referred to as the European Association for the Study of the Liver (EASL) and modified RECIST (mRECIST), and others such as the Choi criteria, have gained increasing utility in determining response to radioembolization.

While a comprehensive discussion of the development and use of the various assessment criteria is outside of the scope of this chapter, it is important to understand their general differences. As previously discussed, the original WHO and RECIST criteria account for only size. Whereas the WHO criteria include bidirectional tumor size, RECIST accounts only for longest tumor dimension. EASL and mRECIST criteria also account for arterial phase tumor enhancement in determining response. A separate assessment method described by Choi et al. for evaluating gastrointestinal stromal tumors (GIST) incorporates changes in both tumor size and mean tumor density (Choi et al., 2007). The Choi criteria have since been applied to the surveillance of HCC and liver metastases following radioembolization (Schlaak, 2013). Each of these criteria can be used for both CT and MRI, and each is outlined in Figure 14.4 and Table 14.1. The primary question asked regarding these various assessment methods is "which one is most accurate in predicting response and time to progression?" This question is not trivial as palliation of liver malignancy hinges on month-to-month changes in disease status so as to tailor an effective treatment regimen to each patient. Fortunately, the use of each of the aforementioned criteria has been the subject of at least some comparative research. The basis for use of enhancement criteria (mRECIST and EASL) derives from experience with chemoembolization. In a large cohort of patients undergoing TACE, mRECIST and EASL more accurately predicted survival than WHO and RECIST criteria (Shim et al., 2012). Consequently, tumor enhancement gained favor over tumor size as a primary determinant of treatment response for chemoembolization and was then applied to radioembolization (Shim et al., 2012). The Choi criteria, which incorporate mean tumor enhancement and tumor size, have been shown

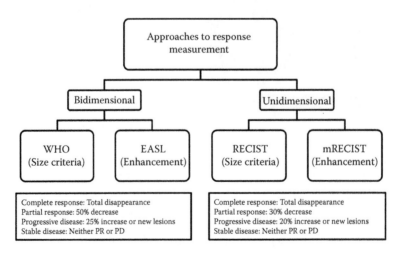

Figure 14.4 Summary of assessment criteria for oncologic reporting.

Table 14.1 Comparison of WHO, RECIST, mRECIST, EASL, and Choi criteria

Response	WHO	RECIST	EASL	mRECIST	Choi
Complete	Disappearance of all known disease	Disappearance of all target lesions	Disappearance of all enhancing disease	Disappearance of any intratumoral arterial enhancement in all target lesions	Disappearance of all target lesions
Partial	At least 50% decrease in tumor size from baseline	At least a 30% decrease in the sum of diameters of target lesions, taking as reference the baseline sum of the diameters of target lesions	At least 50% decrease in tumor enhancement from baseline	At least a 30% decrease in the sum of diameters of viable (enhancement in the arterial phase) target lesions, taking as reference the baseline sum of the diameters of target lesions	Decrease in tumor size ≥10% or decrease in tumor density ≥15% on CT
Stable disease	Any cases that do not qualify for either partial response or progressive disease	Any cases that do not qualify for either partial response or progressive disease	Any cases that do not qualify for either partial response or progressive disease	Any cases that do not qualify for either partial response or progressive disease	Any cases that do not qualify for either partial response or progressive disease
Progressive disease	At least 25% increase of one or more lesions, or appearance of new lesions	An increase of at least 20% in the sum of the diameters of target lesions, taking as reference the smallest sum of the diameters of target lesions recorded since treatment started	At least 25% increased enhancement of one or more lesions, or appearance of new lesions	An increase of at least 20% in the sum of the diameters of viable (enhancing) target lesions, taking as reference the smallest sum of the diameters of viable (enhancing) target lesions recorded since treatment started	Increase in tumor size ≥10% and does not meet partial response (PR) criteria by tumor density

Notes: WHO, World Health Organization; RECIST, Response Evaluation Criteria in Solid Tumors; mRECIST, modified RECIST; EASL, European Association for the Study of the Liver; CT, computed tomography.

to be superior to mRECIST in evaluating HCC response to therapy in at least one single-center review (Schlaak, 2013). However, it should be noted that the Choi criteria were initially developed for CT and include quantitative density analysis specific to CT (Bonekamp et al., 2013) that may be somewhat imprecise when used in MRI follow-up.

14.2.4.1 The Role of ^{18}Fdg-Pet/Ct

Although tumor enhancement is now primarily emphasized in evaluating effects of locoregional therapy, follow-up assessment with ^{18}FDG-PET/CT also plays a significant role after ^{90}Y treatment. PET/CT is particularly useful in surveillance of metastatic disease to liver, which is generally metabolically active (Singh and Anil, 2013; Vouche et al., 2015). For many metastatic tumors such as CRC, ^{18}FDG-PET/CT has shown superiority to CT and MRI for lesional detection and is similarly helpful in determining residual viable tumor after ^{90}Y (Sacks et al., 2011). Further, ^{18}FDG-PET/CT may provide important prognostic information in the early posttreatment period after ^{90}Y. Prior research has shown that quantitative changes in SUV_{max} on ^{18}FDG-PET/CT at 6 weeks after therapy predict early response to radioembolization (Soydal et al., 2013). However, it should be emphasized that PET/CT should interpreted with caution prior to 2 months, so as to not confuse postprocedural inflammation with residual/progressive tumor not seen on preprocedural imaging. The reporting scheme for ^{18}FDG-PET/CT is considerably less controversial than for CT and MRI, and the PET Response Criteria in Solid Tumors (PERCIST) criteria are

widely used. A description of PERCIST is provided in Table 14.2.

14.2.4.2 Investigational Follow-Up Methods

While postradioembolization surveillance usually involves evaluation of tumor size/enhancement and FDG avidity when applicable, a number of additional methods have been described on small patient series. CT perfusion (CTp) involves serial acquisition of CT images after administration of iodinated contrast to characterize and quantify enhancement. The general principle of its use relates to the importance of tumor vascularity as a prognostic factor after locoregional therapy (Shim et al., 2012), as previously discussed. However, CTp allows for more accurate quantification of enhancement than simple visual estimation. Preliminary studies using CTp have shown that significant reduction in post-^{90}Y perfusion translates to treatment response of metastatic disease (Reiner et al., 2014). Interestingly however, these findings were not corroborated in patient with HCC (Reiner et al., 2014).

In some situations, inflammatory changes from radioembolization and viable tumor may be difficult to distinguish. One potential troubleshooting method could be DWI-MRI. In a cohort of 20 patients with HCC, quantitative analysis of the ADC map on DWI-MRI predicted response within 42 days of radioembolization, 2 months prior to reduction in tumor size (Rhee et al., 2008) (Figure 14.5). It remains unclear how these methods compare with mRECIST and EASL in larger patient cohorts and in variable tumor histology.

Table 14.2 PERCIST criteria for oncologic reporting

PERCIST criteria	
Complete response	Disappearance of all metabolically active lesions
Partial response	30% and 0.8-unit decline in (SUL SUV normalized to lean body mass and corrected to background) peak between the most intense lesion before treatment and the most intense lesion after treatment, although not necessarily the same lesion
Progressive disease	30% and 0.8-unit increase in SUL peak or new lesions or 75% increase in total lesion glycolysis
Stable disease	Neither partial response nor progressive disease.

Note: PERCIST, Positron Emission Tomography Response Criteria in Solid Tumors.

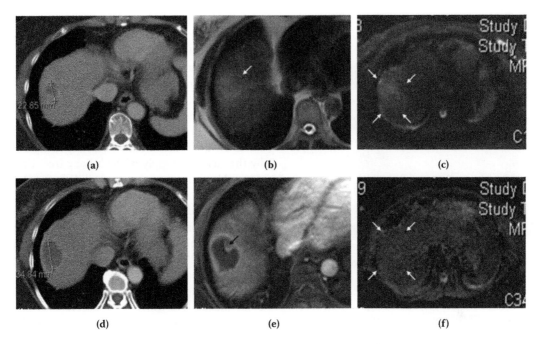

Figure 14.5 Images illustrating the utility of contrast-enhanced and diffusion-weighted imaging on magnetic resonance imaging (DWI-MRI) in the evaluation of treatment response: (a) a metastatic lesion was noted in the right hepatic lobe prior to therapy by CT. (b and c): Lesion has increased in size 1 month after radioembolization. The treatment site manifests as (d) hypodensity on CT and (e) peripheral nodular enhancement by contrast-enhanced MRI. (f) The absence of signal on DWI-MRI confirms the absence of residual disease after therapy. (Reproduced from Singh, P. and Anil, G., *Cancer Imag.*, 13, 645–657, 2013 under the terms of the Creative Commons Attribution 4.0 International License; http://creativecommons.org/licenses/by/4.0/.)

14.3 IMAGING OF COMPLICATIONS

Potential adverse events to the liver and extrahepatic soft tissues from ^{90}Y radioembolization are numerous and are often present with distinct clinical signs/symptoms. This section emphasizes imaging findings, pathogenesis, and epidemiology of the more common complications from ^{90}Y. Complications that are not directly caused by ^{90}Y microspheres such as vascular injury, exposure to ionizing radiation, and iodinated contrast exposure are also omitted from this discussion.

14.3.1 HEPATIC TOXICITY

Radioembolization-induced liver disease (REILD) is a form of hepatic subacute liver injury from radiation exposure. It is conceptually similar to radiation-induced liver disease (RILD) seen in EBRT for liver tumors and in whole-body radiation performed in preparation for allogenic bone marrow transplant (Salem et al., 2013b). As one would expect, REILD presents several weeks after radiation dose delivery, commonly in the 4- to 8-week posttherapy interval (Kuo et al., 2014). Histologically, it is similar to RILD, characterized by sinusoidal congestion, cholestasis, and areas of perivenular necrosis reflecting venoocclusive disease (Sangro et al., 2008). REILD represents a spectrum of disease and its clinical impact is generally determined by pretreatment hepatic functional reserve, the volume of tissue affected, systemic chemotherapy, and presence of clinical intervention. As expected, this phenomenon occurs more frequently in patients treated in a bilobar fashion, those who have undergone prior EBRT, and in sequential radioembolization therapies (Lam et al., 2013b). In the most severe cases,

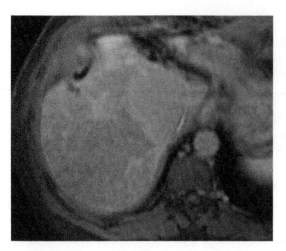

Figure 14.6 Fulminant hepatic failure is largely a clinical diagnosis with nonspecific imaging findings. This patient underwent lobar radio-embolization after left lobectomy for colorectal carcinoma (CRC) metastasis. Eight weeks after therapy, the patient was found to have fulminant hepatic failure, manifested by patchy hypoen-hancement. (Reproduced from Hamoui, N. and Ryu, R., *Semin. Intervent. Radiol.*, 28, 246–251, 2011. With permission.)

REILD can lead to fulminant hepatic failure and death (Hamoui and Ryu, 2011). The overall incidence of REILD from radioembolization ranges from approximately 4% to 9%, although it can reach 50% in patients receiving aggressive chemotherapy regimens (Sangro et al., 2008). REILD is primarily diagnosed by its clinical features; it has been accompanied by ill-defined heterogenous hypoenhancement on short-term follow-up MRI and CT (Hamoui and Ryu, 2011) (Figure 14.6). Hepatic toxicity from ^{90}Y is discussed in more detail in Chapter 8, including a comprehensive discussion of the hepatic dose toxicity models.

14.3.2 BILIARY EFFECTS

While REILD is a serious concern that greatly effects ^{90}Y treatment planning and much technical emphasis is placed on avoiding complications from NTE, biliary toxicity provides some of the most common adverse effects of radioembolization (Atassi et al., 2008b; Singh and Anil, 2013). With current techniques, it is now generally assumed that biliary complications including radiation cholangitis, biloma, biliary stricture,

and cholecystitis are much more common than those related to extrahepatic NTE (Singh and Anil, 2013). In one large series, combined biliary complications occurred in approximately 10% of patients, requiring intervention in ~2% (Atassi et al., 2008b). These rates are generally higher in patients with prior biliary intervention or biliary-enteric anastomosis (Riaz, 2014). This relatively high rate of biliary toxicity is explained by the vascular supply of the biliary tree. Unlike normal hepatic parenchyma, the biliary tree lacks a dual blood supply (Northover and Terblanche, 1979). Instead, the supraduodenal common bile duct and hepatic duct are supplied by either the gastroduodenal artery (68%) or right hepatic artery (32%) (Northover and Terblanche, 1979). The intrahepatic biliary tree is supplied by its accompanying lobar artery and its arterial vascular plexus (Northover and Terblanche, 1979). This paucity of collateral vessels makes the biliary tree particularly susceptible to ischemic injury. Interestingly, peribiliary vascular hypertrophy found in cirrhotic livers is thought to prevent peribiliary microsphere deposition and reduce rates of biliary injury (Yu et al., 2002).

14.3.2.1 Biloma

The potential for biloma formation has been well documented in patients undergoing chemoembolization for hepatic malignancy, occurring in approximately 4% of all cases (Minocha and Lewandowski, 2011). Biloma are most commonly caused by small vessel ischemia resulting in bile duct disruption, although radiation effects may also contribute. Comparative rates of biloma in TACE and ^{90}Y are unclear as are incidence rates among the glass and resin products. However, a large independent observational study reported rates approximating 1% for radioembolization. On imaging, biloma typically manifests as a well-defined intrahepatic collection approximating the biliary tree (Figure 14.7). A direct communication with the bile ducts may not be seen by CT or MRI, but is usually demonstrated by cholangiogram (Figure 14.8). The presence of rim enhancement is common and should be considered suspicious for secondary infection only in the clinical context of fever, right upper quadrant pain, and leukocytosis (Singh and Anil, 2013). In this setting, biloma may difficult to differentiate from intrahepatic abscess.

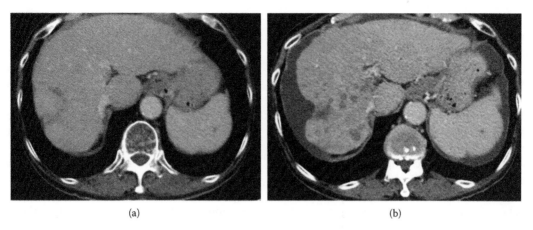

Figure 14.7 **(a)** CT scan prior to radioembolization shows multiple hepatic metastatic lesions. **(b)** Two months after therapy, the right hepatic lobe shows significant volume loss and capsular retraction with several hypodense lesions typical of intrahepatic biloma. (Reproduced from Singh, P. and Anil, G., *Cancer Imag.*, 13, 645–657, 2013 under the terms of the Creative Commons Attribution 4.0 International License; http://creativecommons.org/licenses/by/4.0/.)

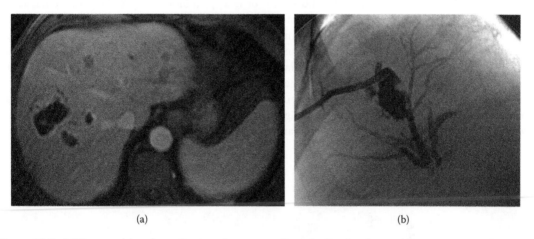

Figure 14.8 A 59-year-old male patient underwent radioembolization for melanoma metastasis to liver. **(a)** Axial T1-weighted, fat-saturated MRI image 25 weeks after therapy shows nonenhancing lesions with thick rim enhancement in the right hepatic lobe and multiple left lobe metastasis. **(b)** Administration contrast through a percutaneous drainage catheter demonstrates communication with the biliary tree, confirming intrahepatic biloma. (Reproduced from Atassi et al., *Radiographics*, 28, 81–99, 2008a. With permission.)

14.3.2.2 Cholangitis and transient hyperbilirubinemia

Radiation cholangitis is a clinical syndrome manifested by jaundice, fever, and right upper quadrant pain. It is rare after radioembolization, with unclear incidence and no clearly defined characteristic imaging findings (Riaz, 2014). While radiation cholangitis may be sterile, it is commonly treated with antibiotics (Riaz, 2014). This should be distinguished from worsening *painless* jaundice. Grade III or higher bilirubin toxicity (total bilirubin >3.0 mg/dL) has been reported in as many as one-third of patients after therapy (Smits et al., 2013). These findings are thought to more likely represent effects of REILD than biliary toxicity and may improve with conservative management (Salem et al., 2013b). It should be emphasized that isolated mild hyperbilirubinemia is commonly asymptomatic and not associated with significant changes in synthetic liver function or impending hepatic decompensation (Salem et al., 2013b; Smits et al., 2013).

14.3.2.3 Biliary stricture

Biliary stricture with upstream dilation often reflects chronic sequela of bile duct injury on the same spectrum biloma, potentially due to radiation and/or ischemia, which most commonly effects intrahepatic biliary radicals and is seen to a lesser degree in cirrhotic livers. Many of these patients will be asymptomatic, only requiring intervention if clinical symptoms present as biliary obstruction or secondary infection (Atassi et al., 2008a, 2008b; Singh and Anil, 2013).

14.3.2.4 Radiation cholecystits

Radiation cholecystitis is a clinical syndrome of cholecystits in patients recently undergoing yttrium-90 radioembolization. Imaging findings of cholecystitis have been reported in approximately 2% of patients, manifested by gallbladder wall thickening and hyperenhancement (Sag et al., 2014) (Figure 14.9). These imaging findings accompany the clinical finding of right upper quadrant pain, with or without fever and leukocytosis. In clinical practice, rates of radiation cholecystitis are highly variable, depending on technical considerations such as prophylactic embolization of the cystic artery and use of antireflux devices (Pasciak et al., 2015). Many interventional radiologists do not routinely embolize the cystic artery to prevent NTE to the gallbladder when it is downstream to the infusion site. This is due in large part to the possible risk of ischemic cholecystits from prophylactic cystic artery embolization, thought by many to outweigh the risk of gallbladder NTE (Hickey and Lewandowski, 2011). In situations where the cystic artery is likely to sump a significant quantity of ^{90}Y activity from the tumor, or when significant deposition of MAA is seen on the pretreatment simulation, prophylactic embolization can be considered. Caution should be taken in embolizing a dominant cystic artery that comprises all or nearly all blood flow to the gallbladder, as this is associated with increased risk of perforation (Hickey and Lewandowski, 2011). Fortunately, radiation cholecystitis is usually self-limited and managed with conservative therapy. Rarely, gallbladder perforation and need for surgical cholecystectomy have been reported, but both are thought to occur in less than 1% of cases (Atassi et al., 2008b).

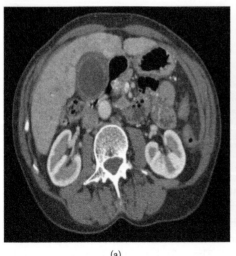

(a)

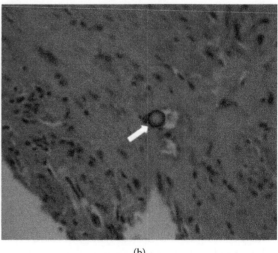

(b)

Figure 14.9 (a) Contrast-enhanced CT scan performed 3 days after radioembolization demonstrates gallbladder wall thickening and pericholecystic fluid suggestive of cholecystitis. The patient underwent laparoscopic cholecystectomy 7 weeks later. (b) Microscopic analysis of the surgically removed gallbladder showed fibrosis, chronic inflammation, and the presence of glass microspheres (arrow). (Reproduced from Hickey, R. and Lewandowski, R., *Semin. Intervent. Radiol.*, 28, 230–233, 2011. With permission.)

14.3.3 HEPATIC ABSCESS

Any rim-enhancing intrahepatic fluid collection should be suspicious for abscess in a patient after radioembolization. The presence of internal gas raises diagnostic probability; however, it should be emphasized that loculated foci of gas may be present in the liver normally after treatment. This is a result of low-volume, intra-arterial gas injection during ^{90}Y infusion, when it can become trapped between microspheres for several days before resorbing (Singh and Anil, 2013). Therefore, imaging findings suggestive of abscess should be taken in context of clinical presentation. Percutaneous aspiration for culture may prove definitive in cases of unclear imaging and clinical findings (Singh and Anil, 2013).

Hepatic abscess may occur after radioembolization as a result of bacterial colonization of necrotic tumor, necrotic nontarget hepatic parenchyma, or biloma. Because ascending infection through the biliary tree is the most common route of infection, rates of hepatic abscess are increased in patients with prior sphincerotomy or biliary-enteric anastomosis. In patients with an intact sphincter of Oddi, abscess occurs in approximately 1%–2% of cases (Brown et al., 2012). However, as many as 15% of patients with prior biliary-enteric anastomosis, sphincerotomy, or indwelling biliary stent experience an abscess, even when periprocedural antibiotic were administered (Brown et al., 2012). An aggressive antibiotic course combined with a bowel prep regimen has been shown to reduce, and potentially eliminate, this risk (Cholapranee et al., 2014). Finally, radioembolization should be deferred in patients with suspicion for bacteremia, as subsequent abscess formation has been reported (Mascarenhas et al., 2011).

14.3.4 GASTROINTESTINAL NONTARGET EMBOLIZATION

Gastrointestinal (GI) tract ulceration from NTE is one of the most feared complications of radioembolization due to its morbidity and the complex nature of surgical intervention in patients containing residual radioactivity. Ulceration is commonly caused by reflux of microspheres into the gastroduodenal (GDA) or right gastric arteries (RGA), where they may deposit and cause hemorrhage, inflammation, and/or ulceration (Veloso et al., 2013; Baumann et al., 2015). Occasionally, peripheral deposition of ^{90}Y within the liver can result in clinically significant dose deposition to adjacent visceral structures (Singh and Anil, 2013). Radiation-induced injury to mucosal stem cells is thought to be the primary causative etiology, with ischemia effects playing a lesser role. Because mucosal stem cells are permanently depopulated by radiation effects, radiation-induced GI tract ulcers rarely resolve spontaneously. In order to mitigate ulceration risks, prophylactic embolization of the GDA and RGA, antacid medicine, and/or use of antireflux devices may employed (Pasciak et al., 2015). However, the practice of routine GDA embolization has been called into question in light of recent research and is performed based on institution-specific preference (Haydar et al., 2010). The exact toxicity thresholds to the small bowel and stomach from ^{90}Y are unclear, although concurrent systemic chemotherapy likely sensitizes mucosal cells to radiation, increasing risk of ulceration (Brown et al., 2012).

By imaging, radiation-induced GI tract ulcers are indistinguishable from other forms of ulceration, and may manifest as focal mural thickening, regional inflammatory changes, and/or signs of perforation. Endoscopy is the gold standard for diagnosis and should be considered in patients with any level of suspicion for NTE to bowel (Singh and Anil, 2013).

Acute pancreatitis is another rare sequela of NTE observed in fewer than 1% of patients (Brown et al., 2012; Riaz, 2014). Pancreatitis manifests similarly when caused by radioembolization as it does when caused by other etiologies, with elevated serum amylase/lipase and intense epigastric pain. Imaging findings of acute pancreatitis related to ^{90}Y are usually localized to the pancreatic head, rather than diffuse in nature (Singh and Anil, 2013). Postprocedural ^{90}Y single-photon emission computed tomography (SPECT/CT) or PET/CT imaging may be helpful to confirm microsphere deposition in the pancreas and exclude other potential causative etiologies.

14.3.5 RADIATION PNEUMONITIS AND OTHER SITES OF NONTARGET EMBOLIZATION

Radiation pneumonitis (RP) is the development of pulmonary fibrosis and restrictive lung disease from exposure to sufficient radiation dose. It manifests in a similar fashion to organizing pneumonia or chronic eosinophilic pneumonia on CT, with patchy areas of ground-glass opacity, mild volume loss, and traction bronchiectasis (Singh and Anil, 2013) (Figure 14.10). Making the diagnosis RP is particularly important in patients being considered for repeat radioembolization as the increased cumulative dose could worsen pulmonary function (Wright et al., 2012). Avoiding RP is a significant emphasis in pretherapy treatment planning, and is the basis for MAA administration and calculation of shunt fraction. Dosimetric and radiobiologic considerations regarding RP are discussed in further detail in Chapter 4. Although the avoidance of RP is greatly emphasized in ^{90}Y treatment planning, its incidence is exceedingly rare. It could be argued that rigorous treatment planning helps to avoid RP; however, it is rarely reported even in cases in which the lungs receive the assumed toxicity threshold of 50 Gy. In one cohort of 58 patients receiving a lung dose above 50 Gy no

cases of radiation pneumonitis were seen (Salem et al., 2008). The rarity of RP makes it suboptimally studied with respect to risk factors and optimal prevention strategies.

Nontarget embolization has been described in other unusual locations such as colon and kidney (Kao, 2014). These represent rare sites of NTE, with poorly understood clinical outcomes as a result of their rarity. Of particular interest, NTE to the abdominal wall from a falciform artery has been reported. The falciform artery is a small terminal arterial branch off the left hepatic or proper hepatic arteries that extends to the umbilicus and communicates with the epigastric arteries (Bhalani and Lewandowski, 2011) (Figure 14.11). Failure to recognize and embolize potential NTE to abdominal wall from a falciform artery can result in periumbilical abdominal pain, cutaneous burning sensation, or skin necrosis (Bhalani and Lewandowski, 2011; Smith et al., 2015). However, it has recently been shown that topically applied ice to the abdominal wall in the periprocedural period can induce sufficient superficial vasoconstriction to eliminate this risk.

14.4 POSTTREATMENT ^{90}Y IMAGING

14.4.1 EVALUATING NONTARGET EMBOLIZATION WITH POSTTREATMENT ^{90}Y IMAGING

Recall that there are no reliable angiographic methods to ensure target delivery of microspheres and exclude nontarget embolization. Both glass and resin microspheres are radiolucent and thus not visualized on fluoroscopy. Although dilute iodinated contrast is commonly used to flush the delivery catheter and confirm antegrade flow during infusion of resin microspheres, this method provides limited evaluation for NTE. It is also known that ^{90}Y microspheres have the potential to distribute in an unpredictable fashion compared with that of the preprocedural Tc99m-MAA injection, owing to differences in particle size, catheter position, injection rate, presence of stasis, and changes in tumor vascular

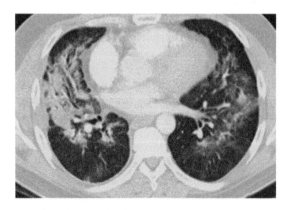

Figure 14.10 Axial thoracic CT showed diffuse ground glass opacity with traction bronchiectasis in a patient following radioembolization. Given this appearance and progressive nature, this was determined to be a result of radiation pneumonitis. (Reproduced from Wright, C.L. et al., *J. Vasc. Interv. Radiol.*, 23, 2012. With permission.)

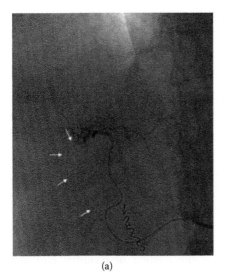
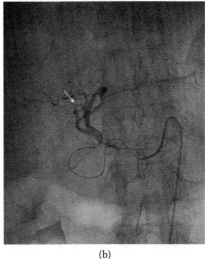

(a)　　　　　　　　　　　　　　　(b)

Figure 14.11 **(a)** Pretreatment angiography prior to radioembolization in a 57-year-old patient with CRC metastases to liver. Selective angiogram of the segment 4 branch demonstrated the presence of a falciform artery (arrows). However, its diminutive size prevented coil embolization. After treatment, the patient experienced localized supraumbilical pain thought to be related to nontarget emboliza-tion of yttrium-90. **(b)** Left hepatic arteriogram prior to therapy in the same patient. (Reproduced from Bhalani, S. and Lewandowski, R., *Semin. Intervent. Radiol.*, 28, 234–239, 2011. With permission.)

dynamics (Lam, 2013c; Wondergem, 2013). Consequently, methods of postprocedural imaging that allow for visualization of ^{90}Y distribution play an indispensible role in confirming technical success and ensuring patient safety.

Identification of GI and other NTE on posttreatment imaging is covered in detail in Chapter 13, in particular, the use of bremsstrahlung SPECT and ^{90}Y PET/CT to gauge technical success and detect NTE, which is not straightforward due to significant differences between ^{90}Y and conventional diagnostic radionuclides.

14.4.2 PROGNOSTICATION WITH ^{90}Y PET/CT

Survival and response to radioembolization is multifactorial, depending on nontreatment-related factors such as tumor burden, histologic subtype, systemic radiosensitizers, and others (Gunduz et al., 2014). When solely focusing on tumor response, the amount of radiation delivered to the tumor (tumor absorbed dose) plays a primary role. However, despite nearly two decades of experience with radioembolization,

exact tumor toxicity threshold remains incompletely understood. Many factors contribute to this uncertainty, including variable methods of dose calculation, tumor heterogeneity, and intrinsic differences among the two ^{90}Y microsphere products (Kao et al., 2013). Much of the prior dose–response data for HCC and metastases derives from predictive dosimetry from Tc^{99}M-MAA SPECT/CT (Eaton et al., 2014; Kokabi et al., 2014). These studies have shown tumor-response threshold ranging from 120 Gy in HCC to 50 Gy in certain metastases (Dezarn et al., 2011). While predictive dosimetry is feasible and commonly used, it is hindered by the differences between ^{90}Y microspheres and MAA discussed previously. Therefore, dosimetry based on Tc^{99}M-MAA simulation provides only an *estimation* of dose distribution that is likely to have been delivered to a tumor (Kao et al., 2013). Other dosimetry methods predicated on bremsstrahlung SPECT are further limited in their accuracy.

The most exciting aspect of ^{90}Y PET/CT is the ability to obtain accurate tumor dosimetry, as previously discussed in Chapters 11 and 13. ^{90}Y PET/CT has the capability to yield potentially important dose–response data and prognostic information. According to phantom studies, quantification of ^{90}Y activity could be performed with accuracy ranging from 0.4% to 10% depending on scan parameters, significantly better than SPECT/CT in comparative studies (Lhommel et al., 2009; Elschot et al., 2013; Gates et al., 2013). A recent international, multicenter trial confirmed the consistent accuracy of ^{90}Y dosimetry using modern PET/CT imaging with time-of-flight (ToF) on a number of scanners with variable scan parameters (Willowson et al., 2015). As quantitative dosimetry with PET/CT has increased in scope, a more in-depth understanding of tumor response will likely come to light. Such information, combined with ^{90}Y PET/CT dosimetry, will allow for prognostication of

outcomes immediately following ^{90}Y radioembolization. Figure 14.12 describes an example of the utilization of PET/CT in the prognostication of a poor clinical response to therapy. Such outcomes are not uncommon in radioembolization and can stem from relative hypovascularity of tumor, poorly optimized treatment planning, and numerous technical and patient-specific challenges occurring during infusion. However, early prognostication of these outcomes can allow for consideration of alternative and adjuvant therapies several months earlier than was previously possible. The potential benefit of using quantitative posttreatment ^{90}Y imaging in the clinical management of disease, particularly in the context of terminal patients, cannot be overstated.

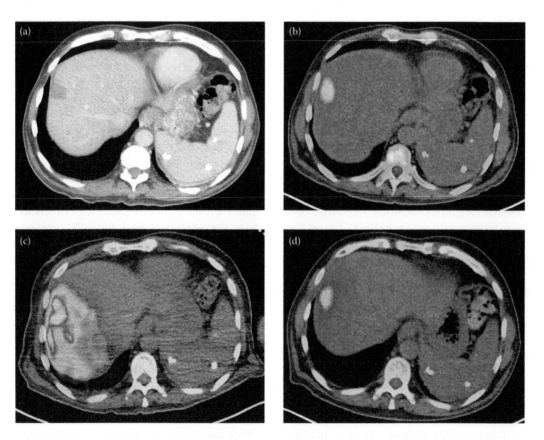

Figure 14.12 A patient with multifocal CRC liver metastases is treated with 1300 MBq of ^{90}Y resin microspheres. **(a)** Lesion appears hypodense on pretreatment contrast-enhanced CT (CECT), with an SUV$_{max}$ of 7.8 2 weeks prior to treatment as shown in **(b)**. **(c)** Posttreatment quantitative ^{90}Y positron emission tomography/CT (PET/CT) reveals an average lesion dose of 64 Gy. **(d)** Follow-up PET/CT at 8 weeks following treatment shows an SUV$_{max}$ of 8.0 with no apparent therapeutic effect owing to the low absorbed dose delivered to the tumor.

14.5 TREATMENT MODIFICATION WITH ^{90}Y PET/CT

14.5.1 INTERPROCEDURAL TREATMENT MODIFICATION

Yttrium-90 PET/CT dosimetry data have also been used in clinical practice to tailor treatments in various case reports. The first example of this was published by Chang et al. (2013). In this case, a patient with cholangiocarcinoma was treated with right lobar infusion of resin microspheres for unresectable recurrent disease after Whipple and radiofrequency ablation. Treatment planning was performed using the standard body surface area approach. The patient experienced minimal (<25%) response to radioembolization on follow-up FDG-PET/CT. Review of the ^{90}Y PET/CT performed immediately after radioembolization yielded a tumor-absorbed dose of 70 Gy and tumor to normal uptake ratio (T/N) of 2.5:1. For a subsequent left lobar infusion, this low calculated T/N was used to justify increased activity infusion that would approach toxicity thresholds reported for HCC (Strigari et al., 2010). The left lobe tumor received 110 Gy on postprocedural ^{90}Y PET/CT and achieved a complete response to therapy. While this dosimetry method is somewhat time consuming, it is similar to partition model dosimetry using Tc^{99}m-MAA and represents a step toward the refinement of dosimetry calculation and treatment personalization.

14.5.2 INTERPROCEDURAL TREATMENT MODIFICATION

Another technique of treatment modification using ^{90}Y PET/CT was performed in a similar fashion, except both infusions targeted a single lesion and the two infusions were performed on the same day (Bourgeois et al., 2015). In this case, a patient with HCC was treated with resin ^{90}Y microspheres via right lobar infusion using the body surface area (BSA) dosimetry method. However, due to intraprocedural technical complications, only a fraction of the ^{90}Y was administered. The patient was promptly transferred to have a ^{90}Y PET/CT, which showed 52 Gy average tumor dose. A simple arithmetic conversion was used to calculate the needed activity to reach 120 Gy tumor dose and the patient returned for a second infusion on the same day. The net result was a robust partial treatment response on follow-up imaging. While the logistics of routinely using this method make it prohibitive for routine use, it demonstrated that ^{90}Y PET/CT may also have value in allowing salvage of a procedural with similar technical complications.

14.6 CONCLUSIONS

Imaging plays an important role in the planning, delivery, follow-up, and clinical management of patients undergoing radioembolization. Familiarity with common complications and associated imaging findings may improve clinical outcomes. Postprocedural imaging with bremsstrahlung SPECT and PET/CT can gauge early treatment response and identify sites of nontarget embolization. PET/CT is likely to play an important role in refining the safety and efficacy profile of radioembolization.

REFERENCES

Annunziata, S., Treglia, G., Caldarella, C., Galiandro, F. (2014). The role of 18F-FDG-PET and PET/CT in patients with colorectal liver metastases undergoing selective internal radiation therapy with yttrium-90: a first evidence-based review. *Sci World J* 2014 :1–8, Article ID 879469. doi: 10.1155/2014/879469. eCollection 2014.

Atassi, B. et al. (2008a). Multimodality imaging following 90Y radioembolization: A comprehensive review and pictorial essay. *Radiographics* 28:81–99.

Atassi, B. et al. (2008b). Biliary sequelae following radioembolization with yttrium-90 microspheres. *J Vasc Interv Radiol* 19:691–697.

Attenberger, U.I. et al. (2015). Fifty years of technological innovation: Potential and limitations of current technologies in abdominal magnetic resonance imaging and computed tomography. *Invest Radiol* 50:584–593.

Barabasch, A. et al. (2015). Diagnostic accuracy of diffusion-weighted magnetic resonance imaging versus positron emission tomography/

computed tomography for early response assessment of liver metastases to y90-radioembolization. *Invest Radiol* 50:409–415.

Bargellini, I. (2014). How does selective internal radiation therapy compare with and/or complement other liver-directed therapies. *Future Oncol* 10:105–109.

Baumann, J., Lin, M., Patel, C. (2015). Image of the month. *Clin Gastroenterol Hepatol* 13:A23–A24.

Bester, L. et al. (2014). Transarterial chemoembolisation and radioembolisation for the treatment of primary liver cancer and secondary liver cancer: A review of the literature. *J Med Imaging Radiat Oncol* 58:341–352.

Bhalani, S., Lewandowski, R. (2011). Radioembolization complicated by nontarget embolization to the falciform artery. *Semin Intervent Radiol* 28:234–239.

Boas, F.E. et al. (2015). Optimal imaging surveillance schedules after liver-directed therapy for hepatocellular carcinoma. *J Vasc Interv Radiol* 26:69–73.

Bonekamp, S. et al. (2013). Unresectable hepatocellular carcinoma: MR imaging after intraarterial therapy. Part I. Identification and validation of volumetric functional response criteria. *Radiology* 268:420–430.

Bourgeois, A.C. et al. (2015). Intraprocedural yttrium-90 positron emission tomography/CT for treatment optimization of yttrium-90 radioembolization. *J Vasc Interven Radiol* 25:271–275.

Braat, A.J.A.T. et al. (2014). Hepatic radioembolization as a bridge to liver surgery. *Front Oncol* 4:199.

Brown, K.T. (2014). Superselective yttrium-90 radioembolization for hepatocellular carcinoma in high-risk cases: Another tool in the toolbox. *J Vasc Interv Radiol* 25:1073–1074.

Brown, D.B. et al. (2012). Quality improvement guidelines for transhepatic arterial chemoembolization, embolization, and chemotherapeutic infusion for hepatic malignancy. *J Vasc Interv Radiol* 23:287–294.

Chang, T.T., Bourgeois, A.C., Balius, A.M., Pasciak, A.S. (2013). Treatment modification of yttrium-90 radioembolization based on quantitative positron emission tomography/CT imaging. *J Vasc Interv Radiol* 24:333–337.

Cho, E. et al. (2015). 18F-FDG PET CT as a prognostic factor in hepatocellular carcinoma. *Turk J Gastroenterol* 26:344–350.

Choi, H. et al. (2007). Correlation of computed tomography and positron emission tomography in patients with metastatic gastrointestinal stromal tumor treated at a single institution with imatinib mesylate: Proposal of new computed tomography response criteria. *J Clin Oncol* 25:1753–1759.

Cholapranee, A. et al. (2014). Risk of liver abscess formation in patients with prior biliary intervention following yttrium-90 radioembolization. *Cardiovasc Intervent Radiol* 38:397–400.

Chung, J. et al. (2010). Haemodynamic events and localised parenchymal changes following transcatheter arterial chemoembolisation for hepatic malignancy: Interpretation of imaging findings. *Br J Radiol* 83:71–81.

Deipolyi, A.R. et al. (2014). High lung shunt fraction in colorectal liver tumors is associated with distant metastasis and decreased survival. *J Vasc Interv Radiol* 25:1604–1608.

Dezarn, W.A. et al. (2011). Recommendations of the American Association of Physicists in Medicine on dosimetry, imaging, and quality assurance procedures for 90Y microsphere brachytherapy in the treatment of hepatic malignancies. *Med Phys* 38:4824.

Dhingra, S. et al. (2014). Histological changes in nontumoral liver secondary to radioembolization of hepatocellular carcinoma with yttrium 90-impregnated microspheres: Report of two cases. *Semin Liver Dis* 34:465–468.

Eaton, B.R. et al. (2014). Quantitative dosimetry for yttrium-90 radionuclide therapy: Tumor dose predicts fluorodeoxyglucose positron emission tomography response in hepatic metastatic melanoma. *J Vasc Interv Radiol* 25:288–295.

Eldredge-Hindy, H. et al. (2014). Yttrium-90 microsphere brachytherapy for liver metastases from uveal melanoma. *Am J Clin Oncol* 39:189–195.

Elschot, M. et al. (2013). Quantitative comparison of PET and bremsstrahlung SPECT for imaging the in vivo yttrium-90 microsphere distribution after liver radioembolization, Villa E, ed. *PLoS One* 8:e55742.

Gaba, R.C. et al. (2009). Radiation lobectomy: Preliminary findings of hepatic volumetric response to lobar yttrium-90 radioembolization. *Ann Surg Oncol* 16:1587–1596.

Gates, V.L., Salem, R., Lewandowski, R.J. (2013). Positron emission tomography/CT after yttrium-90 radioembolization: Current and future applications. *J Vasc Interv Radiol* 24:1153–1155.

Gunduz, S. et al. (2014). Yttrium-90 radioembolization in patients with unresectable liver metastases: Determining the factors that lead to treatment efficacy. *Hepatogastroenterology* 61:1529–1534.

Hamoui, N., Ryu, R. (2011). Hepatic radioembolization complicated by fulminant hepatic failure. *Semin Intervent Radiol* 28:246–251.

Haydar, A., Wasan, H., Wilson, C., Tait, P. (2010). 90Y radioembolization: Embolization of the gastroduodenal artery is not always appropriate. *Cardiovasc Intervent Radiol* 33:1069–1071.

Hickey, R., Lewandowski, R. (2011). Hepatic radioembolization complicated by radiation cholecystitis. *Semin Intervent Radiol* 28:230–233.

Higgins, M., Soulen, M. (2013). Combining locoregional therapies in the treatment of hepatocellular carcinoma. *Semin Intervent Radiol* 30:074–081.

Hilgard, P. et al. (2010). Radioembolization with yttrium-90 glass microspheres in hepatocellular carcinoma: European experience on safety and long-term survival. *Hepatology* 52:1741–1749.

Jakobs, T.F. et al. (2008). Fibrosis, portal hypertension, and hepatic volume changes induced by intra-arterial radiotherapy with 90yttrium microspheres. *Dig Dis Sci* 53:2556–2563.

Kao, Y.H. (2014). Non-target activity detection by post-radioembolization yttrium-90 PET/CT: Image assessment technique and case examples. *Front Oncol* 4: 11.

Kao, Y.H. et al. (2013). Post-radioembolization yttrium-90 PET/CT—Part 2: Dose-response and tumor predictive dosimetry for resin microspheres. *EJNMMI Res* 3:1–1.

Kennedy, A. et al. (2007). Recommendations for radioembolization of hepatic malignancies using yttrium-90 microsphere brachytherapy: A Consensus Panel Report from the Radioembolization Brachytherapy Oncology Consortium. *Int J Radiat Oncol Biol Phys* 68:13–23.

Keppke, A.L. et al. (2007). Imaging of hepatocellular carcinoma after treatment with yttrium-90 microspheres. *AJR Am J Roentgenol* 188:768–775.

Khan, A.S. (2014). Current surgical treatment strategies for hepatocellular carcinoma in North America. *WJG* 20:15007.

Kim, H.J. et al. (2015). Incremental value of liver MR imaging in patients with potentially curable colorectal hepatic metastasis detected at CT: A prospective comparison of diffusion-weighted imaging, gadoxetic acid–enhanced MR imaging, and a combination of both MR techniques. *Radiology* 274:712–722.

Kokabi, N. et al. (2014). A simple method for estimating dose delivered to hepatocellular-carcinoma after yttrium-90 glass-based radioembolization therapy: Preliminary results of a proof of concept study. *J Vasc Interv Radiol* 25:277–287.

Kulik, L.M. et al. (2006). Yttrium-90 microspheres (TheraSphere) treatment of unresectable hepatocellular carcinoma: Downstaging to resection, RFA and bridge to transplantation. *J Surg Oncol* 94:572–586.

Kuo, J.C. et al. (2014). Serious hepatic complications of selective internal radiation therapy with yttrium-90 microsphere radioembolization for unresectable liver tumors. *Asia-Pac J Clin Oncol* 10:266–272.

Lam, M.G.E.H., Banerjee, A., Louie, J.D., Sze, D.Y. (2013a). Splenomegaly-associated thrombocytopenia after hepatic yttrium-90 radioembolization. *Cardiovasc Intervent Radiol* 37:1009–1017.

Lam, M.G.E.H. et al. (2013b). Safety of repeated yttrium-90 radioembolization. *Cardiovasc Intervent Radiol* 36:1320–1328.

Lam, M., Smits, M. (2013c). Value of 99mTc-macroaggregated albumin SPECT for radioembolization treatment planning. *J Nucl Med* 54:1681–1682. Available at: http://jnm.snmjournals.org/cgi/content/full/54/9/1681.

Lencioni, R., Llovet, J. (2010). Modified RECIST (mRECIST) assessment for hepatocellular carcinoma. *Semin Liver Dis* 30:052–060.

Lhommel, R. et al. (2009). Yttrium-90 TOF PET scan demonstrates high-resolution biodistribution after liver SIRT. *Eur J Nucl Med Mol Imaging* 36:1696.

Mascarenhas, N., Ryu, R., Salem, R. (2011). Hepatic radioembolization complicated by abscess. *Semin Intervent Radiol* 28:222–225.

Miller, F.H. et al. (2007). Response of liver metastases after treatment with yttrium-90 microspheres: Role of size, necrosis, and PET. *AJR Am J Roentgenol* 188:776–783.

Miller, A.B., Hoogstraten, B., Staquet, M., Winkler, A. (1981). Reporting results of cancer treatment. *Cancer* 47:207–214.

Minocha, J., Lewandowski, R. (2011). Radioembolization for hepatocellular carcinoma complicated by biliary stricture. *Semin Intervent Radiol* 28:226–229.

Neperud, J. (2013). Can imaging patterns of neuroendocrine hepatic metastases predict response yttruim-90 radioembolotherapy? *WJR* 5:241.

Northover, J.M., Terblanche, J. (1979). A new look at the arterial supply of the bile duct in man and its surgical implications. *Br J Surg* 66:379–384.

Pasciak, A.S. et al. (2015). The impact of an antireflux catheter on target volume particulate distribution in liver-directed embolotherapy: A pilot study. *J Vasc Interv Radiol.* 26:660–669.

Piduru, S.M. et al. (2012). Prognostic value of 18f-fluorodeoxyglucose positron emission tomography-computed tomography in predicting survival in patients with unresectable metastatic melanoma to the liver undergoing yttrium-90 radioembolization. *J Vasc Interv Radiol* 23:943–948.

Poon, R.T.-P., Fan, S.T., Tsang, F.H.-F., Wong, J. (2002). Locoregional therapies for hepatocellular carcinoma: A critical review from the surgeon's perspective. *Ann Surg* 235:466–486.

Reiner, C.S. et al. (2014). Early treatment response evaluation after yttrium-90 radioembolization of liver malignancy with CT perfusion. *J Vasc Interv Radiol* 25:747–759.

Rhee, T.K. et al. (2008). Tumor response after yttrium-90 radioembolization for hepatocellular carcinoma: Comparison of diffusion-weighted functional MR imaging with anatomic MR imaging. *J Vasc Interv Radiol* 19:1180–1186.

Riaz, A. (2014). Side effects of yttrium-90 radioembolization. *Front Oncol* 4:198.

Sacks, A. et al. (2011). Value of PET/CT in the management of liver metastases, Part 1. *Am J Roentgenol* 197:W256–W259.

Sag, A.A., Savin, M.A., Lal, N.R., Mehta, R.R. (2014). Yttrium-90 radioembolization of malignant tumors of the liver: Gallbladder effects. *Am J Roentgenol* 202:1130–1135.

Salem, R. et al. (2002). Yttrium-90 microspheres: Radiation therapy for unresectable liver cancer. *J Vasc Interv Radiol* 13:S223–S229.

Salem, R. et al. (2008). Incidence of radiation pneumonitis after hepatic intra-arterial radiotherapy with yttrium-90 microspheres assuming uniform lung distribution. *Am J Clin Oncol* 31:431–438.

Salem, R. et al. (2010). Radioembolization for hepatocellular carcinoma using yttrium-90 microspheres: A comprehensive report of long-term outcomes. *Gastroenterology* 138:52–64.

Salem, M.E. et al. (2013a). Radiographic parameters in predicting outcome of patients with hepatocellular carcinoma treated with yttrium-90 microsphere radioembolization. *ISRN Oncol* 2013:1–8.

Salem, R., Mazzaferro, V., Sangro, B. (2013b). Yttrium 90 radioembolization for the treatment of hepatocellular carcinoma: Biological lessons, current challenges, and clinical perspectives. *Hepatology* 58:2188–2197.

Salem, R., Thurston, K.G. (2006). Radioembolization with 90yttrium microspheres: A state-of-the-art brachytherapy treatment for primary and secondary liver malignancies. *J Vasc Interven Radiol* 17:1251–1278.

Sangro, B. et al. (2006). Radioembolization using 90Y-resin microspheres for patients with advanced hepatocellular carcinoma. *Radiat Oncol Biol* 66:792–800.

Sangro, B. et al. (2008). Liver disease induced by radioembolization of liver tumors: Description and possible risk factors. *Cancer* 112:1538–1546.

Sato, K. et al. (2006). Treatment of unresectable primary and metastatic liver cancer with yttrium-90 microspheres (TheraSphere®): Assessment of hepatic arterial embolization. *Cardiovasc Intervent Radiol* 29:522–529.

Schlaak, J. (2013). Choi criteria are superior in evaluating tumor response in patients treated with transarterial radioembolization for hepatocellular carcinoma. *Oncol Lett* 6:1707–1712.

Shim, J.H. et al. (2012). Which response criteria best help predict survival of patients with hepatocellular carcinoma following chemoembolization? A validation study of old and new models. *Radiology* 262:708–718.

Singh, P., Anil, G. (2013). Yttrium-90 radioembolization of liver tumors: What do the images tell us? *Cancer Imaging* 13:645–657.

Smith, M.T., Johnson, D.T., Gipson, M.G. (2015). Skin necrosis resulting from nontarget embolization of the falciform artery during transarterial chemoembolization with drug-eluting beads. *Semin Intervent Radiol* 32:22–25.

Smits, M.L.J. et al. (2013). Clinical and laboratory toxicity after intra-arterial radioembolization with 90Y-microspheres for unresectable liver metastases Aravindan N, ed. *PLoS One* 8:e69448.

Soydal, C. et al. (2013). The prognostic value of quantitative parameters of 18F-FDG PET/CT in the evaluation of response to internal radiation therapy with yttrium-90 in patients with liver metastases of colorectal cancer. *Nucl Med Commun* 34:501–506.

Strigari, L. et al. (2010). Efficacy and toxicity related to treatment of hepatocellular carcinoma with 90Y-SIR spheres: Radiobiologic considerations. *J Nucl Med* 51:1377–1385.

Therasse, P. et al. (2000). New guidelines to evaluate the response to treatment in solid tumors. European Organization for Research and Treatment of Cancer, National Cancer Institute of the United States, National Cancer Institute of Canada. *J Natl Cancer Inst* 92:205–216.

Veloso, N. et al. (2013). Gastroduodenal ulceration following liver radioembolization with yttrium-90. *Endoscopy* 45:E108–E109.

Vouche, M. et al. (2015). Clinical imaging. *J Clin Imaging* 39:454–462.

Wang, L.M., Jani, A.R., Hill, E.J., Sharma, R.A. (2013). Anatomical basis and histopathological changes resulting from selective internal radiotherapy for liver metastases. *J Clin Pathol* 66:205–211.

Willowson, K., Tapner, M., Bailey, D. (2015). A multicentre comparison of quantitative 90Y PET/CT for dosimetric purposes after radioembolization with resin microspheres. *Eur J Nucl Med Mol Imaging*:1–21.

Wondergem, M. et al. (2013). 99mTc-macroaggregated albumin poorly predicts the intrahepatic distribution of 90Y resin microspheres in hepatic radioembolization. *J Nucl Med* 54:1294–1301. Available at: http://jnm.snmjournals.org/cgi/doi/10.2967/jnumed.112.117614.

Wright, C.L. et al. (2012). Radiation pneumonitis following yttrium-90 radioembolization: Case report and literature review. *J Vasc Interv Radiol* 23:669–674.

Yu, J.-S. et al. (2002). Predisposing factors of bile duct injury after transcatheter arterial chemoembolization (TACE) for hepatic malignancy. *Cardiovasc Intervent Radiol* 25:270–274.

PART 5

New Horizons

15

Future directions in radioembolization

ALEXANDER S. PASCIAK, J. MARK MCKINNEY, AND YONG C. BRADLEY

15.1 INTRODUCTION

From the perspective of medical physicists, nuclear medicine radiologists, and radiation oncologists involved in radioembolization, the ideal direction of the future of this therapy may appear different than it does to interventional radiologists. This is primarily due to the differences in the common procedures in which each of these groups normally participate. Interventional radiologists focus on procedures that, while technically demanding, normally do not involve the physics, mathematics, or the related precision commonly employed in radiation oncology and nuclear medicine. Further, while the technical considerations involved in a complex vascular intervention necessitate careful review of pretreatment structural and angiographic imaging, most nuclear medicine procedures are completely imaging based, focusing on both function and structure to diagnose and treat the patient. The distinction between these two groups and the ownership that interventional radiologists

normally have over patients undergoing radioembolization therapy have largely driven the radioembolization technical process in the past. Evidence for this can be seen in Chapters 4 and 5, where the manufacturer-recommended treatment-planning methods for radioembolization are largely based on simple empiric calculations that are only patient specific at the simplest level. For example, recall the specifics of the body surface area (BSA) model often employed for treatment planning using resin microspheres. In this model, the impact of height and weight on the prescribed dosage exceed the impact of relative tumor burden. Additionally, the tumor type, vascularity, prior treatment, homogeneity of uptake, and any other patient-specific factors are not considered. While BSA is widely used in medicine for determining dosages for medications such as chemotherapy, it certainly makes more sense for a systemic administration than for a local brachytherapy such as radioembolization. It is for these reasons that many medical physicists, nuclear medicine radiologists, and radiation oncologists new to radioembolization find themselves immediately frustrated with the seeming lack of precision associated with the current standard of care in radioembolization treatment planning. It should be noted that this brief discussion of the commonly employed BSA treatment planning method for resin microspheres is not unique in its lack of tumor specificity—the recommended method for glass microspheres is even less tumor specific and does not account for tumor burden at all. Those looking for a more detailed review of these methods should refer to Chapter 5.

While these treatment-planning methodologies may be simplistic, radioembolization is still successful and beneficial for many patients. Nevertheless, improving the treatment-planning process for both glass and resin microspheres is one way that the future of radioembolization may be shaped. As the subject of this text would suggest, radioembolization is an inherently multidisciplinary field requiring not only an interventional radiologist but also the skills of many other medical specialties. There are numerous necessary steps required to perform a safe and effective radioembolization and with so many potential hands involved, the task can sometimes seem overwhelming. While covered in some detail in the previous chapters, a summary list of tasks required and potential specialties involved are shown below, with a particular focus on the tasks necessary for the audience of this text.

- Patient recruitment and selection (interventional radiologist, oncologist)
- Vascular treatment planning (interventional radiologist)
- Evaluating lung shunt fraction (interventional radiologist, nuclear medicine radiologist)
- Dosimetric treatment planning (nuclear medicine radiologist, radiation oncologist, medical physicist, interventional radiologist)
- Safely preparing the radioembolization dosage (technologist, radiopharmacy, medical physicist, health physicist)
- Preparing the angiographic suite (technologist, health physicist)
- Controlling entry and exit into the treatment room (technologist, health physicist, medical physicist)
- Delivering the dosage (interventional radiologist)

- Determining delivered dose (medical physicist, technologist)
- Posttreatment yttrium-90 (^{90}Y) imaging (nuclear medicine radiologist, medical physicist, technologist)
- Surveying and clearing radioactive contamination (technologist, health physicist)
- Releasing the patient and providing release instructions (health physicist, medical physicist, technologist)
- Using posttreatment ^{90}Y imaging to plan future treatments (interventional radiologist, nuclear medicine radiologist, radiation oncologist, medical physicist)

In this chapter, we will discuss how the future of radioembolization will be affected by improvements related to some of the above tasks. The authors acknowledge that radioembolization is a field that has grown tremendously in recent years but is still relatively young compared with alternative treatments for liver cancer.

15.2 BACKGROUND

The use of ^{90}Y and other radionuclides with localized energy deposition in the percutaneous treatment of disease has a more lengthy history than one might initially suspect. Before the widespread use of hepatic radioembolization, simpler percutaneous procedures were being performed clinically for patients with chronic synovitis due, in part, to hypertrophy of the synovial membrane. Ansell et al. (1963) used a gold-198 (^{198}Au) colloid percutaneously injected into the synovial space to destroy the superficial luminal layers of the synovial membrane. However, as this form of therapy expanded, ^{90}Y in various chemical forms soon replaced ^{198}Au (Oka et al., 1971; Prosser et al., 1993; Stucki et al., 1993; Asavatanabodee et al., 1997; Jahangier et al., 1997; Taylor et al., 1997; Jacob et al., 2003; Oztürk et al., 2008; Thomas et al., 2011).

Radiation synovectomy using ^{90}Y is a treatment that has seen some clinical use, particularly for chronic synovitis that is refractory to traditional intra-articular steroid injections. Treatment traditionally has been used for rheumatoid arthritis, osteoarthritis (Taylor et al., 1997), psoriatic arthritis (Stucki et al., 1993), and hemophilic arthritis (Thomas et al., 2011) unresponsive to systemic medical therapy. 5 mCi of ^{90}Y silicate or ^{90}Y resin-colloid injected into the knee is capable of producing an absorbed dose in the synovium at a depth of 1 mm exceeding 50 Gy (Oka et al., 1971). The goal of radiation synovectomy is to create fibrosis in the hypertrophic areas of the synovium. ^{90}Y has been used extensively for knees; however, for smaller joints, that is, elbows and shoulders, ^{186}Re is preferred due to the lower beta energy (Kavakli et al., 2008). However, the downside of ^{186}Re is the gamma component of the decay, which could result in radiation dose to sensitive tissues near the injection site (e.g., lymph nodes).

Moving in the direction of endovascular therapy, vascular disease, one of the most common diseases in the world, has also been treated with internal emitters. Percutaneous transluminal angioplasty (PTA) is one of the most common treatments for vascular stenosis. However, the durability of PTA is largely determined by restenosis rates, which can be high. Prophylactic endovascular brachytherapy (EVBT), as a preventative tool for restenosis due to intimal hyperplasia (Amols, 1999), has been used successfully for a number of years with various radionuclides and in various parts of the body (Minar, 2012). While early uses of EVBT were based on iridium-192 (^{192}Ir), a low-energy gamma emitter (Schopohl et al., 1996; Reynaert et al., 2001; Piermattei et al., 2002), some newer techniques use high-energy beta particles from phosphorus-32 (^{32}P) (Piermattei et al., 2003), rhenium-188 (^{188}Re) (Werner et al., 2012), strontium-90 (^{90}Sr), or ^{90}Y (Coucke, 2009). The advantage of pure beta emitters is, of course, the markedly reduced radiation safety concerns associated with the procedure. However, despite clinical efficacy, the technical difficulty of EVBT has hindered its widespread clinical use in favor of alternatives such as drug-eluting stents for the management of restenosis.

Radioembolization also has a more lengthy clinical history than one might initially expect. Ariel and Padula (1978a, 1978b) in 1978 reported the first cases of the clinical use of ^{90}Y microspheres in the intra-arterial treatment of colorectal metastases to the liver. Ariel combined intra-arterial infusion of resin ^{90}Y microspheres with chemotherapy in the form of 5-fluorouracil. Ariel's patients received relatively large dosages of ^{90}Y, ranging from 100 to 150 mCi (3.7–5.5 GBq) that led to improved response in their 65 patient cohort. Interestingly, Ariel infused ^{90}Y microspheres into these patients using both percutaneous delivery and delivery through open laparotomy with a catheter inserted directly into the hepatic artery. Since this initial experience, the techniques and sophistication involved in the manufacture and treatment with ^{90}Y microspheres have tremendously improved.

The therapeutic percutaneous uses of endovascular brachytherapy and radioembolization using ^{90}Y and other radionuclides have lengthy and interesting histories. However, this chapter will look to the future of radioembolization.

15.3 TOWARD IMPROVED TREATMENT PLANNING

Treatment planning in ^{90}Y radioembolization lacks much of the detail and patient specificity required for external beam radiation therapy. This is logical since there is an element of control in external beam radiation therapy and even in conventional brachytherapy that is lacking in radioembolization—namely, physical flow dynamics that determine the final location of the microspheres. While this process cannot be controlled, it can be predicted to a limited extent.

As discussed extensively in Chapters 4 and 5, technetium-99m (^{99m}Tc)-macroaggregated albumin (MAA) single-photon emission computed tomography/computed tomography (SPECT/CT) can be used as a standard component of treatment planning using the partition model. Many authors have examined the validity of MAA as a radioembolization surrogate, with no clear consensus (Knesaurek et al., 2010; Kao et al., 2012; Lam and Smits, 2013; Lam et al., 2013; Wondergem et al., 2013; Garin et al., 2014; Lam and Sze, 2014) as to its accuracy in the modeling of hepatic distribution. The position of the catheter tip during both infusion of MAA and radioembolization is among the most critical factors to the prognostic utility of tumor to normal uptake ratio (T:N) measurements made from ^{99m}Tc-MAA SPECT/CT. Positioning of the catheter tip becomes especially critical when it is near a bifurcation or when it is positioned in a tortuous vessel (Jiang et al., 2012; Wondergem et al., 2013). However, in spite of variable correlation between MAA and ^{90}Y microspheres in the literature, many authors agree that MAA is an excellent option for treatment planning and predictive dosimetry. This has been discussed in detail in Chapters 4 through 6.

The use of ^{99m}Tc-MAA as a treatment-planning guide certainly has potential utility in the prognostication of lung shunt fraction, intrahepatic dose distribution, and presence of gastroduodenal nontarget embolization (NTE). Due to a handful of publications describing the difficulty of managing patients with ulcerations from gastroduodenal NTE, the majority of radioembolization treatment centers carefully examine post-MAA nuclear imaging to look for the presence of NTE. However, the majority of treatment centers perform only planar imaging of ^{99m}Tc-MAA, as this is the most common method used to determine the lung shunt fraction and is recommended in the package insert for both resin and glass ^{90}Y microspheres. Many institutions can improve both the sensitivity and specificity of ^{99m}Tc-MAA for the determination of NTE with several simple protocol modifications. As discussed in Chapter 14, there is a substantial increase in the prognostic utility of MAA as a predictor for extrahepatic NTE with the use of SPECT/CT compared with planar imaging. A detailed study by Ahmadzadehfar et al. (2010) suggested that the relative sensitivity of detecting extrahepatic NTE increased from 32% to 100% when SPECT/CT was used compared with planar imaging. That said, however, an increase in sensitivity may lead to exclusion of patients from treatment owing to false positives, such as free technetium. In Chapter 4, there is a detailed discussion of the biological half-life of MAA, binding efficiency, and effects of free ^{99m}Tc. Free ^{99m}Tc in the form of pertechnetate (^{99m}TcO$_4^-$) will show a strong uptake in the gastric mucosa, potentially leading to false positives when MAA is used to assess NTE. Standard of care prophylaxis should include oral administration of sodium perchlorate (NaClO$_4$) as a blocking agent to prevent the uptake of ^{99m}TcO$_4^-$ in gastroduodenal tissues. Sodium perchlorate has been shown to substantially increase the specificity of

^{99m}Tc-MAA as a tool for prognostication of gastroduodenal NTE (Sabet et al., 2011).

Stepping away from the standard of care ^{99m}Tc-MAA for evaluating lung shunt fraction and treatment simulation, there have been several attempts to identify alternative tracers that can be applied to this purpose. Mathias and Green (2008) described a simple method for the production of gallium-68 (^{68}Ga) MAA from the elutant of a conventional germanium-68/^{68}Ga generator. The size range of MAA did not change with the binding of ^{68}Ga and binding efficiency exceeded 99% (Mathias and Green, 2008). ^{68}Ga is a convenient positron emission tomography (PET) radiotracer that provides fully quantitative imaging with a superior resolution to ^{99m}Tc SPECT. ^{68}Ga MAA may have utility in conventional pulmonary perfusion studies, evaluation of lung shunt fraction for radioembolization, and, potentially, improved intrahepatic treatment planning given the improved resolution and quantification of PET/CT.

It is likely, however, that the largest source of error in the use of ^{99m}Tc-MAA as a ^{90}Y microsphere surrogate is in particulate deposition differences rather than SPECT resolution limits. Therefore, the greatest advances in pretreatment simulation may come from replacing MAA with a superior particle, rather than identification of a radionuclide with superior imaging characteristics to ^{99m}Tc. Biodegradable human serum albumin (HSA) microspheres have been proposed for use in pulmonary perfusion studies labeled with ^{99m}Tc or ^{86}Y. HSA microspheres have also been proposed as an alternative form of radioembolization when labeled with ^{188}Re or ^{90}Y (Schiller et al., 2008). A potential benefit of HSA microspheres over MAA is that the size and surface characteristics of the microsphere much more closely approximate those of ^{90}Y microspheres, as shown in Figure 1 of Schiller et al. (2008). Naturally, the specific gravity of an HSA microsphere will also closely approximate that of ^{90}Y resin microspheres. As a tool for pretreatment simulation, the biodegradability of HSA microspheres has a favorable characteristic since it may be degraded and removed from the hepatic vasculature prior to ^{90}Y radioembolization.

Currently, there are multiple clinical trials underway to evaluate the utility of alternative microspheres for pretreatment planning, both in biodegradable and permanent physical form. Potential imaging modalities for these microspheres primarily include SPECT or PET, but in some cases have extended to magnetic resonance imaging (MRI). Holmium-166 (^{166}Ho) microspheres (Quiremspheres, Quirem Medical, The Netherlands) are relatively unique in radioembolization for their multimodality imaging and therapy potential. ^{166}Ho decays with the emission of a two medium-energy beta particles at a high yield with average energies of 654 and 691 keV. ^{166}Ho also emits an 80 keV gamma ray with a lower yield (approximately 6.5 per 100 transformations) that can be effectively imaged using SPECT. Finally, Holmium is strongly paramagnetic, resulting in convenient visualization using MRI with high contrast and resolution. Examples of direct imaging of ^{166}Ho by both MRI and SPECT are shown in Figure 1.1, in Chapter 1. The clinical protocol for ^{166}Ho radioembolization takes advantage of multimodality imaging and includes pretreatment infusion of a low-activity dosage of ^{166}Ho radioembolization as a tool for treatment planning and evaluation of lung shunt fraction via nuclear imaging. This process eliminates MAA altogether in the ^{166}Ho radioembolization treatment process.

Many current and previous investigations have focused on improving pretreatment imaging and simulation for radioembolization in an effort to improve treatment planning. However, while these tools can help to predict tumor and normal liver tissue absorbed doses, this is just part of the treatment-planning equation. Patient-specific treatment planning also requires a precise knowledge of tumoricidal dose thresholds, which will vary at a minimum on tumor type, size, vascularity, previous therapy, and use of glass or resin microspheres. The difficulty of obtaining these data has been compounded in the past by the limited availability of postradioembolization quantitative imaging; however, some excellent sources do exist particularly for hepatocellular carcinoma (HCC) (Strigari et al., 2010). Going forward, the discovery of ^{90}Y PET/CT will open up the door to precisely defining toxicity thresholds for radioembolization. An international clinical trial sponsored by SIRTeX medical (SIR-Spheres®, SIRTex Technology Pty, Lane Cove, NSW, Australia) will use postradioembolization ^{90}Y PET/CT to elucidate dose–response thresholds based on tumor type for metastatic breast and colon cancer. Some initial data from this effort have already been published (Willowson et al., 2015).

15.4 IMPROVING EFFICACY AND SAFETY

15.4.1 ENHANCEMENT OF T:N

As discussed in detail in Chapters 5 and 8, the success of radioembolization depends on delivering sufficient dosage to the tumor to elicit a therapeutic response while sparing uninvolved liver tissue from excessive toxicity. In conventional lobar therapy, it is the T:N that in many ways will define the balance between sufficient absorbed dose to the tumor and limitation of dose to normal liver tissue. T:N is defined in Equation 15.1:

$$\text{T:N} = \frac{A_{^{90}\text{Y,tumor}} / V_{\text{tumor}}}{A_{^{90}\text{Y,normal}} / V_{\text{normal}}} \tag{15.1}$$

where, $A_{^{90}\text{Y,tumor}}$ is the ^{90}Y activity (MBq) deposited in the tumor and $A_{^{90}\text{Y,normal}}$ is the activity deposited in uninvolved liver tissue. V_{normal} and V_{tumor} are the respective volumes of each. Tumor type, size, burden, prior treatments, and other patient-specific physiological factors significantly affect T:N, leading to a range of clinical values which can vary from nearly 15:1 to less than 1. Table 5.2 in Chapter 5 lists some typical T:N values from the literature. While there are many factors in addition to T:N that determine the success or failure of a radioembolic therapy, lower T:Ns make a robust clinical response difficult to achieve. The premise behind the difficulty of successful radioembolization in the setting of a low T:N (<2:1) can be linked back to the fundamental fallacy of treating liver cancer with external beam radiation therapy: the radiation toxicity threshold for uninvolved liver tissue is low—potentially lower than the absorbed dose necessary for effective tumor control.

Several methods of prophylactically increasing T:N are available and may improve efficacy in certain classes of patients receiving lobar therapy. Patients with low T:N (<2:1), or patients with moderate T:N (<3:1) in the setting of underlying liver disease and/or tumors that require large absorbed doses, may benefit from these techniques. Both pharmaceutical and physical techniques can be used to prophylactically increase T:N prior to radioembolization in patients who fall into either of these categories.

15.4.1.1 Pharmaceutical methods

Tumor angiogenesis results in the formation of arterioles and capillaries that are structurally and physiologically abnormal and support modification of T:N using both pharmaceutical and physical techniques. Tumor arterioles are abnormal in that they do not have complete formation of smooth muscle coat and lack autonomic innervation and, therefore, autoregulatory response (Mattsson et al., 1977; Ashraf et al., 1996; Burke et al., 2001; Tanaka et al., 2008). Tumor capillaries are abnormal, owing to both incomplete formation of the basement membrane and pericyte detachment (Nagy et al., 2009).

Intra-arterial infusion of the vasoconstrictor angiotensin II has been used in several studies (Sasaki et al., 1985; Goldberg et al., 1991a, 1991b; Burke et al., 2001; Flower et al., 2001) as a means to preferentially constrict arterioles feeding normal liver tissue, while arterioles supplying tumor will remain largely unaffected owing, again, to the lack of innervation and incomplete smooth muscle coat (Mattsson et al., 1977; Hafström et al., 1980; Burke et al., 2001). A recent review by van den Hoven et al. (2014b) summarized the results of all published studies that quantified the degree of change in T:N before and after infusion of angiotensin II. In these publications spanning findings in 71 patients, an increase in T:N ranging from 180% to 310% following infusion of angiotensin II has been reported. These changes, which are substantial, can dramatically affect the distribution of radioactive microspheres. Figure 15.1 shows the absorbed dose to the tumor and to the uninvolved liver as a function of T:N with the pre- and postangiotensin II (van den Hoven et al., 2014b) ranges, highlighted to illustrate the potential benefit.

Angiotensin II is not the only vasoconstrictor that has been evaluated as a tool to improve liver-directed therapies. The effect of intra-arterial infusion of antidiuretic hormone (ADH) has been demonstrated in a porcine model (Durack et al., 2012). Specifically, the benefits of ADH have been quantified in terms of its ability to reduce NTE by preferential constriction of gastroenteric

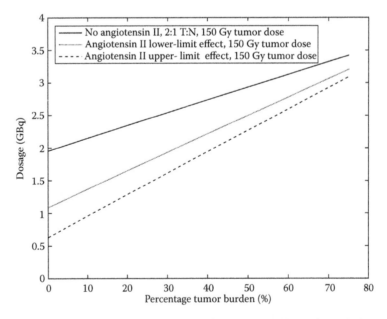

Figure 15.1 The theoretical treatment dosage (GBq) of yttrium-90 (^{90}Y) radioembolization necessary to achieve an average tumor absorbed dose of 150 Gy. Baseline in the absence of intra-arterial antiotensin II administration assumes a 2:1 tumor to normal uptake ratio (T:N) and a 1300 cc total liver volume. The slope of the curve indicates the necessary increase in the ideal treatment dosage as a function of percentage tumor infiltration. With angiotensin II and its related lower and upper limit of efficacy on T:N (van den Hoven et al., 2014b), the treatment dosage required to achieve 150 Gy is decreased substantially at low tumor infiltration. As percentage tumor infiltration rises, the benefit of angiotensin II is reduced. (From van den Hoven, A.F. et al., *PLoS ONE*, 9, e86394, 2014b.)

collaterals. ADH was effective in reducing the ratio of gastric to hepatic activity by a factor of two (Durack et al., 2012). Similar analyses based on infusion of catecholamines (Hafström et al., 1980; Tanaka et al., 2008) have been performed on both rats (Hafström et al., 1980) and humans (Tanaka et al., 2008) with concomitant increases in T:N from norepinephrine (Hafström et al., 1980) and epinephrine (Tanaka et al., 2008). While vasoconstrictors can illicit a positive increase on T:N that may contribute to the improved safety and efficacy of radioembolization in some patients, their use must be weighed against the potential contraindications related to their systemic effects. Median increases in systolic blood pressure of more than 40 mm Hg (Sasaki et al., 1985; Goldberg et al., 1991a) have been reported following hepatic arterial infusion of angiotensin II, and while transient, this increase may be an important safety consideration in some patients.

15.4.1.2 Physical techniques

As described in Chapter 3, pretreatment occlusion of the right gastric artery (RGA) and gastroduodenal arteries (GDA) is often performed prior to radioembolization to prevent extrahepatic NTE. The use of specialty catheters, or antireflux catheters, such as the Surefire Infusion System (Surefire Medical Inc., Westminster, CO), may also protect extrahepatic tissues from NTE by preventing retrograde flow (Arepally et al., 2013; Fischman et al., 2014; van den Hoven et al., 2014a; Morshedi et al., 2015) of microspheres into unprotected collaterals. However, antireflux catheters (Figure 15.2) have recently been shown to alter the hepatic distribution of radioembolization and other liver-directed therapies (Arepally et al., 2013; Pasciak et al., 2015; van den Hoven et al., 2015). Arepally et al. (2013) showed increases in distal arterial penetration of tantalum microspheres following renal artery embolization in a porcine model with the use of an antireflux microcatheter compared with a conventional end-hole catheter, sparking interest in the use of these devices for a purpose other than preventing reflux.

While pharmaceutical techniques for increasing T:N take advantage of the lack of complete smooth muscle coat and innervation in tumor arterioles, physical techniques build upon the related inability of tumor arterioles to autoregulate in the setting of changing arterial pressure.

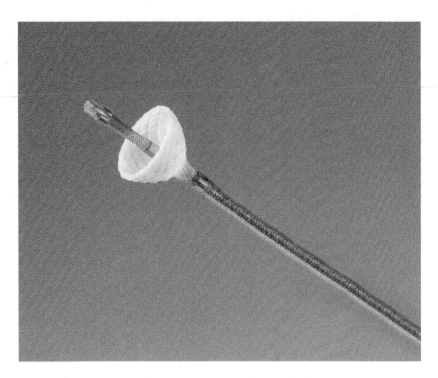

Figure 15.2 The Surefire Precision Infusion system. (Surefire Medical, Denver, CO.)

Rose et al. (2013) made *in vivo* downstream hepatic-arterial blood pressure measurements with the expandable tip (Figure 15.3) of an antireflux catheter open and with it closed. The findings of this study indicated that the open tip of the antireflux catheter reduced downstream systolic and diastolic arterial pressure by nearly 50%. In theory, owing to the autoregulatory response of normal arteries and arterioles, radioembolization performed using an antireflux microcatheter may trigger vasoconstriction of the vessels perfusing uninvolved hepatic parenchyma. At the same time, the structurally abnormal angiogenesis-induced tumor arterioles are not likely to constrict owing to absence of smooth muscle, innervation, and autoregulatory properties (Mattsson et al., 1977; Burke et al., 2001). This process could preferentially shunt microspheres toward the tumor compartment, temporarily increasing the T:N.

The effect of an antireflux catheter on T:N in radioembolization has been shown experimentally using ^{99m}Tc-MAA with a two-step same-day infusion (Figure 15.4) based on a protocol traditionally used for routine renal and cardiac perfusion imaging (Pasciak et al., 2015). Statistically significant increases in tumor uptake, commensurate with decreases in uninvolved liver, support the premise that an antireflux catheter can increase T:N (Pasciak et al., 2015). Differences in hepatic distribution of ^{99m}Tc-MAA with an end-hole catheter and an antireflux catheter are shown in Figure 15.4a and b, respectively. The distribution of ^{90}Y resin microspheres following infusion with the antireflux catheter and imaged directly using ^{90}Y PET/CT is shown in Figure 15.4d and demonstrates excellent agreement to Figure 15.4b.

While it is convenient to cite downstream pressure changes to explain the effect of an antireflux catheter, it is likely that centering of the tip as well as turbulence of flow also contributes to the impact of these devices. For example, the open semiocclusive tip (Figure 15.3) will result in an increase in the turbulence of downstream arterial flow. Increased turbulence of flow may result in a more homogenous cross-sectional distribution of microspheres in downstream arteries, potentially improving homogeneity of deposition (van den Hoven et al., 2015).

15.4.2 ALTERNATIVE METHODS TO ENHANCE TUMOR TARGETING

A more proactive approach to increasing T:N involves the occlusion of arteries supplying uninvolved areas of the liver in an effort to redirect hepatic-arterial blood flow, and thus ^{90}Y microspheres, to the tumor itself. Naturally,

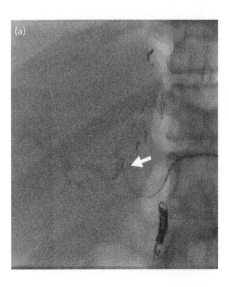
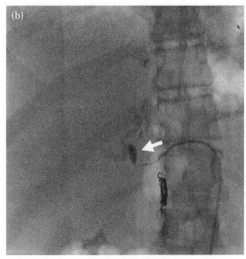

Figure 15.3 Proper hepatic angiogram showing (a) an end-hole catheter and (b) an antireflux catheter with the tip expanded.

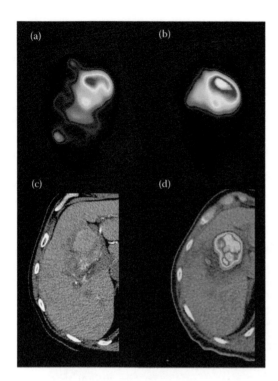

Figure 15.4 (a) Single-photon emission computed tomography (SPECT) following infusion of technetium-99m-macroaggregated albumin (^{99m}Tc-MAA) using a conventional end-hole catheter and **(b)** using an antireflux catheter. The catheter tip position was identical in each case. **(c)** The distribution of ^{99m}Tc-MAA can be compared with contrast-enhanced computed tomography (CT) in this patient with focal hepatocellular carcinoma (HCC). **(d)** ^{90}Y positron emission tomography (PET)/CT following infusion of resin ^{90}Y microspheres with the antireflux catheter. The distribution of ^{90}Y was well predicted with ^{99m}Tc-MAA, as shown in (b).

such efforts would only be indicated with temporary methods of occlusion such as biodegradable starch microspheres or gelatin powder. Degradable starch microspheres (DSM) have been shown in a porcine model to have a short biological half-life with complete re-perfusion in the liver in 30 minutes (Pieper et al., 2015). DSM may have utility in improving safety by protecting normal liver tissue in situations where subselection or repositioning of the microcatheter to avoid microsphere deposition in large volumes of uninvolved liver is not a possibility (Meyer et al., 2013).

15.5 TOWARD IMPROVED POSTTREATMENT IMAGING

15.5.1 NEW TRACERS AND METHODS

In the past, the difficulty of directly imaging ^{90}Y has led to a variety of interesting experimental modifications to ^{90}Y microspheres. Aliva-Rodriguez was able to successfully bind ^{86}Y and ^{89}Zr to the surface of resin ^{90}Y microspheres. Spheres were radiolabeled and the binding stability for *in vivo* applications was confirmed out to 24 hours at a physiological pH temperature (Avila-Rodriguez et al., 2007). This technique had the benefit of direct quantitative imaging of ^{86}Y and ^{89}Zr using PET/CT. Investigations into dual isotope ^{90}Y and ^{177}Lu microspheres have also been performed (Poorbaygi et al., 2011), with effective imaging of the gamma emissions of ^{177}Lu using SPECT/CT. ^{166}Ho radioembolization, as already discussed, is perhaps the best in multimodality direct imaging of radioactive microspheres.

Chapters 10 and 11 discussed in detail techniques for quantification of ^{90}Y bremsstrahlung SPECT and ^{90}Y PET/CT, which have been substantially refined over the past 5 years. Particularly in the case of ^{90}Y PET/CT, many modern PET/CT systems are able to accurately quantify ^{90}Y right out of the box using existing software (Willowson et al., 2015). In light of this information, the driving force behind development of directly imageable microspheres using alternative tracers has decreased substantially. This is not to say, however, that there are no areas for improvement.

As quantitative imaging of ^{90}Y radioembolization becomes more common, routine techniques normally reserved for diagnostic studies can be applied. One such example is respiratory gating, which has been shown to affect both lesion size and quantification in 2-deoxy-2-[fluorine-18]fluoro-D-glucose (^{18}FDG) PET/CT (Suenaga et al., 2013). This technique can be applied to postradioembolization quantitative imaging, including ^{90}Y PET/CT (Pasciak et al., 2014), and may be capable of improving lesion detection and quantitative accuracy. Figure 15.5 visually illustrates the advantages of amplitude-based respiratory gating in postradioembolization imaging.

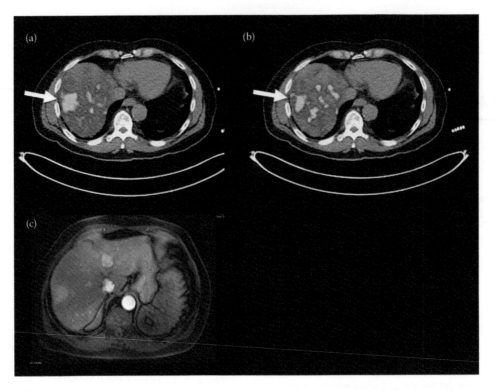

Figure 15.5 (a) Posttreatment ^{90}Y PET/CT on a Siemens mCT Flow without respiratory gating and (b) using amplitude-based gating (Siemens HD•Chest). Figure parts (a) and (b) were both reconstructed from the same acquisition data, making the decreased segment VIII lesion size easy to appreciate when amplitude-based gating is used. Respiratory motion can effect image quality and quantification in hepatic ^{90}Y PET/CT just as it can in 2-deoxy-2-[fluorine-18]fluoro-D-glucose (^{18}FDG) PET/CT. (c) Pretreatment hepatic protocol magnetic resonance imaging (MRI) for comparison.

15.5.2 IMAGE ANALYSIS

The increasing use of quantitative postradioembolization imaging based on either alternative radiotracers, ^{90}Y PET/CT, or quantitative bremsstrahlung SPECT suggests the need for standardized techniques to interpret these images. While there are numerous methods that can be used to convert quantitative ^{90}Y images into three-dimensional representations of absorbed dose as described in Chapter 12, there is a significant ambiguity in the simple process of using this information to determine the absorbed dose to a tumor. Using this interpreted tumor absorbed-dose data to predict treatment efficacy is associated with several pitfalls:

- What "dose metric" to the tumor correlates with published tumoricidal thresholds? Average dose, maximum dose, something else?

- There are many ways to contour the tumor. Contouring based on ^{90}Y uptake on postradioembolization imaging or contouring on pretreatment ^{18}FDG PET/CT, contrast-enhanced CT, or hepatic protocol MRI?

- Published dose–response thresholds are based on a particular method of dose characterization that may or may not be reproducible or even specified in the literature.

- Finally, the quantitative imaging modality used for postradioembolization imaging will itself effect the dose measurement owing to differences in quantification accuracy and contrast recovery (a function of tumor size).

The future of radioembolization includes the use of quantitative postradioembolization imaging to predict treatment efficacy by comparison to tumoricidal thresholds. This is incredibly powerful since it allows a physician to immediately consider alternative or adjuvant therapy, allowing radioembolization to mesh synergistically with the spectrum of treatment options available to the patient. Several examples in the literature have illustrated the benefits of quantitative postradioembolization imaging as a tool to influence patient care decisions (Chang et al., 2013; Bourgeois et al., 2014; Pasciak et al., 2014). These techniques are also discussed in Chapter 14.

As we progress as a field we must become unified and consistent to avoid the aforementioned pitfalls. To this end, the suggestions in Section 15.5.2.1 may serve as a starting point.

15.5.2.1 Tumor contouring on postradioembolization quantitative imaging

Posttreatment imaging of radioactive microspheres is not a diagnostic test for liver cancer and should not be treated as such. Although contouring tumor based on areas of high ^{90}Y activity concentration may sometimes correlate with active tumor, more often it will not. Therefore, this practice may lead to unreliable correlation with dose–response thresholds and poor prognostic accuracy of quantitative postradioembolization imaging. Instead, the following guidelines should be applied.

1. Tumor contours on quantitative posttreatment imaging should be defined based on pretreatment *diagnostic* scans. Pretreatment diagnostic imaging may include ^{18}FDG PET/CT, three-phase hepatic CT, hepatic protocol MRI, and, for example, specialty procedures such as ^{111}In octreotide SPECT for neuroendocrine tumors. Based on the patient's history, the appropriate standard of care diagnostic imaging procedure should be selected by the care provider and used to define tumor contours.
2. Tumor contours on pretreatment imaging should be translated onto postradioembolization quantitative imaging using count-preserving deformable image-registration software. An example of this process is shown in Figure 15.6.

Certainly, the software in (2) above may not always be available. A qualified radiologist or nuclear medicine physician can still reliably draw contours manually so long as the pretreatment

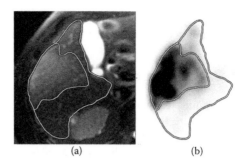

(a)   (b)

Figure 15.6 An example of count-preserving deformable registration using the MIM 6 (MIM Software Inc., Cleveland, OH) software package. **(a)** Semiautomatic contouring of liver and tumor is performed on hepatic protocol MRI. **(b)** Deformable registration algorithm is used to copy contours onto ^{90}Y PET/CT posttreatment imaging. To preserve absorbed dose quantification, contours are deformed to match the previously defined registration.

diagnostic scan is referenced. Those involved in this task must understand that microsphere deposition is a mechanical process and, again, postradioembolization imaging is not diagnostic evaluation of tumor location or activity.

15.5.2.2 Dose metrics of interest

The most common dose metric reported in the literature is the average absorbed dose to the tumor (D_{avg}). In many respects, D_{avg} is flawed despite its wide use for convenience. The failure of radioactive microspheres to penetrate into small areas of tumor will artificially reduce the D_{avg} measurements, even if the majority of the tumor may respond to the therapy. On the other hand, D_{avg} may be artificially inflated if a small portion of the tumor receives a large dose, even if the majority of it is left untreated.

Alternatively, some authors have used standards from external beam radiation therapy such as D_{70} and V_{100} for tumor analysis (Kao et al., 2013). D_{70} is the minimum absorbed dose delivered to 70% of the tumor volume, while V_{100} is the percentage of tumor volume exceeding 100 Gy. D_{70} is not skewed by either of the aforementioned scenarios and is a more reproducible metric for dosimetry based on postradioembolization quantitative imaging. While not reported by some nuclear image analysis software packages,

D_{70} can always be computed from dose–volume histogram data exported from a tumor contour.

15.5.2.3 Standardization

Even if count-preserving deformable image registration software is used to measure a robust dose metric such as D_{70}, dose–response data used for comparison must also have been measured in a similar reproducible way. Standardization as a field is essential to utilizing dose–response data to influence the clinical decision-making process for a patient. In this regard, radioembolization is decades behind external beam radiation therapy. Published guidelines from the Society of Nuclear Medicine, Society of Interventional Radiology, American Association of Physics in Medicine, or the American Society of Therapeutic Radiation Oncology would be helpful in moving the field of radioembolization forward in the establishment of dose–response thresholds.

15.6 TOWARD EXTRAHEPATIC TUMOR RADIOEMBOLIZATION

Several attempts at using radioembolization for extrahepatic tumor control and other endpoints have been investigated. One that has received some attention is renal artery radioembolization for renal cell carcinoma (RCC). While renal artery embolization with bland microspheres prior to surgical resection is a widely used tool that aids in the control of blood loss during surgery, renal artery radioembolization as a treatment modality is much more controversial. Over 20 years ago, selective renal artery radioembolization (RARE) using ^{90}Y microspheres was performed in a porcine model, with the capacity to deliver in excess of 100 Gy to the kidney with greater than 95% dose retention and sparing of extrarenal tissues (Zimmermann et al., 1995). However, its use in humans has never progressed beyond a few case reports. One such example of a large (14.7 × 11.1 cm) focal RCC mass was embolized with ^{90}Y glass microspheres delivering an average tumor dose of 80 Gy (Hamoui et al., 2013). At 8 weeks following RARE, CT imaging revealed patchy tumor hypodensity consistent with necrosis. The

patient lived 23 months following RARE and expired from extrarenal metastases; however, the primary renal tumor remained well controlled and stable (Hamoui et al., 2013). Several small-scale clinical trials are underway to continue the investigation into the utility of RARE for RCC.

Radioembolization of lung metastases via the bronchial artery has also been performed in a small patient series by Ricke et al. (2013). Two patients, one with colorectal cancer and the another with RCC lung metastasis who had failed chemotherapy and were not surgical candidates, were treated using radioembolization via the bronchial artery (Ricke et al., 2013). A conservative treatment plan was assumed based on a T:N of 1:1 owing to the differing arterial anatomy in the lung and liver. 200 MBq of ^{90}Y resin microspheres was infused into both patients through a branch of the bronchial artery perfusing multiple lung segments. The authors performed both posttreatment bremsstrahlung SPECT and ^{90}Y PET/CT and found, surprisingly, that pulmonary deposition of microspheres was limited only to active tumor (Ricke et al., 2013). In fact, the T:N for bronchial artery radioembolization seems far higher than typical T:N for hepatic radioembolization. This finding is critical in that it supports future investigations into the utility of radioembolization for lung metastases.

A nononcologic extrahepatic use of radioembolization has also recently been explored. Pasciak et al. (2016) infused 90Y radioembolization into the gastric fundus in a porcine model to evaluate the potential of radioembolization in the management of obesity. Although the animal cohort was small, decreased weight gain was noted. The mechanism of action was thought to be both a decrease in ghrelin producing cells in the gastric mucosa and a decrease in stomach size and volume. While this is an interesting application of radioembolization, additional animal studies are needed before a human trial is considered.

15.7 CONCLUSIONS

Many of the ideas provided in this chapter on the future of radioembolization stand upon research done in the past. In some cases, interesting and novel trials related to improving radioembolization were performed years ago, before radioembolization was widely used. However, the failure to widely adopt these methods may simply be due to the fact that radioembolization has shown strong clinical efficacy and safety as-is. This has made drastic changes in the procedure or treatment difficult to justify. That said, an improved understanding of radioembolization dose–response may be within reach due to the availability of new imaging techniques such as ^{90}Y PET/CT. These data could have a broad impact on both treatment planning, patient selection, and patient follow-up. However, this information cannot effectively be elucidated by individual researchers; instead, an international effort should be organized in order to obtain useful and reproducible results, ideally through professional societies.

As more is understood about the biological effects of radioembolization, it will in turn will further the multidisciplinary nature of radioembolization, expanding the role of medical physicists, radiation oncologists, and nuclear medicine radiologists in this treatment modality. Because of the unique nature of this therapy, we believe this can only result in the improved patient outcomes.

REFERENCES

Ahmadzadehfar, H. et al. (2010). The significance of 99mTc-MAA SPECT/CT liver perfusion imaging in treatment planning for 90Y-microsphere selective internal radiation treatment. *J Nucl Med* 51:1206–1212.

Amols, H.I. (1999). Review of endovascular brachytherapy physics for prevention of restenosis. *Cardiovasc Radiat Med* 1:64–71.

Ansell, B.M., Crook, A., Mallard, J.R. (1963). Evaluation of intra-articular colloidal gold Au 198 in the treatment of persistent knee effusions. *Ann Rheum Dis* 22:435–439.

Arepally, A., Chomas, J., Kraitchman, D., Hong, K. (2013). Quantification and reduction of reflux during embolotherapy using an antireflux catheter and tantalum microspheres: Ex vivo analysis. *J Vasc Interv Radiol* 24:575–580.

Ariel, I.M., Padula, G. (1978a). Treatment of symptomatic metastatic cancer to the liver from primary colon and rectal cancer by the intra-arterial administration of chemotherapy and radioactive isotopes. *Prog Clin Cancer* 7:247–254.

Ariel, I.M., Padula, G. (1978b). Treatment of symptomatic metastatic cancer to the liver from primary colon and rectal cancer by the intraarterial administration of chemotherapy and radioactive isotopes. *J Surg Oncol* 10:327–336.

Asavatanabodee, P., Sholter, D., Davis, P. (1997). Yttrium-90 radiochemical synovectomy in chronic knee synovitis: A one year retrospective review of 133 treatment interventions. *J Rheumatol* 24:639–642.

Ashraf, S. et al. (1996). The absence of autonomic perivascular nerves in human colorectal liver metastases. *Br J Cancer* 73:349–359.

Avila-Rodriguez, M.A. et al. (2007). Positron-emitting resin microspheres as surrogates of 90Y SIR-Spheres: A radiolabeling and stability study. *Nucl Med Biol* 34:585–590.

Bourgeois, A.C. et al. (2014). Intra-procedural 90Y PET/CT for treatment optimization of 90Y radioembolization. *J Vasc Interv Radiol* 25:271–275.

Burke, D. et al. (2001). Continuous angiotensin II infusion increases tumour: Normal blood flow ratio in colo-rectal liver metastases. *Br J Cancer* 85:1640–1645.

Chang, T.T., Bourgeois, A.C., Balius, A.M., Pasciak, A.S. (2013). Treatment modification of yttrium-90 radioembolization based on quantitative positron emission tomography/CT imaging. *J Vasc Interv Radiol* 24:333–337.

Coucke, P. (2009). Basic rules of dosimetry in endovascular brachytherapy. *J Interv Cardiol* 13:425–429.

Durack, J.C. et al. (2012). Intravenous vasopressin for the prevention of nontarget gastrointestinal embolization during liver-directed cancer treatment: Experimental study in a porcine model. *J Vasc Interv Radiol* 23:1505–1512.

Fischman, A.M. et al. (2014). Prospective, randomized study of coil embolization versus surefire infusion system during yttrium-90 radioembolization with resin microspheres. *J Vasc Interv Radiol* 25:1709–1716.

Flower, M.A. et al. (2001). 62Cu-PTSM and PET used for the assessment of angiotensin II-induced blood flow changes in patients with colorectal liver metastases. *Eur J Nucl Med* 28:99–103.

Garin, E., Boucher, E., Rolland, Y. (2014). 99mTc-MAA-based dosimetry for liver cancer treated using 90Y-loaded microspheres: Known proof of effectiveness. *J Nucl Med* 55:1391–1392.

Goldberg, J.A. et al. (1991a). The use of angiotensin II as a potential method of targeting cytotoxic microspheres in patients with intrahepatic tumour. *Br J Cancer* 63:308–310.

Goldberg, J.A. et al. (1991b). Angiotensin II as a potential method of targeting cytotoxic-loaded microspheres in patients with colorectal liver metastases. *Br J Cancer* 64:114–119.

Hafström, L., Nobin, A., Persson, B., Sundqvist, K. (1980). Effects of catecholamines on cardiovascular response and blood flow distribution to normal tissue and liver tumors in rats. *Cancer Res* 40:481–485.

Hamoui, N. et al. (2013). Radioembolization of renal cell carcinoma using yttrium-90 microspheres. *J Vasc Interv Radiol* 24:298–300.

Jacob, R., Smith, T., Prakasha, B., Joannides, T. (2003). Yttrium90 synovectomy in the management of chronic knee arthritis: A single institution experience. *Rheumatol Int* 23:216–220.

Jahangier, Z.N., Jacobs, J.W., van Isselt, J.W., Bijlsma, J.W. (1997). Persistent synovitis treated with radiation synovectomy using yttrium-90: A retrospective evaluation of 83 procedures for 45 patients. *Br J Rheumatol* 36:861–869.

Jiang, M., Fischman, A., Nowakowski, F.S. (2012). Segmental perfusion differences on paired Tc-99m macroaggregated albumin (MAA) hepatic perfusion imaging and yttrium-90 (Y-90) bremsstrahlung imaging studies in SIR-Sphere radioembolization: Associations with angiography. *J Nucl Med Radiat Ther* 3:122.

Kao, Y.H. et al. (2012). Image-guided personalized predictive dosimetry by artery-specific SPECT/CT partition modeling for safe and effective 90Y radioembolization. *J Nucl Med* 53:559–566.

Kao, Y.H. et al. (2013) Post-radioembolization yttrium-90 PET/CT—Part 2: Dose-response and tumor predictive dosimetry for resin microspheres. *EJNMMI Res* 3:1–1.

Kavakli, K. et al. (2008). Radioisotope synovectomy with rhenium186 in haemophilic synovitis for elbows, ankles and shoulders. *Haemophilia* 14:518–523.

Knesaurek, K. et al. (2010). Quantitative comparison of yttrium-90 (90Y)-microspheres and technetium-99m (99mTc)-macroaggregated albumin SPECT images for planning 90Y therapy of liver cancer. *Technol Cancer Res Treatm* 9:253–261.

Lam, M.G.E.H. et al. (2013). Prognostic utility of 90Y radioembolization dosimetry based on fusion 99mTc-macroaggregated albumin-99mTc-sulfur colloid SPECT. *J Nucl Med* 54:2055–2061.

Lam, M.G.E.H., Smits, M.L.J. (2013). Value of 99mTc-macroaggregated albumin SPECT for radioembolization treatment planning. *J Nucl Med* 54:1681–1682.

Lam, M.G.E.H., Sze, D.Y. (2014). Reply: 99mTc-MAA-based dosimetry for liver cancer treated using 90Y-loaded microspheres: Known proof of effectiveness. *J Nucl Med* 55:1392–1393.

Mathias, C.J., Green, M.A. (2008). A convenient route to [68Ga]Ga-MAA for use as a particulate PET perfusion tracer. *Appl Radiat Isot* 66:1910–1912.

Mattsson, J., Appelgren, L., Hamberger, B., Peterson, H.I. (1977). Adrenergic innervation of tumour blood vessels. *Cancer Lett* 3:347–351.

Meyer, C. et al. (2013). Feasibility of temporary protective embolization of normal liver tissue using degradable starch microspheres during radioembolization of liver tumours. *Eur J Nucl Med Mol Imaging* 41:231–237.

Minar, E. (2012). Commentary: Resuscitation of endovascular brachytherapy owing to improved logistics. *J Endovasc Ther* 19:476–479.

Morshedi, M.M., Bauman, M., Rose, S.C., Kikolski, S.G. (2015). Yttrium-90 resin microsphere radioembolization using an antireflux catheter: An alternative to traditional coil embolization for nontarget protection. *Cardiovasc Intervent Radiol* 38:381–388.

Nagy, J.A., Chang, S.-H., Dvorak, A.M., Dvorak, H.F. (2009). Why are tumour blood vessels abnormal and why is it important to know? *Br J Cancer* 100:865–869.

Oka, M., Rekonen, A., Ruotsi, A., Seppälä, O. (1971). Intra-articular injection of Y-90 resin colloid in the treatment of rheumatoid knee joint effusions. *Acta Rheumatol Scand* 17:148–160.

Oztürk, H., Oztemür, Z., Bulut, O. (2008). Treatment of skin necrosis after radiation synovectomy with yttrium-90: A case report. *Rheumatol Int* 28:1067–1068.

Pasciak, A.S. et al. (2014). Radioembolization and the dynamic role of (90)Y PET/CT. *Front Oncol* 4:38.

Pasciak, A.S. et al. (2015). The impact of an anti-reflux catheter on target volume particulate distribution in liver-directed embolotherapy: A pilot study. *J Vasc Interv Radiol* 26:660–669.

Pasciak, A.S., Bourgeois, A.C., Paxton, B.E., Nodit, L., Coan, P.N., Kraitchman, D., Stinnett, S., Patel, V.M., Fu, Y., Adams, J.K., Tolbert, M.K., Lux, C.N., Arepally, A., Bradley, Y.C. (2016). Bariatric radioembolization: A pilot study on technical feasibility and safety in a porcine model. *J Vasc Interv Radiol,* in press.

Pieper, C.C. et al. (2015). Temporary arterial embolization of liver parenchyma with degradable starch microspheres (EmboCept®S) in a swine model. *Cardiovasc Intervent Radiol* 38:435–441.

Piermattei, A. et al. (2002). A standard dosimetry procedure for ^{192}Ir sources used for endovascular brachytherapy. *Phys Med Biol* 47:4205–4221.

Piermattei, A. et al. (2003). Experimental dosimetry of a ^{32}P catheter-based endovascular brachytherapy source. *Phys Med Biol* 48:2283–2296.

Poorbaygi, H. et al. (2011). Applied radiation and isotopes. *Appl Radiat Isot* 69:1407–1414.

Prosser, J.S. et al. (1993). Induction of micronuclei in peripheral blood lymphocytes of patients treated for rheumatoid or osteo-arthritis of the knee with dysprosium-165 hydroxide macroaggregates or yttrium-90 silicate. *Cytobios* 73:7–15.

Reynaert, N., Van Eijkeren, M., Taeymans, Y. (2001) Dosimetry of ^{192}Ir sources used for endovascular brachytherapy. *Phys Med Biol* 46:499–516.

Ricke, J., Großer, O., Amthauer, H. (2013). Y^{90}-radioembolization of lung metastases via the bronchial artery: A report of 2 cases. *Cardiovasc Intervent Radiol* 36:1664–1669.

Rose, S.C., Kikolski, S.G., Chomas, J.E. (2013). Downstream hepatic arterial blood pressure changes caused by deployment of the surefire antireflux expandable tip. *Cardiovasc Intervent Radiol* 36:1262–1269.

Sabet, A. et al. (2011). Significance of oral administration of sodium perchlorate in planning liver-directed radioembolization. *J Nucl Med* 52:1063–1067.

Sasaki, Y. et al. (1985). Changes in distribution of hepatic blood flow induced by intra-arterial infusion of angiotensin II in human hepatic cancer. *Cancer* 55:311–316.

Schiller, E. et al. (2008). Yttrium-86-labelled human serum albumin microspheres: Relation of surface structure with in vivo stability. *Nucl Med Biol* 35:227–232.

Schopohl, B. et al. (1996). 192Ir endovascular brachytherapy for avoidance of intimal hyperplasia after percutaneous transluminal angioplasty and stent implantation in peripheral vessels: 6 years of experience. *Int J Radiat Oncol Biol Phys* 36:835–840.

Strigari, L. et al. (2010). Efficacy and toxicity related to treatment of hepatocellular carcinoma with 90Y-SIR spheres: Radiobiologic considerations. Journal of nuclear medicine: Official publication. *Soc Nucl Med* 51:1377–1385.

Stucki, G. et al. (1993). Efficacy and safety of radiation synovectomy with yttrium-90: A retrospective long-term analysis of 164 applications in 82 patients. *Br J Rheumatol* 32:383–386.

Suenaga, Y. et al. (2013). Respiratory-gated ^{18}F-FDG PET/CT for the diagnosis of liver metastasis. *Eur J Radiol* 82:1696–1701.

Tanaka, O. et al. (2008). Epinephrine-infused CTHA for HCCs. *Abdom Imaging* 33:308–312.

Taylor, W.J., Corkill, M.M., Rajapaske, C.N. (1997) A retrospective review of yttrium-90 synovectomy in the treatment of knee arthritis. *Br J Rheumatol* 36:1100–1105.

Thomas, S. et al. (2011). Radioactive synovectomy with Yttrium^{90} citrate in haemophilic synovitis: Brazilian experience. *Haemophilia* 17:e211–e216.

van den Hoven, A.F. et al. (2014a). Posttreatment PET-CT-confirmed intrahepatic radioembolization performed without coil embolization, by using the antireflux Surefire Infusion System. *Cardiovasc Intervent Radiol* 37:523–528.

van den Hoven, A.F. et al. (2014b). The effect of intra-arterial angiotensin II on the hepatic tumor to non-tumor blood flow ratio for radioembolization: A systematic review, Morishita R, ed. *PLoS ONE* 9:e86394.

van den Hoven, A.F. et al. (2015). Innovation in catheter design for intra-arterial liver cancer treatments results in favorable particle-fluid dynamics. *J Exp Clin Cancer Res* 34:74.

Werner, M. et al. (2012). Endovascular brachytherapy using liquid beta-emitting rhenium-188 for the treatment of long-segment femoropopliteal in-stent stenosis. *J Endovasc Ther* 19:467–475.

Willowson, K.P., Tapner, M.; QUEST Investigator Team, Bailey, D.L. (2015). A multicentre comparison of quantitative (90)Y PET/CT for dosimetric purposes after radioembolization with resin microspheres: The QUEST Phantom Study. *Eur J Nucl Med Mol Imaging* 42:1202–1222.

Wondergem, M. et al. (2013). ^{99}mTc-macroaggregated albumin poorly predicts the intrahepatic distribution of 90Y resin microspheres in hepatic radioembolization. *J Nucl Med* 54:1294–1301.

Zimmermann, A. et al. (1995). Renal pathology after arterial yttrium-90 microsphere administration in pigs. A model for superselective radioembolization therapy. *Invest Radiol* 30:716–723.

Index

Printed and bound by CPI Group (UK) Ltd, Croydon, CR0 4YY

01/11/2024

01782601-0006